IF FOUND, please notify and arrange return to owner. The
the Certified Management Accountant Examination and/or the
Examination.

Name of CMA/CFM Candidate ______________________________

Address ______________________________

City, State, Zip ______________________________

Telephone (______) ______________________ E-mail ______________________

Groundwood Paper and Highlighters -- This book is printed on high quality groundwood paper. It is lightweight and easy-to-recycle. We recommend that you purchase a highlighter specifically designed to be non-bleed-through (e.g., Avery *Glidestick*™) at your local office supply store.

REVIEWERS AND CONTRIBUTORS

Mark E. Fraser, C.P.A., J.D., Duke University, contributed questions and answer explanations for Study Unit 7, Information Systems I.

Karen Hom, B.A., University of Florida, is our book production coordinator. Ms. Hom coordinated the production staff and provided production assistance throughout the project.

Grady M. Irwin, J.D., University of Florida Holland Law Center, has taught in the University of Florida College of Business. Mr. Irwin provided many outline sections, answer explanations and extensive editorial assistance throughout.

Travis Moore, M.B.A., University of Florida, provided technical and editorial assistance, reviewed the final manuscript, and prepared the page layout.

Jan Strickland is our production assistant. Ms. Strickland reviewed the final manuscript and prepared the page layout.

A PERSONAL THANKS

This manual would not have been possible without the extraordinary effort and dedication of Jim Collis and Terry Hall who typed the entire manuscript and all revisions as well as prepared the camera-ready pages.

The authors appreciate the proofreading and production assistance of Kim Blackburn, Mark Fraser, Rob Gallardo, Jaime Gentile, Melissa Guinand, Jeff King, Lynn Li, Jessica Medina, Shane Rapp, McKenzie Rogers, and Jenna Zirkel.

The authors also appreciate the editorial assistance of Jennifer Forrester, Nefertiti Griffin, Ryan Kroll, Walter Mansfield, and Jose Martinez.

Finally, we appreciate the encouragement and tolerance of our families throughout the project.

NONDISCLOSED EXAM

The CMA/CFM is a nondisclosed exam, and you will encounter questions which may be totally unfamiliar to you. That is the nature of nondisclosed exams. Please follow the study suggestions on pages 13 through 15. We have the best and most efficient CMA/CFM exam prep system for success.

TENTH EDITION

CMA/CFM REVIEW

for PART 4
DECISION ANALYSIS AND INFORMATION SYSTEMS

by Irvin N. Gleim, Ph.D., CPA, CIA, CMA, CFM

and

Dale L. Flesher, Ph.D., CPA, CIA, CMA, CFM

ABOUT THE AUTHORS

Irvin N. Gleim is Professor Emeritus in the Fisher School of Accounting at the University of Florida and is a member of the American Accounting Association, Academy of Legal Studies in Business, American Institute of Certified Public Accountants, Association of Government Accountants, Florida Institute of Certified Public Accountants, Institute of Internal Auditors, and the Institute of Management Accountants. He has had articles published in the *Journal of Accountancy, The Accounting Review,* and *The American Business Law Journal* and is author/coauthor of numerous accounting and aviation books and CPE courses.

Dale L. Flesher is the Arthur Andersen Alumni Professor in the School of Accountancy at the University of Mississippi and has written over 180 articles for business and professional journals, including *Management Accounting, Journal of Accountancy,* and *The Accounting Review,* as well as numerous books. He is a member of the Institute of Management Accountants, American Institute of Certified Public Accountants, Institute of Internal Auditors, American Accounting Association, and American Taxation Association. He is a past editor of *The Accounting Historians' Journal* and is a trustee and past president of the Academy of Accounting Historians.

Gleim Publications, Inc.
P.O. Box 12848
University Station
Gainesville, Florida 32604
(352) 375-0772
(800) 87-GLEIM or (800) 874-5346
FAX: (352) 375-6940
E-mail: admin@gleim.com
Internet: www.gleim.com

This is the first printing of the tenth edition of ***CMA/CFM Review: Part 4***. Please e-mail update@gleim.com with CMA/CFM Part 4 10-1 included in the subject or text. You will receive our current update as a reply.

EXAMPLE:

To: update@gleim.com
From: your e-mail address
Subject: CMA/CFM Part 4 10-1

ISSN: 1093-2577

ISBN 1-58194-127-7

First Printing: February 2001

ACKNOWLEDGMENTS

The authors are indebted to the Institute of Certified Management Accountants for permission to use problem materials from past CMA examinations. Questions and unofficial answers from the Certified Management Accountant Examinations, copyright © 1973 through 1997 by the Institute of Certified Management Accountants, are reprinted and/or adapted with permission.

The authors are also indebted to The Institute of Internal Auditors, Inc. for permission to use Certified Internal Auditor Examination Questions and Suggested Solutions, copyright © 1977 through 1997 by The Institute of Internal Auditors, Inc.

The authors would also like to thank the following professors for contributing questions to this book: C.J. Skender of North Carolina State University and A. Wilson of Auburn University.

This publication was printed and bound by Corley Printing Company, St. Louis, MO, a registered ISO-9002 company. More information about Corley Printing Company is available at www.corleyprinting.com or by calling (314) 739-3777.

PREFACE FOR CMA/CFM CANDIDATES

The purpose of this book is to help YOU prepare YOURSELF to pass Part 4 of the CMA/CFM examinations. The overriding consideration is an inexpensive, effective, and easy-to-use study program. This manual

1. Defines topics tested on Part 4 of the CMA/CFM exams.
2. Explains how to optimize your exam score by analyzing how the CMA/CFM exams are constructed, administered, and graded.
3. Outlines all of the subject matter tested on Part 4 of the CMA/CFM exams in 10 easy-to-use study units.
4. Presents multiple-choice questions to prepare you to answer questions on Part 4 of the CMA/CFM exams. The answer explanations are presented to the immediate right of the questions for your convenience. You MUST cover these answers until you commit to the correct answer. Use the bookmark provided at the back of this book.
5. Illustrates **individual question answering techniques** to minimize selecting incorrect answers and to maximize your exam score.
6. Suggests **exam-taking techniques** to help you maintain control and achieve success.

This is the Tenth Edition of *CMA/CFM Review: Part 4*, which covers only Part 4 of the CMA/CFM exams. Recall that the only difference between the CFM exam and the CMA exam is Part 2. Part 2CFM is "Corporate Financial Management." Part 2CMA is "Financial Accounting and Reporting."

To maximize the efficiency of your review program, begin by **studying** (not reading) the introduction in this book. "Preparing for and Taking the CMA/CFM Exams" is very short but very important. It has been carefully organized and written to provide you with important information to assist you in successfully completing Part 4 of the CMA/CFM examination.

Thank you for your interest in our review materials. We very much appreciate the thousands of letters and suggestions received from CIA, CMA/CFM, and CPA candidates since 1974. Please give us feedback concerning this book. Do NOT disclose information about individual questions beyond subject matter not covered in our books. Tell us which part and which topics were NOT covered or were inadequately covered. The last two pages have been designed to help you note corrections and suggestions throughout your study process. Please tear them out and mail or fax them to us with your comments.

Good Luck on the Exam,

Irvin N. Gleim
Dale L. Flesher

February 2001

TABLE OF CONTENTS

PUBLISHER'S NOTE:

This is the first printing of the tenth edition of ***CMA/CFM Review: Part 4***. Please help. Use the form on pages 503 and 504 to send us corrections and suggestions for improvement (please send after you take Part 4 of the CMA/CFM exams). FAX the support request form on page vii or e-mail irvin@gleim.com for inquiries about errors, omissions, etc.

For updates and information on this and other Gleim books, please see our web page (www.gleim.com/updates.html), or send e-mail to update@gleim.com with "CMA/CFM Part 4 10-1" in the subject line.

Gleim's *CMA/CFM Review* Support via Fax

If you fax us inquiries about errors, omissions, etc., before 1:00 p.m. eastern time, we will respond by fax the following business day. If we have trouble faxing our response, it will be mailed. Technical support is also available via e-mail (support@gleim.com). Please include your e-mail address on the fax form below if you wish to receive a response to your request by e-mail (required if you are outside the United States).

Please photocopy this *CMA/CFM Review* Support Request fax form. It must be completed as requested so we can address the issues and questions you have. All items should refer to a specific page number and outline letter/number or question number.

Wait until after you take the CMA/CFM exam to send us the separate evaluation form provided on pages 503 and 504.

Gleim Publications, Inc. FAX (352) 375-6940

***CMA/CFM REVIEW* SUPPORT REQUEST**

Complete this form and attach additional pages as necessary.

Your name ______________________ Your Fax # ( ) ______-______

Address: ____

____

City: ______________________ State: ______ Zip: ________

E-mail: ____

Inquiry: ____

____

____

____

____

____

____

____

____

____

PREPARING FOR AND TAKING THE CMA/CFM EXAMS

ABOUT THE CMA/CFM EXAMS

INTRODUCTION

CMA is the acronym for Certified Management Accountant. CFM is an acronym for Certified in Financial Management. The CFM exam is an offshoot of the CMA. The two exams are identical except that the Part 2CFM exam is "Corporate Financial Management," while the Part 2CMA exam is "Financial Accounting and Reporting."

The CMA examination has been, and the CMA/CFM exams will continue to be, developed and offered by the Institute of Certified Management Accountants (ICMA) in approximately 200 locations in the U.S. and an additional 200 international locations.

CMA/CFM Review: Part 4 contains this 22-page Introduction and 10 study units of outlines and multiple-choice questions that cover all of the material tested on Part 4 of the CMA/CFM exams. This Introduction discusses exam content, pass rates, administration, organization, background information, preparing for the CMA/CFM exams, and taking the CMA/CFM exams. We urge you to read the next 20 pages carefully because they will help you dramatically improve your study and test-taking procedures.

CORPORATE MANAGEMENT ACCOUNTANTS AND FINANCIAL MANAGERS

1. Objective: Maximize the value of the firm by optimizing
   a. Long-term investment strategies
   b. The capital structure, i.e., how these long-term investments are funded
   c. Short-term cash flow management
2. Corporate financial management involves a financial manager, usually a vice-president/chief financial officer, who is assisted by the
   a. Treasurer -- cash, credit, capital outlay management
   b. Controller -- financial, cost, tax accounting
3. All accounting and finance personnel are beneficiaries of CFM and/or CMA participation.
4. The CMA/CFM tools are set forth in the ICMA's Content Specification Outlines for the CMA/CFM exams.
5. The diagram below illustrates an entity combining the factors of production into finished goods (arrows to the right) with money flowing to the left.

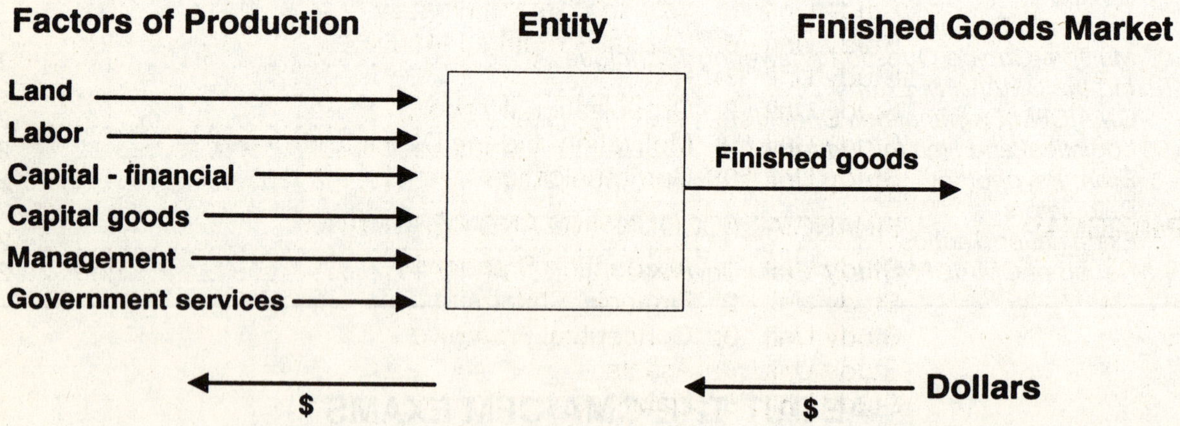

6. Put CMA and CFM in perspective of goods or services production in our capitalistic society. An entity combines the factors of production into finished goods.
   a. Note that the CMA/CFM programs focus on financial capital and the other factors of production as well as the finished goods market. The CEO (chief executive officer) has overall responsibility for the entity's operations.

OBJECTIVES AND CONTENT OF THE CMA/CFM EXAMINATIONS

The primary purpose of the CMA/CFM examination programs is "to establish an objective measure of an individual's knowledge and competence in the fields of management accounting and financial management." Three other objectives set forth by the ICMA are to

1. "Establish management accounting and financial management as recognized professions by identifying the role of the professional, the underlying body of knowledge, and a course of study by which such knowledge is acquired"
2. "Encourage higher educational standards in the management accounting and financial management fields"
3. "Encourage continued professional development"

The exams test the candidates' knowledge and ability with respect to the current state of the art in the fields of management accounting and financial management.

We have arranged the subject matter tested on the CMA/CFM examinations into 10 study units for each part. Each part is presented in a separate book. For CMA/CFM Parts 1, 3, and 4, study the corresponding *CMA/CFM Review* book. For Part 2CMA, study *CMA Review: Part 2CMA* and, for Part 2CFM, study *CFM Review: Part 2CFM*. All contain review outlines and prior CMA/CFM exam questions and answers. The 50 study units are organized as follows:

CMA/CFM Part 1: ECONOMICS, FINANCE, AND MANAGEMENT*

- Study Unit 1: Microeconomics
- Study Unit 2: Macroeconomics
- Study Unit 3: International Business Environment
- Study Unit 4: Domestic Institutional Environment of Business
- Study Unit 5: Working Capital Finance
- Study Unit 6: Capital Structure Finance
- Study Unit 7: Risk
- Study Unit 8: Organizational Theory
- Study Unit 9: Motivation and the Directing Process
- Study Unit 10: Communication

Part 2CMA: FINANCIAL ACCOUNTING AND REPORTING

- Study Unit 1: Accounting Standards
- Study Unit 2: Financial Statements
- Study Unit 3: Conceptual Framework
- Study Unit 4: Assets
- Study Unit 5: Liabilities
- Study Unit 6: Equity
- Study Unit 7: Other Income Items
- Study Unit 8: Other Reporting Issues
- Study Unit 9: Financial Statement Analysis
- Study Unit 10: External Auditing

***WARNING!!!**

About 25 to 30% of Part 1 test questions will require mathematical calculations. All other parts are approximately 35 to 40% computational questions. Practice computational questions to prepare for exam success!

Part 2CFM:	CORPORATE FINANCIAL MANAGEMENT	
	Study Unit 1:	Financial Statements and Annual Reports
	Study Unit 2:	Financial Statements: Special Topics
	Study Unit 3:	Long-Term Capital Financing
	Study Unit 4:	Financial Markets and Interest Rates
	Study Unit 5:	Investment Banking and Commercial Banking
	Study Unit 6:	Financial Statement Analysis
	Study Unit 7:	Business Combinations and Restructurings
	Study Unit 8:	Risk Management
	Study Unit 9:	External Financial Environment
	Study Unit 10:	Accounting Standard Setting
CMA/CFM Part 3:	MANAGEMENT REPORTING, ANALYSIS, AND BEHAVIORAL ISSUES	
	Study Unit 1:	Cost and Managerial Accounting Definitions
	Study Unit 2:	Product Costing and Related Topics
	Study Unit 3:	Cost Behavior
	Study Unit 4:	Statements on Management Accounting
	Study Unit 5:	Planning
	Study Unit 6:	Budgeting
	Study Unit 7:	The Controlling Process
	Study Unit 8:	Standard Costs and Variance Analysis
	Study Unit 9:	Responsibility Accounting
	Study Unit 10:	Behavioral Issues
CMA/CFM Part 4:	DECISION ANALYSIS AND INFORMATION SYSTEMS	
	Study Unit 1:	Decision Analysis
	Study Unit 2:	Cost-Volume-Profit Analysis
	Study Unit 3:	Capital Budgeting
	Study Unit 4:	Decision Making under Uncertainty
	Study Unit 5:	Quantitative Methods I
	Study Unit 6:	Quantitative Methods II
	Study Unit 7:	Information Systems I
	Study Unit 8:	Information Systems II
	Study Unit 9:	Internal Control
	Study Unit 10:	Internal Auditing

Recap: The only difference between the CFM exam and the CMA exam is Part 2. This book covers Part 4 of the CMA/CFM exam.

The *CMA/CFM Review* study unit titles and organization differ somewhat from the subtopic titles used by the ICMA in its content specification outlines for the CMA/CFM exams. The detailed ICMA Content Specification Outline for Part 4 is presented in Appendix D beginning on page 483. The selection of study units in *CMA/CFM Review: Part 4* is based on the types and number of questions that have appeared on past CMA and CFM exams as well as the extensiveness of past and expected future exam coverage.

CONTENT SPECIFICATION OUTLINES

The ICMA has developed content specification outlines and has committed to follow them on each examination. A complete CSO is presented in Appendix D beginning on page 483. Thus, each examination will cover the major topics specified below; e.g., investment decision analysis will constitute 20% to 30% of Part 4 on each examination.

Candidates for the CMA/CFM designations are expected to have a minimum level of business knowledge that transcends all examination parts. This minimum level includes knowledge of basic financial statements, time value of money concepts, and elementary statistics. Specific discussion of the ICMA's Levels of Performance (A, B, and C) is provided on the next page.

CMA/CFM Part 1: Economics, Finance, and Management

A. Microeconomics--10%-15% (Level B)
B. Macroeconomics--10%-15% (Level A)
C. International Business Environment--10%-15% (Level B)
D. Domestic Institutional Environment of Business--10%-15% (Level A)
E. Working Capital Policy and Management--10%-15% (Level C)
F. Long-term Finance and Capital Structure--10%-15% (Level B)
G. Organizational Structures, Management, and Communication--20%-30% (Level B)

Part 2CMA: Financial Accounting and Reporting

A. Development of Accounting Standards--10%-15% (Level A)
B. Advanced Topics on the Preparation of Financial Statements--50%-70% (Level C)
C. Interpretation and Analysis of Financial Statements--10%-15% (Level C)
D. External Auditing--10%-15% (Level A)

Part 2CFM: Corporate Financial Management

A. Use of Financial Statements--10%-20% (Level B)
B. Advanced Topics in Corporate Financial Management--50%-70% (Level C)
C. Risk Management--10%-15% (Level C)
D. External Financial Environment--10%-20% (Level A)
E. Accounting Standard Setting Environment--5%-10% (Level A)

CMA/CFM Part 3: Management Reporting, Analysis, and Behavioral Issues

A. Cost Measurement--20%-30% (Level C)
B. Planning--20%-30% (Level C)
C. Control and Performance Evaluation--20%-30% (Level C)
D. Behavioral Issues--20%-30% (Level B)

CMA/CFM Part 4: Decision Analysis and Information Systems

A. Decision Theory and Operational Decision Analysis--20%-30% (Level C)
B. Investment Decision Analysis--20%-30% (Level C)
C. Quantitative Methods for Decision Analysis--10%-15% (Level B)
D. Information Systems--20%-30% (Level B)
E. Management Controls--10%-15% (Level B)

LEVEL OF PERFORMANCE REQUIRED

All parts of the exams appear to be tested at the skill level of a final examination for the appropriate course at a good school of business. The ICMA has specified three levels of coverage as reproduced below and indicated in its content specification outlines. You will evaluate and compare the difficulty of each part of the CMA/CFM exams as you work the questions in this book.

Authors' Note: Rely on the questions at the back of each study unit in each *CMA/CFM Review* book and *CMA/CFM Test Prep* software.

Level A: Requiring an appreciation of the broad nature and fundamentals, the ability to recognize the existence of special features and problems in various business transactions, and the ability to apply this knowledge in new and concrete situations, i.e., requiring the skill levels of knowledge and comprehension.

Level B: Requiring a sound understanding of the broad aspects of practices and procedures and awareness of the problems relating to more detailed aspects, the ability to apply such broad knowledge in new and concrete situations, the ability to analyze the components of the material and their relationships, and the ability to reformulate the components into a new whole, i.e., requiring the skill levels of knowledge, comprehension, application and analysis.

Level C: Requiring a sound understanding of principles, practices, and procedures, the ability to analyze material so that its organizational structure can be understood, the proficiency to apply such knowledge to situations likely to be encountered, the ability to deal with all aspects of the situation without extensive recourse to technical research and assistance, and the ability to judge the value of material for a given purpose, i.e., requiring all six skill levels.

HOW ETHICS AND TAXES ARE TESTED

Ethics will be tested on every exam within the context of a specific subject area tested on that part. Appendix A beginning on page 455 covers ethics.

Federal income taxes are tested as (1) accounting for income taxes and (2) tax implications for decisions. Accounting for income taxes including the proper treatment of deferred income taxes is tested in Part 2CMA, Financial Accounting and Reporting, and Part 2CFM, Corporate Financial Management.

The tax code provisions that affect business decisions may be tested in (1) questions that require decision analysis regarding debt versus equity issues in CMA/CFM Part 1, Economics, Finance, and Management; (2) questions in Part 2CFM, Corporate Financial Management; and (3) the decision analysis questions in CMA/CFM Part 4, Decision Analysis and Information Systems.

REQUIREMENTS TO ATTAIN THE CMA/CFM DESIGNATIONS

The CMA and/or the CFM designations are granted only by the ICMA. Candidates must complete five steps to become a CMA and/or CFM:

1. Become a member of the Institute of Management Accountants.
2. File an application for admission with the ICMA and register for the CMA/CFM exams.

3. Pass all four parts of the CMA or CFM examination. To attain both designations, CMAs in good standing need only pass Part 2CFM; CFMs in good standing need only pass Part 2CMA. "Continuous candidacy" is required to retain credit after successful completion of one or more parts. "Continuous candidacy" includes IMA membership, 20 hours of CPE annually, and at least one examination part attempted every 2 years.
4. Meet the CMA/CFM experience requirement.
5. Comply with the Standards of Ethical Conduct.
6. Meet the CMA/CFM education requirements (see below).

Credit can be retained indefinitely as long as these requirements are fulfilled.

Once a designation is earned, the CMA and/or CFM is a member of the Institute of Certified Management Accountants and must comply with the program's CPE requirement and maintain IMA membership in good standing. The certificate of a CMA or CFM delinquent in these requirements will be subject to recall.

ADMISSION TO THE CMA/CFM PROGRAMS

In addition to the above requirements, candidates seeking admission to the CMA/CFM programs must also meet one of the following educational credentials:

1. Hold a baccalaureate degree, in any area, from an accredited college or university. Students may apply when they attain senior standing and will be permitted to take the examination pending receipt of a degree. Degrees from foreign institutions must be evaluated by an independent agency approved by the ICMA; or
2. Hold a CPA license to practice or other professional qualification issued in a foreign country that is comparable to the CPA, CMA, CFM, etc., or
3. Achieve a score in the 50th percentile or higher on either the Graduate Management Admissions Test (GMAT) or the Graduate Record Examination (GRE).

NOTE: Educational credentials may be submitted at a later date, i.e., a degree or senior standing is not required to sit for the exam but must be satisfied before you are certified.

Also, the applicant must

1. Be a member of the Institute of Management Accountants or submit an IMA application when applying to the ICMA
2. Fulfill or expect to fulfill the educational requirements
3. Be employed or expect to be employed in a position that meets the experience requirement
4. Submit the names of two character references (one from employer, one from a person who is neither employed by the same firm nor a family member)
5. Be of good moral character
6. Abide by the Standards of Ethical Conduct

CMA/CFM WORK EXPERIENCE REQUIRED

Two continuous years of professional experience in financial management and/or management accounting are required any time prior to, or within 7 years of, passing the examination.

1. Professional experience shall be defined as full-time continuous experience at a level where judgments are regularly made that employ the principles of financial management and/or management accounting, e.g.,
   a. Financial analysis
   b. Budget preparation
   c. Management information systems analysis
   d. Management accounting in government, finance, or industry
   e. Auditing in government, finance, or industry
   f. Management consulting
   g. Audit work in public accounting (3 years required)
   h. Research, teaching, or consulting related to management accounting (for teaching, a significant portion required to be above the principles level)
2. Employment in functions that require the occasional application of financial management or management accounting principles but are not essentially management accounting oriented will not satisfy the requirement, e.g.,
   a. Computer operations
   b. Sales and marketing
   c. Manufacturing
   d. Engineering
   e. Personnel
   f. Employment in internships, trainee, clerical, or nontechnical positions

If you have any questions about the acceptability of your work experience or baccalaureate degree, please write or call the ICMA. Include a complete description of your situation. You will receive a response from the ICMA as soon as your request is evaluated.

Institute of Certified Management Accountants
10 Paragon Drive
Montvale, NJ 07645-1759
(201) 573-9000
(800) 638-4427

NOTE: If you have passed the U.S. CPA exam, you can waive Part 2CMA. There is an $80 fee for the processing of this waiver. To receive a waiver, you must request that a letter from your state board be sent directly to the ICMA confirming your licensure or passing of the U.S. CPA exam; copies are not acceptable.

THE INSTITUTE OF MANAGEMENT ACCOUNTANTS (IMA)

Conceived as an educational organization to develop the individual management accountant professionally and to provide business management with the most advanced techniques and procedures, the IMA was founded as the National Association of Accountants in 1919 with 37 charter members. It grew rapidly, with 2,000 applications for membership in the first year, and today it is the largest management accounting association in the world, with approximately 75,000 members and more than 300 chapters in the U.S. and abroad.

The IMA has made major contributions to business management through its continuing education program, with courses and seminars conducted in numerous locations across the country; its two magazines, *Strategic Finance*, which is a new monthly publication, and *Management Accounting*, which is now a quarterly; other literature, including research reports, monographs, and books; a technical inquiry service; a library; the annual international conference; and frequent meetings at chapter levels.

Membership in the IMA is open to all persons interested in advancing their knowledge of accounting or financial management. It is required for CMA/CFM candidates and CMAs/CFMs.

IMA Dues in the USA and Canada

1. **Regular**: 1 year, $140
2. **Associate**: $47.00 (2nd year, $93.00); must apply within 2 years of completing fulltime studies
3. **Academic member**: $70; must be a full-time faculty member and reside in the U.S. or Canada
4. **Student**: $28.00; must have 6 or more equivalent hours per semester and reside in the U.S. or Canada

Membership application forms may be obtained by writing the Institute of Management Accountants, 10 Paragon Drive, Montvale, NJ 07645-1760, or calling (201) 573-9000 or (800) 638-4427. A sample of the two-page form appears in Appendix C on pages 478 and 479.

THE INSTITUTE OF CERTIFIED MANAGEMENT ACCOUNTANTS (ICMA)

The ICMA is located at the IMA headquarters in Montvale, New Jersey. The only function of the ICMA is to offer and administer the CFM and CMA designations. The staff consists of the senior director, the director of examinations, and three associates (all of whom are professional accountants), plus support personnel. The ICMA occupies about 4,000 square feet of office space in the IMA headquarters. This office is where the examinations are prepared and graded and where all records are kept.

ICMA BOARD OF REGENTS AND STAFF

The ICMA Board of Regents is a special committee of the IMA established to direct the CFM and CMA programs for financial managers and management accountants through the ICMA.

The Board of Regents consists of 16 regents, one of whom is designated as chair by the president of the IMA. The regents are appointed by the president of the IMA to serve 3-year terms. Membership on the Board of Regents rotates, with one-third of the regents being appointed each year. The regents usually meet two times a year for 1 or 2 days.

The managing director of the ICMA, the director of examinations, and the ICMA staff are located at the ICMA office in Montvale, NJ. They undertake all of the day-to-day work with respect to the CFM and CMA programs.

HOW TO (1) APPLY AND (2) REGISTER FOR THE CMA/CFM EXAMS

First, you are required to **apply** both for membership in the IMA (see sample application form in Appendix C on pages 478 and 479) and for admission into the ICMA (see sample application form in Appendix C on pages 480 and 481). Thus, currently two applications are required if you are not already an IMA member. The ICMA is also phasing in a single combined IMA and ICMA application. Contact the ICMA for more information.

Apply to join the IMA and the ICMA **today** -- it only takes a few minutes. Application to the ICMA requires education, employment, and reference data. The educational experience requirements are discussed on page 7. You must provide two references: one from your employer and the second from someone other than a family member or fellow employee. Character reference forms are sent by the ICMA with your application forms. An official transcript providing proof of graduation is also required. There is a $50 credentialing fee (waived for students). Once a person has become a candidate, there is no participant's fee other than IMA membership dues.

Second, it is necessary to **register** each time you wish to sit for the exams. The exam registration form (see page 482) is very simple (it takes about 2 minutes to complete). The registration fee for each part of the exam is $80. There is an additional $40 processing fee per part for candidates taking the examination at an international site making the fee $120 per part. Graduating seniors and full-time graduate students are charged a one-time special rate of $40 per part. Full-time faculty are permitted to take the exam one time at no cost and thereafter pay $40 per part.

A sample of the first page of the ICMA registration form available to candidates from the ICMA appears in Appendix C. Order a registration booklet and IMA membership application form from the ICMA at (800) 638-4427, extension 510. The IMA encourages candidates to view information and complete IMA and ICMA applications and exam registration forms online. Visit the IMA's web site at www.imanet.org for more information.

SPECIAL STUDENT EXAMINATION FEE

U.S. college students may take five examination parts at $40 per part (versus the normal $80). These discounts must be used within the year following application or they will be forfeited. To be eligible for this discount, students must

1. Provide the name of someone who can verify student status
2. Apply to the ICMA while enrolled in school
3. Be undergraduates who have attained senior status or graduate students who are enrolled for at least 6 credit hours in the current term, or undergraduate or graduate students who are taking less than 6 hours but have taken 6 credit hours for the prior two semesters or three quarters and can be expected to graduate at the end of the current term

FEES FOR FULL-TIME PROFESSORS

Full-time faculty members are permitted to take five examination parts once at no charge. The fee for any parts that must be retaken is 50% of the normal fee. The $50 credentialing fee is no longer waived. To qualify, a faculty member must submit a letter on school stationery affirming his/her full-time status. Faculty should sit for the CMA/CFM examinations because a professor's status as a CMA/CFM encourages students to enter the program. Full-time doctoral students who plan to pursue a teaching career are treated as faculty members for purposes of qualifying for the free examination.

ICMA REFUND POLICY

In general, the fees paid to the ICMA discussed above are nonrefundable. However, a candidate withdrawing from the examination within 30 days of registration may receive a credit for the examination fee less a $25 processing charge. Failure to take the examination within the assigned 120-day authorization period will result in the loss of examination fees; there will be no refund or credit available.

PASS/FAIL AND GRADE REPORTS

The passing score is approximately 60-70%. You will receive your test results immediately after completion of the computerized exam. In addition to pass or fail, you are told a "scaled score" between 200 and 700, with 500 being the passing mark. This provides information about how close, above or below, you were to passing. If you fail the exam, there is a 90-day waiting period before you may retake the part failed.

MAINTAINING YOUR CMA/CFM DESIGNATIONS

Membership in the IMA is required to maintain your CMA/CFM certificates. The general membership fee is $140. There is no additional participant fee.

Continuing professional education is required of CMAs/CFMs to maintain their proficiency in the fields of managerial accounting and financial management. Every year, 30 hours of CPE must be completed, which is about 4 days per year. Qualifying topics include management accounting, financial management, corporate taxation, statistics, computer science, systems analysis, management skills, marketing, business law, and insurance.

Credit for hours of study will be given for participation in programs sponsored by businesses, educational institutions, or professional and trade associations at either the national or local level. Programs conducted by an individual's employer must provide for an instructor or course leader. There must be formal instructional training material. On-the-job training does not qualify. An affidavit from the employer is required to attest to the hours of instruction. The programs may be seminars, workshops, technical meetings, or college courses under the direction of an instructor. The method of instruction may include lecture, discussion, case studies, and teaching aids such as training films and cassettes.

Credit for hours of study may be given for technical articles published in business, professional, or trade journals, and for major technical talks given for the first time before business, professional, or trade organizations. The specific hours of credit in each case will be determined by the Institute.

CMA/CFM TEST ADMINISTRATION AND GRADING

The ICMA will send you authorization to take the exam, which will have a 120-day window. The authorization will instruct you to call the Prometric registration office 1(800) 479-6370 and register for your test at a local Prometric testing center. Your test will be graded by the computer as soon as you complete the test.

COMPUTER TESTING PROCEDURES

When you arrive at the computer testing center, you will be required to check in. Be sure to bring your authorization letter and two forms of identification, one with a photograph, both with your signature. If you have any questions, please call the IMA at 1(800) 638-4427.

Next, you will be taken into the testing room and seated at a computer terminal. A person from the testing center will assist you in logging on the system, and you will be asked to confirm your personal data. Then you will be prompted and given an online introduction to the computer testing system and you will view a tutorial.

If you have used our ***CMA/CFM Test Prep*** software, you will be conversant with the computer testing methodology and environment, and you will probably want to skip the tutorial and begin the actual test immediately. Once you begin your test, you will be allowed 3 hrs. to complete the actual test. This is just over 1.6 minutes per question. You may take a break during the exam, BUT the clock continues to run during your break. Before you leave the testing center, you will be required to check out of the testing center.

CONCEPTUAL VS. CALCULATION QUESTIONS

About 25 to 30% of CMA/CFM Part 4 test questions will be calculations in contrast to conceptual questions. When you take the test, it will appear that more than 40% of the questions are calculation-type because they take longer and are "more difficult."

PREPARING FOR THE CMA/CFM EXAMS

HOW MANY PARTS TO TAKE

We suggest that you take one part at a time.

CMA/CFM Part 1:	Economics, Finance, and Management
Part 2CMA:	Financial Accounting and Reporting
Part 2CFM:	Corporate Financial Management
CMA/CFM Part 3:	Management Reporting, Analysis, and Behavioral Issues
CMA/CFM Part 4:	Decision Analysis and Information Systems

Parts 2CMA, 2CFM, and 3 cover many of the accounting and finance topics included in typical undergraduate accounting/finance programs. Part 1 covers the material found in management, microeconomics, macroeconomics, and corporate finance courses. Part 4 covers computer systems, quantitative methods, and statistics courses.

There is no time limit for taking and passing all of the parts. Candidates can maintain credit for passed parts as long as they maintain continuous candidacy. To maintain candidacy, individuals must continue their membership in the IMA, complete 20 hours of continuing education each year beginning the calendar year after they passed their first part, and take (not necessarily pass) an exam part every 2 years. Note that a candidate receives 12 hours of continuing professional education for each exam part passed.

HOW TO STUDY A STUDY UNIT (books only*)

1. Gain an overview of the study unit -- familiarity with the topic, number of pages of outline, number of multiple-choice questions -- and estimate the time you will invest.
2. Answer five to ten multiple-choice questions. Choose one or two questions from each subunit.

   The purpose of answering multiple-choice questions before working through the study outline is to understand the standards to which you will be held. This will motivate you to concentrate on the study outline. You will also practice answering questions for which you are unprepared. Become an educated guesser when you encounter questions in doubt; you will outperform the inexperienced exam candidate.
3. Work through the study outline. Learn and understand the concepts. Remember, you are aiming toward the analysis, synthesis, and evaluation levels of knowledge, not rote memorization. Study the outlines with the objective of being able to explain the subject matter to third parties.
4. After you are comfortable with the study outlines, apply your multiple-choice question answering technique (see page 18) to answer all of the multiple-choice questions by marking the correct answer before consulting the answer and answer explanation. It is essential to mark your answer choice before looking at the answer and answer explanation. Use the bookmark at the back of each Gleim ***CMA/CFM Review*** book.
5. Develop a 90% proficiency level within each study unit. You will achieve this proficiency level by studying the outlines and answering multiple-choice questions.

   Learning from questions you answer incorrectly is very important. Each question you answer incorrectly is an opportunity to avoid missing actual test questions on your CMA/CFM exam. Thus, you should carefully study the answer explanations provided until you understand why the original answer you chose is wrong, as well as why the correct answer indicated is correct. This study technique may prove to be the difference between passing and failing for many candidates.

* We recommend using *CMA/CFM Test Prep* software discussed on page 14. It will give you a definite advantage.

You **must** determine why you answered questions incorrectly and learn how to avoid the same error in the future. Reasons for missing questions include

a. Misreading the requirement (stem)
b. Failing to understand what is required
c. Making a math error
d. Applying the wrong rule or concept
e. Being distracted by one or more of the answers
f. Incorrectly eliminating answers from consideration
g. Lacking any knowledge of the topic tested
h. Employing bad intuition (Why?) when guessing

6. Gleim ***CMA/CFM Test Prep*** software will significantly benefit your study efforts, especially when using the 20-question test routine suggested below or the Prometric test emulation contained in the software.

ADDING CMA/CFM TEST PREP SOFTWARE

Using ***CMA/CFM Test Prep*** really works! The software forces you to commit to your answer choice before looking at answer explanations. It also keeps track of your time and the results of your effort. Each study session and each test session are kept in the performance history and are viewable in either a table or a graphical format. You can know exactly what to expect when you go to take your exam by using the Prometric test emulation contained within the software.

Each Test Prep disk covers a different part of the exam and includes over 1,000 questions. All questions have been updated to reflect the current subject matter.

Read and study the following six steps regarding how to use the software with the books. Using ***CMA/CFM Test Prep*** will greatly facilitate your study and success on the exam! DO NOT neglect diagnosing the reasons for answering questions incorrectly; i.e., learn from your mistakes while studying so you avoid making mistakes on the exam.

The best way to prepare to PASS:

1. In test mode, answer a 20-question test from each study unit before studying the study unit.
2. Study the knowledge transfer outline for the study unit in your Gleim book.
3. Take two or three 20-question tests in test mode after studying knowledge transfer outlines.
4. After EACH test session, immediately switch to study mode and select questions "missed on last session" so you can reanswer these questions AND analyze why you answered each question incorrectly.
5. Continue the process until you approach a 75% proficiency level.
6. Modify the process to suit your individual learning process.

> Avoid studying Gleim questions to learn the correct answers. Use Gleim questions to help you learn how to answer CMA/CFM questions under exam conditions. Expect the unexpected and be prepared to deal with the unexpected. Always take one 20-question test in test mode *before* studying the material in each study unit. These test sessions will allow you to practice answering questions you have not seen before. Become an educated guesser when you encounter questions in doubt; you will outperform the inexperienced exam taker.

STUDY PLAN AND TIME BUDGET

Complete one study unit at a time. Initially, budget 3 to 4 hours per study unit (1 to 2 hours studying the outline and 1 to 2 minutes each on all the multiple-choice questions).

This Introduction	2
10 study units at 3.5 hours each	35
General review	3
Total Hours	40

Each week you should evaluate your progress and review your preparation plans for the time remaining prior to the exam. Review your commitments, e.g., out-of-town assignments, personal responsibilities, etc., and note them on your calendar. This precaution will assist you in keeping to your schedule.

CONTROL

You have to be in control to be successful during exam preparation and execution. Control can also contribute greatly to your personal and other professional goals. Control is a process whereby you

1. Develop expectations, standards, budgets, and plans.
2. Undertake activity, production, study, and learning.
3. Measure the activity, production, output, and knowledge.
4. Compare actual activity with expected, or budgeted.
5. Modify the activity, behavior, or production to better achieve the desired outcome.
6. Revise expectations and standards in light of actual experience.
7. Continue the process.

The objective is to be confident that the best possible performance is being generated. Most accountants study this process in relation to standard costs, i.e., establish cost standards and compute cost variances.

Every day you rely on control systems implicitly. When you groom your hair, you have expectations about the desired appearance of your hair and the time required to style it. You monitor your progress and make adjustments as appropriate, e.g., brush it a different way or speed up.

Develop and enforce standards in all of your endeavors. Exercise control, implicitly or explicitly. Most endeavors will improve with explicit control. This is particularly true with certification examinations and other academic tests.

1. Practice your question answering techniques (and develop control) as you prepare answers/solutions to practice questions/problems during your study program.
2. Develop explicit control over your study programs based on the control process discussed previously.
3. Think about using more explicit control systems over any and all of your endeavors.
4. Seek continuous improvement to meet your needs given a particular situation or constraint. Additional practice will result in further efficiencies.

PROMETRIC TESTING CENTERS (Domestic Only): Part of the Prometric Learning Center Network

ALABAMA
Birmingham
Decatur
Dothan
Mobile
Montgomery

ALASKA
Anchorage

ARIZONA
Chandler
Phoenix
Tucson

ARKANSAS
Ft. Smith
Little Rock

CALIFORNIA
Anaheim Hills
Atascadero
Brea
Culiver City
Diamond Bar
Fair Oaks
Fremont
Gardena
Irvine
La Jolla
Piedmont
Rancho Cuca
Redlands
Riverside
San Francisco
San Jose
Santa Rosa
Walnut Creek
Westlake

COLORADO
Boulder
Colorado Springs
Littleton
Pueblo

CONNECTICUT
Glastonbury
Hamden

FLORIDA
Davie
Ft. Myers
Gainesville
Jacksonville
Miami Lakes
Sarasota
Tallahassee
Temple Terr.
Winter Park

GEORGIA
Augusta
Jonesboro
Macon
Savannah
Smyrna
Valdosta

HAWAII
Kailua/Oahu

IDAHO
Boise

ILLINOIS
Carbondale
Carpentersville
Chicago
Homewood
Normal
Northbrook
Springfield
Westchester

INDIANA
Ft. Wayne
Indianapolis
Layfayette
Merrillville
Mishawaka

IOWA
Bettendorf
Cedar Rapids
W. Des Moines

KANSAS
Topeka
Wichita

KENTUCKY
Lexington
Louisville

LOUISIANA
Baton Rouge
Bossier City
New Orleans

MAINE
Pikesville
South Portland

MARYLAND
Bethesda
Columbia
Pikesville

MASSACHUSETTS
Boston
E. Longmeadow
Worcester

MICHIGAN
Ann Arbor
Grand Rapids
Lansing
Livonia
Troy
Utica

MINNESOTA
Bloomington
Duluth
Rochester
Woodbury

MISSISSIPPI
Jackson

MISSOURI
Ballwin
Creve Coeur
Gladstone
Springfield

MONTANA
Billings
Helena

NEBRASKA
Lincoln
Omaha

NEVADA
Las Vegas
Reno

NEW HAMPSHIRE
Concord

NEW JERSEY
East Brunswick
Fairlawn
Hamilton Town
Verona

NEW MEXICO
Albuquerque

NEW YORK
Albany
Amherst
East Syracuse
Forest Hills
Garden City
Ithaca
Penn Plaza
Rochester
Staten Island
Vestal
Wappingers Falls
White Plains

NORTH CAROLINA
Charlotte
Gastonia
Greensboro
Greenville
Raleigh

NORTH DAKOTA
Bismarck

OHIO
Akron
Centerville
Cincinnati-North
Columbus
Lima
Mentor
Niles
Reynoldsburg
Solon
Strongsville
Toledo

OKLAHOMA
Oklahoma City
Tulsa

OREGON
Eugene
Milwaukie
Portland
Salem

PENNSYLVANIA
Allentown
Center City
Erie
Harrisburg
Lancaster
North Hills
North Wales
Pittsburgh
Plymouth Meeting
Scranton
York

RHODE ISLAND
Cranston

SOUTH CAROLINA
Charleston
Greenville
Irmo

SOUTH DAKOTA
Sioux Falls

TENNESSEE
Chattanooga
Clarksville
Franklin
Knoxville
Madison

TEXAS
Abilene
Amarillo
Austin
Beaumont
Bedford
Corpus Christi
Dallas Metro/
Arlington
El Paso
Houston
Lubbock
Midland
San Antonio
Waco

UTAH
Orem
Salt Lake City

VERMONT
Williston

VIRGINIA
Arlington
Newport News
Richmond
Roanoke

WASHINGTON
Lynnwood
Puyallup
Spokane

WASHINGTON, DC

WEST VIRGINIA
Charleston
Morgantown

WISCONSIN
Brookfield

WYOMING
Casper

CAUTION: Since actual sites change often, you should contact the Prometric registration office (1(800) 479-6370) to verify the location of your local testing center or obtain international test sites.

After you register to take each part of the CMA/CFM exams and pay $80 ($120 internationally), the ICMA will send you a letter with an authorization number and instructions on contacting Prometric to schedule your test. Call your local Prometric testing center a day or two before your test to confirm your time and obtain directions to the testing center or visit the Prometric web site at www.prometric.com. A listing of Prometric testing centers is presented on page 16.

DOMESTIC INSTRUCTIONS TO CANDIDATES

The letter accompanying these instructions is your authorization to schedule the taking of your examination part(s) with Prometric Technologies. Please note that you have a separate authorization number for each part that you registered to take. Questions regarding your registration with IMA should be directed to 800-638-4427, ext. 301 or 303.

Scheduling with Prometric Technologies

The interval dates shown on your authorization letter represent the time/authorization period during which you are authorized to take CMA/CFM examination part(s).

- Before calling for your appointment, select two or three dates during your authorization period that are convenient for you to take the examination in case your first choice is not available. Please be aware that Saturdays fill quickly and you may not be able to get a Saturday appointment.
- Choose a Prometric Technology Center from the accompanying list of locations.
- Be sure you have your authorization number(s) from the accompanying letter handy when you call as you will be required to provide this information.
- To schedule your examination, please call Prometric Technologies National Registry at 800-479-6370 at your earliest convenience. You also have the option of calling directly to the Prometric Technology Center of your choice.

You must schedule your examination during the dates provided as your authorization will expire at the end of the period, and your fees will be forfeited.

Cancellation of a Scheduled Appointment

If you find that you are unable to keep a scheduled appointment at Prometric, you must cancel 48 business hours before the appointment. To cancel, call 800-479-6370 at any time; if you cannot speak to a Prometric representative, you may leave a voice mail message as these messages are marked with the date and time. Be sure to leave sufficient information for identification such as your name, your authorization number, and the site location where you are scheduled. If you cancel an appointment, you must wait 72 hours before calling to reschedule. If you do not comply with this 48-hour policy, you will be considered a "no-show" and you will need to reregister with ICMA and repay the examination fee.

General Instructions

- Be sure to arrive at the Prometric Technology Center on time. If you are more than 30 minutes late, you may lose your scheduled sitting and be required to reschedule at a later date at an additional cost.
- You will be required to sign the Prometric Log Book when you enter the center.
- Be sure to take your authorization letter with you as you will be asked to show this document. You may wish to make a copy of this letter if you are taking multiple parts.
- For admission to the examination, you will be required to present two forms of identification, one with a photograph, both with your signature. Approved IDs are a passport, driver's license, military ID, credit card with photo, or company ID. Student IDs are **not** acceptable. You will not be permitted into the examination without proper photo identification.
- Small lockers are available at the test centers for personal belongings. Items such as purses, briefcases, and jackets will not be allowed in the testing room.
- Small battery or solar powered electronic calculators restricted to a maximum of six functions - addition, subtraction, multiplication, division, square root, and percent - are allowed. The calculator must be non-programmable and must not use any type of tape. Memory must erase when the calculator is turned off. Candidates **will not** be allowed to use calculators that do not comply with these restrictions.
- On request, candidates will be provided with scrap paper which will be counted by the test center personnel when given to candidates and will also be counted when collected from candidates. Candidates should bring pencils or pens to the test center for use in making calculations, etc., on the provided scrap paper.
- The staff at the Prometric Technology Centers is not involved in the development of the examination or the procedures governing the evaluation of your performance. Questions or comments on the examination content or performance evaluation should be directed only to the ICMA as this is a nondisclosed examination. You may also include comments on the survey at the end of the exam.
- At the beginning of your test administration, you will be given the opportunity to take a tutorial that introduces the testing screens; the tutorial is not part of your testing time and may be repeated if the candidate wishes; however, total tutorial time is limited.
- You will have three hours to complete the examination; your time remaining will be displayed on the screen.
- Upon completion of each examination part, your performance results will be displayed on the screen, and you will also receive a printed copy of your results before leaving the testing center.
- If you pass your final part, you will receive further certification information from the ICMA. If you do not pass an examination, you will receive a new registration form from the ICMA; it should be noted that you may retake a part at any time, but you are limited to taking the part three times in a 12-month period.

MULTIPLE-CHOICE QUESTION ANSWERING TECHNIQUE

1. **Budget your time.**
   a. We make this point with emphasis. Just as you should fill up your gas tank prior to reaching empty, you should finish your exam before time expires.
   b. You have 180 minutes to answer 110 questions, i.e., 1.6 minutes per question. We suggest you attempt to answer eight questions every 10 minutes, which is 1.25 minutes per question. This would result in completing 110 questions in 137 minutes to give you almost 45 minutes to review questions that you have marked. (See 3.c.2) for a brief discussion of marking questions at Prometric.)
   c. On your Prometric computer screen, the time remaining (starting with 180 minutes) appears at the upper right of your screen.
2. **Answer the questions in chronological order.**
   a. Do **not** agonize over any one question. Stay within your time budget.
   b. It is not necessary to "mark" unanswered questions. Prometric provides options to review both unanswered questions and marked questions.
3. **For each question**
   a. **Read the question** stem carefully (the part of the question that precedes the answer choices) to determine the precise requirement.
      1) Focusing on what is required enables you to ignore extraneous information and to proceed directly to determining the correct answer.
         a) Be especially careful to note when the requirement is an **exception**; e.g., "All of the following statements regarding a company's internal rate of return are true **except**:."
   b. **Determine the correct answer** before reading the answer choices. The objective is to avoid allowing the answer choices to affect your reading of the question.
      1) When four answer choices are presented, three of them are incorrect. They are called **distractors** for a very good reason.
      2) **Read each answer choice** with close attention.
         a) Even if answer (A) appears to be the correct choice, do **not** skip the remaining answer choices. Answer (B), (C), or (D) may be better.
         b) Treat each answer choice as a true-false question.
   c. **Select the best answer**. The answer is selected by either pressing the answer letter on your keyboard or by using your mouse. Select the most likely or best answer choice. If you are uncertain, make an educated guess.
      1) The CMA/CFM does not penalize guessing because your score is determined by the number of correct responses. Thus, you should answer every question.
      2) As you answer a question, you can mark it by pressing "M" or unmark a marked question by pressing "M." After you have answered, marked, or looked at and not answered all 110 questions, you will be presented with a summary screen that shows how many questions you did not answer and how many you marked. You then have the option of revisiting all of the unanswered questions and "marked" questions.

4. Prometric Computer Screen Layout

Question 10 of 120 YOUR NAME Time remaining: 02:51

The sky is ______________

O A. blue

O B. brown

O C. green

O D. silver

Directions: Select the best answer

← Previous Next → Mark ? Help

NOTE: A menu offering a number of options is displayed at the bottom of the question screen. The options enable you to view the previous question or the next question, mark a question to be revisited, or request help. You may select an option by pressing the appropriate symbol key or highlighted letter on your keyboard or by clicking on the symbol or letter with your mouse.

1) View the previous question by indicating the letter **P** or the left arrow key.
2) View the next question by indicating the letter **N** or the right arrow key.
3) Mark a question by indicating the letter **M**.
4) Request help by indicating the question mark.

USE GLEIM *CMA/CFM TEST PREP* SOFTWARE. It emulates the Prometric testing procedures and environment including computer screen layout, software operation, etc.

TAKING THE CMA/CFM EXAMS

CMA/CFM EXAMINATIONS CHECKLIST

1. Acquire your study materials. Rely on this book and *CMA/CFM Test Prep* software. It's all you need!
2. **Apply** for membership in the IMA (see pages 478 and 479) and for admission to the ICMA (pages 480 and 481).
3. **Register** to take the desired part of the exam using the examination registration form (page 482) and send it with your applications to the ICMA. Take one part at a time. When you register and pay $80 for a part ($120 for exams at international sites), you have 120 days to take the part. Upon receipt of authorization to take the exam, call Prometric to schedule your test.
4. Plan your preparation process. It's easy. You have 10 study units to complete.
5. Orderly, controlled preparation builds confidence, reduces anxiety, and produces success!
6. PASS THE EXAMINATION (study this Introduction)!

LOGISTICAL AND HEALTH CONCERNS

As soon as the ICMA sends you a computer-based test authorization, call and schedule your test at a convenient time and convenient Prometric testing center. In almost all cases, you should be able to drive to your testing site, take the test, and return home in one day. If the exam is not being given within driving distance of your home, call Prometric Technology Centers to inquire about accommodations. Stay by yourself at a hotel to be assured of avoiding distractions. The hotel should be soundproof and have a comfortable bed and desk suitable for study. If possible, stay at a hotel with recreational facilities you normally use, e.g., a swimming pool.

Proper exercise, diet, and rest in the weeks before you take your exam are very important. High energy levels, reduced tension, and an improved attitude are among the benefits. A good aerobic fitness program, a nutritious and well-balanced diet, and a regular sleep pattern will promote your long-term emotional and physical well-being as well as contribute significantly to a favorable exam result. Of course, the use of health-undermining substances should be avoided.

EXAM PSYCHOLOGY

Plan ahead and systematically prepare. Then go to the exam and give it your best: neither you nor anyone else can expect more. Having undertaken a systematic preparation program, you will do fine.

Maintain a positive attitude and do not become depressed if you encounter difficulties either before or during the exam. An optimist will usually do better than an equally well-prepared pessimist. Remember, you have reason to be optimistic because you will be competing with many less qualified persons who have not prepared as well as you have.

CALCULATORS

Only simple six-function calculators are permitted (i.e., addition, subtraction, multiplication, division, square root, percent). Candidates are responsible for providing their own calculators. You should be thoroughly experienced in the operations of your calculator. Make sure it has fresh batteries just prior to the examination.

1. Consider bringing a backup calculator with you.
2. The calculator must be small, quiet, and battery- or solar-powered so it will not be distracting to other candidates.
3. The calculator may have a memory. However, the memory must be temporary and erase when the memory is cleared or the calculator is turned off.
4. The calculator must not use any type of tape.
5. The calculator must be nonprogrammable.
6. Nonconforming calculators and calculator instruction books are **not** permitted.

EXAMINATION TACTICS

1. Remember to bring your authorization and appropriate identification to the exam site. The photo ID requirement is **strictly** enforced.
2. Arrive at the test center at least 30 minutes prior to the scheduled exam time to allow for orientation and check-in procedures. Your appointment may be canceled if you are 30 minutes late.
3. Dressing for exam success means emphasizing comfort, not appearance. Be prepared to adjust for changes in temperature, e.g., to remove or put on a sweater.
4. Do not bring notes, this text, other books, etc., into the Prometric testing center. You will only make yourself nervous and confused by trying to cram the last 5 minutes before the exam. Books are not allowed in the testing room, anyway. You should, however, bring an appropriate calculator.
5. Adequate scratch paper and pencils are provided. You must turn in your scratch paper as you leave the exam site. Any breath mints, gum, etc., should be in your pocket as they may distract other persons taking the test.
6. As soon as you complete the exam, we would like you to e-mail, fax, or write to us with your comments on our books and software. We are particularly interested in which topics need to be added or expanded in our books and software. We are NOT asking about specific CMA/CFM questions; rather we are asking for feedback on our books and software.

RECAP OF CMA/CFM PART 4 EXAM COVERAGE

ICMA Major Topics and Percent Coverage	Gleim/Flesher Study Units
A. Decision Theory and Operational Decision Analysis--20%-30% (Level C)	1. Decision Analysis 2. Cost-Volume-Profit Analysis
B. Investment Decision Analysis--20%-30% (Level C)	3. Capital Budgeting
C. Quantitative Methods for Decision Analysis--10%-15% (Level B)	4. Decision Making under Uncertainty 5. Quantitative Methods I 6. Quantitative Methods II
D. Information Systems--20%-30% (Level B)	7. Information Systems I 8. Information Systems II
E. Management Controls--10%-15% (Level B)	9. Internal Control 10. Internal Auditing

GO FOR IT!
IT'S YOURS TO PASS!

***WARNING!!!**

About 25 to 30% of Part 1 test questions will require mathematical calculations. All other parts are approximately 35 to 40% computational questions. Practice computational questions to prepare for exam success!

STUDY UNIT 1: DECISION ANALYSIS

23 pages of outline
66 multiple-choice questions

A. Decision-Making Process
B. Marginal Analysis Applications
C. Pricing
D. Inventory Control Models
E. Government Contracts and the CASB

This study unit concerns decision-making principles and their operational applications to managerial accounting, especially differential analysis, pricing, and inventory management. The content specification outline assigns a relative weight range of 20% to 30% to decision theory and operational decision analysis. We cover this major topic here and in Study Unit 2, Cost-Volume-Profit Analysis.

A. Decision-Making Process

1. The decision-making process is an integral part of the planning process. The consideration of choices and the selection of the choice expected to give the best results are a key part of the planner's job. Rational decision making involves
   a. Defining the problem
   b. Identifying choices
   c. Evaluating choices
   d. Making the choice
   e. Following up
2. Decision making may be programmed or nonprogrammed.
   a. The **programmed decision process** is used for routine or repetitive decisions. This process is based upon decision rules. They are the parameters within which a routine decision is based.
   b. The **nonprogrammed decision process** is used for complex, nonroutine situations, calling for the use of the problem-solving process. Four questions need to be answered before making a nonprogrammed decision.
      1) What decision needs to be made?
      2) When does it need to be made?
      3) Who makes the decision?
      4) Who needs to be consulted before implementation of the decision?
3. Awareness of the existence of a problem, i.e., realization that a gap exists between planned results and reality (between where the enterprise is and where it wants to be), precedes defining the problem.
   a. **Bounded rationality** (the inability to see all aspects of the situation) may prevent accurate problem definition. Faulty perception may preclude the decision maker from seeing the correct problem.
      1) We see what we **expect** to see.
      2) We see what we are **trained** to see.
      3) Our **needs** influence what we observe.
      4) **Group pressure** may influence what problems we perceive or how we define them.

b. The most common difficulty in problem solving is defining the problem. Difficulties in defining problems include

1) Defining the problem according to a solution

a) Do not rule out other solutions.
b) Focus on "bringing the actual situation closer to the desired situation."

2) Concentrating on nonpertinent areas

a) Define minor problems.
b) Concentrate on problems that are most important.
c) Rank the priority of the problems in an organized manner.

3) Defining problems by their symptoms

a) Focus on the causes of the gap between "the actual and the desired situations."

4. **Searching for Solutions**

a. The search can be primary (e.g., surveys, samples, attitude questionnaires) or secondary (e.g., reading scholarly papers).

b. It can be creative.

1) **Brainstorming** -- an unstructured approach that relies on the spontaneous contribution of ideas from all members of a group. This technique breaks down broadly based problems into their essentials.

2) **Synectics** -- a highly structured group approach to problem statement and solution based on creative thinking. It involves free use of analogies and metaphors in informal exchange within a carefully selected small group of individuals of diverse personality and areas of specialization.

3) **Delphi technique** -- an approach in which the manager solicits opinions on a problem from experts in the field, summarizes the opinions, and feeds the summaries back to the experts (without revealing any of the participants to each other). The process is reiterated until the opinions converge on an optimal solution.

4) **Value analysis** -- a methodical approach primarily designed to optimize performance of an activity at a minimum cost. Emphasis is on the cost-benefit criterion.

5) **Free association** -- an approach to idea generation that reports the first thought to come to mind in response to a given stimulus, for example, a symbol or analogy pertaining to a product for which an advertising slogan is sought. The objective is to express the content of consciousness without censorship or control.

6) **Forced relationship** -- a structured adaptation of free association. The elements of a problem are analyzed and the associations among them are identified so as to detect patterns that may suggest new ideas.

7) **Blast! then refine** -- a U.S. Navy problem-solving methodology. It completely disregards (blasts out of one's consciousness) any existing approach. An entirely new problem solution is then sought that will attain the original objectives.

8) **Edisonian approach** -- a trial-and-error experimental method. It should usually not be applied unless other approaches have been unsuccessful.

9) **Morphological matrix analysis** -- a structured technique that plots decision variables along the axes of a matrix. The relationships of these variables are found in the squares within the chart.

10) **Attribute listing** -- a technique applied primarily to improve a tangible object. It lists the parts and essential features of the object and systematically analyzes modifications intended as improvements.

11) **Trial-and-error** -- an approach that generates new ideas based on failed ones

12) **Creative leap** -- a process that formulates an ideal solution and then works back to a feasible one

c. In open systems, the search cannot feasibly consider all possibilities.

d. The search ordinarily ends for most people when they find a solution they consider not optimal but good enough (satisficing).

e. Information processing for decision making is characterized by two styles.

1) The thinking or **analytic** style is objective, deductive, precise, detailed, logical, and repetitive. It tends to prevail in the traditional, pyramidal organization.

2) The **intuitive** style is subjective, inductive, nonroutine, and unstructured. Intuitive people may rely on hunches and reach decisions without explicitly applying a formal process.

f. **The Limiting Factor Principle**. The crucial factor is the one that limits or prevents achievement of the desired goal. The mass of alternative choices can be reduced if managers can identify the limiting factors to a problem.

1) EXAMPLE: If a company directs a division to turn a loss into a profit, the immediate means might be purchasing new equipment, and a limiting factor might be lack of cash or good credit.

a) Seeking alternative sources for the equipment may reveal someone willing to swap, to take a lien, etc.

b) Alternatives calling for purchases and loans can be discarded.

5. **Evaluation of Solutions**. The analysis of facts and the development of solutions have been improved in recent years through the adoption of mathematical techniques. These techniques are not a panacea -- they will not provide useful solutions for poorly formulated problems -- but they will provide valuable results for well-defined quantitative problems. Some of these quantitative tools or approaches are defined below and on the following page.

a. **Operations research** is an overall approach that applies a variety of scientific and mathematical techniques. It involves the

1) Recognition of the complete system, including interrelationships, in which the problem operates

2) Formulation of a mathematical model

3) Testing of the model

4) Implementation of the optimal solution as produced by the model

b. **Simulation** is an approach that is economically feasible because of the availability of high-speed computers.

   1) Simulation is used when an optimization model is not possible.

      a) For example, the number of variables may exceed the number of equations describing the variables.

   2) Simulation requires the construction of a quantitative model that incorporates most of the features of the problem situation.

   3) The model is then tested in a variety of hypothetical situations.

      a) For example, if certain conditions hold true, what performance can be expected from the system being modeled?

   4) The modeled response to the situations can then be measured.

c. **Linear programming** is a tool for allocating scarce resources in the presence of constraints.

   1) The problem must be described by a system of linear equations. An objective function (usually cost or revenue) is maximized or minimized subject to constraint equations. This description is seldom strictly accurate but provides useful approximations.

   2) The simultaneous solution to the system of equations maximizes profit (or minimizes cost).

d. **Cost-benefit analysis** is a simple test for possible solutions. Cost should be less than the benefits realized.

e. **Queuing theory** is a technique to analyze waiting-line problems through the use of probability theory. It minimizes the sum of the cost of waiting lines and the cost of reducing or eliminating queues.

f. **Decision trees (decision theory)** map out possible actions given probabilistic events. Probabilities are assigned, and the expected value (payoff or loss) for each decision choice and the events that might follow from that choice is calculated.

   1) A current decision may entail a choice among multiple options, each of which may lead to different future events (states of nature).

      a) EXAMPLE: If one invests in a fast-food franchise and the economy improves, investing in the franchise is the action, and the improvement in the economy is the future event.

g. **Monte Carlo method**. Random behavior may be added to otherwise deterministic models to simulate the uncertainty inherent in real-world situations. The model is then run a large number of times, and the variance of the probability distribution of the results and the average performance are determined.

h. **Certainty, risk, and uncertainty**. Assigning probabilities or assuming away uncertainty with precise mathematical formulas is not always feasible.

   1) **Certainty** is a condition in which a decision maker can predict the outcome of a decision with perfect accuracy.

2) **Risk** is commonly defined as a condition in which a decision maker, using incomplete but reliable information, can estimate the probability of a given outcome of a decision.

a) **Probability** is a number between 0 and 1 inclusive that measures the likelihood of the occurrence of some event. The less likely the event, the closer the number is to 0. The two types of probabilities are objective and subjective.

i) Objective probabilities are derived mathematically from historical data.

ii) Subjective probabilities are estimated based on past experience and judgment.

3) **Uncertainty** may be used in two senses. It is most often defined as an absence of certainty. Some technical writers, however, define it narrowly as the condition in which no reliable factual evidence is available to support a probability estimate. Thus, an objective probability cannot be stated for a given outcome of a decision. Intuition and creative decision-making may be necessary.

6. **Making the Choice**. The selection of the best available solution is the objective of the decision-making process.

a. The optimal solution is selected.

1) If the complete decision-making process has been followed, the result will be profit-maximizing behavior rather than satisficing behavior.

a) This result should be attained if cost-benefit analysis is applied to the overall decision-making process as well as to its individual steps.

b. Human aspects of the choice must be considered. A decision is useful only if those responsible for its implementation are convinced of its value.

1) Decisions are made in a multi-dimensional environment and therefore have many, sometimes unexpected, effects.

2) An otherwise optimal solution may fail if the opinion and attitudes of the people involved are not considered in the decision-making process.

3) The multi-step process of decision making must be followed when making organizational changes.

a) Implementing change is a critical part of the decision-making process because people are crucial to the effectiveness of the choice.

7. **Following Up on the Effects of the Action**. The proper delegation of authority is made to the individual responsible for implementing the action to be taken. Performance measurement is then provided for.

a. This process puts the solution into a feasible framework, which assures it will be carried out.

b. The failure to provide for practical implementation causes many otherwise correct decisions to fail to produce the desired results.

c. Decisions cannot be fully implemented until they are effectively communicated.

8. **Three Decision-Making Problems**
   a. **Framing error** is the human tendency to judge positively presented information favorably and negatively presented information unfavorably.
   b. **Escalation** is the human tendency to become committed to failing courses of action instead of quitting or admitting error.
   c. **Overconfidence** can expose the decision maker to unreasonable risks.

9. The foregoing discussions of problem solving and decision making are based on the optimizing model and the concept of rationality. According to Robbins, *Organizational Behavior* (6th ed., Prentice-Hall, 1993), rationality has the same assumptions as the optimizing (outcome-maximizing) model for decision making. Rational decision making is fully objective and logical. The model assumes that
   a. A single, well-defined goal is to be maximized.
   b. All relevant criteria and feasible options are known.
   c. The criteria and options can be assigned numerical values and are ranked.
   d. Preferences are constant (criteria and their assigned weights do not change).
   e. The decision maker will choose the option with the highest rank (the maximum benefits).

10. Effective decision making entails sound reasoning. Decisions are inferences that should be derived from carefully selected premises. Inferences are of two basic types:
   a. **Inductive reasoning** is the inference of a generalized conclusion from particular instances. It is the basis of experimental science.
      1) For example, a financial audit involves gathering specific evidence to arrive at a general conclusion about the fair presentation of the financial statements.
   b. **Deductive reasoning** is the inference of a conclusion about particulars from general or universal premises.
      1) For example, an auditor may apply the principles of statistical analysis to arrive at a specific conclusion regarding the likelihood that an account balance is not misstated by more than a stated amount.

11. Stop and review! You have completed the outline for this subunit. Study multiple-choice questions 1 through 9 beginning on page 46.

B. Marginal Analysis Applications

1. The typical problem for which marginal (differential or incremental) analysis can be used involves choices among courses of action. The quantitative analysis emphasizes the ways in which costs and benefits vary with the option chosen. Thus, the focus is on incremental, not total, costs.
   a. EXAMPLE: Assume that a firm produces a product for which it incurs the following unit costs:

Raw materials	$2.00
Direct labor	3.00
Variable overhead	.50
Fixed overhead	2.50
Total cost	$8.00

1) The product normally sells for $10 per unit. An application of marginal cost analysis is necessary if a foreign buyer, who has never before been a customer, offers to pay $5.75 per unit for a **special order** of the firm's product. The immediate reaction might be to refuse the offer because the selling price is less than the average cost of production by a considerable amount. However, marginal cost analysis results in a different decision.

2) Under marginal cost analysis, assuming that the firm has idle capacity, only the additional costs should be considered. In this example, the only marginal costs are for raw materials, direct labor, and variable overhead. No additional fixed overhead costs would be incurred. Because the $5.75 selling price (marginal revenue) exceeds the $5.50 of marginal costs ($2 materials + $3 labor + $.50 variable OH), accepting the special order will be profitable.

2. In addition to special-order situations, marginal costing can be used in other cases.

   a. Make-or-buy (insourcing or outsourcing) decisions
   b. Adding or dropping product lines
   c. Sell-or-process-further decisions
   d. Acquiring or divesting a company segment
   e. Selecting a market channel

3. Caution must always be used in applying marginal cost analysis because of the many nonquantitative factors that must be considered, including

   a. Will special price concessions place the firm in violation of the price discrimination provisions of the Robinson-Patman Act of 1936?

   b. What is the effect of government contract pricing regulations (see subunit E. on page 44)?

   c. Will sales to the special customer affect sales in the firm's regular market?

   d. Will regular customers find out about the special price and demand equal terms?

   e. In the case of dropping a product line, will sales be hurt in the other product lines (e.g., the dropped product may have been an unintended loss leader)?

4. Additional cost terms may appear in decision analysis situations.

   a. An **avoidable cost** can be saved by not adopting a particular option.

   b. An **imputed cost** requires no dollar outlay and is not recognized in the accounting records but must be considered in an investment decision to give recognition to economic reality. An opportunity cost is an example.

   c. A **postponable cost** may be shifted to the future with little effect on the efficiency of current operations. Routine maintenance is an example.

   d. A **relevant cost** should be considered in decision making because it is a future cost that will vary with the option chosen.

   e. A **sunk cost** cannot be avoided because the expenditure has occurred or an irrevocable decision to incur the cost has been made. Sunk costs are irrelevant to management decision making because they cannot vary with the option selected.

   f. **Taxes**. See Appendix B, which begins on page 467. Taxes are usually a significant cost component of every business decision.

5. **Make-or-Buy Decisions**

   a. The concept behind a make-or-buy decision is to use available resources as efficiently as possible before outsourcing. Often, an array of products can be produced efficiently if production capacity is available. If not enough capacity is available to produce them all, the products that are produced least efficiently should be outsourced (or capacity should be expanded).

   b. In a make-or-buy decision, the manager considers only the costs relevant to the investment decision. If the total relevant costs of production are less than the cost to buy the item, it should be insourced. The key variable is relevant costs, not total costs.

      1) Past (sunk) costs are irrelevant. Hence, a production plant's $100,000 of repairs last year is irrelevant to this year's make-or-buy decision.

      2) Book value of old equipment is also irrelevant because it is a past (sunk) cost.

      3) Opportunity costs must be considered when idle capacity is not available. They are of primary importance because they represent the forgone opportunities of the firm. The **opportunity cost** is the maximum benefit forgone by using a scarce resource for a given purpose. It is the benefit, for example, the contribution to income, provided by the best alternative use of that resource.

   c. EXAMPLE: Should a company make or buy an item?

	Make	Buy
Total VC	$10	
Allocation of FC	5	
Total unit costs	$15	$13

      1) At excess capacity, the decision should be to produce the item. Total variable cost ($10) is less than the purchase price. However, if the plant is already running at 100% capacity, the opportunity cost of displaced production becomes a relevant cost that might alter the decision in favor of purchasing the item from a supplier.

6. **Disinvestment**

   a. Disinvestment decisions are the opposite of capital budgeting decisions, i.e., to terminate rather than start an operation. Four steps should be taken in making a disinvestment decision:

      1) Identify fixed expenses that would be curtailed by the disinvestment decision, e.g., depreciation and insurance on equipment used.

      2) Determine the revenue needed to justify continuing operations (variable cost of production).

      3) Establish the opportunity cost of funds that will be received upon disinvestment (e.g., salvage value).

4) Determine if the book value of the assets is equal to the economic value of the capital. If not, reevaluate the decision using current fair value rather than the book value.

b. When a firm disinvests in a project, excess capacity exists unless another project uses this capacity immediately. The cost of idle capacity should be treated as a relevant cost.

c. In general, if the marginal cost of a project is greater than the marginal revenue, the firm should disinvest.

7. Stop and review! You have completed the outline for this subunit. Study multiple-choice questions 10 through 32 beginning on page 49.

C. Pricing

1. Pricing Objectives

a. Profit maximization. Classical economic theory assumes all firms always select the price that results in the highest profit.

b. Target margin maximization. This objective is stated as a percentage ratio of profits to sales.

c. Volume-oriented objectives set prices to meet target sales volumes or market shares.

d. Image-oriented objectives set prices to enhance the consumer's perception of the firm's merchandise mix.

e. Stabilization objectives set prices to maintain a stable relationship between the firm's prices and the industry leader's prices.

2. Price-Setting Factors

a. Internal factors

1) Marketing objectives may include survival, current profit maximization, market-share leadership, or product-quality leadership.

2) Marketing-mix strategy

3) Costs (variable, fixed, and total costs)

4) Organizational locus of pricing decisions

5) Capacity

a) For example, under **peak-load pricing**, prices vary directly with capacity usage. Thus, when idle capacity is available, that is, when demand falls, the price of a product or service tends to be higher given a peak-load pricing approach.

b. External factors

1) The market and demand

a) Type of market (pure competition, monopolistic competition, oligopolistic competition, or pure monopoly)

b) Consumer perceptions of price and value

c) The price-demand relationship

i) The demand curve for normal goods is ordinarily downward sloping to the right (quantity demanded increases as the price decreases).

ii) However, over some intermediate range of prices, the reaction to a price increase for prestige goods is an increase, not a decrease, in the quantity demanded. Within this range, the demand curve is upward sloping. The reason is that consumers interpret the higher price to indicate a better or more desirable product. Above some price level, the relation between price and quantity demanded will again become negatively sloped.

d) Price elasticity of demand. If demand is price elastic (inelastic), the ratio of the percentage change in quantity demanded to the percentage change in price is greater (less) than 1.0.

2) Competitors' costs, prices, and offers

3. **General Pricing Approaches**

a. Cost-based pricing (See sections 4. and 5. beginning on page 34.)

b. Buyer-based pricing involves basing prices on the product's perceived value rather than on the seller's cost. Nonprice variables in the marketing mix augment the perceived value. For example, a cup of coffee may have a higher price at an expensive restaurant than at a fast-food outlet.

c. Competition-based pricing

1) Going-rate pricing bases price largely on competitors' prices.

2) Sealed-bid pricing bases price on a company's perception of its competitors' prices.

d. New product pricing

1) Price skimming is the practice of setting an introductory price relatively high to attract buyers who are not concerned about price and to recover research and development costs.

2) Penetration pricing is the practice of setting an introductory price relatively low to gain deep market penetration quickly.

e. Pricing by intermediaries

1) Using markups tied closely to the price paid for a product
2) Using markdowns -- a reduction in the original price set on a product

f. Price adjustments

1) Geographical pricing

a) FOB-origin pricing charges each customer its actual freight costs.

b) A seller that uses uniform delivered pricing charges the same price, inclusive of shipping, to all customers regardless of their location.

i) Thus, both nearby and distant customers are charged the same amount. This policy is easy to administer, permits the company to advertise one price nationwide, and facilitates marketing to faraway customers.

c) Zone pricing sets differential freight charges for customers on the basis of their location. Customers are not charged actual average freight costs.

d) Basing-point pricing charges each customer the freight costs incurred from a specified city to the destination regardless of the actual point of origin of the shipment.

e) A seller that uses freight-absorption pricing absorbs all or part of the actual freight charges. Customers are not charged actual delivery costs.

2) Discounts and allowances

a) Cash discounts encourage prompt payment, improve cash flows, and avoid bad debts.

b) Quantity discounts encourage large volume purchases.

c) Trade (functional) discounts are offered to other members of the marketing channel for performing certain services, such as selling.

d) Seasonal discounts are offered for sales out of season. They help smooth production.

e) Allowances (e.g., trade-in and promotional allowances) reduce list prices.

3) Discriminatory pricing adjusts for differences among customers, the forms of a product, or locations.

4) Psychological pricing is based on consumer psychology. For example, consumers who cannot judge quality may assume higher prices correlate with higher quality.

5) Promotional pricing temporarily reduces prices below list or even cost to stimulate sales.

6) Value pricing entails redesigning products to improve quality without raising prices or offering the same quality at lower prices.

7) International pricing adjusts prices to local conditions.

g. Product-mix pricing

1) Product-line pricing sets price steps among the products in the line based on costs, consumer perceptions, and competitors' prices.

2) Optional-product pricing requires the firm to choose which products to offer as accessories and which as standard features of a main product.

3) Captive-product pricing involves products that must be used with a main product, such as razor blades with a razor. Often the main product is relatively cheap, but the captive products have high markups.

4) By-product pricing usually sets prices at any amount in excess of storing and delivering by-products. Such prices allow the seller to reduce the costs and therefore the prices of the main products.

5) Product-bundle pricing entails selling combinations of products at a price lower than the combined prices of the individual items. This strategy promotes sales of items consumers might not otherwise buy if the price is low enough. An example is season tickets for sports events.

h. Certain pricing tactics are illegal. For example, pricing products below cost to destroy competitors (**predatory pricing**) is illegal.

1) The U.S. Supreme Court has held that a price is predatory if it is below an appropriate measure of costs and the seller has a reasonable prospect of recovering its losses in the future through higher prices or greater market share.

2) Also illegal is price **discrimination** among customers. The **Robinson-Patman Act** makes such pricing illegal if it has the effect of lessening competition, although price discrimination may be permissible if the competitive situation requires it and if costs of serving some customers are lower. The Robinson-Patman Act applies to manufacturers, not service entities.

3) Another improper form of pricing is **collusive pricing**. Companies may not conspire to restrict output and set artificially high prices. Such behavior violates antitrust laws.

4) Still another inappropriate pricing tactic is selling below cost in other countries (**dumping**), which may trigger retaliatory tariffs and other sanctions.

4. A **cost-plus price** equals the cost plus a markup. Cost may be defined in many ways. Most companies use either absorption manufacturing cost or total cost when calculating the price. Variable costs may be used as the basis for cost, but then fixed costs must be covered by the markup.

a. Following are four commonly used cost-plus pricing formulas:

1) $$\textit{Price} = \textit{Total cost} + (\textit{Total cost} \times \textit{Markup percentage})$$

2) $$\textit{Price} = \begin{array}{c}\textit{Absorption}\\ \textit{manufacturing}\\ \textit{cost}\end{array} + \left(\begin{array}{c}\textit{Absorption}\\ \textit{manufacturing}\\ \textit{cost}\end{array} \times \textit{Markup percentage}\right)$$

3) $$\textit{Price} = \begin{array}{c}\textit{Variable}\\ \textit{manufacturing}\\ \textit{cost}\end{array} + \left(\begin{array}{c}\textit{Variable}\\ \textit{manufacturing}\\ \textit{cost}\end{array} \times \textit{Markup percentage}\right)$$

4) $$\textit{Price} = \begin{array}{c}\textit{Total}\\ \textit{variable}\\ \textit{cost}\end{array} + \left(\begin{array}{c}\textit{Total}\\ \textit{variable}\\ \textit{cost}\end{array} \times \textit{Markup percentage}\right)$$

b. The costs of unused capacity in production facilities, distribution channels, marketing organizations, etc., are ordinarily not assignable to products or services on a cause-and-effect basis, so their inclusion in overhead rates may distort pricing decisions. Including the fixed costs of unused capacity in a cost-based price results in higher prices and in what is known as the **downward (black hole) demand spiral**.

1) As higher prices depress demand, unused capacity costs increase, the fixed costs included in prices increase, and demand will continue to spiral downward. One way to avoid this problem is not to assign unused capacity costs to products or services. The result should be better operating decisions and better evaluation of managerial performance.

5. A **target price** is the expected market price for a product, given the company's knowledge of its consumers and competitors.
   a. Subtracting the unit target profit margin determines the long-term **target cost** of the product.
      1) Because it may be lower than the full cost of the product, the target cost may not be achievable unless the company adopts comprehensive cost-reduction measures.
      2) The Japanese concept of **Kaizen** is relevant to target costing. A policy of seeking continuous improvement in all phases of company activities facilitates cost reduction, often through numerous minor changes.
      3) **Value engineering** is a means of reaching targeted cost levels. It is a systematic approach to assessing all aspects of the value chain cost buildup for a product: R&D, design of products, design of processes, production, marketing, distribution, and customer service. The purpose is to minimize costs without sacrificing customer satisfaction.
         a) Value engineering requires distinguishing between cost incurrence and locked-in costs. Cost incurrence is the actual use of resources, whereas locked-in (designed-in) costs will result in use of resources in the future as a result of past decisions. Traditional cost accounting focuses on budget comparisons, but value engineering emphasizes controlling costs at the design stage, that is, before they are locked in.
6. **Life-cycle costing** is sometimes used as a basis for cost planning and product pricing. Life-cycle costing estimates a product's revenues and expenses over its expected life cycle. The result is to highlight upstream and downstream costs in the cost planning process that often receive insufficient attention. Emphasis is on the need to price products to cover all costs, not just production costs.
   a. A concept related to life-cycle cost that is relevant to pricing is **whole-life cost**, which equals life-cycle costs plus after-purchase costs (operating, support, repair, and disposal) incurred by customers. Reduction of whole-life costs is a strong competitive weapon. Customers may pay a premium for a product with low after-purchase costs.
7. Stop and review! You have completed the outline for this subunit. Study multiple-choice questions 33 through 41 beginning on page 58.

D. Inventory Control Models

1. **Inventory Fundamentals**
   a. An organization carries inventories because of the difficulty in predicting the amount, timing, and location of both supply and demand. The purpose of inventory control is to determine the optimal level of inventory necessary to minimize costs.
   b. Although the traditional approach to inventory management has been to minimize inventory and the related holding costs, many companies find inventory a good hedge against inflation.
      1) Stockpiles also guarantee future availability of inventory.

c. Inventory carrying costs are sometimes transferred to suppliers or customers.

1) If a manufacturer has good enough control of production schedules to know exactly when materials are needed, orders can be placed so that materials arrive no earlier than when actually needed.

a) This practice relies on a supplier who takes the responsibility for storing the needed inventory and shipping it to arrive on time.

b) Suppliers are more willing to provide this service when they have many competitors.

2) Customers can sometimes be persuaded to carry large quantities of inventory by allowing them special quantity discounts or extended credit terms.

3) If customers are willing to accept long lead times, inventory can be manufactured to order to avoid storing large quantities.

4) Although these measures can reduce inventory carrying costs, additional costs might be incurred by adopting them.

a) Stockout costs may increase (e.g., sales may be lost) because customers may not always be willing to wait for goods to be produced.

b) Production shutdowns and additional shipping costs may also prevent minimization of carrying costs from being cost effective.

d. **Inventory costs**

1) **Order costs** include all costs associated with preparing a purchase order.

2) **Carrying costs** include rent, insurance, taxes, security, depreciation, and opportunity cost (i.e., the cost incurred by investing in inventory rather than making an income-earning investment). Carrying costs may also include a charge for spoilage of perishable items or for obsolescence.

3) **Shortage (stockout) costs** include those costs incurred when an item is out of stock. These include the lost contribution margin on sales, customer ill will, and production interruptions.

e. Inventory policies should consider the types of costs and any limitations the firm may have, such as storage space.

1) Constraints may also be imposed by suppliers.
2) The cost of maintaining inventory records should also be considered.

2. **Economic Order Quantity (EOQ)**

a. **Inventory models** are quantitative models designed to control inventory costs by determining the optimal time to place an order (or begin production) and the optimal order quantity (production run).

1) The timing of an order can be periodic (placing an order every X days) or perpetual (placing an order whenever the inventory declines to X units).

a) **Periodic order systems** place minimal emphasis on record keeping. However, a risk of substantial overstock or understock may arise unless inventories are checked for assurance that the model is still appropriate.

b) **Perpetual systems** detect an inventory decline to the reorder point by entering every addition and withdrawal on a perpetual record displaying the balance.

i) An alternative is the **two-bin method** for physical storage. In this system, the reorder level amount is stored separately from the balance of the items. When the stock clerk removes the last item from the balance bin, an order should be placed. The reorder level bin is then used until the order is received. A **red-line system** uses only one bin, but a red line is drawn at the reorder level.

c) Physical inventories should be taken to reconcile records and verify models in either a periodic or a perpetual system. Some companies take a single, comprehensive year-end inventory. Others use a cycle inventory count; they count a different segment of the inventory periodically, for example, every week.

b. The basic EOQ model minimizes the sum of ordering (setup) costs and carrying costs.

1) The following are the characteristics of this model:

a) Demand is known and uniform throughout the period.

b) The fixed costs of ordering are eliminated when the total cost equation is differentiated to arrive at the EOQ.

c) Cost per order (setup) and unit carrying cost are constant.

i) Thus, the model is based on variable costs.

d) Full replenishment occurs instantly when the last item is used, stockout costs are zero, and no safety stock is held.

2) The EOQ is the square root of twice the periodic demand multiplied by the order (setup) cost, divided by the periodic unit carrying cost.

$$X = \sqrt{\frac{2aD}{k}}$$

If: X = EOQ
a = variable cost per order (setup)
D = periodic demand in units
k = unit periodic carrying cost

3) EXAMPLE: If periodic demand is uniform at 1,000 units, the cost to place an order is \$4, and the cost to carry one unit in inventory for a period is \$2, the EOQ is 63.25 units.

$$EOQ = \sqrt{\frac{2(\$4)(1{,}000)}{\$2}} = 63.25 \text{ units}$$

4) The average level of inventory for this model will be one-half of the EOQ. The formula shows that the EOQ varies directly with demand and order (setup) costs, but inversely with carrying costs. Thus, if demand quadruples, the EOQ will only double.

5) The EOQ is a periodic model. The number of orders (production runs) per period is given by the periodic demand divided by the EOQ.

6) The EOQ results from differentiating the total cost with regard to order (production) quantity. It is the minimum point on the total cost curve. It also corresponds to the intersection of the variable carrying cost and variable ordering cost curves. The fundamental EOQ model is based on variable costs. As explained on the previous page, order (setup) cost and unit carrying cost are variable, and fixed costs are eliminated when the total cost equation is differentiated.

$$c = \left(\frac{D}{x}\right)a + \frac{xk}{2} + F$$

$$dc/dx = \frac{-Da}{x^2} + \frac{k}{2}$$

If:
- c = total cost
- x = EOQ
- a = variable cost per order (setup)
- D = periodic demand in units
- k = unit periodic carrying cost
- F = fixed costs

Setting dc/dx = 0 at minimum total cost,

$$\frac{-Da}{x^2} + \frac{k}{2} = 0$$

$$\frac{k}{2} = \frac{Da}{x^2}$$

$$x^2 = \frac{2Da}{k}$$

$$x = \sqrt{\frac{2aD}{k}}$$

c. **Reorder points, safety stock, stockout cost**

1) In addition to minimizing the sum of order and carrying costs, an organization should minimize the cost of holding safety stock and the cost of stockouts.

a) Safety stock is the amount of extra inventory that is kept to guard against stockouts. It is the amount at the time of reordering minus the expected usage while the new goods are in transit. The problem may be diagrammed as follows:

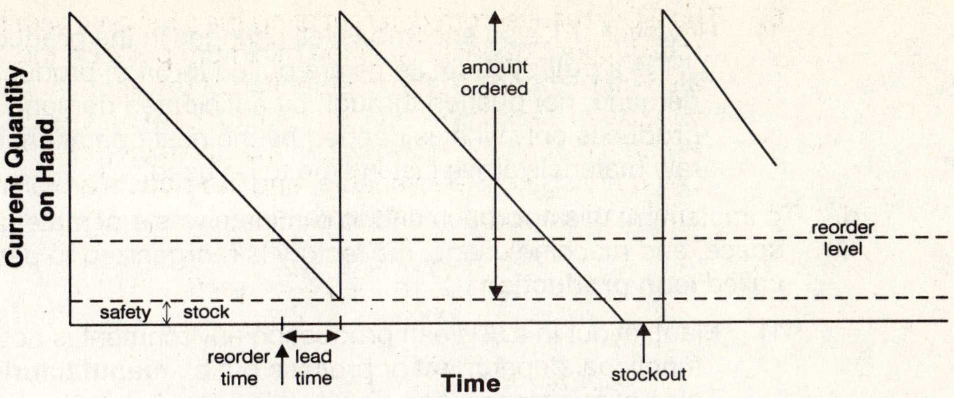

i) The EOQ determines order size.

ii) The reorder point is the intersection of the reorder level and the downward-sloping total inventory line that allows sufficient lead time for an order to be placed and received.

3. **Just-in-Time Systems**

a. Modern inventory management has departed from the EOQ approach in favor of the **just-in-time (JIT)** model. U.S. companies have traditionally built parts and components for subsequent operations on a preset schedule. Such a schedule provides a cushion of inventory so that the next operation will always have parts to work with -- a **just-in-case method**. In contrast, JIT limits output to the demand of the subsequent operation.

1) Reductions in inventory levels result in less money invested in idle assets; reduction of storage space requirements; and lower inventory taxes, pilferage, and obsolescence risks. High inventory levels often mask production problems because defective parts can be overlooked when plenty of good parts are available. If only enough parts are made for the subsequent operation, however, any defects will immediately halt production.

a) The focus of quality control under JIT shifts from the discovery of defective parts to the prevention of quality problems, so total preventive maintenance and zero defects are ultimate goals. Higher quality and lower inventory go together.

b. JIT is a reaction to the trends of global competition and rapid technological progress that have resulted in shorter product life-cycles and greater consumer demand for product diversity. Higher productivity, reduced order costs as well as carrying costs, faster and cheaper setups, shorter manufacturing cycle times, better due date performance, improved quality, and more flexible processes are goals of JIT methods. The ultimate objectives are increased competitiveness and higher profits.

c. JIT systems are based on a manufacturing philosophy popularized by the Japanese that combines purchasing, production, and inventory control. Minimization of inventory is a goal because many inventory-related activities are viewed as nonvalue-added. Indeed, carrying inventory is regarded as a symptom of correctable problems such as poor quality, long cycle times, and lack of coordination with suppliers.

1) However, JIT also encompasses changes in the production process itself. JIT is a **pull system**; items are pulled through production by current demand, not pushed through by anticipated demand. Thus, one operation produces only what is needed by the next operation, and components and raw materials arrive just in time to be used.

d. To implement this approach and to eliminate waste of materials, labor, factory space, and machine usage, the factory is reorganized to permit what is often called **lean production**.

1) Plant layout in a JIT/lean production environment is not arranged by functional department or process but by **manufacturing cells**. Cells are sets of machines, often grouped in semicircles, that produce a given product or product type. Each worker in a cell must be able tc operate all machines and, possibly, to perform support tasks, such as setup activities, preventive maintenance, movement of work-in-process within the cell, and quality inspection.

a) In a pull system, workers might often be idle if they are not multi-skilled. Hence, central support departments are reduced or eliminated, space is saved, fewer and smaller factories may be required, and materials and tools are brought close to the point of use. Manufacturing cycle time and setup time are also reduced. As a result, on-time delivery performance and response to changes in markets are enhanced, and production of customized goods in small lots becomes feasible.

e. A cellular organization requires workers to operate as effective teams, so **employee empowerment** is crucial in a JIT/lean production system. Greater participation by employees is needed to achieve continuous improvement and zero defects goals, so they may, for example, have the power to stop production to correct a problem, be consulted about changes in processes, or become involved in hiring co-workers. Thus, managers in such a system usually play more of a facilitating rather than a support role.

f. The Japanese term **kanban** and JIT have often been confused. JIT is the total system of purchasing, production, and inventory control. Kanban is one of the many elements in the JIT system as it is used in Japan.

1) The word kanban means ticket. Tickets (also described as cards or markers) control the flow of production or parts so that they are produced or obtained in the needed amounts at the needed times. A basic kanban system includes a **withdrawal kanban** that states the quantity that a later process should withdraw from its predecessor, a **production kanban** that states the output of the preceding process, and a **vendor kanban** that tells a vendor what, how much, where, and when to deliver.

a) U.S. companies have not been comfortable with controlling production using tickets on the production floor. Computerized information systems have been used for many years, and U.S. companies have been reluctant to give up their computers in favor of the essentially manual kanban system. Instead, U.S. companies have integrated their existing systems, which are complex computerized planning systems, with the JIT system.

g. Another feature of the lower inventory levels in a JIT system is elimination of the need for several traditional internal controls.

1) Frequent receipt of deliveries from suppliers often means less need for a sophisticated inventory control system and for control personnel.

2) JIT also may eliminate central receiving areas, hard copy receiving reports, and storage areas. A central warehouse is not needed because deliveries are made by suppliers directly to the area of production.

3) The quality of parts provided by suppliers is verified by use of statistical controls rather than inspection of incoming goods. Because storage, counting, and inspecting are eliminated in an effort to perform only value-adding work, the supplier's dependability is crucial.

h. Organizations that adopt JIT systems develop close relationships with a few carefully chosen suppliers who are extensively involved in the buyer's processes, and long-term contracts are typically negotiated to reduce order costs.

1) Indeed, some major retailers have agreed to **continuous replenishment** arrangements whereby a supplier with superior demand forecasting ability essentially tells the buyer when and how much to reorder.

2) Buyer-supplier relationships are further facilitated by **electronic data interchange (EDI)**, a technology that allows the supplier access to the buyer's online inventory management system. Thus, electronic messages replace paper documents (purchase orders and sales invoices), and the production schedules and deliveries of the parties can be more readily coordinated.

4. **Other Inventory Control Techniques**

a. **The ABC system**

1) This method controls inventories by dividing items into three groups.

a) Group A -- high-dollar-value items, which account for a small portion (perhaps 10%) of the total inventory usage

b) Group B -- medium-dollar-value items, which may account for about 20% of the total inventory items

c) Group C -- low-dollar-value items, which account for the remaining 70% of sales or usage

2) The ABC system permits the proper degree of managerial control to be exercised over each group. The level of control reflects cost-benefit concerns.

a) Group A items are reviewed on a regular basis.

b) Group B items may not have to be reviewed as often as group A items, but more often than group C items.

c) For group C, extensive use of models and records is not cost effective. It is cheaper to order large quantities infrequently.

b. **Materials requirements planning (MRP)**

1) MRP is a computer-based information system designed to plan and control raw materials used in a production setting.

a) It assumes that the demand for materials is typically dependent upon some other factor, which can be programmed into the computer.

b) The timing of deliveries is vital to avoid production delays.

c) EXAMPLE: An auto manufacturer need only tell a computer how many autos of each type are to be manufactured. The MRP system determines how many of every component part will be needed. The computer will generate a complete list of every part and component needed.

2) MRP, in effect, creates schedules of when items of inventory will be needed in the production departments. If parts are not in stock, the computer will automatically generate a purchase order on the proper date (considering lead times) so that deliveries will arrive on time.

c. **Manufacturing resource planning (MRP-II)**

1) MRP-II is a closed-loop manufacturing system that integrates all facets of a manufacturing business, including production, sales, inventories, schedules, and cash flows. The same system is used for both the financial reporting and managing operations (both use the same transactions and numbers).

a) MRP-II uses a master production schedule (MPS), which is a statement of the anticipated manufacturing schedule for selected items for selected periods.

b) MRP also uses the MPS. Thus, MRP is a component of an MRP-II system.

d. **Enterprise resource planning (ERP)** is the latest phase in the development of computerized systems for managing organizational resources. ERP is intended to integrate enterprise-wide information systems. An ERP application consists of database workflow software that connects all organizational operations (personnel, the financial accounting system, production, marketing, distribution, etc.) and also connects the organization with its suppliers and customers.

e. Flexible manufacturing is the capacity of computer-controlled machinery to perform many different programmed functions. By eliminating machine setup time, strengthening control, and automating handling processes, computer-aided manufacturing permits the efficient production of small numbers of different products by the same machines.

1) A company can therefore more accurately match output with consumer tastes and avoid long production runs of identical goods. A **flexible manufacturing system** consists of two or more computer-controlled machines linked by automated handling devices such as robots and transport systems.

f. **Computer-aided design and manufacturing (CAD/CAM)**

1) CAD/CAM models and predicts the outcomes of alternative product decisions.

a) CAD/CAM is essentially a system that combines

i) Storage and retrieval of drawings and parts attributes
ii) Computer graphics for drawing and display
iii) Data acquisition and control
iv) Mathematical modeling and control

b) Alternatively, an MRP system can be based primarily on database management and exclude the other components.

c) The database created in the design phase (CAD) can be used to produce a bill of materials for the manufacturing phase (CAM). CAM can use CAD drawings and information to give specific instructions to individual machines.

d) CAD/CAM helps engineers examine more options, conduct sophisticated simulations and tests, and make themselves more effective.

i) Product value and quality are improved.
ii) The need for physical prototypes is reduced.

e) CAD/CAM systems are usually expensive, high-risk, high-payback investments.

g. **Computer-integrated manufacturing (CIM)**

1) Modern inventory management is often accomplished in the context of automated manufacturing, whether involving one piece of equipment, a cell, or an integrated plant. A computer-integrated manufacturing (CIM) system involves designing products using computer-aided design (CAD), testing the design using computer-aided engineering (CAE), manufacturing products using computer-aided manufacturing (CAM), and integrating all components with a computerized information system.

2) Accordingly, CIM entails a holistic approach to manufacturing in which design is translated into product by centralized processing and robotics. The concept also includes materials handling.

a) The advantages of CIM include flexibility, integration, and synergism.

b) Flexibility is a key advantage. A traditional manufacturing system might become disrupted from an emergency change, but CIM will reschedule everything in the plant when a priority requirement is inserted into the system. The areas of flexibility include

i) Varying production volumes during a period
ii) Handling new parts added to a product
iii) Changing the proportion of parts being produced
iv) Adjusting to engineering changes of a product
v) Adapting the sequence in which parts come to the machinery
vi) Adapting to changes in materials
vii) Rerouting parts as needed because of machine breakdowns or other production delays
viii) Allowing for defects in materials

c) CIM integrates all production machinery using one computer system.

d) Benefits of CIM include improved product quality (less rework), better customer service, faster response to market changes, greater product variety, lower production costs, and shorter product development times.

e) JIT is sometimes adopted prior to CIM because JIT simplifies production processes and provides a better understanding of actual production flow, which are essential factors for CIM success.

f) The flexibility offered by CIM is almost a necessity for JIT suppliers. For example, a company that provides JIT deliveries to automobile plants cannot adapt to changing customer production schedules with a manual system unless a high inventory level is maintained.

g) The emphasis is on materials control rather than the direct labor control that is dominant in most cost systems.

h) CIM is an addition to, not a substitute for, other types of manufacturing concepts such as JIT. In other words, JIT should already be in place for CIM to work most effectively.

5. Stop and review! You have completed the outline for this subunit. Study multiple-choice questions 42 through 64 beginning on page 62.

E. Government Contracts and the CASB

1. Congress has granted the **Cost Accounting Standards Board** "exclusive authority to make, promulgate, amend, and rescind cost accounting standards and interpretations thereof." Its purpose is "to achieve uniformity and consistency in the cost accounting standards governing measurement, assignment, and allocation of costs to contracts within the United States." CASB pronouncements apply to negotiated contracts and subcontracts over $500,000.

   a. The CASB is an independent entity within the Office of Federal Procurement Policy.

   b. CASB pronouncements apply to all kinds of government contracts, not just those related to defense procurement.

2. **CASB Standards Incorporated into the Federal Acquisition Regulations**

   a. Standard 402 -- Costs incurred for the same purpose and in like circumstances shall be allocated as either direct costs or indirect costs, but not both.

      1) Each type of cost is to be allocated only once. Hence, double-counting is prohibited.

   b. Standard 403 -- Home office expenses shall be allocated on the basis of the relationship between supporting and receiving activities. Expenses not directly allocable on a beneficial or causal basis shall be grouped into homogeneous pools and indirectly allocated. The preferable allocation base is some measure of an input activity consumed. The next best choice is a measure of the output of the supporting function.

   c. Standard 404 -- Acquisition costs of tangible capital assets shall be capitalized. Contractors shall provide written policies that are reasonable and consistently applied.

      1) Minimum life for capitalization shall not exceed 2 years.
      2) Minimum cost for capitalization shall not exceed $1,000.
      3) Acquisition cost equals

a) Purchase price and preparation costs

b) Fair market value of donated assets

c) Direct and indirect costs, including general and administrative expenses, of self-constructed assets

d. Standard 405 -- Costs determined to be unallowable by agreement shall be identified and excluded from billings, claims, and proposals.

1) If the contracting officer determines that a cost is unallowable, it shall be identified if used in a billing, claim, or proposal.

e. Standard 406 -- A contractor shall use the fiscal year as the cost accounting period unless another cost accounting period is an established practice of the contractor.

f. Standard 409 -- Depreciation shall be computed and allocated as follows:

1) Depreciable basis is cost minus salvage value.

2) Method of depreciation is to be based on the pattern of consumption of services.

a) Method used for financial accounting may be used if it is accepted for income tax purposes.

3) Gain or loss on disposition is allocated to period of disposition.

g. Standard 410 -- General and administrative expenses shall be grouped in a separate indirect cost pool and allocated only to final cost objectives.

h. Standards 412 and 413 -- Pension costs shall be assigned to the period for which computed.

i. Standard 414 -- Amortization of intangible assets shall be allocated to periods based on the cost of money (interest rates) of the period; i.e., if the cost of money is 12%, amortization shall be 12% of the cost of the amortizable assets.

j. Standard 415 -- Deferred compensation shall be allocated to periods in which obligation to pay arises (earned) and computed as the present value of future benefits.

k. Standard 416 -- Insurance costs shall be allocated to periods based on projected average loss for the period plus administration expenses for the period.

l. Standard 417 -- The imputed cost of money is an element of the cost of capital assets under construction and is recovered as depreciation.

m. Standard 418 -- Indirect costs must be accumulated in homogeneous pools and allocated to cost objectives based on some reasonable relationship of the costs and the objectives.

n. Standard 420 -- Independent research and development costs and bid and proposal costs

1) Shall be aggregated to the project for which they are incurred
2) Consist of all allocable costs except general and administrative costs
3) Shall be allocated to the period incurred

3. Stop and review! You have completed the outline for this subunit. Study multiple-choice questions 65 and 66 on page 70.

MULTIPLE-CHOICE QUESTIONS

A. Decision-Making Process

1. A company is about to introduce a new service and wishes to develop a new slogan and logo to be used in advertising and on company publications. You have been chosen to participate in this process and to look at past slogans, logos, and suggestions given by the advertising agency. You are not limited to the suggested ideas and have been encouraged to suggest original ideas of your own. Which of the following techniques is most appropriate to use?

A. Brainstorming.

B. Value analysis.

C. Free association.

D. Attribute listing.

The correct answer is (C). *(CIA, adapted)*

REQUIRED: The technique that considers suggested ideas but also encourages original thinking.

DISCUSSION: Free association is an approach to idea generation that reports the first thought to come to mind in response to a given stimulus, for example, a symbol or analogy pertaining to a product for which an advertising slogan is sought. The objective is to express the content of consciousness without censorship or control.

Answer (A) is incorrect because brainstorming breaks down broadly based problems into their essentials. It is an unstructured approach that relies on the spontaneous contribution of ideas from all members of a group.. Answer (B) is incorrect because value analysis is a methodical approach primarily designed to optimize performance of an activity at a minimum cost. Answer (D) is incorrect because attribute listing is applied primarily to improve a tangible object.

2. A company has a computer that it no longer needs because of a discontinued operation. Currently, there are several computer projects that may be able to use the machine, but some modification will be necessary if such new application is to be successful. Which of the following techniques is most appropriate to use?

A. Attribute listing.

B. Operations research.

C. Morphological matrix analysis.

D. Synectics.

The correct answer is (A). *(CIA, adapted)*

REQUIRED: The technique useful in developing new uses for a computer.

DISCUSSION: Attribute listing is applied primarily to improve a tangible object. It lists the parts and essential features of the object and systematically analyzes modifications intended as improvements.

Answer (B) is incorrect because operations research attempts to find optimal solutions using classical concepts such as statistics, simulation, logical thinking, and other scientific and mathematical techniques to develop and test hypotheses. This application closely fits the problem and charge given. Answer (C) is incorrect because morphological matrix analysis is a structured technique that plots decision variables along the axes of a matrix. The relationships of these variables are found in the squares within the chart. Answer (D) is incorrect because synectics is a highly structured group approach to problem statement and solution based on creative thinking. It involves free use of analogies and metaphors in informal exchange within a carefully selected small group of individuals of diverse personality and areas of specialization.

3. A company has experienced numerous complaints because a passenger door has not opened smoothly. All attempts to repair the door have failed. The item is no longer under warranty, and the manufacturer has informed the company that a very expensive new door must be purchased and installed to solve the problem. You have been asked to supervise an attempt by your engineering department to come up with less expensive alternative solutions. Which of the following techniques is most appropriate to use?

A. Free association.

B. Operations research.

C. Blast! then refine.

D. Edisonian.

The correct answer is (D). *(CIA, adapted)*

REQUIRED: The technique that would be applied by an engineering department to come up with less expensive alternative solutions.

DISCUSSION: The Edisonian approach is a trial-and-error experimental method. It should usually not be applied unless other approaches have been unsuccessful.

Answer (A) is incorrect because free association is an approach to idea generation that reports the first thought to come to mind in response to a given stimulus, for example, a symbol or analogy pertaining to a product for which an advertising slogan is sought. The objective is to express the content of consciousness without censorship or control. Answer (B) is incorrect because operations research attempts to find optimal solutions using classical concepts such as statistics, simulation, logical thinking, and other scientific and mathematical techniques to develop and test hypotheses. This application closely fits the problem and charge given. Answer (C) is incorrect because "blast! then refine" is a U.S. Navy problem-solving methodology. It completely disregards (blasts out of one's consciousness) any existing approach. An entirely new problem solution is then sought that will attain the original objectives.

4. A company is concerned that spare parts inventories are too large. It has attempted to keep critical parts for its fleet in stock so that equipment will have minimal downtime. Management wants to know what the optimal spare parts inventory should be if downtime is estimated to cost $150 per day. Carrying cost and order cost have not been measured. You have been asked to make a formal recommendation on spare parts stocking levels. Which of the following techniques is most appropriate to use?

A. Operations research.

B. Value analysis.

C. Attribute listing.

D. Brainstorming.

The correct answer is (A). *(CIA, adapted)*

REQUIRED: The technique for determining the optimal spare parts inventory.

DISCUSSION: Operations research attempts to find optimal solutions using classical concepts such as statistics, simulation, logical thinking, and other scientific and mathematical techniques to develop and test hypotheses. This application closely fits the problem and charge given.

Answer (B) is incorrect because value analysis is a methodical approach primarily designed to optimize performance of an activity at a minimum cost. Emphasis is on the cost-benefit criterion. Answer (C) is incorrect because attribute listing is applied primarily to improve a tangible object. It lists the parts and essential features of the object and systematically analyzes modifications intended as improvements. Answer (D) is incorrect because brainstorming breaks down broadly based problems into their essentials. It is an unstructured approach that relies on the spontaneous contribution of ideas from all members of a group.

5. As part of the decision-making process, alternative solutions ordinarily should be screened for the most appealing balance of effectiveness and efficiency in view of relevant constraints and intangibles. When a problem is resolved by selecting a course of action that meets the minimum constraints, a manager is said to be

A. Satisficing.

B. Optimizing.

C. Idealizing.

D. Maximizing.

The correct answer is (A). *(CIA, adapted)*

REQUIRED: The term for resolving a problem by selecting a course of action that meets the minimum constraints.

DISCUSSION: Satisficing is the decision-making process in which the decision maker uses a minimal amount of information in an effort to reduce to a minimum the amount of time it takes to make a decision. The result is a decision that is not the best but is good enough to give satisfactory resolution.

Answer (B) is incorrect because optimizing means selecting the best possible solution. Answer (C) is incorrect because idealizing involves changing the situation so that the problem no longer exists. Answer (D) is incorrect because maximizing is a synonym for optimizing and involves selecting the best solution.

6. In making management decisions, the concept of probability is most closely related to which of the following?

A. Management's confidence in its control systems.

B. The likelihood that a given event will occur.

C. The availability of factual information about alternatives.

D. Management's ability to cope with subjective information.

The correct answer is (B). *(CIA, adapted)*

REQUIRED: The item most closely related to the concept of probability.

DISCUSSION: Probability is a number between 0 and 1 inclusive that measures the likelihood of the occurrence of some event. The less likely the event, the closer the number is to 0. The two types of probabilities are objective and subjective. Objective probabilities are derived mathematically from historical data. Subjective probabilities are estimated based on past experience and judgment. Thus, probabilities are used to determine the likelihood that a given event will occur.

Answer (A) is incorrect because management's confidence in its control systems is not closely related to the concept of probability. Answer (C) is incorrect because the availability of factual information helps in the decision-making process, but it is not closely related to the concept of probability. Answer (D) is incorrect because management's ability to cope with subjective information may be a factor in decision making, but it is not closely related to the concept of probability.

7. Internal auditors are often very good at identifying problems that have occurred in an organizational unit. However, the auditee may not always react to the suggestions made by the auditor in a positive manner. One behavioral approach that has been documented has been termed "escalation of commitment." The term "escalation of commitment" refers to

A. The process of continuing to fund old projects because of inadequate information or time to formally analyze the costs and benefits associated with continued investment.

B. The decision maker increasing the resources to a new course of action if the auditor can recommend alternatives such that the decision maker will view the new course of action as his/her own initiative.

C. Committing to projects that have been shown to be successful and limiting the additional commitment of resources to projects that have been unsuccessful.

D. The decision maker increasing the resources to the previous course of action in an effort to demonstrate that the previous course of action was appropriate.

The correct answer is (D). *(CIA, adapted)*

REQUIRED: The definition of escalation of commitment.

DISCUSSION: Consistency may be dysfunctional because it may result in inflexibility. The desire to avoid the inner conflict resulting from inconsistency may have adverse effects on decision making. For example, when a decision maker feels responsible for the failure of a course of action, (s)he may want to prove that the original decision was sound despite increasing evidence to the contrary. Thus, his/her commitment to a failed policy may escalate, and additional resources may be squandered.

Answer (A) is incorrect because escalation of commitment is a psychological phenomenon unrelated to availability of information or time. Answers (B) and (C) are incorrect because increasing resources to a new course of action and limiting resources to unsuccessful projects are antithetical to escalation of commitment.

8. Auditors must use both inductive and deductive reasoning when gathering audit evidence and reaching audit conclusions. Auditors must also understand the differences between the two types of reasoning in order to assess the strengths and weaknesses of each. Which of the following audit procedures uses deductive reasoning?

A. The internal auditor uses generalized audit software to select a sample of items for confirmation. Based on the confirmation responses, the auditor concludes that accounts receivable as recorded exist.

B. The internal auditor performs analytical procedures to estimate the accuracy of the sales account balance. No material differences are noted. Based on this, the auditor assumes that the underlying record keeping must be correct.

C. The auditor observes the client's physical inventory process and records test counts of inventory. Those test counts are traced to the year-end inventory compilation and no exceptions are noted. The auditor concludes that the compiled inventory exists.

D. The auditor examines sales transactions recorded during January 1995 and finds that none of those sales represent 1994 sales. The auditor concludes that 1994 sales are recorded properly.

The correct answer is (B). *(CIA, adapted)*

REQUIRED: The audit procedure using deductive reasoning.

DISCUSSION: Deductive reasoning is the inference of a conclusion about particulars from general or universal premises. In contrast, inductive reasoning is the inference of a generalized conclusion from particular instances. Analytical procedures are based on the general premise that plausible relationships among the data may reasonably be expected to continue in the absence of known conditions to the contrary. For example, absent unusual circumstances, the sales balance is relatively predictable given its known relationship to other factors, such as cost of sales. Thus, the process is one of deductive reasoning because a general premise (the known relationship of sales to other data) is applied to draw a conclusion about a particular instance (the sales balance).

Answers (A), (C), and (D) are incorrect because sampling receivables, test counting inventory, and examining sales recorded after the balance sheet date are examples of inductive processes. Each entails gathering empirical data about particular items and then determining the general principle that the evidence supports.

9. Which of the following is deductive reasoning?

A. Reasoning from one or more particular facts to reach a general conclusion.

B. Reasoning on a post hoc basis.

C. Reasoning from the general to the particular.

D. Reasoning through the development of a carefully constructed hypothesis.

The correct answer is (C). *(CIA, adapted)*

REQUIRED: The definition of deductive reasoning.

DISCUSSION: Deduction is a process of reasoning that applies general or universal premises to reach a conclusion about particulars. For example, the laws of physics may be used to predict the duration of a particular orbit of a satellite.

Answer (A) is incorrect because reasoning from one or more particular facts to reach a general conclusion is induction. Answer (B) is incorrect because post hoc reasoning is a logical fallacy. It assumes that one event causes another because the first precedes the second. Answer (D) is incorrect because inductive reasoning is more closely associated with constructing hypotheses.

B. Marginal Analysis Applications

10. In a decision analysis situation, which one of the following costs is not likely to contain a variable cost component?

A. Labor.

B. Overhead.

C. Depreciation.

D. Selling.

The correct answer is (C). *(CMA, adapted)*

REQUIRED: The cost not likely to contain a variable component.

DISCUSSION: Most types of costs either are variable or contain some mix of fixed and variable components. For example, labor and materials are primarily variable. Their amounts fluctuate with the relevant cost drivers. Selling expenses and overhead costs can be either fixed or variable. Depreciation, however, is strictly a fixed cost when viewed in a decision analysis situation because it requires no cash outlay. Instead, the initial acquisition of the long-lived asset requires the cash outlay.

Answers (A), (B), and (D) are all incorrect because labor, overhead, and selling costs are either variable or mixed.

11. An important concept in decision making is described as "the contribution to income that is forgone by not using a limited resource in its best alternative use." This concept is called

A. Marginal cost.

B. Incremental cost.

C. Potential cost.

D. Opportunity cost.

The correct answer is (D). *(CMA, adapted)*

REQUIRED: The contribution to income that is forgone by not adopting the best alternative use of a limited resource.

DISCUSSION: An opportunity cost is the maximum benefit sacrificed by employing a scarce productive resource in a specified manner. In other words, it is the value or worth of that resource in its next best alternative use.

Answer (A) is incorrect because marginal cost is an incremental or differential cost. Answer (B) is incorrect because incremental cost is the difference in total cost between two decision choices. Answer (C) is incorrect because a potential cost may arise in the future.

12. American Coat Company estimates that 60,000 special zippers will be used in the manufacture of men's jackets during the next year. Reese Zipper Company has quoted a price of $.60 per zipper. American would prefer to purchase 5,000 units per month, but Reese is unable to guarantee this delivery schedule. In order to ensure availability of these zippers, American is considering the purchase of all 60,000 units at the beginning of the year. Assuming American can invest cash at 8%, the company's opportunity cost of purchasing the 60,000 units at the beginning of the year is

A. $1,320

B. $1,440

C. $1,500

D. $2,640

The correct answer is (A). *(CMA, adapted)*

REQUIRED: The opportunity cost of purchasing the total annual requirement at the beginning of the year.

DISCUSSION: The cost of 60,000 zippers is $36,000 (60,000 x $.60). The monthly cost is $3,000 (5,000 x $.60). Because the company wants to purchase the items monthly, it will invest at least $3,000 in January. Accordingly, the zippers to be used in January will be purchased at the first of the year even if no special purchase is made. Thus, the incremental advance purchase will be only $33,000. The incremental investment will decline at a constant rate of $3,000 per month, so the amount for February will be $30,000, for March $27,000, etc. Hence, the incremental investment for December will be $0. Accordingly, assuming consumption is uniform, the average incremental investment during the year will be $16,500 ($33,000 ÷ 2) if the onetime purchase is made. Total opportunity cost is therefore $1,320 (8% cost of capital x $16,500).

Answers (B), (C), and (D) are incorrect because total opportunity cost is $1,320.

13. In a decision analysis situation, which one of the following costs is generally not relevant to the decision?

A. Incremental cost.

B. Differential cost.

C. Avoidable cost.

D. Historical cost.

The correct answer is (D). *(CMA, adapted)*

REQUIRED: The cost ordinarily not relevant to a decision.

DISCUSSION: Management decision analysis is based on the concept of relevant costs. Relevant costs differ among decision choices. Thus, incremental (differential or avoidable) costs are always relevant. Opportunity cost is also relevant because it is the benefit forgone by selecting one choice instead of another. Historical costs, because they occurred in the past, are sunk costs and not relevant to most management decisions.

Answers (A), (B), and (C) are all incorrect because incremental cost, differential cost, and avoidable cost are relevant to management decision analysis.

14. All of the following costs are relevant to a decision to accept or reject an order except

A. Differential costs.

B. Out-of-pocket costs.

C. Replacement costs.

D. Sunk costs.

The correct answer is (D). *(CMA, adapted)*

REQUIRED: The irrelevant costs.

DISCUSSION: A sunk cost cannot be avoided because either the expenditure has occurred, or an irrevocable decision to incur the cost has been made. Sunk costs are irrelevant to management decision-making because they cannot vary with the option selected.

Answer (A) is incorrect because a differential (incremental) cost is relevant. It is the difference in total cost between two decisions. Answer (B) is incorrect because out-of-pocket costs (outlay costs) are relevant. They require negative cash flows (expenditures) currently or in the future. Answer (C) is incorrect because a replacement cost is relevant. It is the cash or equivalent that would be paid for a current acquisition of the same or an equivalent asset.

15. Which one of the following is most relevant to a manufacturing equipment replacement decision?

A. Original cost of the old equipment.

B. Disposal price of the old equipment.

C. Gain or loss on the disposal of the old equipment.

D. A lump-sum write-off amount from the disposal of the old equipment.

The correct answer is (B). *(CMA, adapted)*

REQUIRED: The item most relevant to a manufacturing equipment replacement decision.

DISCUSSION: Management decision analysis is based on the concept of relevant costs. Relevant costs differ among decision choices. Thus, incremental or differential costs are always relevant. Historical costs, such as the original cost of the equipment, because they were incurred in the past, are not relevant. Similarly, any gain or loss on the old equipment is not relevant because this amount is based on the historical cost. However, the disposal price of the old equipment is relevant because it involves a future cash inflow that will not occur unless the equipment is disposed of.

Answer (A) is incorrect because the original cost of the old equipment is a sunk cost with no relevance to future decision making. Answer (C) is incorrect because gain or loss is based on historical cost, which is a sunk cost. Answer (D) is incorrect because a lump-sum write-off of a sunk cost is not relevant to a future decision.

Questions 16 through 18 are based on the following information. Kator Co. is a manufacturer of industrial components. One of their products that is used as a subcomponent in auto manufacturing is KB-96. This product has the following financial structure per unit:

Selling price	$150
Direct materials	$ 20
Direct labor	15
Variable manufacturing overhead	12
Fixed manufacturing overhead	30
Shipping and handling	3
Fixed selling and administrative	10
Total costs	$ 90

16. Kator Co. has received a special, one-time, order for 1,000 KB-96 parts. Assuming Kator has excess capacity, the minimum price that is acceptable for this one-time special order is in excess of

A. $47

B. $50

C. $60

D. $77

The correct answer is (B). *(CMA, adapted)*

REQUIRED: The minimum acceptable price.

DISCUSSION: A company must cover the incremental costs of a special order when it has excess capacity. The incremental costs for product KB-96 are $50 ($20 direct materials + $15 direct labor + $12 variable overhead + $3 shipping and handling). The fixed costs will not change as a result of the special order, so they are not relevant. Thus, any price is excess of $50 per unit is acceptable.

Answer (A) is incorrect because $47 ignores the shipping and handling costs. Answer (C) is incorrect because $60 includes fixed selling and administrative costs. Answer (D) is incorrect because $77 includes fixed manufacturing overhead but omits shipping and handling costs.

17. During the next year, KB-96 sales are expected to be 10,000 units. All of the costs will remain the same except that fixed manufacturing overhead will increase by 20% and direct materials will increase by 10%. The selling price per unit for next year will be $160. Based on this data, the contribution margin from KB-96 for next year will be

A. $620,000

B. $750,000

C. $1,080,000

D. $1,110,000

The correct answer is (C). *(CMA, adapted)*

REQUIRED: The contribution margin for next year.

DISCUSSION: Contribution margin equals sales minus variable costs. All variable costs will remain the same except that direct materials will increase to $22 per unit (1.1 x $20). Thus, total unit variable costs will be $52 ($22 + $15 + $12 + $3), and the contribution margin will be $1,080,000 [10,000 units ($160 unit selling price – $52)].

Answer (A) is incorrect because $620,000 includes all fixed costs. Answer (B) is incorrect because $750,000 includes all manufacturing costs. Answer (D) is incorrect because $1,110,000 assumes that the fixed costs and shipping and handling are the only relevant costs.

18. Kator Co. has received a special, one-time, order for 1,000 KB-96 parts. Assume that Kator is operating at full capacity and that the contribution margin of the output that would be displaced by the special order is $10,000. The minimum price that is acceptable, using the original data, for this one-time special order is in excess of

A. $60

B. $70

C. $87

D. $100

The correct answer is (A). *(CMA, adapted)*

REQUIRED: The minimum acceptable price.

DISCUSSION: Given no excess capacity, the price must cover the incremental costs. The incremental costs for KB-96 equal $50 ($20 direct materials + $15 direct labor + $12 variable overhead + $3 shipping and handling). Opportunity cost is the benefit of the next best alternative use of scarce resources. Because acceptance of the special order would cause the company to forgo a contribution margin of $10,000, that amount must be reflected in the price. Hence, the minimum unit price is $60 [$50 unit incremental cost + ($10,000 lost CM ÷ 1,000 units)].

Answer (B) is incorrect because $70 includes fixed selling and administrative costs. Answer (C) is incorrect because $87 includes fixed manufacturing overhead but omits shipping and handling costs. Answer (D) is incorrect because $100 is based on full absorption cost.

Questions 19 and 20 are based on the following information. Whitehall Corporation produces chemicals used in the cleaning industry. During the previous month, Whitehall incurred $300,000 of joint costs in producing 60,000 units of AM-12 and 40,000 units of BM-36. Whitehall uses the units-of-production method to allocate joint costs. Currently, AM-12 is sold at split-off for $3.50 per unit. Flank Corporation has approached Whitehall to purchase all of the production of AM-12 after further processing. The further processing will cost Whitehall $90,000.

19. Concerning AM-12, which one of the following alternatives is most advantageous?

A. Whitehall should process further and sell to Flank if the total selling price per unit after further processing is greater than $3.00, which covers the joint costs.

B. Whitehall should continue to sell at split-off unless Flank offers at least $4.50 per unit after further processing, which covers Whitehall's total costs.

C. Whitehall should process further and sell to Flank if the total selling price per unit after further processing is greater than $5.00.

D. Whitehall should process further and sell to Flank if the total selling price per unit after further processing is greater than $5.25, which maintains the same gross profit percentage.

The correct answer is (C). *(CMA, adapted)*

REQUIRED: The processing and selling alternative that is most advantageous.

DISCUSSION: The unit price of the product at the split-off point is known to be $3.50, so the joint costs are irrelevant. The additional unit cost of further processing is $1.50 ($90,000 ÷ 60,000 units). Consequently, the unit price must be at least $5.00 ($3.50 opportunity cost + $1.50).

Answer (A) is incorrect because the joint costs are irrelevant. Answer (B) is incorrect because the unit price must cover the $3.50 opportunity cost plus the $1.50 of additional costs. Answer (D) is incorrect because any price greater than $5 will provide greater profits, in absolute dollars, even though the gross profit percentage declines.

20. Assume that Whitehall Corporation agreed to sell AM-12 to Flank Corporation for $5.50 per unit after further processing. During the first month of production, Whitehall sold 50,000 units with 10,000 units remaining in inventory at the end of the month. With respect to AM-12, which one of the following statements is correct?

A. The operating profit last month was $50,000, and the inventory value is $15,000.

B. The operating profit last month was $50,000, and the inventory value is $45,000.

C. The operating profit last month was $125,000, and the inventory value is $30,000.

D. The operating profit last month was $200,000, and the inventory value is $30,000.

The correct answer is (B). *(CMA, adapted)*

REQUIRED: The operating profit and inventory value after specified sales of a product.

DISCUSSION: Joint costs are allocated based on units of production. Accordingly, the unit joint cost allocated to AM-12 is $3.00 [$300,000 ÷ (60,000 units of AM-12 + 40,000 units of BM-36)]. The unit cost of AM-12 is therefore $4.50 [$3.00 joint cost + ($90,000 additional cost ÷ 60,000 units)]. Total inventory value is $45,000 (10,000 units x $4.50), and total operating profit is $50,000 [($5.50 unit price – $4.50 unit cost) x 50,000 units sold].

Answer (A) is incorrect because the $3 unit joint cost should be included in the inventory value. Answer (C) is incorrect because the $1.50 unit additional cost should be included in total unit cost. Answer (D) is incorrect because the $3 unit joint cost should be included in the cost of goods sold, and inventory should include the $1.50 unit additional cost.

Questions 21 through 23 are based on the following information. Condensed monthly operating income data for Korbin Inc. for May follow:

	Urban Store	Suburban Store	Total
Sales	$80,000	$120,000	$200,000
Variable costs	32,000	84,000	116,000
Contribution margin	$48,000	$ 36,000	$ 84,000
Direct fixed costs	20,000	40,000	60,000
Store segment margin	$28,000	$ (4,000)	$ 24,000
Common fixed cost	4,000	6,000	10,000
Operating income	$24,000	$ (10,000)	$ 14,000

Additional information regarding Korbin's operations follows:

- One-fourth of each store's direct fixed costs would continue if either store is closed.
- Korbin allocates common fixed costs to each store on the basis of sales dollars.
- Management estimates that closing Suburban Store would result in a 10% decrease in Urban Store's sales, whereas closing Urban Store would not affect Suburban Store's sales.
- The operating results for May are representative of all months.

21. A decision by Korbin to close Suburban Store would result in a monthly increase (decrease) in Korbin's operating income of

A. $(10,800)

B. $(6,000)

C. $(1,200)

D. $4,000

The correct answer is (A). *(CMA, adapted)*

REQUIRED: The effect on operating income of closing Suburban Store.

DISCUSSION: If Suburban Store is closed, one-fourth of its direct fixed costs will continue. Thus, the segment margin that should be used to calculate the effect of its closing on Korbin's operating income is $6,000 {$36,000 contribution margin – [$40,000 direct fixed costs x (1.0 – .25)]}. In addition, the sales (and contribution margin) of Urban Store will decline by 10% if Suburban Store closes. A 10% reduction in Urban's $48,000 contribution margin will reduce income by $4,800. Accordingly, the effect of closing Suburban Store is to decrease operating income by $10,800 ($6,000 + $4,800).

Answer (B) is incorrect because $(6,000) overlooks the decline in profitability at Urban Store. Answer (C) is incorrect because $(1,200) assumes that the effect on Urban Store is a $4,800 increase in contribution margin. Answer (D) is incorrect because profits will decline.

22. Korbin is considering a promotional campaign at Suburban Store that would not affect Urban Store. Increasing annual promotional expense at Suburban Store by $60,000 in order to increase this store's sales by 10% would result in a monthly increase (decrease) in Korbin's operating income during the year (rounded) of

A. $(5,000)

B. $(1,400)

C. $487

D. $7,000

The correct answer is (B). *(CMA, adapted)*

REQUIRED: The effect on monthly income of an advertising campaign.

DISCUSSION: The $60,000 advertising campaign will increase direct fixed costs by $5,000 per month ($60,000 ÷ 12). Sales and contribution margin will also increase by 10%. Hence, the contribution margin for Suburban Store will increase by $3,600 (10% x $36,000), and income will decline by $1,400 ($5,000 – $3,600).

Answer (A) is incorrect because $(5,000) is the monthly advertising cost. Answer (C) is incorrect because the contribution margin of Suburban Store will increase by $3,600, which is $1,400 less than the increased advertising cost. Answer (D) is incorrect because $7,000 omits the 10% increase in variable costs from the calculation.

23. One-half of Suburban Store's dollar sales are from items sold at variable cost to attract customers to the store. Korbin is considering the deletion of these items, a move that would reduce Suburban Store's direct fixed expenses by 15% and result in a 20% loss of Suburban Store's remaining sales volume. This change would not affect Urban Store. A decision by Korbin to eliminate the items sold at cost would result in a monthly increase (decrease) in Korbin's operating income of

A. $(5,200)

B. $(1,200)

C. $(7,200)

D. $2,000

The correct answer is (B). *(CMA, adapted)*

REQUIRED: The effect on monthly income of eliminating sales made at variable cost.

DISCUSSION: If 50% of Suburban Store's sales are at variable cost, its contribution margin (sales – variable costs) must derive wholly from sales of other items. However, eliminating sales at variable cost reduces other sales by 20%. Thus, the effect is to reduce the contribution margin to $28,800 ($36,000 x .8). Moreover, fixed costs will be reduced by 15% to $34,000 ($40,000 x .85). Consequently, the new segment margin is $(5,200) ($34,000 direct fixed costs – $28,800 contribution margin), a decrease of $1,200 [$(5,200) – $(4,000)].

Answer (A) is incorrect because $(5,200) is the new segment margin. Answer (C) is incorrect because $(7,200) is the reduction in Suburban Store's contribution margin. Answer (D) is incorrect because operating income must decrease.

24. Polar Company sells refrigeration components both in the U.S. and to a subsidiary located in France. One of the components, Part No. 456, has a variable manufacturing cost of $30. The part can be sold domestically or shipped to the French subsidiary for use in the manufacture of a residential subassembly. Relevant data with regard to Part No. 456 are shown below.

	Part No. 456
Domestic selling price	$ 65
Shipping charges to France	15
Cost of acquiring Part No. 456 in France	75
French residential subassembly:	
Sales price	170
Other additional manufacturing costs	55
Units shipped to France	150,000*

*If deemed preferable, these units could be sold in the U.S.

Polar's applicable income tax rates are 40% in the U.S. and 70% in France.

Polar will transfer Part No. 456 to the French subsidiary at either variable manufacturing cost or the domestic market price. On the basis of this information, which one of the following strategies should be recommended to Polar's management?

A. Transfer 150,000 units at $30 and the French subsidiary pays the shipping costs.

B. Transfer 150,000 units at $65 and the French subsidiary pays the shipping costs.

C. Sell 150,000 units in the U.S. and the French subsidiary obtains Part No. 456 in France.

D. Transfer 150,000 units at $65 and have the U.S. company absorb the shipping costs.

The correct answer is (C). *(CMA, adapted)*

REQUIRED: The best strategy for use of a component if a company has the option of selling it domestically or including it in a product sold by a foreign subsidiary.

DISCUSSION: If the 150,000 components are sold separately in the U.S., the after-tax profit per unit will be $21 [($65 – $30) x .6]. The French subsidiary will then purchase 150,000 components domestically to include in the subassembly. The per-unit profit on sales of the subassembly will be $12 [($170 – $75 – $55) x .3]. The total per-unit profit for the consolidated entity from these transactions will therefore be $33 ($21 + $12).

Answer (A) is incorrect because, if the components are transferred at cost, no tax will be paid in the U.S., and the per-unit tax on French sales of the subassembly will be $49 [($170 – $30 – $15 – $55) x .7]. Thus, the consolidated entity's per-unit profit will be $21 ($170 – $30 – $15 – $55 – $49). Answer (B) is incorrect because, if the transfer is at $65, U.S. tax per unit will be $14 [($65 – $30) x .4], and the French tax per unit will be $24.50 [($170 – $65 – $15 – $55) x .7]. The consolidated entity's per-unit profit will therefore be $31.50 ($170 – $30 – $15 – $55 – $14 – $24.50). Answer (D) is incorrect because, if the transfer is at $65 and the U.S. parent pays the shipping costs, the U.S. tax per unit will be $8 [($65 – $30 – $15) x .4], and the French tax per unit will be $35 [($170 – $65 – $55) x .7]. The consolidated entity's per-unit profit will be $27 ($170 – $30 – $15 – $55 – $8 – $35).

25. In a make-versus-buy decision, the relevant costs include variable manufacturing costs as well as

A. Factory management costs.

B. General office costs.

C. Avoidable fixed costs.

D. Depreciation costs.

The correct answer is (C). *(CMA, adapted)*

REQUIRED: The relevant costs in a make-versus-buy decision.

DISCUSSION: The relevant costs in a make-versus-buy decision are those that differ between the two decision choices. These costs include any variable costs plus any avoidable fixed costs. Avoidable fixed costs will not be incurred if the "buy" decision is selected.

Answers (A), (B), and (D) are incorrect because factory management costs, general office costs, and depreciation costs are unlikely to differ regardless of which decision is selected.

26. Listed below are a company's monthly unit costs to manufacture and market a particular product.

Manufacturing costs:	
Direct materials	$2.00
Direct labor	2.40
Variable indirect	1.60
Fixed indirect	1.00
Marketing costs:	
Variable	2.50
Fixed	1.50

The company must decide to continue making the product or buy it from an outside supplier. The supplier has offered to make the product at the same level of quality that the company can make it. Fixed marketing costs would be unaffected, but variable marketing costs would be reduced by 30% if the company were to accept the proposal. What is the maximum amount per unit that the company can pay the supplier without decreasing operating income?

A. $8.50

B. $6.75

C. $7.75

D. $5.25

The correct answer is (B). *(CMA, adapted)*

REQUIRED: The maximum amount paid to an outside supplier without decreasing operating income.

DISCUSSION: The company's avoidable unit costs total $6.75 [($2.00 DM + $2.40 DL + $1.60 VOH) + (.3 x $2.50 variable marketing costs)]. Hence, it will at least break even by paying no more than $6.75.

Answer (A) is incorrect because $8.50 assumes that all variable marketing costs are avoidable. Answer (C) is incorrect because $7.75 assumes that fixed manufacturing costs of $1 are avoidable. Answer (D) is incorrect because $5.25 results from subtracting the savings in marketing costs from the manufacturing savings.

27. When a multiproduct plant operates at full capacity, quite often decisions must be made as to which products to emphasize. These decisions are frequently made with a short-run focus. In making such decisions, managers should select products with the

A. Highest sales price per unit.

B. Highest individual unit contribution margin.

C. Highest sales volume potential.

D. Highest contribution margin per unit of the constraining resource.

The correct answer is (D). *(CMA, adapted)*

REQUIRED: The products to be made in the short run.

DISCUSSION: In the short run, many costs are fixed. Hence, contribution margin (revenues – all variable costs) becomes the best measure of profitability. Moreover, certain resources are also fixed. Accordingly, when deciding which products to produce at full capacity, the criterion should be the contribution margin per unit of the most constrained resource. This approach maximizes total contribution margin.

Answers (A) and (C) are incorrect because the highest sales price or volume does not consider costs and constraints. Answer (B) is incorrect because the product with the highest UCM may require greater usage of the constrained resource than another product contributing a lesser amount per unit.

28. Laurel Corporation has its own cafeteria with the following annual costs:

Food	$100,000
Labor	75,000
Overhead	110,000
Total	$285,000

The overhead is 40% fixed. Of the fixed overhead, $25,000 is the salary of the cafeteria supervisor. The remainder of the fixed overhead has been allocated from total company overhead. Assuming the cafeteria supervisor will remain and that Laurel will continue to pay his/her salary, the maximum cost Laurel will be willing to pay an outside firm to service the cafeteria is

A. $285,000

B. $175,000

C. $219,000

D. $241,000

The correct answer is (D). *(CMA, adapted)*

REQUIRED: The maximum that could be paid.

DISCUSSION: Given that overhead is 40% fixed, $66,000 (60% x $110,000) is variable, and $44,000 is fixed. Of the latter amount, $25,000 is attributable to the supervisor's salary. The $19,000 remainder is allocated from total company overhead and is unavoidable. Assuming the company will continue to pay the supervisor's salary if an outside firm services the cafeteria, the total fixed overhead is not an avoidable (incremental) cost. Thus, the total avoidable cost of the cafeteria's operation is $241,000 ($100,000 food + $75,000 labor + $66,000 VOH). This amount is the savings from hiring an outside firm. Accordingly, it is also the maximum that Laurel should be willing to pay the outside firm.

Answer (A) is incorrect because $285,000 is greater than the avoidable cost of operating the cafeteria. Answer (B) is incorrect because the company can pay more than $175,000. Its overhead costs are avoidable. Answer (C) is incorrect because the company can pay more than $219,000.

29. Power Systems, Inc. manufactures jet engines for the United States armed forces on a cost-plus basis. The cost of a particular jet engine the company manufactures is shown below.

Direct materials	$200,000
Direct labor	150,000
Overhead:	
Supervisor's salary	20,000
Fringe benefits on direct labor	15,000
Depreciation	12,000
Rent	11,000
Total cost	$408,000

If production of this engine were discontinued, the production capacity would be idle, and the supervisor would be laid off. When asked to bid on the next contract for this engine, the minimum unit price that Power Systems should bid is

A. $408,000

B. $365,000

C. $397,000

D. $385,000

The correct answer is (D). *(CMA, adapted)*

REQUIRED: The minimum unit price that should be bid.

DISCUSSION: The company will need to cover its variable costs and any other incremental costs. Thus, direct materials ($200,000), direct labor ($150,000), the supervisor's salary ($20,000), and fringe benefits on direct labor ($15,000) are the incremental unit costs of manufacturing the engines. The breakeven price is therefore $385,000 ($200,000 + $150,000 + $20,000 + $15,000).

Answer (A) is incorrect because depreciation and rent are allocated costs that will be incurred even if the contract is lost. Answer (B) is incorrect because the supervisor's salary will have to be covered. The $20,000 salary is an avoidable cost. Answer (C) is incorrect because depreciation is a cost that cannot be avoided.

30. Listed below are a company's monthly unit costs to manufacture and market a particular product.

Manufacturing Costs:	
Direct materials	$4.00
Direct labor	4.80
Variable indirect	3.20
Fixed indirect	2.00
Marketing Costs:	
Variable	$5.00
Fixed	3.00

The company must decide whether to continue making the product or to buy it from an outside supplier. The supplier has offered to make a product of the same quality as that manufactured in-house. If the company were to accept the proposal, fixed marketing costs would be unaffected, but variable marketing costs would be reduced by 30%. Moreover, if the company outsources the product, the capacity formerly used for its manufacture will be idle. What is the maximum amount per unit that the company can pay the supplier without decreasing its operating income?

A. $17.00

B. $13.50

C. $15.50

D. $10.50

The correct answer is (B). *(Publisher)*

REQUIRED: The maximum amount the company can pay the outside supplier without decreasing its operating income.

DISCUSSION: The company is currently incurring variable costs of $4.00 for materials, $4.80 for direct labor, and $3.20 for variable indirect costs, all of which will be eliminated if the purchase is made from the outside supplier. These variable manufacturing costs total $12.00 per unit. In addition, the company's variable marketing costs will be decreased by $1.50 (30% x $5). Hence, given that the resulting idle capacity cannot be used for other profitable activities (opportunity cost is zero), the company can pay $13.50 per unit ($12.00 + $1.50) to the outside supplier without reducing its operating income.

Answer (A) is incorrect because $17.00 assumes that all variable marketing costs will be eliminated. Answer (C) is incorrect because $15.50 assumes that fixed manufacturing costs of $2 will be eliminated. Answer (D) is incorrect because $10.50 subtracts the savings in marketing costs from the manufacturing savings.

Questions 31 and 32 are based on the following information. Pontotoc Industries manufactures a product that is used as a subcomponent by other manufacturers. It has the following price and cost structure:

Selling price		$300
Costs		
Direct materials	40	
Direct labor	30	
Variable manufacturing overhead	24	
Fixed manufacturing overhead	60	
Variable selling	6	
Fixed selling and administrative	20	(180)
Operating margin		$120

31. Pontotoc received a special, one-time order for 1,000 of the above parts. Assuming Pontotoc has excess capacity, the minimum unit price for this special, one-time order is in excess of

A. $180

B. $120

C. $100

D. $160

The correct answer is (C). *(Publisher)*

REQUIRED: The minimum unit price for a special order when a company has excess manufacturing capacity.

DISCUSSION: In a special order situation, a company with excess capacity has a $0 opportunity cost of filling the special order. Accordingly, it should be willing to sell the product at a price that exceeds its incremental costs. The incremental (relevant) costs for Pontotoc equal the variable costs of $100 ($40 direct materials + $30 direct labor + $24 variable overhead + $6 variable selling costs). Thus, if the selling price is in excess of $100, the company should be willing to accept the order.

Answer (A) is incorrect because, given excess capacity, the total absorption cost of $180 per unit is not relevant. Answer (B) is incorrect because $120 is the normal operating margin, not a cost. Answer (D) is incorrect because $160 includes the fixed manufacturing overhead, an irrelevant cost.

32. Pontotoc received a special, one-time order for 1,000 units of its product. However, Pontotoc has an alternative use for this capacity that will result in a contribution of $20,000. The minimum unit price for this special, one time order is in excess of

A. $200

B. $180

C. $140

D. $120

The correct answer is (D). *(Publisher)*

REQUIRED: The minimum unit price for a special order when the company does not have idle capacity.

DISCUSSION: Incremental (relevant) costs include the variable costs of $100 ($40 direct materials + $30 direct labor + $24 variable overhead + $6 variable selling costs). Furthermore, if the company does not have idle capacity, the unit price must cover the opportunity costs as well as the variable costs. Given that the alternative use will generate a $20,000 contribution, the 1,000 special-order units will have to generate at least $20 per unit ($20,000 ÷ 1,000 units) above their variable costs, a total of $120 per unit.

Answer (A) is incorrect because $200 is based on a $20 increment over the total cost on an absorption costing basis. Answer (B) is incorrect because $180 is the total absorption cost. Answer (C) is incorrect because $140 is based on a $20 increment over the operating margin.

C. Pricing

33. Which one of the following graphs depicts the demand curve for prestige goods?

A.

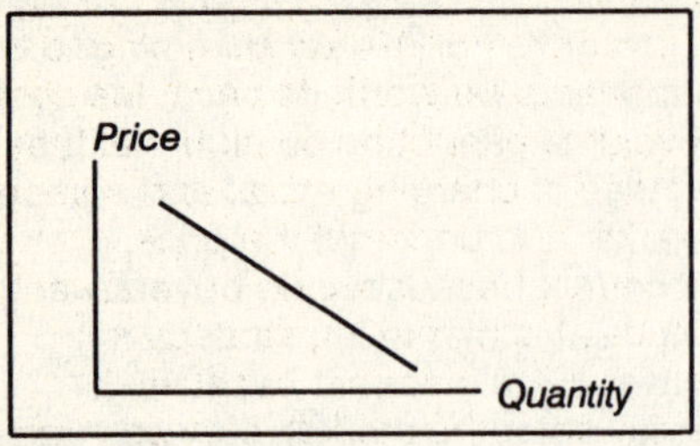

B.

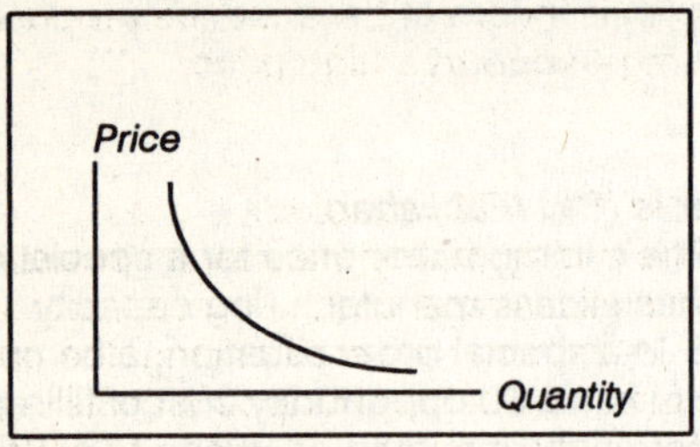

C.

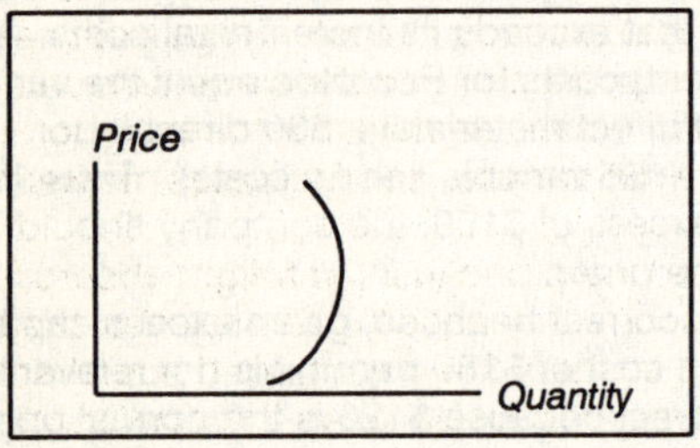

D.

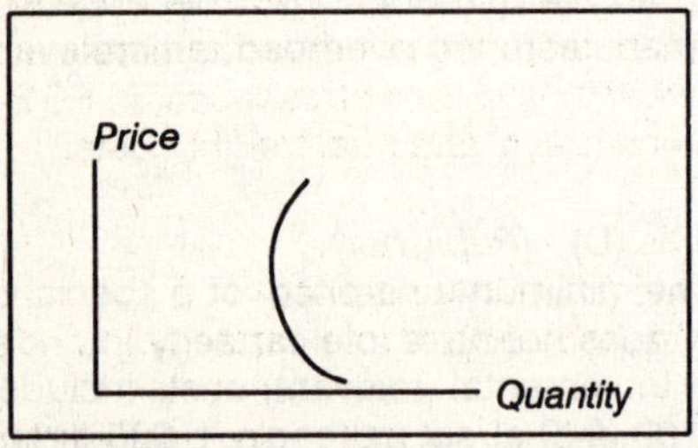

The correct answer is (C). *(CIA, adapted)*

REQUIRED: The graph depicting the demand curve for prestige goods.

DISCUSSION: Over some intermediate range of prices, the reaction to a price increase for prestige goods is an increase, not a decrease, in the quantity demanded. Within this range, the demand curve is upward sloping. The reason is that consumers interpret the higher price to indicate a better or more desirable product. Above some price level, the relation between price and quantity demanded will again become negatively sloped.

Answers (A) and (B) are incorrect because these graphs describe the familiar, negatively sloped relation between price charged and the resulting demand level for normal goods. The demand curve can be linear or curvilinear. Answer (D) is incorrect because this demand curve has the same basic shape as the demand curve for prestige goods, but it bends the wrong way. As prices increase, quantity demanded first falls and then rises in this graph.

34. Buyer-based pricing involves

A. Adding a standard markup to the cost of the product.

B. Determining the price at which the product will earn a target profit.

C. Basing prices on the product's perceived value.

D. Basing prices on competitors' prices.

The correct answer is (C). *(CIA, adapted)*

REQUIRED: The definition of buyer-based pricing.

DISCUSSION: Buyer-based pricing involves basing prices on the product's perceived value rather than on the seller's cost. Nonprice variables in the marketing mix augment the perceived value. For example, a cup of coffee may have a higher price at an expensive restaurant than at a fast-food outlet.

Answer (A) is incorrect because adding a standard markup to the cost of the product is cost-plus pricing. Answer (B) is incorrect because determining the price at which the product will earn a target profit is target profit pricing. Answer (D) is incorrect because basing prices on competitors' prices is going-rate pricing.

35. Market-skimming pricing strategies could be appropriate when

A. No buyers want the product at a high price.

B. The costs of producing a small volume are low.

C. Competitors can easily enter the market.

D. The product is of poor quality.

The correct answer is (B). *(CIA, adapted)*

REQUIRED: The circumstances under which market-skimming pricing strategies are appropriate.

DISCUSSION: Market-skimming pricing is used when a new product is introduced at the highest price possible given the benefits of the product. For market skimming to work, the product must appear to be worth its price, the costs of producing a small volume cannot be so high that they eliminate the advantage of charging more, and competitors cannot enter the market and undercut the price.

Answer (A) is incorrect because, if no buyers want the product at a high price, this marketing strategy is inappropriate. Answer (C) is incorrect because, if competitors can easily enter the market, they can undercut the price. Answer (D) is incorrect because the product quality and image must support a high price.

36. Which of the following pricing policies involves the selling company setting freight charges to customers at the actual average freight cost?

A. Freight absorption pricing.

B. Uniform delivered pricing.

C. Zone pricing.

D. FOB-origin pricing.

The correct answer is (B). *(CIA, adapted)*

REQUIRED: The pricing policy that results in setting freight charges at actual average cost.

DISCUSSION: In uniform delivered pricing, the company charges the same price, inclusive of shipping costs, to all customers regardless of their location. This price is the company's average actual freight cost. Thus, both nearby and distant customers are charged the same amount. This policy is easy to administer, permits the company to advertise one price nationwide, and facilitates marketing to faraway customers.

Answer (A) is incorrect because, in freight absorption pricing, the selling company absorbs all or part of the actual freight charges. Customers are not charged actual delivery costs. Answer (C) is incorrect because, in zone pricing, differential freight charges are set for customers on the basis of their location. Customers are not charged actual average freight costs. Answer (D) is incorrect because, in FOB-origin pricing, each customer pays its actual freight costs.

37. Which of the following price adjustment strategies is designed to stabilize production for the selling firm?

A. Cash discounts.

B. Quantity discounts.

C. Functional discounts.

D. Seasonal discounts.

The correct answer is (D). *(CIA, adapted)*

REQUIRED: The price adjustment strategy intended to stabilize production.

DISCUSSION: Seasonal discounts are designed to smooth production by the selling firm. For example, a ski manufacturer offers seasonal discounts to retailers in the spring and summer to encourage early ordering.

Answer (A) is incorrect because cash discounts encourage prompt payment. Answer (B) is incorrect because quantity discounts encourage large volume purchases. Answer (C) is incorrect because functional or trade discounts are provided to channel members in return for the performance of certain functions, such as selling, storing, and record keeping.

38. In which product-mix pricing strategy is it appropriate for the seller to accept any price that exceeds the storage and delivery costs for the product?

A. By-product pricing.

B. Optional-product pricing.

C. Captive-product pricing.

D. Product-bundle pricing.

The correct answer is (A). *(CIA, adapted)*

REQUIRED: The pricing strategy that accepts any price greater than storage and delivery costs.

DISCUSSION: A by-product is a product of relatively minor importance generated during the production of one or more other products. Its production entails no additional costs. Any amount received above the storage and delivery costs for a by-product allows the seller to reduce the main product's price to make it more competitive.

Answer (B) is incorrect because optional products are offered for sale along with the main product. They are unlikely to have a zero production cost, so the seller must receive a price above their storage and delivery costs. Answer (C) is incorrect because captive products must be used along with the main product, such as film for use with a camera. Sellers often make their profits on the captive products rather than on the main product, which is sold at a low price. The captive products therefore will be priced well above the storage and delivery costs. Answer (D) is incorrect because product bundles are combinations of products sold together at a reduced price, such as season tickets for a theater. Products are bundled to promote the sale of certain items that consumers might not otherwise purchase. The combined price of the bundle must be low enough to encourage consumers to buy the bundle but must recover production costs and provide some profit for the seller, so the price must exceed storage and delivery costs.

39. Several surveys point out that most managers use full product costs, including unit fixed costs and unit variable costs, in developing cost-based pricing. Which one of the following is least associated with cost-based pricing?

A. Price stability.

B. Price justification.

C. Target pricing.

D. Fixed-cost recovery.

The correct answer is (C). *(CMA, adapted)*

REQUIRED: The concept least associated with cost-based pricing.

DISCUSSION: A target price is the expected market price of a product, given the company's knowledge of its customers and competitors. Hence, under target pricing, the sales price is known before the product is developed. Subtracting the unit target profit margin determines the long-term unit target cost. If cost-cutting measures do not permit the product to be made at or below the target cost, it will be abandoned.

Answer (A) is incorrect because full-cost pricing promotes price stability. It limits the ability to cut prices. Answer (B) is incorrect because full-cost pricing provides evidence that the company is not violating antitrust laws against predatory pricing. Answer (D) is incorrect because full-cost pricing has the advantage of recovering the full long-term costs of the product. In the long term, all costs are relevant.

40. If a U.S. manufacturer's price in the U.S. market is below an appropriate measure of costs and the seller has a reasonable prospect of recovering the resulting loss in the future through higher prices or a greater market share, the seller has engaged in

A. Collusive pricing.

B. Dumping.

C. Predatory pricing.

D. Price discrimination.

The correct answer is (C). *(Publisher)*

REQUIRED: The pricing strategy characterized by charging a price that is below an appropriate measure of costs when the seller reasonably expects to recover the loss through higher prices or a greater market share.

DISCUSSION: Predatory pricing is intentionally pricing below cost to eliminate competition and reduce supply. Federal statutes and many state laws prohibit the practice. The U.S. Supreme Court has held that pricing is predatory when two conditions are met: (1) the seller's price is below "an appropriate measure of its costs," and (2) it has a reasonable prospect of recovering the resulting loss through higher prices or greater market share.

Answer (A) is incorrect because collusive pricing involves a conspiracy to set higher prices. Answer (B) is incorrect because dumping is defined under U.S. law as sale by a non-U.S. company in the U.S. market of a product below its market value in the country where it was produced. Such sale is illegal if it threatens material injury to a U.S. industry. Answer (D) is incorrect because price discrimination entails charging different prices to different customers for essentially the same product if the effect is to lessen competition substantially; to tend to create a monopoly; or to injure, destroy, or prevent competition.

41. A company's product has an expected 4-year life cycle from research, development, and design through its withdrawal from the market. Budgeted costs are

Upstream costs (R&D, design)	$2,000,000
Manufacturing costs	3,000,000
Downstream costs (marketing, distribution, customer service)	1,200,000
After-purchase costs	1,000,000

The company plans to produce 200,000 units and price the product at 125% of the whole-life unit cost. Thus, the budgeted unit selling price is

A. $15

B. $31

C. $36

D. $45

The correct answer is (D). *(Publisher)*

REQUIRED: The unit selling price.

DISCUSSION: Whole-life costs include after-purchase costs (operating, support, repair, and disposal) incurred by customers as well as life-cycle costs (R&D, design, manufacturing, marketing, distribution, and research). Hence, the budgeted unit whole-life cost is $36 [($2,000,000 + $3,000,000 + $1,200,000 + $1,000,000) ÷ 200,000 units], and the budgeted unit selling price is $45 (125% × $36).

Answer (A) is incorrect because $15 is the budgeted unit manufacturing cost. Answer (B) is incorrect because $31 is the budgeted unit life-cycle cost. Answer (C) is incorrect because $36 is the budgeted unit whole-life cost.

D. Inventory Control Models

Questions 42 and 43 are based on the following information. Candman Company is a wholesale distributor of candy. The company leases space in a public warehouse and is charged according to the square feet occupied. Candman has decided to employ the economic order quantity (EOQ) method to determine the optimum number of cases of candy to order.

The company placed 2,400 orders the last year. Data for the high-activity month, the low-activity month, and the year for the purchasing and warehouse operations appear in the next column.

The annual charges for the warehouse totaled $12,750 last year. In addition, the annual insurance and property taxes on the candy stored in the warehouse amounted to $1,500 and $2,250, respectively. The average monthly inventory last year was $75,000.

Long-term capital investments are expected to earn 12% after income taxes. Candman is subject to an effective income tax rate of 40%.

	High-Activity Month (160 Orders)	Low-Activity Month (100 Orders)	Annual Costs
Purchasing Dept.			
Manager	$ 1,600	$1,600	$ 23,400
Clerks	1,750	1,250	18,000
Supplies	400	260	3,500
Warehouse			
Supervisor	1,550	1,550	18,600
Rcvg. clerks	2,200	1,700	20,500
Ship. clerks	2,800	2,500	31,200
Totals	$10,300	$8,860	$115,200

42. The incremental cost of placing an order that would be used in the EOQ model is

A. $48

B. $35

C. $24

D. $19

The correct answer is (D). *(CMA, adapted)*

REQUIRED: The order cost used in the EOQ model.

DISCUSSION: The incremental cost of additional orders equals the variable costs incurred in purchasing and receiving inventory (but not costs associated with shipping). Using the high-low method of analysis, these variable costs can be computed by determining the apparent variable costs for the two given levels of activity. For 160 orders, the relevant costs with a variable element include purchasing clerks, $1,750; supplies, $400; and receiving clerks, $2,200. The total is $4,350. For 100 orders, these costs are $1,250, $260, and $1,700, respectively, for a total of $3,210. As the number of orders increased by 60, the costs increased by $1,140. Consequently, the variable or incremental costs per order must have been $19 ($1,140 ÷ 60 orders).

Answer (A) is incorrect because $48 equals total annual cost divided by total orders from the previous year. Answer (B) is incorrect because $35 equals total annual cost minus the costs associated with the shipping clerks, divided by total orders from the previous year. Answer (C) is incorrect because $24 includes the costs associated with shipping.

43. The annual carrying cost, stated as a percentage, that would be used in the EOQ model is

A. 42%

B. 37%

C. 34%

D. 22%

The correct answer is (A). *(CMA, adapted)*

REQUIRED: The percentage annual carrying cost used in the EOQ model.

DISCUSSION: The annual carrying costs are $12,750 for the warehouse, $1,500 for insurance, and $2,250 for property taxes. These costs total $16,500. In addition, the company desires a 12% after-tax return on investments. Because the tax rate is 40%, the 12% after-tax return equals a 20% before-tax return (12% ÷ 60%). A 20% return on the $75,000 average investment in inventory is $15,000. Hence, the total carrying cost is $31,500 ($16,500 + $15,000). This amount is 42% of the $75,000 investment in inventory.

Answer (B) is incorrect because 37% does not consider insurance and property taxes. Answer (C) is incorrect because 34% is based on a 12% before-tax return. Answer (D) is incorrect because 22% equals carrying costs other than return on investment divided by average inventory.

44. The calculation of an economic order quantity (EOQ) considers

A. The purchasing manager's salary.

B. A corporate charge for advertising expenses.

C. The shipping costs to deliver the product to the customer.

D. Capital costs.

The correct answer is (D). *(CMA, adapted)*

REQUIRED: The true statement about the calculation of the economic order quantity.

DISCUSSION: The determination of the economic order quantity balances the variable costs of ordering and carrying inventory. Factors in the equation include the cost of placing an order, unit carrying cost, and annual demand in units. Carrying costs include storage costs, handling costs, insurance, property taxes, obsolescence, and the opportunity cost of investing capital in inventory. Thus, the return on capital that is forgone when it is invested in inventory should be considered.

Answer (A) is incorrect because the purchasing manager's salary is a fixed cost. The EOQ model includes variable costs only. Answer (B) is incorrect because advertising is not an ordering or carrying cost. Answer (C) is incorrect because the cost of shipping to customers is a selling expense.

45. A decrease in inventory order costs will

A. Decrease the economic order quantity.

B. Increase the reorder point.

C. Increase the economic order quantity.

D. Decrease the holding cost percentage.

The correct answer is (A). *(CMA, adapted)*

REQUIRED: The effect of a decrease in inventory ordering costs.

DISCUSSION: A decrease in inventory ordering costs should decrease the economic order quantity. The effect is that more orders can be made (of smaller quantities) without increasing costs. Accordingly, in the EOQ model, ordering cost is a numerator value.

Answer (B) is incorrect because the reorder point is based on lead time, not the EOQ model. Answer (C) is incorrect because the EOQ should decline. Answer (D) is incorrect because the holding cost percentage will always be identical to the ordering cost percentage in accordance with the fundamental calculus underlying the EOQ model.

46. Which one of the following items is not directly reflected in the basic economic order quantity (EOQ) model?

A. Interest on invested capital.

B. Inventory obsolescence.

C. Public warehouse rental charges.

D. Quantity discounts lost on inventory purchases.

The correct answer is (D). *(CMA, adapted)*

REQUIRED: The item not reflected in the basic EOQ model.

DISCUSSION: The basic EOQ model minimizes the sum of ordering (or setup) and carrying costs. Included in the formula are annual demand, ordering (or setup) costs, and carrying costs. Carrying costs include warehousing costs, insurance, spoilage, obsolescence, and interest on invested capital. The cost of the inventory itself and any quantity discounts lost on inventory purchases are not components of the EOQ model.

Answers (A), (B), and (C) are incorrect because interest on invested capital, inventory obsolescence, and public warehouse rental charges are all items considered in the basic EOQ model.

47. A characteristic of the basic economic order quantity (EOQ) model is that it

A. Is relatively insensitive to error.

B. Should not be used when carrying costs are large in relation to procurement costs.

C. Is used when product demand, lead-time, and ordering costs are uncertain.

D. Should not be used in conjunction with computerized perpetual inventory systems.

The correct answer is (A). *(CMA, adapted)*

REQUIRED: The characteristic of the basic EOQ model.

DISCUSSION: The basic EOQ model equals the square root of the quotient of (1) the product of twice the demand times the cost per order, (2) divided by the periodic carrying cost. Hence, the model is relatively insensitive to error. A given percentage error in a value results in a lower percentage change in the EOQ.

Answer (B) is incorrect because the EOQ model can be used regardless of the relationship between carrying and holding costs. Answer (C) is incorrect because product demand and ordering costs must be known with some certainty. Answer (D) is incorrect because an EOQ model can be used with any type of system.

48. Companies that adopt just-in-time purchasing systems often experience

A. An increase in carrying costs.

B. A reduction in the number of suppliers.

C. A greater need for inspection of goods as the goods arrive.

D. Less need for linkage with a vendor's computerized order entry system.

The correct answer is (B). *(CMA, adapted)*

REQUIRED: The true statement about companies that adopt just-in-time (JIT) purchasing systems.

DISCUSSION: The objective of JIT is to reduce carrying costs by eliminating inventories and increasing the deliveries made by suppliers. Ideally, shipments of raw materials are received just in time to be incorporated into the manufacturing process. The focus of quality control under JIT is the prevention of quality problems. Quality control is shifted to the supplier. JIT companies typically do not inspect incoming goods; the assumption is that receipts are of perfect quality. Suppliers are limited to those who guarantee perfect quality and prompt delivery.

Answer (A) is incorrect because carrying costs typically decline in JIT companies. Less inventory is on hand. Answer (C) is incorrect because, in a JIT system, materials are delivered directly to the production line ready for insertion in the finished product. Answer (D) is incorrect because the need for communication with the vendor is greater. Orders and deliveries must be made on short notice, sometimes several times a day.

49. In production management, product breakdown into component parts and lead times for procuring these parts is necessary for

A. A critical path method (CPM) system.

B. A materials requirements planning (MRP) system.

C. An economic order quantity (EOQ) system.

D. An ABC system.

The correct answer is (B). *(CMA, adapted)*

REQUIRED: The production management system that requires breaking a product down into its components.

DISCUSSION: Materials requirements planning (MRP) is usually considered a computer-based information system designed to plan and control raw materials used in a production setting. It assumes that estimated demand for materials is reasonably accurate and that suppliers can deliver based upon this accurate schedule. It is crucial that delivery delays be avoided because, under MRP, production delays are almost unavoidable if the materials are not on hand. An MRP system uses a parts list, often called a bill of materials, and lead times for each type of material to obtain materials just as they are needed for planned production.

Answer (A) is incorrect because a CPM system is a project management and scheduling technique that determines the longest time path from the first to the last event for a project. Answer (C) is incorrect because the EOQ model does not adjust for the dependent demand for the components of an inventory item. Answer (D) is incorrect because the main purpose of the ABC system is to increase control over items accounting for the most profit.

50. A major justification for investments in computer-integrated manufacturing (CIM) projects is

A. Reduction in the costs of spoilage, reworked units, and scrap.

B. Lower book value and depreciation expense for factory equipment.

C. Increased working capital.

D. Stabilization of market share.

The correct answer is (A). *(CIA, adapted)*

REQUIRED: The major justification for investments in CIM.

DISCUSSION: Automating and computerizing production processes requires a substantial investment in fixed assets and an increase in risk because of greater fixed costs. CIM also necessitates an increase in software costs and extensive worker retraining. However, the costs of spoilage, rework, and scrap are reduced along with labor costs. The qualitative advantages of CIM are increased flexibility, shorter manufacturing lead time, quicker development of new products, better product delivery and service, faster response to market changes, and improved competitiveness.

Answer (B) is incorrect because an increase in fixed assets results in higher book value and depreciation expense. Answer (C) is incorrect because working capital normally is reduced as investments shift from current to fixed assets. Answer (D) is incorrect because actual or potential market share changes may trigger investments in CIM.

51. Increased competition, technological innovation, and a shift from mass production of standardized products to custom-produced products in many industries have increased the need for productivity improvement and flexibility of production systems. In response to these demands, organizations have increased their reliance on automation and the use of advanced technologies in their operations. Which of the following is an example of the use of automation and advanced technologies?

A. Flexible manufacturing system (FMS).

B. Just-in-time (JIT) system.

C. Master budgeting system (MBS).

D. Economic order quantity (EOQ).

The correct answer is (A). *(CIA, adapted)*

REQUIRED: The example of the use of automation and advanced technologies.

DISCUSSION: Flexible manufacturing is the capacity of computer-controlled machinery to perform many different programmed functions. By eliminating machine setup time, strengthening control, and automating handling processes, computer-aided manufacturing permits the efficient production of small numbers of different products by the same machines. A company can therefore more accurately match output with consumer tastes and avoid long production runs of identical goods. A flexible manufacturing system consists of two or more computer-controlled machines linked by automated handling devices such as robots and transport systems.

Answer (B) is incorrect because a JIT system involves the purchase of materials and production of components immediately preceding their use. Answer (C) is incorrect because a master budget is the detailed financial plan for the next period. Answer (D) is incorrect because the EOQ is the quantity that minimizes total costs.

52. Traditionally, large manufacturers have believed that economies of scale gained through large production runs of like, or similar, products are the best way to keep production costs down and remain competitive. Select the most appropriate response to whether this theory is still valid.

A. Yes, larger economies of scale continue to accrue from ever larger production runs.

B. Yes, lower-per-unit costs for standard products continue to guarantee a competitive advantage.

C. No, economies of scale can no longer be gained from long production runs.

D. No, production flexibility and diversity of products are needed to remain competitive.

The correct answer is (D). *(CIA, adapted)*

REQUIRED: The most appropriate evaluation of the belief that economies of scale are the best means of reducing costs.

DISCUSSION: Global competition and more rapidly changing consumer tastes are lessening the need for long production runs and increasing the need for diversity of products and flexibility of manufacturing. The economies of scale concept is being replaced by economies of scope, which result when flexible manufacturing methods (use of computerized machines to perform many programmed functions) permit the economical output of small numbers of many different products with the same machines.

Answers (A) and (B) are incorrect because economies of scale cannot continue to be achieved if the products are not salable. Answer (C) is incorrect because economies of scale can still be achieved from long production runs if the goods are salable.

53. The inventory model that follows the concept that 80% of the value of an inventory is in 20% of the inventory items is the

A. ABC system.

B. Economic order quantity (EOQ) model.

C. Just-in-time inventory system.

D. Materials requirements planning (MRP) system.

The correct answer is (A). *(CMA, adapted)*

REQUIRED: The inventory system that follows the concept that 80% of the value of the inventory is in 20% of the items.

DISCUSSION: The ABC method of inventory control requires management to exert greatest control over the A classification items, which usually include a relatively small percentage of total items but a high percentage of the dollar volume. This method is analogous to the 80/20 rule, which says, for instance, that 20% of the customers account for 80% of the profit.

Answer (B) is incorrect because the EOQ model is intended to minimize the sum of holding and ordering costs. Answer (C) is incorrect because a just-in-time system attempts to reduce holding costs by scheduling deliveries of materials as closely as possible to when they are needed in production. Answer (D) is incorrect because an MRP system recognizes that the demand for one inventory item creates a dependent demand for the components of that item.

54. When a specified level of safety stock is carried for an item in inventory, the average inventory level for that item

A. Decreases by the amount of the safety stock.

B. Is one-half the level of the safety stock.

C. Increases by one-half the amount of the safety stock.

D. Increases by the amount of the safety stock.

The correct answer is (D). *(CMA, adapted)*

REQUIRED: The effect on the average inventory level of carrying safety stock.

DISCUSSION: If there is no safety stock, the average level of inventory will be one-half of the economic order quantity. For example, if the EOQ is 500, the basic model predicts that the average inventory level will be 250. The company will have 500 units immediately after a purchase and zero immediately before the receipt of the next purchase (replenishment is assumed to be instantaneous). However, safety stock increases the average inventory level by the amount of the safety stock. The modified EOQ model assumes that safety stock will never be used. Thus, if a safety stock of 100 is carried by the company in the example, the average inventory level will increase to 350. Inventory will be 600 units immediately upon receipt of a purchase and 100 units immediately before the receipt of the next purchase.

Answers (A), (B), and (C) are incorrect because the average inventory level increases by the amount of the safety stock.

55. Arnold Enterprises uses the EOQ model for inventory control. The company has an annual demand of 50,000 units for part number 191 and has computed an optimal lot size of 6,250 units. Per-unit carrying costs and stockout costs are $13 and $3, respectively. The following data have been gathered in an attempt to determine an appropriate safety stock level:

Units Short Because of Excess Demand During the Lead Time Period	Number of Times Short in the Last 40 Reorder Cycles
200	6
300	12
400	6

The annual cost of establishing a 200-unit safety stock is expected to be

A. $2,600

B. $4,040

C. $4,260

D. $5,200

The correct answer is (B). *(CMA, adapted)*

REQUIRED: The annual cost of establishing a given safety stock.

DISCUSSION: The annual cost consists of the carrying cost of the 200 units of safety stock at $13 each, or $2,600, plus the stockout costs incurred when 200 units are insufficient. The stockout cost per unit is $3. The excess demand has been 100 units (300 – 200) greater than the proposed safety stock 30% of the time (12 ÷ 40). The cost per stockout was $300 ($3 x 100). Demand has exceeded the safety stock by 200 units (400 – 200) 15% of the time (6 ÷ 40). The cost per stockout was $600 ($3 x 200). Given 30% and 15% probabilities of $300 and $600 stockout costs, respectively, the expected stockout cost for a 200-unit safety stock for one reorder cycle is $180 [(30% x $300) + (15% x $600)]. Thus, the annual expected stockout cost for a 200-unit safety stock is $180 times the number of reorder cycles per year, or $1,440 [$180 x (50,000 ÷ 6,250 EOQ). The annual cost of a 200-unit safety stock is $4,040 ($2,600 + 1,440).

Answer (A) is incorrect because $2,600 is the carrying cost of the safety stock. Answer (C) is incorrect because $4,260 is the annual cost of establishing a 300-unit safety stock. Answer (D) is incorrect because $5,200 is the annual cost of establishing a 400-unit safety stock.

Questions 56 through 58 are based on the following data. The diagram presented below represents the economic order quantity (EOQ) model.

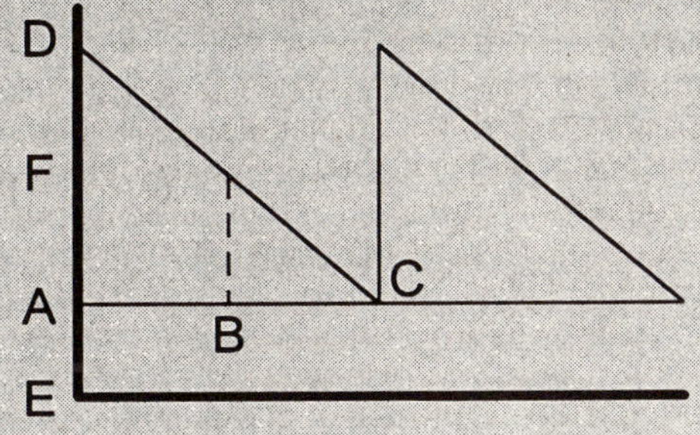

56. Which line segment represents the reorder lead time?

A. AB.

B. AE.

C. AF.

D. BC.

The correct answer is (D). *(CMA, adapted)*

REQUIRED: The line segment representing the reorder lead time.

DISCUSSION: The quantity of inventory on hand is represented by the y axis and time by the x axis. The reorder lead time is represented by the line segment BC.

Answer (A) is incorrect because AB is the time between receipt of the last order and the placing of the next order. Answer (B) is incorrect because AE is the safety stock. Answer (C) is incorrect because AF represents the quantity of inventory that will be used during the reorder lead time.

57. Which line segment identifies the quantity of safety stock maintained?

A. AB.

B. AE.

C. AC.

D. BC.

The correct answer is (B). *(CMA, adapted)*

REQUIRED: The line segment representing the quantity of safety stock maintained.

DISCUSSION: Quantities of inventory are shown along the y axis. Safety stock is represented by the line AE.

Answer (A) is incorrect because AB is the time between the receipt of the last order and the placing of the next order. Answer (C) is incorrect because AC is the time to consume the EOQ. Answer (D) is incorrect because BC is the reorder lead time.

58. Which line segment represents the length of time to consume the total quantity of materials ordered?

A. DE.

B. BC.

C. AC.

D. AE.

The correct answer is (C). *(CMA, adapted)*

REQUIRED: The line representing the time to consume all the materials ordered.

DISCUSSION: Time is shown along the x axis. The line segment AC depicts the time to consume an entire order (to reduce the inventory to the safety stock).

Answer (A) is incorrect because DE represents the total inventory on hand just after an order has been received. Answer (B) is incorrect because BC is the reorder lead time. Answer (D) is incorrect because AE is the safety stock.

59. Each stockout of a product sold by A.W. Inn Co. costs $1,750 per occurrence. The carrying cost per unit of inventory is $5 per year, and the company orders 1,500 units of product 24 times a year at a cost of $100 per order. The probability of a stockout at various levels of safety stock is

Units of Safety Stock	Probability of a Stockout
0	.50
100	.30
200	.14
300	.05
400	.01

The optimal safety stock level for the company is

A. 0 units.

B. 100 units.

C. 300 units.

D. 400 units.

The correct answer is (D). *(CMA, adapted)*

REQUIRED: The optimal level of safety stock.

DISCUSSION: The total expected cost of safety stock equals the sum of the expected annual stockout cost and the expected annual carrying cost. Annual expected stockout cost equals the cost per occurrence ($1,750), times the probability of a stockout per cycle, times the number of cycles (24). Annual expected carrying cost of a safety stock equals the unit carrying cost ($5) times the number of units. Hence, a safety stock of 400 units has the lowest total expected cost.

Units Held	Carrying Cost	Expected Stockout Cost Per Cycle	Expected Stockout Cost for 24 Cycles	Total Expected Cost
0	$ 0	$875.00	$21,000	$21,000
100	500	525.00	12,600	13,100
200	1,000	245.00	5,880	6,880
300	1,500	87.50	2,100	3,600
400	2,000	17.50	420	2,420

Answer (A) is incorrect because a safety stock of 0 units has a total expected cost of $21,000. Answer (B) is incorrect because a safety stock of 100 units has a total expected cost of $13,100. Answer (C) is incorrect because a safety stock of 30 units has a total expected cost of $3,600.

Questions 60 and 61 are based on the following information. The Huron Corporation purchases 60,000 headbands per year. The average purchase lead time is 20 working days. Maximum lead time is 27 working days. The corporation works 240 days per year.

60. Huron Corporation should carry safety stock of

A. 5,000 units.

B. 6,750 units.

C. 1,750 units.

D. 250 units.

The correct answer is (C). *(CMA, adapted)*

REQUIRED: The amount of safety stock to be carried.

DISCUSSION: Safety stock is defined as the amount of extra stock that is kept to guard against stockouts. It is the inventory level at the reorder point minus the expected usage during the lead time. Daily usage is 250 units (60,000 per year ÷ 240 days). Given a maximum lead time of 27 days and a normal lead time of 20 days, a safety stock for 7 days (27 – 20) should be maintained. Hence, safety stock is 1,750 units (7 x 250 units).

Answer (A) is incorrect because 5,000 units represents the quantity expected to be sold during the normal lead time. Answer (B) is incorrect because 6,750 units includes not only the safety stock but also the units expected to be sold during the normal lead time. Answer (D) is incorrect because 250 units is 1 day's sales.

61. Huron Corporation should reorder headbands when the quantity in inventory reaches

A. 5,000 units.

B. 6,750 units.

C. 1,750 units.

D. 5,250 units.

The correct answer is (B). *(CMA, adapted)*

REQUIRED: The reorder point in units.

DISCUSSION: The reorder point is the quantity on hand when an order is placed. With a 20-day normal lead time, a 7-day safety stock (see preceding question), and usage of 250 units per day (see preceding question), an order should be placed when 27 days of inventory are on hand, a total of 6,750 units (27 x 250).

Answer (A) is incorrect because 5,000 units does not allow for safety stock. Answer (C) is incorrect because 1,750 units covers only safety stock. Answer (D) is incorrect because 5,250 units includes only 1 day of safety stock.

Questions 62 through 64 are based on the following information. Ryerson Computer Furniture Inc. (RCF) manufactures a line of office computer chairs. The annual demand for the chairs is estimated to be 5,000 units. The annual cost to hold one unit in inventory is $10 per year, and the cost to initiate a production run is $1,000. There are no computer chairs on hand, and RCF has scheduled four equal production runs of computer chairs for the coming year, the first of which is to be run immediately. RCF has 250 business days per year, sales occur uniformly throughout the year, and production start-up is within one day. RCF is considering using the following formula for determining the economic order quantity (EOQ):

$$EOQ = \sqrt{\frac{2AD}{K}}$$

If: A = cost to initiate a production run per purchase order
D = annual unit demand
K = cost of carrying one unit per year

62. The number of production runs per year of computer chairs that would minimize the sum of carrying and setup costs for the coming year is

A. 1

B. 2

C. 4

D. 5

The correct answer is (D). *(CMA, adapted)*

REQUIRED: The number of production runs that minimizes the sum of setup and carrying costs.

DISCUSSION: The EOQ minimizes the sum of carrying and setup costs. The EOQ is the amount at which carrying costs are equal to setup costs. Thus, plugging the data into the EOQ formula results in the following:

$$EOQ = \sqrt{\frac{2(\$1{,}000)(5{,}000)}{\$10}} = 1{,}000 \textit{ units}$$

Thus, if each lot consists of 1,000 units, five production runs per year are needed to meet the 5,000-unit demand. At this level, setup costs will total $5,000 (5 x $1,000). Carrying costs will also equal $5,000 ($10 per unit carrying cost x average inventory of 500 units). Accordingly, total costs are minimized at $10,000.

Answer (A) is incorrect because a single production run indicates an EOQ of 5,000 units. The carrying costs of $25,000 [$10 x ($5,000 ÷ 2)] would exceed the $1,000 of setup costs. Answer (B) is incorrect because two production runs correspond to an EOQ of 2,500 units and an average inventory of 1,250 units. The resulting $12,500 of carrying costs would exceed the $2,000 of setup costs. Answer (C) is incorrect because four production runs correspond to an EOQ of 1,250 units and an average inventory of 625 units. The resulting $6,250 of carrying costs would exceed the $4,000 of setup costs.

63. If RCF does not maintain a safety stock, the estimated total carrying costs for the computer chairs for the coming year based on their current schedule is

A. $4,000

B. $5,000

C. $6,250

D. $12,500

The correct answer is (C). *(CMA, adapted)*

REQUIRED: The estimated total annual carrying costs.

DISCUSSION: Given four production runs and an annual demand of 5,000 units, each production run must generate 1,250 units. Inventory will total 1,250 units at the completion of each run but will decline to zero just prior to the next run. Thus, the average inventory is 625 units (1,250 ÷ 2), and the total carrying cost is $6,250 ($10 x 625 units).

Answer (A) is incorrect because the cost of maintaining an average inventory of 625 units is $6,250. Answer (B) is incorrect because $5,000 is based upon an EOQ of 1,000 units and an average inventory of 500 units. Answer (D) is incorrect because $12,500 is based on the maximum inventory level.

64. A safety stock of a 5-day supply of computer chairs would increase RCF's planned average inventory in units (rounded) by

A. Zero units.

B. 20 units.

C. 50 units.

D. 100 units.

The correct answer is (D). *(CMA, adapted)*

REQUIRED: The increase in planned average inventory.

DISCUSSION: The annual sales of 5,000 units are made over 250 days, or 20 units per day. Thus, a 5-day safety stock equals 100 units (5 days x 20 units per day).

Answer (A) is incorrect because average inventory must increase if the company maintains a safety stock. Answer (B) is incorrect because 20 units is the safety stock per day (5,000 units ÷ 250 business days). Answer (C) is incorrect because 50 units is the average safety stock assuming that it will decrease to zero.

E. Government Contracts and the CASB

65. The Cost Accounting Standards Board (CASB) was established by

A. The Financial Accounting Standards Board (FASB).

B. The General Accounting Office (GAO).

C. The U.S. Congress.

D. The Securities and Exchange Commission (SEC).

The correct answer is (C). *(CMA, adapted)*

REQUIRED: The source of authority for the CASB.

DISCUSSION: The CASB was established by the U.S. Congress in 1970 to develop cost accounting standards applicable to defense contractors and subcontractors. Other government agencies also use the standards. When the CASB was eliminated in 1980, the standards remained in effect. In 1988, Congress reestablished the CASB.

Answer (A) is incorrect because the FASB is a private sector organization concerned with financial accounting and reporting issues. Answer (B) is incorrect because the GAO is the audit agency of the Congress. Answer (D) is incorrect because the SEC is a government agency concerned with public disclosure of financial and other data by publicly held companies.

66. The Cost Accounting Standards Board's (CASB) purpose is to

A. Develop accounting principles and standard practices for industry.

B. Achieve uniformity and consistency in cost accounting standards for contracts and subcontracts with the U.S. government.

C. Work in conjunction with the Securities and Exchange Commission (SEC) in examining registration forms and statements filed by corporations.

D. Administer all contracts and subcontracts with federal agencies.

The correct answer is (B). *(CMA, adapted)*

REQUIRED: The purpose of the CASB.

DISCUSSION: The CASB has "exclusive authority to make, promulgate, amend, and rescind cost accounting standards and interpretations thereof" for negotiated contracts and subcontracts over $500,000. Its purpose is "to achieve uniformity and consistency in the cost accounting standards governing measurement, assignment, and allocation of costs to contracts within the United States."

Answer (A) is incorrect because the CASB does not develop accounting principles or set standards other than for contractors doing business with the U.S. Government. Answer (C) is incorrect because the CASB has no connection with the SEC and does not examine registration statements. Answer (D) is incorrect because the CASB does not administer contracts.

STUDY UNIT 2: COST-VOLUME-PROFIT ANALYSIS

6 pages of outline
81 multiple-choice questions

A. Cost-Volume-Profit (CVP) Analysis
B. Applications

Cost-volume-profit (CVP) analysis, also known as breakeven analysis, is tested in Part 4. It is a methodology useful in operational decision analysis. CVP analysis concerns cost behavior, that is, whether costs are fixed, variable, or some combination thereof. Cost behavior must be understood to plan for firm profitability.

The tax consequences of business decisions are considered briefly in Appendix B, Taxes as Tested on the CMA/CFM Exams. Operating income subject to income taxes, gains and losses subject to taxes, tax credits, tax shields from sunk costs, transaction timing, tax elections, etc., are among the relevant issues.

CVP analysis is also a subunit of Study Unit 3, "Cost Behavior," in Part 3. We believe, however, that the topic will be more heavily tested in Part 4.

A. Cost-Volume-Profit (CVP) Analysis

1. CVP analysis is a means of predicting the relationships among revenues, variable costs, and fixed costs at various production levels. It allows management to discern the probable effects of changes in sales volume, sales price, product mix, etc. A key calculation in CVP analysis is the breakeven point.
2. The **variables** include
   a. Revenue as a function of price per unit and quantity produced
   b. Fixed costs
   c. Variable cost per unit or as a percentage of sales
   d. Profit per unit or as a percentage of sales
3. The inherent simplifying **assumptions** of CVP analysis are the following:
   a. Cost and revenue relationships are predictable and linear. These relationships are true over a **relevant range** of activity and specified time span. For example, reductions in prices are not necessary to increase revenues, and no learning curve effect operates to reduce unit variable labor costs at higher output levels.
   b. Total **variable costs** change proportionally with volume, but unit variable costs are constant over the relevant range.
   c. Changes in inventory are insignificant in amount.
   d. **Fixed costs** remain constant over the relevant range of volume, but unit fixed costs vary indirectly with volume.

e. Unit selling prices and market conditions are constant.

f. Production equals sales.

g. The **revenue (sales) mix** is constant, or the firm makes and sells only one product.

h. All costs are either fixed or variable relative to a given cost object.

i. Technology and productive efficiency are constant.

j. Revenues and costs vary only with changes in physical unit volume. Hence, volume is the sole revenue driver and cost driver.

k. The breakeven point is directly related to costs and inversely related to the budgeted margin of safety and the contribution margin.

l. The time value of money is ignored.

4. The assumptions under which CVP analysis operates primarily hinge on **certainty**. However, many decisions must be made even though uncertainty exists. Assigning probabilities to the various outcomes and sensitivity ("what-if") analysis are important approaches to dealing with uncertainty.

5. **Definitions**

a. The **breakeven point** is the level of output at which total revenues equal total expenses, that is, the point at which operating income is zero.

b. The **margin of safety** is a measure of risk. It is the excess of budgeted revenues over breakeven revenues (or budgeted units over breakeven units).

c. **Mixed costs (or semivariable costs)** are costs with both fixed and variable elements.

d. The **revenue (sales) mix** is the composition of total revenues in terms of various products, i.e., the percentages of each product included in total revenues. It is maintained for all volume changes.

e. **Sensitivity analysis** examines the effect on the outcome of not achieving the original forecast or of changing an assumption.

f. **Unit contribution margin (UCM)** is the unit selling price minus the unit variable cost. It is the contribution from the sale of one unit to cover fixed costs (and possibly a targeted profit).

1) It is expressed as either a percentage of the selling price (**contribution margin ratio**) or a dollar amount.

2) The UCM is the slope of the total cost curve plotted so that volume is on the x axis and dollar value is on the y axis.

6. The **breakeven formula** equates operating income with revenues (quantity sold x unit price), minus fixed costs, minus variable costs (quantity sold x unit variable cost).

   a. $OI = R - FC - VC$
      $R = PQ$

      If: OI = operating income ($0 at breakeven)
      R = revenues
      FC = fixed costs
      VC = variable costs
      Q = quantity of units sold
      P = unit sales price

   b. EXAMPLE: A product is sold for $.60 per unit, and variable costs are $.20 per unit. If fixed costs are $10,000, what is the breakeven point?

$$\begin{aligned} Q &= \textit{Units produced and sold} \\ \$.60Q &= \$10{,}000 \textit{ of FC} + \$.20Q \textit{ of VC} \\ \$.40Q &= \$10{,}000 \\ Q &= 25{,}000 \textit{ units} \end{aligned}$$

      1) A simpler calculation is to divide fixed costs by the unit contribution margin (the unit contribution to coverage of fixed costs).

$$\frac{\textit{Fixed costs}}{\textit{UCM}} = \textit{Breakeven point (in units)}$$

      2) The UCM is $.40 ($.60 sales price – $.20 variable cost). Thus, to cover $10,000 of fixed costs, 25,000 units ($10,000 ÷ $.40 UCM) must be sold to break even.

      3) The breakeven point in dollars can be calculated by dividing fixed costs by the contribution margin ratio {[$10,000 ÷ ($.40 ÷ $.60)] = $15,000}.

7. The **contribution income statement** is useful for confirming the breakeven calculation.

   a.
Revenues	
(Variable costs)	
Contribution margin	
(Fixed costs)	
Operating income	

8. Stop and review! You have completed the outline for this subunit. Study multiple-choice questions 1 through 13 beginning on page 77.

B. Applications

1. The basic problem equates revenues with the sum of fixed costs and variable costs.
2. An amount of operating income, either in dollars or as a percentage of sales, is frequently required.
   a. EXAMPLE: If units are sold at \$6.00 and variable costs are \$2.00, how many units must be sold to realize operating income of 15% (\$6.00 x .15 = \$.90 per unit) before taxes, given fixed costs of \$37,500?

$$\begin{aligned} R &= FC + VC + OI \\ \$6.00Q &= \$37{,}500 + \$2.00Q + \$.90Q \\ \$3.10Q &= \$37{,}500 \\ Q &= 12{,}097 \textit{ units at breakeven } + 15\% \textit{ profit} \end{aligned}$$

      1) The target operating income of \$.90 per unit is treated as a variable cost. If it were stated in total dollars rather than as a percentage, it would be treated as a fixed cost.
      2) Selling 12,097 units results in \$72,582 of revenues. Variable costs are \$24,194 and operating income is \$10,888 (\$72,582 x 15%). The proof is that fixed costs of \$37,500, plus variable costs of \$24,194, plus operating income of \$10,888 equals \$72,582 of revenues.
      3) A variation of this problem asks for net income (an after-tax amount) instead of operating income (a pretax amount). In this case, the computation requires converting the target net income to target operating income by dividing the target net income by one minus the income tax rate. Income tax liability does not change the breakeven point because, at that output level, operating income and therefore income tax expense are zero. However, other taxes, such as sales tax, do change the breakeven point.
3. Multiple products may be involved in calculating a breakeven point.
   a. EXAMPLE: A and B account for 60% and 40% of total revenues, respectively. The variable costs of A and B are 60% and 85% of individual product revenues, respectively. What is the breakeven point, given fixed costs of \$150,000?

$$\begin{aligned} R &= FC + VC \\ R &= \$150{,}000 + .6(.6R) + .85(.4R) \\ R &= \$150{,}000 + .36R + .34R \\ .30R &= \$150{,}000 \\ R &= \$500{,}000 \end{aligned}$$

      1) In effect, the result is obtained by calculating a weighted-average contribution margin ratio (100% – 36% – 34% = 30%) and dividing it into the fixed costs to arrive at the breakeven point in revenue dollars.
      2) Another approach to multiproduct breakeven problems is to divide fixed costs by the UCM for a composite unit (when unit prices are known) to determine the number of composite units. The number of individual units can then be calculated based on the stated mix.
         a) EXAMPLE: If 150,000 units of X and 300,000 units of Y are expected to be sold, the composite unit consists of 1 unit of X and 2 units of Y. If X and Y have UCMs of \$5 and \$7, respectively, the composite UCM is \$19 (\$5 + \$7 + \$7). Dividing \$19 into fixed costs gives the breakeven point in composite units. The units of X and Y equal the number of composite units and twice the number of composite units, respectively.

b. When resources are limited, a company may produce only a single product. A breakeven analysis of the point where the same operating income or loss will result, regardless of the product selected, is calculated by setting the breakeven formulas of the individual products equal to each other.

1) EXAMPLE: Assume a lessor can rent property to either of two lessees. One lessee offers a rental fee of $100,000 per year plus 2% of revenues. The other lessee offers $20,000 per year plus 5% of revenues. The optimal solution depends on the level of revenues. A typical CMA/CFM question asks at what level the lessor will be indifferent. The solution is to equate the two formulas as follows:

$$\$100{,}000 + .02R = \$20{,}000 + .05R$$
$$\$80{,}000 = .03R$$
$$\$80{,}000 \div .03 = R$$
$$\$2{,}666{,}667 = R$$

Thus, if revenues are expected to be less than $2,666,667, the lessor would prefer the larger fixed rental of $100,000 and the smaller variable rental.

4. Sometimes CVP analysis is applied to special orders. This application is essentially contribution margin analysis.

a. EXAMPLE: What is the effect of accepting a special order for 10,000 units at $8.00, given the following operating data?

	Per Unit	Total
Revenues	$12.50	$1,250,000
Manufacturing costs -- variable	$ 6.25	$ 625,000
-- fixed	1.75	175,000
-- total	$ 8.00	$ 800,000
Gross profit	$ 4.50	$ 450,000
Selling expenses -- variable	$ 1.80	$ 180,000
-- fixed	1.45	145,000
-- total	$ 3.25	$ 325,000
Operating income	$ 1.25	$ 125,000

1) Because the variable cost of manufacturing is $6.25, the UCM is $1.75 ($8.00 – $6.25), and the increase in operating income is $17,500 ($1.75 x 10,000 units).

2) The assumptions are that idle capacity is sufficient to manufacture 10,000 extra units, that sale at $8.00 per unit will not affect the price or quantity of other units sold, and that no additional selling expenses are incurred.

5. The foregoing discussion of breakeven applications assumes that **variable costing** is used. Under variable costing, the breakeven level of unit sales does not vary with production. It is a function solely of total fixed costs and the UCM.

a. Under **absorption costing**, however, the breakeven level of unit sales also depends on units produced and the amount of the denominator value used to assign fixed manufacturing costs to inventory. As the formula below indicates, the greater the units produced, the lower the breakeven point in units sold. This inverse relationship arises because some fixed manufacturing costs are inventoried under absorption costing. Accordingly, the breakeven point (BEP) in units sold under absorption costing equals total fixed costs (TFC) plus the product of the fixed manufacturing cost application rate (FMCAR) and the difference between the BEP in units sold and total units produced (UP), all divided by the UCM:

$$BEP = \frac{TFC + [FMCAR(BEP - UP)]}{UCM}$$

6. The **degree of operating leverage (DOL)** is the change in operating income (earnings before interest and taxes) resulting from a percentage change in revenues. It measures the extent to which a firm incurs fixed rather than variable costs in operations. Thus, the greater the DOL, the greater the risk of loss when revenues decline and the greater the reward when revenues increase.

   a. $$\textit{Operating leverage} = \frac{\textit{Percentage change in operating income}}{\textit{Percentage change in revenues}}$$

   b. The assumption is that organizations with larger investments (and greater fixed costs) will have higher contribution margins and more operating leverage.

      1) Thus, as they invest in better and more expensive equipment, their variable production costs should decrease.

      2) EXAMPLE: If revenues increase by 40% and operating income increases by 50%, the operating leverage is 1.25 (50% ÷ 40%).

   c. Given that Q equals the number of units sold, P is unit price, V is unit variable cost, and F is fixed cost, the DOL can also be calculated from the formula below, which equals total contribution margin divided by operating income (total contribution margin – fixed cost). This formula is derived from the formula above, but the derivation is not given.

$$\frac{Q(P - V)}{Q(P - V) - F} = \frac{\textit{Total Contribution Margin (TCM)}}{\textit{Operating Income}}$$

   d. The DOL is calculated with respect to a given base level of revenues. The significance of the DOL is that a given percentage increase in revenues yields a percentage increase in operating income equal to the DOL for the base revenues level times the percentage increase in revenues.

   e. Furthermore, the DOL is related to the **margin of safety percentage (MSP)**. The DOL equals 1 divided by the MSP, and the MSP equals 1 divided by the DOL. The reason for this relationship is that the margin of safety (operating income) is the excess of the CM over fixed costs (the amount by which the TCM may be reduced without incurring a loss). Thus, the MSP equals operating income divided by the TCM, which is the inverse of the DOL (TCM ÷ operating income).

7. Stop and review! You have completed the outline for this subunit. Study multiple-choice questions 14 through 81 beginning on page 81.

MULTIPLE-CHOICE QUESTIONS

A. Cost-Volume-Profit (CVP) Analysis

1. Cost-volume-profit (CVP) analysis is a key factor in many decisions, including choice of product lines, pricing of products, marketing strategy, and use of productive facilities. A calculation used in a CVP analysis is the breakeven point. Once the breakeven point has been reached, operating income will increase by the

A. Gross margin per unit for each additional unit sold.

B. Contribution margin per unit for each additional unit sold.

C. Fixed costs per unit for each additional unit sold.

D. Variable costs per unit for each additional unit sold.

The correct answer is (B). *(CMA, adapted)*

REQUIRED: The amount by which operating income will increase once the breakeven point has been reached.

DISCUSSION: At the breakeven point, total revenue equals the fixed cost plus the variable cost. Beyond the BEP, each unit sale will increase operating income by the unit contribution margin (unit sales price – unit variable cost) because fixed cost will already have been recovered.

Answer (A) is incorrect because the gross margin equals sales price minus cost of goods sold, including fixed cost. Answers (C) and (D) are incorrect because operating income will increase by the UCM.

2. If inventories are expected to change, the type of costing that provides the best information for breakeven analysis is

A. Job-order costing.

B. Variable (direct) costing.

C. Joint costing.

D. Absorption (full) costing.

The correct answer is (B). *(CMA, adapted)*

REQUIRED: The type of costing system that provides the best information for breakeven analysis if inventories are expected to change.

DISCUSSION: A variable (direct) costing system is best for providing the information needed for CVP analysis because both techniques separate the fixed costs from variable costs. CVP analysis calculates a variable cost per unit and deducts it from unit sales price to determine the unit contribution margin. The total contribution margin from a given level of sales measures the extent of recovery of fixed costs and the profit earned. Direct costing is likewise oriented toward determination of the contribution margin because it treats fixed manufacturing overhead as a period, not a product, cost. Thus, it facilitates CVP analysis by isolating variable manufacturing costs.

Answers (A), (C), and (D) are incorrect because job-order costing, joint costing, and absorption (full) costing do not separate fixed costs from variable costs.

3. One of the major assumptions limiting the reliability of breakeven analysis is that

A. Efficiency and productivity will continually increase.

B. Total variable costs will remain unchanged over the relevant range.

C. Total fixed costs will remain unchanged over the relevant range.

D. The cost of production factors varies with changes in technology.

The correct answer is (C). *(CMA, adapted)*

REQUIRED: The major assumption limiting the use of breakeven analysis.

DISCUSSION: The inherent simplifying assumptions used in CVP analysis are the following: costs and revenues are predictable and are linear over the relevant range; variable costs change proportionally with activity level; changes in inventory are insignificant in amount; fixed costs remain constant over the relevant range of volume; prices remain fixed; production equals sales; there is a relevant range in which the various relationships are true for a given time span; all costs are either fixed or variable; productive efficiency is constant; costs vary only with changes in sales volume; and there is a constant mix of products (or only one product).

Answer (A) is incorrect because breakeven analysis assumes no changes in efficiency and productivity. Answer (B) is incorrect because one limiting assumption is that unit variable cost, not total variable cost, is constant. Answer (D) is incorrect because the cost of production factors is assumed to be stable.

4. In working on a CVP analysis, the accountant is unsure of the exact results and/or assumptions under which to operate. What can the accountant do to help management in this CVP decision?

A. Nothing. It is not the responsibility of the accountant to be concerned with the ambiguity of the results and/or assumptions.

B. Ascertain the probabilities of various outcomes and work with management on understanding those probabilities in reference to the CVP decision.

C. Calculate the probabilities of various outcomes and make the decision for management.

D. Use a random number table to generate a decision model and make the decision for management.

The correct answer is (B). *(Publisher)*

REQUIRED: The approach to take in a probabilistic CVP decision.

DISCUSSION: The assumptions under which CVP analysis operates primarily hinge on certainty. Once uncertainty enters the situation, the results are not nearly so clear. Thus, the accountant should make an appropriate effort to ascertain the probabilities of various outcomes. The accountant can then work with management to help make the appropriate decision.

Answer (A) is incorrect because it is the responsibility of the accountant to be involved with management and aid in their decision making. Answer (C) is incorrect because it is not appropriate for the accountant to make the decision. Management should make the decision with the accountant's advice and help. Answer (D) is incorrect because, although using a random number table may be part of the decision process, the decision should not rest with the accountant alone.

5. The margin of safety is a key concept of CVP analysis. The margin of safety is

A. The contribution margin rate.

B. The difference between budgeted contribution margin and breakeven contribution margin.

C. The difference between budgeted sales and breakeven sales.

D. The difference between the breakeven point in sales and cash flow breakeven.

The correct answer is (C). *(CMA, adapted)*

REQUIRED: The definition of the margin of safety.

DISCUSSION: The margin of safety measures the amount by which sales may decline before losses occur. It is the excess of budgeted or actual sales over sales at the BEP. It may be stated in either units sold or sales revenue.

Answer (A) is incorrect because the contribution margin rate is computed by dividing contribution margin by sales. The contribution margin equals sales minus total variable costs. Answer (B) is incorrect because the margin of safety is expressed in revenue or units, not contribution margin. Answer (D) is incorrect because cash flow breakeven is not relevant.

6. When used in cost-volume-profit analysis, sensitivity analysis

A. Determines the most profitable mix of products to be sold.

B. Allows the decision maker to introduce probabilities in the evaluation of decision alternatives.

C. Is done through various possible scenarios and computes the impact on profit of various predictions of future events.

D. Is limited because in cost-volume-profit analysis, costs are not separated into fixed and variable components.

The correct answer is (C). *(CMA, adapted)*

REQUIRED: The true statement about sensitivity analysis.

DISCUSSION: Sensitivity analysis permits the decision maker to measure the effects of errors in certainty equivalents, which are estimated amounts developed by the best means available and assumed for purposes of a given decision model to be certain. The decision model then may be evaluated by changing certain data variables (certainty equivalents) critical to the success of the entity and observing the outcomes. This analysis allows the decision maker to quantify the effects of forecasting or prediction errors and to identify the most critical variables. For example, with respect to breakeven analysis, a firm might make computations using several different estimates of what fixed costs are expected to be. These calculations indicate how sensitive the results are to changes in fixed costs.

Answer (A) is incorrect because CVP analysis assumes a constant product mix or only one product. Answer (B) is incorrect because expected value analysis allows the decision maker to introduce probabilities in the evaluation of decision alternatives. Answer (D) is incorrect because, in cost-volume-profit analysis, costs are separated into fixed and variable components.

7. Which of the following would decrease unit contribution margin the most?

A. A 15% decrease in selling price.

B. A 15% increase in variable expenses.

C. A 15% decrease in variable expenses.

D. A 15% decrease in fixed expenses.

The correct answer is (A). *(CMA, adapted)*

REQUIRED: The change in a CVP variable causing the greatest decrease in UCM.

DISCUSSION: The plausible options are a decrease in selling price or an increase in variable expenses because UCM equals unit sales price minus unit variable cost. However, a given percentage change in unit sales price must have a greater effect than an equal but opposite percentage change in unit variable cost because the former is greater than the latter. The example below demonstrates this point.

Original: SP =	$100		
VC =	50		
CM =	$ 50		
SP – 15%: SP =	$ 85		
VC =	50		
CM =	$ 35	($15 decrease)	
VC + 15%: SP =	$100.00		
VC =	57.50		
CM =	$ 42.50	($7.50 decrease)	

Answer (B) is incorrect because a 15% increase in variable expenses will not decrease the CM as much as a 15% decrease in sales price. Answer (C) is incorrect because a decrease in variable expenses would increase UCM. Answer (D) is incorrect because a decrease in fixed expenses has no effect on the UCM.

8. Marston Enterprises sells three chemicals: petrol, septine, and tridol. Petrol is the company's most profitable product; tridol is the least profitable. Which one of the following events will definitely decrease the firm's overall breakeven point for the upcoming accounting period?

A. The installation of new computer-controlled machinery and subsequent layoff of assembly-line workers.

B. A decrease in tridol's selling price.

C. An increase in the overall market for septine.

D. An increase in anticipated sales of petrol relative to sales of septine and tridol.

The correct answer is (D). *(CMA, adapted)*

REQUIRED: The event that will decrease the firm's overall breakeven point.

DISCUSSION: A company's breakeven point will be reduced if fixed costs are lowered or the average unit contribution margin is increased. Given that petrol is the company's most profitable product, and assuming that it has a higher unit contribution margin than septine and tridol, an increase in sales of petrol relative to the other products will result in a higher average unit contribution margin and a lower breakeven point (fixed costs ÷ average UCM).

Answer (A) is incorrect because the acquisition of new machinery will result in greater fixed costs and the possibility of a higher breakeven point. Answer (B) is incorrect because a decrease in selling price reduces the unit contribution margin, which in turn increases the breakeven point. Answer (C) is incorrect because an increase in the market for septine has an indeterminate effect. The facts given do not indicate whether its unit contribution margin is greater or less than the average unit contribution margin for all products.

9. Product A accounts for 75% of a company's total sales revenue and 60% of its variable costs. Product B accounts for 25% of total sales revenue and 85% of variable costs. What is the breakeven point given fixed costs of $150,000?

A. $375,000

B. $444,444

C. $500,000

D. $545,455

The correct answer is (B). *(Publisher)*

REQUIRED: The breakeven point in a multiproduct firm.

DISCUSSION: Sales = FC + VC. The proportion of each product's sales must be considered.

$$S = \$150{,}000 + 0.75S(0.60) + 0.25S(0.85)$$
$$S = \$150{,}000 + 0.45S + 0.2125S$$
$$S - 0.6625S = \$150{,}000$$
$$0.3375S = \$150{,}000$$
$$S = \$444{,}444$$

Answer (A) is incorrect because $375,000 is based on the contribution margin of Product A only rather than a weighted average. Answer (C) is incorrect because $500,000 is based on half of the required sales at B's contribution margin. Answer (D) is incorrect because $545,455 is based on an unweighted average of the two contribution margins.

10. Which of the following will result in raising the breakeven point?

A. A decrease in the variable cost per unit.

B. An increase in the semivariable cost per unit.

C. An increase in the contribution margin per unit.

D. A decrease in income tax rates.

The correct answer is (B). *(CIA, adapted)*

REQUIRED: The change in a CVP factor that will raise the breakeven point.

DISCUSSION: The BEP equals fixed cost divided by the UCM (selling price – unit variable cost). An increase in semivariable costs increases fixed costs and/or variable costs. An increase in either will raise the BEP. If fixed costs increase, more units must be sold, assuming the same UCM, to cover the greater fixed costs. If variable costs increase, the UCM will decrease and again more units must be sold to cover the fixed costs.

Answer (A) is incorrect because, if other factors are constant, an increase in sales price or a decrease in unit variable cost increases the CM and lowers the BEP. Answer (C) is incorrect because an increase in the CM decreases the BEP. Answer (D) is incorrect because, if income taxes are taken into account, they are treated as variable costs. A decrease in variable costs lowers the BEP.

11. A company's breakeven point in sales dollars may be affected by equal percentage increases in both selling price and variable cost per unit (assume all other factors are constant within the relevant range). The equal percentage changes in selling price and variable cost per unit will cause the breakeven point in sales dollars to

A. Decrease by less than the percentage increase in selling price.

B. Decrease by more than the percentage increase in the selling price.

C. Increase by the percentage change in variable cost per unit.

D. Remain unchanged.

The correct answer is (D). *(CIA, adapted)*

REQUIRED: The effect of equal percentage changes in selling price and variable cost on the breakeven point in sales dollars.

DISCUSSION: The BEP in sales dollars is equal to the fixed cost divided by the CMR. Accordingly, equal percentage changes in selling price and variable cost per unit will not affect the BEP in sales dollars. For example, assume the unit price of a product is $1 and its unit variable cost is $.60. The CMR equals 40% [($1 – $.60)UCM ÷ $1 unit price]. If fixed cost is $100, the BEP in sales dollars is $250 ($100 ÷ 40%). Raising the selling price and variable cost by 20% to $1.20 and $.72, respectively, leaves the CMR at 40% ($.48 ÷ $1.20). Similarly, lowering the selling price and variable cost to $.80 and $.48, respectively, also leaves the CMR at 40% ($.32 ÷ $.80).

Answer (A) is incorrect because the breakeven point in sales dollars will not change. Answer (B) is incorrect because the CMR will remain the same. Therefore, the breakeven point in sales dollars will remain unchanged. Answer (C) is incorrect because the breakeven point in sales dollars will not change.

12. Two companies produce and sell the same product in a competitive industry. Thus, the selling price of the product for each company is the same. Company 1 has a contribution margin ratio of 40% and fixed costs of $25 million. Company 2 is more automated, making its fixed costs 40% higher than those of Company 1. Company 2 also has a contribution margin ratio that is 30% greater than that of Company 1. By comparison, Company 1 will have the <List A> breakeven point in terms of dollar sales volume and will have the <List B> dollar profit potential once the indifference point in dollar sales volume is exceeded.

	List A	List B
A.	Lower	Lesser
B.	Lower	Greater
C.	Higher	Lesser
D.	Higher	Greater

The correct answer is (A). *(CIA, adapted)*

REQUIRED: The breakeven points and dollar profit potentials for two manufacturers.

DISCUSSION: Company 1's breakeven point is lower because its fixed costs are lower. Company 1's breakeven point is $62,500,000 ($25,000,000 ÷ 40%). Company 2's breakeven point is $67,307,692 [($25,000,000 x 1.4) ÷ (40% x 1.3)]. The indifference point, at which dollar profits are equal, is $83,000,000 ($25,000,000 + .60X = $35,000,000 + .48X). Once the indifference point is passed, Company 1 will make lower profits than Company 2 because Company 2 has a higher contribution margin.

Answers (B), (C), and (D) are incorrect because Company 1's breakeven point is lower, but its profits will be lower past the indifference point.

13. Which of the following is a characteristic of a contribution income statement?

A. Fixed and variable expenses are combined as one line.

B. Fixed expenses are listed separately from variable expenses.

C. Fixed and variable manufacturing costs are combined as one line item, but fixed operating expenses are shown separately from variable operating expenses.

D. Fixed and variable operating expenses are combined as one line item, but fixed manufacturing expenses are shown separately from variable manufacturing expenses.

The correct answer is (B). *(CIA, adapted)*

REQUIRED: The characteristic of a contribution income statement.

DISCUSSION: The contribution income statement emphasizes the distinction between fixed and variable costs. Making this distinction facilitates determination of CVP relationships and the effects of changes in sales volume on income. Thus, fixed manufacturing costs and other fixed costs are separated from variable manufacturing costs and other variable costs. The basic categories in the contribution income statement are variable costs, contribution margin, fixed costs, and operating income.

Answer (A) is incorrect because the contribution income statement shows the contribution margin (sales – variable costs) before it subtracts the fixed costs. Fixed costs are not combined with variable costs. Answer (C) is incorrect because fixed costs are not combined with variable costs on contribution income statements. Answer (D) is incorrect because fixed costs are not combined with variable costs on contribution income statements.

B. Applications

14. BE&H Manufacturing is considering dropping a product line. It currently produces a multipurpose woodworking clamp in a simple manufacturing process that uses special equipment. Variable costs amount to $6.00 per unit. Fixed overhead costs, exclusive of depreciation, have been allocated to this product at a rate of $3.50 a unit and will continue whether or not production ceases. Depreciation on the special equipment amounts to $20,000 a year. If production of the clamp is stopped, the special equipment can be sold for $18,000; if production continues, however, the equipment will be useless for further production at the end of 1 year and will have no salvage value. The clamp has a selling price of $10 a unit. Ignoring tax effects, the minimum number of units that would have to be sold in the current year to break even on a cash flow basis is

A. 4,500 units

B. 5,000 units

C. 20,000 units

D. 36,000 units

The correct answer is (A). *(CMA, adapted)*

REQUIRED: The breakeven point in units on a cash flow basis.

DISCUSSION: The BEP in units is equal to fixed costs divided by the difference between unit selling price and unit variable cost (UCM). The $18,000 salvage value, the cash flow to be received if production is discontinued, is treated here as a fixed cost. Hence, continuation of the product line will permit the firm to break even or make a profit only if the total CM is $18,000 or more.

$$BEP = \frac{\$18{,}000}{\$10 - \$6} = 4{,}500 \textit{ units}$$

Fixed overhead allocated is not considered in this calculation because it is not a cash flow and will continue regardless of the decision.

Answer (B) is incorrect because the BEP is equal to the salvage value (not depreciation) divided by the UCM of $4 ($10 – $6). Depreciation is a noncash flow and therefore should not be considered in the cash flow breakeven point calculation. Answer (C) is incorrect because the BEP is equal to the salvage value (not depreciation) divided by the UCM of $4 ($10 – $6). Depreciation is a noncash flow and therefore should not be considered in the cash flow breakeven point calculation. Answer (D) is incorrect because unit fixed costs should not be subtracted in determining the unit contribution margin. The fixed costs will continue regardless so they are not included in the calculation. Therefore, the $18,000 salvage value will be divided by the $4 unit contribution margin in determining the cash flow breakeven point in units.

Questions 15 and 16 are based on the following information. Delphi Company has developed a new project that will be marketed for the first time during the next fiscal year. Although the Marketing Department estimates that 35,000 units could be sold at $36 per unit, Delphi's management has allocated only enough manufacturing capacity to produce a maximum of 25,000 units of the new product annually. The fixed costs associated with the new product are budgeted at $450,000 for the year, which includes $60,000 for depreciation on new manufacturing equipment.

Data associated with each unit of product are presented below. Delphi is subject to a 40% income tax rate.

	Variable Costs
Direct material	$ 7.00
Direct labor	3.50
Manufacturing overhead	4.00
Total variable manufacturing cost	14.50
Selling expenses	1.50
Total variable cost	$16.00

15. The maximum after-tax profit that can be earned by Delphi Company from sales of the new product during the next fiscal year is

A. $30,000

B. $50,000

C. $110,000

D. $66,000

The correct answer is (A). *(CMA, adapted)*

REQUIRED: The maximum after-tax profit.

DISCUSSION: The maximum output (maximum sales level) is 25,000 units (given). Thus, the breakeven point in units is 22,500 ($450,000 FC ÷ $20). The unit contribution margin is $20 ($36 selling price – $16 unit VC). At the breakeven point, all fixed costs have been recovered. Hence, pretax profit equals the unit contribution margin times unit sales in excess of the breakeven point, or $50,000 [(25,000 unit sales – 22,500 BEP) x $20 UCM]. After-tax profit is $30,000 [$50,000 x (1.0 – .4 tax rate)].

Answer (B) is incorrect because $50,000 is the pretax profit. Answer (C) is incorrect because $110,000 fails to include depreciation as a fixed cost and ignores income taxes. Answer (D) is incorrect because $66,000 fails to include depreciation as a fixed cost.

16. Delphi Company's management has stipulated that it will not approve the continued manufacture of the new product after the next fiscal year unless the after-tax profit is at least $75,000 the first year. The unit selling price to achieve this target profit must be at least

A. $37.00

B. $36.60

C. $34.60

D. $39.00

The correct answer is (D). *(CMA, adapted)*

REQUIRED: The unit selling price to achieve a targeted after-tax profit.

DISCUSSION: If X represents the necessary selling price, 25,000 equals maximum sales volume, $16 is the variable cost per unit, $450,000 is the total fixed cost, and $125,000 [$75,000 target after-tax profit ÷ (1.0 – .4 tax rate)] is the desired pretax profit, the following formula may be solved to determine the requisite unit price:

$$\begin{aligned} 25{,}000\,(X - \$16) - \$450{,}000 &= \$125{,}000 \\ 25{,}000X - \$400{,}000 - \$450{,}000 &= \$125{,}000 \\ 25{,}000X &= \$975{,}000 \\ X &= \$39 \end{aligned}$$

Answer (A) is incorrect because $37.00 does not consider income taxes. Answer (B) is incorrect because $36.60 excludes depreciation. Answer (C) is incorrect because $34.60 does not include depreciation or taxes.

Questions 17 through 19 are based on the following information. Bruell Electronics Co. is developing a new product, surge protectors for high-voltage electrical flows. The cost information in the opposite column relates to the product:

	Unit Costs
Direct materials	$3.25
Direct labor	4.00
Distribution	.75

The company will also be absorbing $120,000 of additional fixed costs associated with this new product. A corporate fixed charge of $20,000 currently absorbed by other products will be allocated to this new product.

17. If the selling price is $14 per unit, the breakeven point in units (rounded to the nearest hundred) for surge protectors is

A. 8,500 units.

B. 10,000 units.

C. 15,000 units.

D. 20,000 units.

The correct answer is (D). *(CMA, adapted)*

REQUIRED: The breakeven point in units.

DISCUSSION: The breakeven point in units equals total additional fixed costs divided by the unit contribution margin. Unit variable costs total $8 ($3.25 + $4.00 + $.75). Thus, UCM is $6 ($14 unit selling price – $8 unit VC), and the breakeven point is 20,000 units ($120,000 FC ÷ $6).

Answer (A) is incorrect because a breakeven point of 8,500 units ignores variable costs. Answer (B) is incorrect because the breakeven point is 20,000 units when the contribution margin is $6 per unit. Answer (C) is incorrect because 15,000 units equals fixed costs divided by unit variable cost.

18. How many surge protectors (rounded to the nearest hundred) must Bruell Electronics sell at a selling price of $14 per unit to gain $30,000 additional income before taxes?

A. 10,700 units.

B. 12,100 units.

C. 20,000 units.

D. 25,000 units.

The correct answer is (D). *(CMA, adapted)*

REQUIRED: The number of units to be sold to generate a targeted pretax income.

DISCUSSION: The number of units to be sold to generate a specified pretax income equals the sum of total fixed costs and the targeted pretax income, divided by the unit contribution margin. Unit variable costs total $8 ($3.25 + $4.00 + $.75), and UCM is $6 ($14 unit selling price – $8). Thus, the desired unit sales level equals 25,000 units [($120,000 + $30,000) ÷ $6].

Answer (A) is incorrect because 10,700 units is based on a UCM equal to selling price. Answer (B) is incorrect because a contribution margin of $6 per unit necessitates sales of 25,000 units to produce a $30,000 before-tax profit. Answer (C) is incorrect because 20,000 units is the breakeven point.

19. How many surge protectors (rounded to the nearest hundred) must Bruell Electronics sell at a selling price of $14 per unit to increase after-tax income by $30,000? Bruell Electronics' effective income tax rate is 40%.

A. 10,700 units.

B. 12,100 units.

C. 20,000 units.

D. 28,300 units.

The correct answer is (D). *(CMA, adapted)*

REQUIRED: The number of units to be sold to generate a specified after-tax income.

DISCUSSION: The number of units to be sold to generate a specified pretax income equals the sum of total fixed costs and the targeted pretax income, divided by the unit contribution margin. Given a desired after-tax income of $30,000 and a tax rate of 40%, the targeted pretax income must be $50,000 [$30,000 ÷ (1.0 – .4)]. Unit variable costs total $8 ($3.25 + $4.00 + $.75), and UCM is $6 ($14 unit selling price – $8). Hence, the desired unit sales level is 28,333 [($120,000 + $50,000) ÷ $6]. Rounded to the nearest hundred, the answer is 28,300.

Answer (A) is incorrect because 10,700 units is based on a UCM equal to selling price and $30,000 of pretax income. Answer (B) is incorrect because a $6 UCM necessitates sales of 28,300 units to produce a $30,000 after-tax profit. Answer (C) is incorrect because 20,000 units is the breakeven point.

Questions 20 through 22 are based on the following information. Siberian Ski Company recently expanded its manufacturing capacity, which will allow it to produce up to 15,000 pairs of cross-country skis of the mountaineering model or the touring model. The Sales Department assures management that it can sell between 9,000 pairs and 13,000 pairs of either product this year. Because the models are very similar, Siberian Ski will produce only one of the two models.

The following information was compiled by the Accounting Department.

	Per-Unit (Pair) Data	
	Mountaineering	Touring
Selling price	$88.00	$80.00
Variable costs	52.80	52.80

Fixed costs will total $369,600 if the mountaineering model is produced but will be only $316,800 if the touring model is produced. Siberian Ski is subject to a 40% income tax rate.

20. The total sales revenue at which Siberian Ski Company would make the same profit or loss regardless of the ski model it decided to produce is

A. $880,000

B. $422,400

C. $924,000

D. $686,400

The correct answer is (A). *(CMA, adapted)*

REQUIRED: The sales revenue resulting in the same profit or loss regardless of which model is produced.

DISCUSSION: The sales revenue at which the same profit or loss will be made equals the unit price times the units sold for each kind of skis. Accordingly, if M is the number of units sold of mountaineering skis and T is the number of units sold of touring skis, this level of sales revenue may be stated as $88M or $80T, and M is therefore equal to ($80 ÷ $88)T. Moreover, given the same profit or loss, the difference between sales revenue and total costs (variable + fixed) will also be the same for the two kinds of skis. Solving the equation below by substituting for M yields sales revenue of $880,000 [(11,000 x $80) or (10,000 x $88)].

$$\begin{aligned} Sales_M - VC_M - FC_M &= Sales_T - VC_T - FC_T \\ \$88M - \$52.80M - \$369{,}600 &= \$80T - \$52.80T - \$316{,}800 \\ \$35.2M - \$52{,}800 &= \$27.2T \\ \$35.2(\$80 \div \$88)T &= \$27.2T + \$52{,}800 \\ T &= 11{,}000 \text{ units} \\ M &= 10{,}000 \text{ units} \end{aligned}$$

Answers (B), (C), and (D) are incorrect because the same profit or loss is earned at a sales revenue of $880,000.

21. If the Siberian Ski Company Sales Department could guarantee the annual sale of 12,000 pairs of either model, Siberian Ski would

A. Produce 12,000 pairs of touring skis because they have a lower fixed cost.

B. Be indifferent as to which model is sold because each model has the same variable cost per unit.

C. Produce 12,000 pairs of mountaineering skis because they have a lower breakeven point.

D. Produce 12,000 pairs of mountaineering skis because they are more profitable.

The correct answer is (D). *(CMA, adapted)*

REQUIRED: The model that should be produced if sales are exactly 12,000 pairs.

DISCUSSION: Preparing income statements determines which model will produce the greater profit at a sales level of 12,000 pairs. Thus, as indicated below, the mountaineering skis should be produced.

	Mountain	Touring
Sales	$1,056,000	$960,000
Variable costs	(633,600)	(633,600)
Fixed costs	(369,600)	(316,800)
Net Income	$ 52,800	$ 9,600

Answer (A) is incorrect because, at a sales volume of 12,000 pairs, the higher contribution margin of the mountaineering skis results in a greater profit. Answer (B) is incorrect because the lower selling price of the touring skis results in a lower contribution margin per unit sold. Answer (C) is incorrect because breakeven point is not a consideration. A sales volume of 12,000 units is above the breakeven point for both models.

22. If Siberian Ski Company desires an after-tax net income of $24,000, how many pairs of touring model skis will the company have to sell?

A. 13,118 pairs.

B. 12,529 pairs.

C. 13,853 pairs.

D. 4,460 pairs.

The correct answer is (A). *(CMA, adapted)*

REQUIRED: The units of touring model skis the company will have to sell to produce a given after-tax profit.

DISCUSSION: The breakeven sales volume equals total fixed costs divided by the unit contribution margin (UCM). In the breakeven formula, the desired profit should be treated as a fixed cost. Because the UCM is stated in pretax dollars, the targeted profit must be adjusted for taxes. Hence, the targeted after-tax net income of $24,000 is equivalent to a pretax profit of $40,000 [$24,000 ÷ (1.0 – .4 tax rate)]. The sum of the pretax profit and the fixed costs is $356,800 ($316,800 + $40,000). Consequently, the desired sales volume is 13,118 pairs of touring skis [$356,800 ÷ ($80 price – $52.80 unit VC)].

Answers (B), (C), and (D) are incorrect because 13,118 pairs of touring skis must be sold.

23. Austin Manufacturing, which is subject to a 40% income tax rate, had the following operating data for the period just ended:

Selling price per unit	$ 60
Variable cost per unit	22
Fixed costs	504,000

Management plans to improve the quality of its sole product by (1) replacing a component that costs $3.50 with a higher-grade unit that costs $5.50 and (2) acquiring a $180,000 packing machine. Austin will depreciate the machine over a 10-year life with no estimated salvage value by the straight-line method of depreciation. If the company wants to earn after-tax income of $172,800 in the upcoming period, it must sell

A. 19,300 units.

B. 21,316 units.

C. 22,500 units.

D. 23,800 units.

The correct answer is (C). *(CMA, adapted)*

REQUIRED: The number of units to be sold to generate a targeted after-tax profit.

DISCUSSION: The units to be sold equal fixed costs plus the desired pretax profit, divided by the unit contribution margin. In the preceding year, the unit contribution margin was $38 ($60 price – $22 unit VC). That amount will decrease by $2 to $36 in the upcoming year because of use of a higher-grade component. Fixed costs will increase from $504,000 to $522,000 as a result of the $18,000 ($180,000 ÷ 10 years) increase in fixed costs attributable to depreciation on the new machine. Dividing the $172,800 of desired after-tax income by 60% (the complement of the tax rate) produces a desired before-tax income of $288,000. Hence, the breakeven point in units is 22,500 [($522,000 + $288,000) ÷ $36].

Answer (A) is incorrect because 19,300 units does not take income taxes into consideration. Answer (B) is incorrect because 21,316 units fails to consider the increased variable costs from the introduction of the higher-priced component. Answer (D) is incorrect because 23,800 units does not take income taxes into consideration, and it includes the entire cost of the new machine as a fixed cost.

Questions 24 and 25 are based on the following information. A company that sells its single product for $40 per unit uses cost-volume-profit analysis in its planning. The company's after-tax net income for the past year was $1,188,000 after applying an effective tax rate of 40%. The projected costs for manufacturing and selling its single product in the coming year are in the next column.

Variable costs per unit:	
Direct material	$ 5.00
Direct labor	4.00
Manufacturing overhead	6.00
Selling and administrative costs	3.00
Total cost per unit	$18.00
Annual fixed operating costs:	
Manufacturing overhead	$6,200,000
Selling and administrative costs	3,700,000
Total annual fixed cost	$9,900,000

24. The dollar sales volume required in the coming year to earn the same after-tax net income as the past year is

A. $20,160,000

B. $21,600,000

C. $23,400,000

D. $26,400,000

The correct answer is (B). *(CIA, adapted)*

REQUIRED: The dollar sales volume required in the coming year to earn the same after-tax net income as the past year.

DISCUSSION: The desired after-tax net income is $1,188,000 (the past year's amount). Given a 40% tax rate, the pretax equivalent is $1,980,000 [$1,188,000 ÷ (1.0 – .40)]. Pretax net income equals dollar sales (unit sales x $40), minus total fixed costs, minus total variable costs (unit sales x unit variable cost). Hence, the contribution margin (sales – variable costs) is equated with the sum of fixed costs and the targeted pretax net income. Unit sales (S) equal 540,000, and sales dollars equal $21,600,000 ($40 x 540,000 units).

$$\$40S - \$9{,}900{,}000 - \$18S = \$1{,}980{,}000$$
$$\$22S = \$11{,}880{,}000$$
$$S = 540{,}000 \text{ units}$$

Answer (A) is incorrect because $20,160,000 does not adjust after-tax net income to pretax net income. Answer (C) is incorrect because $23,400,000 results from adjusting after-tax net income by dividing by the tax rate rather than one minus the tax rate. Answer (D) is incorrect because $26,400,000 results from equating the sum of the desired pretax net income and total fixed costs with total variable costs instead of the contribution margin.

25. The company has learned that a new direct material is available that will increase the quality of its product. The new material will increase the direct material costs by $3 per unit. The company will increase the selling price of the product to $50 per unit and increase its marketing costs by $1,575,000 to advertise the higher-quality product. The number of units the company has to sell in order to earn a 10% before-tax return on sales would be

A. 337,500 units.

B. 346,875 units.

C. 425,000 units.

D. 478,125 units.

The correct answer is (D). *(CIA, adapted)*

REQUIRED: The number of units the company has to sell to earn a 10% before-tax return on sales.

DISCUSSION: Pretax net income (10% of dollar sales) equals dollar sales (unit sales x $50), minus total fixed costs (increased by $1,575,000 of marketing costs), minus total variable costs (increased by $3 per unit). Unit sales (S) therefore equal 478,125 units.

$$.10(\$50S) = \$50S - (\$9{,}900{,}000 + \$1{,}575{,}000) - (\$18 + \$3)S$$
$$\$24S = \$11{,}475{,}000$$
$$S = 478{,}125 \text{ units}$$

Answer (A) is incorrect because 337,500 units results from using the wrong sign for the pretax net income. Answer (B) is incorrect because 346,875 units results from subtracting, not adding, the incremental marketing costs to determine total fixed costs. Answer (C) is incorrect because 425,000 units fails to adjust for the increase in direct materials costs.

26. A company allocates its variable factory overhead based on direct labor hours. During the past 3 months, the actual direct labor hours and the total factory overhead allocated were as follows:

	January	February	March
Direct labor hours	1,000	3,000	5,000
Total factory overhead allocated	$80,000	$140,000	$200,000

Based upon this information, monthly fixed factory overhead was

A. $50,000

B. $46,667

C. $33,333

D. $30,000

The correct answer is (A). *(CIA, adapted)*

REQUIRED: The monthly fixed overhead given direct labor hours and the total overhead allocation.

DISCUSSION: This question requires solving simultaneous equations because both the variable overhead per direct labor hour (Y) and the fixed overhead (X) are unknown.

$$\begin{aligned} \$80{,}000 &= X + 1{,}000Y \\ \$140{,}000 &= X + 3{,}000Y \\ \$60{,}000 &= 2{,}000Y \\ Y &= \$30 \text{ per direct labor hour} \end{aligned}$$

Substituting,

$$\begin{aligned} \$80{,}000 &= X + 1{,}000(\$30) \\ X &= \$50{,}000 \end{aligned}$$

Answer (B) is incorrect because $46,667 is the average total factory overhead applied per month, a portion of which is variable. Answer (C) is incorrect because the monthly fixed factory overhead is calculated by dividing the difference between total factory overhead allocated in January and February by the change in direct labor hours. This equals the variable overhead per direct labor hour. Subtracting the quantity of direct labor hours times variable overhead rate per hour from total factory overhead equals fixed overhead. Answer (D) is incorrect because $30,000 is the variable factory overhead per 1,000 hours.

27. Green Company produces Product A and sells it for $18.00. The following cost data apply:

Type of Cost	Per Unit
Direct materials (3 lb. × $1.50)	$ 4.50
Direct labor	6.45
Variable overhead	1.35
Fixed overhead	1.50
Variable selling expense	1.10
Fixed selling expense	2.20
	$17.10

Green has thought of marketing a new Product B with the same cost structure as Product A except that the price will be $15.60. Green Company currently has the plant capacity necessary for this expansion. Because of the cost structure, Green Company will find the production and sale of Product B in the short run to be

A. Not profitable unless the price can be raised to $17.10.

B. Not profitable at any price.

C. Not profitable at $15.60 because the fixed selling expense and fixed manufacturing overhead will not be covered by the price.

D. Profitable to produce and sell Product B in the short run at the price of $15.60.

The correct answer is (D). *(CIA, adapted)*

REQUIRED: The short-run profitability of a product given cost data and excess capacity.

DISCUSSION: With excess capacity, production is profitable if the incremental revenues are greater than the incremental costs. Here, the incremental costs equal total costs minus any fixed costs ($17.10 – $1.50 – $2.20 = $13.40). If Product B can be sold for a price greater than $13.40, short-run production will be profitable. Long-run profitability, however, will depend on fixed costs as well as variable costs and sales price.

Answer (A) is incorrect because the cost for Product B in the short run will be lower than $17.10 because the fixed costs of $3.70 would not be included in the incremental cost of Product B. Answer (B) is incorrect because Product B would be profitable at a $15.60 selling price. Answer (C) is incorrect because, in the short run, the fixed expenses and overhead do not need to be included in the cost of Product B because they are all covered by Product A.

28. Orange Company's controller developed the following direct-costing income statement for year 1:

			Per Unit
Sales (150,000 units at $30)		$4,500,000	$30
Variable costs:			
Direct materials	$1,050,000		$ 7
Direct labor	1,500,000		10
Mfg. overhead	300,000		2
Selling & mkg.	300,000		2
		(3,150,000)	$21
Contribution margin		$1,350,000	$ 9
Fixed costs:			
Mfg. overhead	$ 600,000		$ 4
Selling & mkg.	300,000		2
		(900,000)	$ 6
Net income		$ 450,000	$ 3

Orange Co. based its next year's budget on the assumption that fixed costs, unit sales, and the sales price would remain as they were in year 1, but with net income being reduced to $300,000. By July of year 2, the controller was able to predict that unit sales would increase over year 1 levels by 10%. Based on the year 2 budget and the new information, the predicted year 2 net income would be

A. $300,000

B. $330,000

C. $420,000

D. $585,000

The correct answer is (C). *(CIA, adapted)*

REQUIRED: The projected net income given constant FC and sales price and different estimates of unit sales.

DISCUSSION: Projected net income is estimated total revenue minus estimated total costs. Given the original assumption that FC, unit sales, and sales price remain the same, and that net income will be reduced, the variable costs must increase. If the July, year 2 prediction that unit sales will increase by 10% from 150,000 to 165,000 is based on the budgeted FC, sales price, and VC, predicted net income will increase from $300,000 to $420,000.

TR – FC – VC = $300,000
TR – $900,000 – VC = $300,000
TR – VC = $1,200,000
CM = $1,200,000

Budgeted UCM = $1,200,000 ÷ 150,000 units = $8

Total CM =	165,000 units × $8 =	$1,320,000
	FC =	(900,000)
Net income		$ 420,000

Answer (A) is incorrect because $300,000 is the projected net income before the increase in sales. Answer (B) is incorrect because $330,000 has taken into account a 10% increase in fixed costs to $990,000 when the fixed costs did not increase. Answer (D) is incorrect because the UCM for year 2 is $8 per unit, not $9.

29. A company has just completed the final development of its only product, general recombinant bacteria, which can be programmed to kill most insects before dying themselves. The product has taken 3 years and $6,000,000 to develop. The following costs are expected to be incurred on a monthly basis for the normal production level of 1,000,000 pounds of the new product:

	1,000,000 pounds
Direct materials	$ 300,000
Direct labor	1,250,000
Variable factory overhead	450,000
Fixed factory overhead	2,000,000
Variable selling, general, and administrative expenses	900,000
Fixed selling, general, and administrative expenses	1,500,000
Total	$6,400,000

At a sales price of $5.90 per pound, the sales in pounds necessary to ensure a $3,000,000 profit the first year would be (to the nearest thousand pounds)

A. 13,017,000 pounds.

B. 14,000,000 pounds.

C. 15,000,000 pounds.

D. 25,600,000 pounds.

The correct answer is (C). *(CIA, adapted)*

REQUIRED: The sales in pounds necessary to ensure a specified profit in year one.

DISCUSSION: In breakeven analysis, total revenue equals fixed costs plus variable costs. If a given profit is desired, it is treated as a fixed cost. Exclusive of profit, the annual fixed cost is $42,000,000 (12 months x $3,500,000 per month of fixed factory overhead and SG&A expense). The variable cost per pound is $2.90 [($300,000 + $1,250,000 + $450,000 + $900,000) ÷ 1,000,000 lbs.]. If X equals sales in pounds, the level of sales needed to earn a $3,000,000 profit is

$5.90X = $3,000,000 + $42,000,000 + $2.90X
$3X = $45,000,000
X = 15,000,000 pounds

This analysis assumes that no additional costs are incurred when production exceeds the normal level of 1,000,000 pounds per month.

Answer (A) is incorrect because the sales necessary to ensure a $3,000,000 profit is 15,000,000 pounds. Answer (B) is incorrect because 14,000,000 pounds is the breakeven point. Answer (D) is incorrect because the sales necessary to ensure a $3,000,000 profit is 15,000,000 pounds.

Questions 30 through 32 are based on the following information. Moorehead Manufacturing Company produces two products for which the following data have been tabulated. Fixed manufacturing cost is applied at a rate of $1.00 per machine hour.

Per Unit	XY-7	BD-4
Selling price	$4.00	$3.00
Variable manufacturing cost	$2.00	$1.50
Fixed manufacturing cost	$ .75	$ .20
Variable selling cost	$1.00	$1.00

The sales manager has had a $160,000 increase in the budget allotment for advertising and wants to apply the money to the most profitable product. The products are not substitutes for one another in the eyes of the company's customers.

30. Suppose the sales manager chooses to devote the entire $160,000 to increased advertising for XY-7. The minimum increase in sales units of XY-7 required to offset the increased advertising is

A. 640,000 units.

B. 160,000 units.

C. 128,000 units.

D. 80,000 units.

The correct answer is (B). *(CMA, adapted)*

REQUIRED: The minimum increase in sales units to offset increased advertising costs.

DISCUSSION: The contribution margin for XY-7 is $1 per unit ($4 sales price – $3 variable costs). Thus, 160,000 units of XY-7 will generate an additional $160,000 of CM, which is sufficient to cover the increase in advertising costs.

Answer (A) is incorrect because 640,000 units would be the result if the unit contribution margin for XY-7 were $.25 instead of $1.00 (640,000 x $.25 = $160,000). Fixed manufacturing costs are not included in determining unit contribution margin. Answer (C) is incorrect because 128,000 units implies a $1.25 unit contribution margin. Variable selling costs are included and fixed manufacturing costs are not included in determining unit contribution margin. Answer (D) is incorrect because 80,000 units implies a $2.00 unit contribution margin. The correct UCM of $1.00 is found by subtracting all variable costs from the selling price.

31. Suppose the sales manager chooses to devote the entire $160,000 to increased advertising for BD-4. The minimum increase in sales dollars of BD-4 required to offset the increased advertising would be

A. $160,000

B. $320,000

C. $960,000

D. $1,600,000

The correct answer is (C). *(CMA, adapted)*

REQUIRED: The minimum increase in sales dollars to offset increased advertising costs.

DISCUSSION: Sales dollars must increase sufficiently to cover the $160,000 increase in advertising. The unit contribution margin for BD-4 is $.50 ($3 – $2.50 variable costs), and the CMR is 1/6 (UCM ÷ $3 sales price). Dividing the $160,000 by 1/6 gives the sales dollars necessary to generate a CM of $960,000 ($160,000 ÷ 1/6 = $960,000).

Answer (A) is incorrect because a $1 increase in sales does not result in a $1 increase in profits. Variable costs of producing the units must be deducted in order to determine the contribution margin derived from each unit sold. Answer (B) is incorrect because 320,000 is the number of sales units, not sales dollars, needed to offset the increased advertising costs. Answer (D) is incorrect because fixed manufacturing costs are not included in determining unit contribution margin.

32. Refer to the information preceding question 30 on page 89. Suppose Moorehead has only 100,000 machine hours that can be made available to produce additional units of XY-7 and BD-4. If the potential increase in sales units for either product resulting from advertising is far in excess of this production capacity, which product should be advertised and what is the estimated increase in contribution margin earned?

A. Product XY-7 should be produced, yielding a contribution margin of $75,000.

B. Product XY-7 should be produced, yielding a contribution margin of $133,333.

C. Product BD-4 should be produced, yielding a contribution margin of $187,500.

D. Product BD-4 should be produced, yielding a contribution margin of $250,000.

The correct answer is (D). *(CMA, adapted)*

REQUIRED: The more profitable product and the estimated increase in contribution margin.

DISCUSSION: The machine hours are a scarce resource that must be allocated to the product(s) in a proportion that maximizes the total contribution margin. Given that potential additional sales of either product are in excess of production capacity, only the product with the greater contribution margin per unit of scarce resource should be produced. XY-7 requires .75 hours; BD-4 requires .2 hours of machine time (given fixed manufacturing cost applied at $1 per machine hour of $.75 for XY-7 and $.20 for BD-4).

XY-7 has a CM of $1.33 per machine hour ($1 UCM ÷ .75 hours), and BD-4 has a CM of $2.50 per machine hour ($.50 ÷ .2 hours). Thus, only BD-4 should be produced, yielding a CM of $250,000 (100,000 x $2.50). The key to the analysis is CM per unit of scarce resource.

Answer (A) is incorrect because product XY-7 actually has a contribution margin of $133,333, which is lower than the $250,000 contribution margin for product BD-4. Answer (B) is incorrect because product BD-4 has a higher contribution margin at $250,000. Answer (C) is incorrect because product BD-4 has a contribution margin of $250,000.

Questions 33 through 35 are based on the following information. MultiFrame Company has the following revenue and cost budgets for the two products it sells:

	Plastic Frames	Glass Frames
Sales price	$10.00	$15.00
Direct materials	(2.00)	(3.00)
Direct labor	(3.00)	(5.00)
Fixed overhead	(3.00)	(4.00)
Net income per unit	$ 2.00	$ 3.00
Budgeted unit sales	100,000	300,000

The budgeted unit sales equal the current unit demand, and total fixed overhead for the year is budgeted at $975,000. Assume that the company plans to maintain the same proportional mix. In numerical calculations, MultiFrame rounds to the nearest cent and unit.

33. The total number of units MultiFrame needs to produce and sell to break even is

A. 150,000 units.

B. 354,545 units.

C. 177,273 units.

D. 300,000 units.

The correct answer is (A). *(CMA, adapted)*

REQUIRED: The total units sold at the breakeven point.

DISCUSSION: The calculation of the breakeven point is to divide the fixed costs by the contribution margin per unit. This determination is more complicated for a multiproduct firm. If the same proportional product mix is maintained, one unit of plastic frames is sold for every three units of glass frames. Accordingly, a composite unit consists of four frames: one plastic and three glass. For plastic frames, the unit contribution margin is $5 ($10 – $2 – $3). For glass frames, the unit contribution margin is $7 ($15 – $3 – $5). Thus, the composite unit contribution margin is $26 ($5 + $7 + $7 + $7), and the breakeven point is 37,500 packages ($975,000 FC ÷ $26). Because each composite unit contains four frames, the total units sold equal 150,000.

Answers (B), (C), and (D) are incorrect because the total units sold at the breakeven point is 150,000.

34. The total number of units needed to break even if the budgeted direct labor costs are $2 for plastic frames instead of $3 is

A. 154,028 units.

B. 144,444 units.

C. 156,000 units.

D. 146,177 units.

The correct answer is (B). *(CMA, adapted)*

REQUIRED: The breakeven point in units if labor costs for the plastic frames are reduced.

DISCUSSION: If the labor costs for the plastic frames are reduced by $1, the composite unit contribution margin will be $27 ($26 as calculated in the preceding question + $1). Hence, the new breakeven point is 144,444 units [($975,000 FC ÷ $27) x 4 units].

Answers (A), (C), and (D) are incorrect because the new breakeven point is 144,444 units.

35. The total number of units needed to break even if sales are budgeted at 150,000 units of plastic frames and 300,000 units of glass frames with all other costs remaining constant is

A. 171,958 units.

B. 418,455 units.

C. 153, 947 units.

D. 365,168 units.

The correct answer is (C). *(CMA, adapted)*

REQUIRED: The total number of units needed to break even if the product mix is changed.

DISCUSSION: The unit contribution margins for plastic frames and glass frames are $5 ($10 – $2 – $3) and $7 ($15 – $3 – $5), respectively. If the number of plastic frames sold is 50% of the number of glass frames sold, a composite unit will contain one plastic frame and two glass frames. Thus, the composite unit contribution margin will be $19 ($5 + $7 + $7), and the breakeven point in units will be 153,947 [($975,000 ÷ $19) x 3 units].

Answers (A), (B), and (D) are incorrect because the breakeven point in units will be 153,947.

36. Mason Enterprises has prepared the following budget for the month of July:

	Selling Price Per Unit	Variable Cost Per Unit	Unit Sales
Product A	$10.00	$4.00	15,000
Product B	15.00	8.00	20,000
Product C	18.00	9.00	5,000

Assuming that total fixed costs will be $150,000 and the mix remains constant, the breakeven point (rounded to the next higher whole unit) will be

A. 20,455 units.

B. 21,429 units.

C. 21,819 units.

D. 6,818 units.

The correct answer is (C). *(CMA, adapted)*

REQUIRED: The breakeven point in units given fixed costs and a constant product mix.

DISCUSSION: Given the constant product mix of 3:4:1 established by the budgeted unit sales, a composite unit consists of eight individual units (3 of A, 4 of B, and 1 of C). The unit contribution margins for A, B, and C are $6 ($10 price – $4 VC), $7 ($15 – $8 VC), and $9 ($18 price – $9 VC), respectively. Hence, the contribution margin for a composite unit is $55 [(3 x $6) + (4 x $7) + (1 x $9)], and the breakeven point is 2,727.2727 composite units ($150,000 FC ÷ $55). This amount equals 21,819 (rounded up) individual units (8 x 2,727.2727).

Answers (A) and (B) are incorrect because the breakeven point is 21,819 units. Answer (D) is incorrect because 6,818 is based on an average contribution margin of $22 per unit.

Questions 37 and 38 are based on the following information. Barnes Corporation manufactures skateboards and is in the process of preparing next year's budget. The pro forma income statement for the current year is presented in the next column.

Sales		$1,500,000
Cost of sales:		
Direct materials	$250,000	
Direct labor	150,000	
Variable overhead	75,000	
Fixed overhead	100,000	575,000
Gross profit		$ 925,000
Selling and G&A		
Variable	$200,000	
Fixed	250,000	450,000
Operating income		$ 475,000

37. The breakeven point (rounded to the nearest dollar) for Barnes Corporation for the current year is

A. $146,341

B. $636,364

C. $729,730

D. $181,818

The correct answer is (B). *(CMA, adapted)*

REQUIRED: The breakeven point in dollars.

DISCUSSION: Fixed costs total $350,000. Variable costs total $675,000. Given sales of $1,500,000, the contribution margin is $825,000 ($1,500,000 – $675,000). Thus, the contribution margin percentage is 55% ($825,000 ÷ $1,500,000). Dividing the $350,000 of fixed costs by 55% produces a breakeven point of $636,364.

Answer (A) is incorrect because $146,341 does not cover fixed costs. Answer (C) is incorrect because $729,730 of sales results in a small profit. Answer (D) is incorrect because $181,818 does not cover the $350,000 of fixed costs.

38. For the coming year, the management of Barnes Corporation anticipates a 10% increase in sales, a 12% increase in variable costs, and a $45,000 increase in fixed expenses. The breakeven point for next year will be

A. $729,027

B. $862,103

C. $214,018

D. $474,000

The correct answer is (A). *(CMA, adapted)*

REQUIRED: The breakeven point following an increase in sales, variable costs, and fixed expenses.

DISCUSSION: Sales are expected to be $1,650,000 (1.10 x $1,500,000), variable costs $756,000 (1.12 x 675,000), and fixed expenses $395,000 ($350,000 + $45,000). Thus, the contribution margin will be $894,000 ($1,650,000 – $756,000), and the contribution margin percentage is 54.1818%. The breakeven point is therefore $729,027 ($395,000 fixed expenses ÷ 54.1818%).

Answer (B) is incorrect because the contribution margin percentage is computed by dividing the contribution margin (total sales – total variable costs) by total sales, not by dividing the total variable costs by total sales. Answer (C) is incorrect because $214,018 does not cover fixed costs. Answer (D) is incorrect because $474,000 barely covers fixed costs.

Questions 39 through 41 are based on the following information. Kim is thinking of organizing a fundraiser to support a local charity. She has planned to rent a banquet hall and provide the guests with food, entertainment, and various party favors. She has decided to charge $500 a person. After researching around town, Kim has discovered the following costs:

Fixed Costs	
Rental fee of banquet hall	$150,000
Advertising	50,000
Entertainment	4,000
Variable Costs	**Per Guest**
Food	$12
Other miscellaneous costs	8

39. What is Kim's contribution margin?

A. $400

B. $450

C. $480

D. $500

The correct answer is (C). *(Publisher)*

REQUIRED: The contribution margin.

DISCUSSION: The contribution margin in a breakeven analysis can be found by taking the revenues per person and subtracting the variable cost per person. In this case, it would be $500 – $20 = $480.

Answers (A), (B), and (D) are incorrect because $480 is the contribution margin.

40. How many guests must attend in order for Kim to break even?

A. 408

B. 417

C. 425

D. 443

The correct answer is (C). *(Publisher)*

REQUIRED: The number needed to break even.

DISCUSSION: The breakeven formula equates operating income with revenues minus fixed costs minus variable costs. By setting operating income equal to 0, the number needed to breakeven can be found.

OI = R – FC – VC, or

$0 = $500x – $204,000 – $20x
$480x = $204,000
x = 425

Answer (A) is incorrect because 408 is found by not subtracting variable costs from revenues to get the contribution margin. Answer (B) is incorrect because 417 is found by not including the $4,000 of entertainment expenses in the fixed cost. Answer (D) is incorrect because 443 is found by subtracting variable cost from the contribution margin.

41. If Kim's goal is to raise $10,000 for her charity, how many people must attend the banquet?

A. 404

B. 425

C. 428

D. 446

The correct answer is (D). *(Publisher)*

REQUIRED: The minimum number of guests required to raise a certain income in a breakeven formula.

DISCUSSION: The breakeven formula equates operating income with revenues minus fixed costs minus variable costs. By setting operating income equal to the desired amount of income, the number of guests required can be discovered.

OI = R – FC – VC, or

$10,000 = $500x – $204,000 – $20x
$480x = $214,000
x = 445.83 or 446 (cannot have .83 of a person)

Answer (A) is incorrect because 404 guests is found by subtracting the $10,000 of desired income from the fixed costs. Answer (B) is incorrect because 425 guests is the amount needed to break even. Answer (C) is incorrect because 428 guests is found by not subtracting the variable costs per person from the revenues per person to get the contribution margin.

42. The following information relates to Clyde Corporation, which produced and sold 50,000 units during a recent accounting period:

Sales	$850,000
Manufacturing costs	
Fixed	210,000
Variable	140,000
Selling and administrative costs	
Fixed	300,000
Variable	45,000
Income tax rate	40%

For the next accounting period, if production and sales are expected to be 40,000 units, the company should anticipate a contribution margin per unit of

A. $1.86

B. $3.10

C. $7.30

D. $13.30

The correct answer is (D). *(CMA, adapted)*

REQUIRED: The contribution margin per unit.

DISCUSSION: Unit contribution margin is the difference between unit selling price and unit variable cost. Unit selling price is $17 ($850,000 ÷ 50,000 units), and unit variable cost is $3.70 [($140,000 variable manufacturing cost + $45,000 variable S&A cost) ÷ 50,000 units sold]. Accordingly, unit contribution margin is $13.30 ($17 – $3.70).

Answer (A) is incorrect because $1.86 is an after-tax amount based on the inclusion of all fixed costs in the calculation. Answer (B) is incorrect because $3.10 is erroneously based on the inclusion of all fixed costs in the calculation of the UCM. Answer (C) is incorrect because $7.30 includes the $300,000 of fixed S&A costs in the calculation of the UCM.

43. A company manufactures a single product. Estimated cost data regarding this product and other information for the product and the company are as follows:

Sales price per unit	$40
Total variable production cost per unit	$22
Sales commission (on sales)	5%
Fixed costs and expenses:	
Manufacturing overhead	$5,598,720
General and administrative	$3,732,480
Effective income tax rate	40%

The number of units the company must sell in the coming year in order to reach its breakeven point is

A. 388,800 units.

B. 518,400 units.

C. 583,200 units.

D. 972,000 units.

The correct answer is (C). *(CIA, adapted)*

REQUIRED: The number of units to reach the break-even point.

DISCUSSION: The breakeven point is determined by dividing total fixed costs by the unit contribution margin. The total fixed costs are $9,331,200 ($5,598,720 manufacturing overhead + $3,732,480 general and administrative). The contribution margin is $16.00 ($40 sales price – $22 variable production cost – $2.00 commission). Thus, the breakeven point is 583,200 units ($9,331,200 ÷ $16).

Answer (A) is incorrect because 388,800 units results from not subtracting the variable costs per unit from sales price. Answer (B) is incorrect because 518,400 units does not reflect the sales commissions in the total variable costs. Answer (D) is incorrect because 972,000 units includes taxes in the total variable costs, which understates the unit contribution margin.

44. A company makes a product that sells for $30. During the coming year, fixed costs are expected to be $180,000, and variable costs are estimated at $26 per unit. How many units must the company sell to break even?

A. 6,000

B. 6,924

C. 45,000

D. 720,000

The correct answer is (C). *(CIA, adapted)*

REQUIRED: The breakeven point in units.

DISCUSSION: The contribution to overhead per unit is $4.00. Fixed costs of $180,000 divided by the contribution of $4.00 gives breakeven sales volume of 45,000 units.

Answer (A) is incorrect because 6,000 units will cover the fixed costs for the year but not the variable costs ($180,000 ÷ $30 = 6,000). Answer (B) is incorrect because 6,924 units is found by dividing the total fixed costs by the variable cost per unit. FC should be divided by the unit contribution margin. Answer (D) is incorrect because the total fixed costs should be divided (not multiplied) by the unit contribution margin.

Questions 45 and 46 are based on the following information. The data available for the current year are given below.

	Whole Company	Division 1	Division 2
Variable manufacturing cost of goods sold	$ 400,000	$220,000	$ 80,000
Unallocated costs (e.g., president's salary)	100,000		
Fixed costs controllable by division managers (e.g., advertising, engineering supervision costs)	90,000	50,000	40,000
Net revenue	1,000,000	600,000	400,000
Variable selling and administrative costs	130,000	70,000	60,000
Fixed costs controllable by others (e.g., depreciation, insurance)	120,000	70,000	50,000

45. Based upon the information presented, the contribution margin for the company was

A. $400,000

B. $470,000

C. $530,000

D. $600,000

The correct answer is (B). *(CIA, adapted)*

REQUIRED: The contribution margin for the company.

DISCUSSION: Contribution margin is sales minus variable costs. Direct costing considers only variable costs as product costs, so contribution margin appears in a direct costing income statement. Absorption costing treats both variable and fixed costs as product costs. Thus, variable costs are not stated separately, and contribution margin would not appear in the income statement. Accordingly, the CM is $470,000 ($1,000,000 net revenues – $400,000 variable CGS – $130,000 variable S&A costs).

Answer (A) is incorrect because $400,000 is the total variable manufacturing cost of goods sold. Contribution margin equals net revenue minus variable costs. Answer (C) is incorrect because $530,000 is the total variable costs for the year. Contribution margin equals net revenue minus variable costs. Answer (D) is incorrect because variable selling and administrative costs must also be subtracted from net revenue in determining contribution margin.

46. Using the information presented, the contribution by Division 1 was

A. $190,000

B. $260,000

C. $310,000

D. $380,000

The correct answer is (A). *(CIA, adapted)*

REQUIRED: The contribution by Division 1.

DISCUSSION: The contribution margin for Division 1 is $310,000 ($600,000 net revenue – $290,000 total variable costs). The contribution controllable by Division 1's manager is $260,000 ($310,000 CM – $50,000 controllable fixed cost). The total contribution by Division 1 equals its net revenue minus all costs traceable to it. Accordingly, the total contribution is $190,000 ($260,000 controllable contribution – $70,000 allocated but controllable by others). Unallocated costs are excluded from the calculation. If separate amounts are determined for the division's contribution and the controllable contribution, the difference between the division's and the manager's performance may be ascertained (assuming controllability of fixed costs can be assigned).

Answer (B) is incorrect because $260,000 is the contribution controllable by the Division 1 manager. All the costs incurred in Division 1 are subtracted from net revenue in determining the contribution by Division 1. Answer (C) is incorrect because $310,000 is the contribution margin for Division 1. All the costs incurred in Division 1 are subtracted from net revenue in determining the contribution by Division 1. Answer (D) is incorrect because all costs (not variable manufacturing costs only) incurred by Division 1 must be subtracted from Division 1 net revenue in order to find the contribution.

Questions 47 through 49 are based on the following information. Harper and his band want to put on a concert. They have looked at two venues, a small one and a large one, and have compiled the following information:

	Small	Large
Capacity of venue	400	1,200
Fixed costs	$2,000	$4,800

The variable cost per customer for both venues is $2. The band will charge $10 per customer for the small venue or $14 for the large venue.

47. What is the breakeven point of the small venue?

A. 200

B. 250

C. 305

D. 315

The correct answer is (B). *(Publisher)*

REQUIRED: The number of ticket sales required to break even in the small venue.

DISCUSSION: In a breakeven analysis, set the operating income equal to 0 and equate it with revenues minus fixed costs minus variable costs. The breakeven point is therefore

$$\begin{aligned} \$0 &= \$10x - \$2{,}000 - \$2x \\ \$8x &= \$2{,}000 \\ x &= 250 \end{aligned}$$

Answer (A) is incorrect because 200 ticket sales is found by not subtracting the variable cost per person from the revenues per person. Answers (C) and (D) are incorrect because 250 ticket sales is the amount needed to break even.

48. What is the breakeven point of the large venue?

A. 343

B. 375

C. 381

D. 400

The correct answer is (D). *(Publisher)*

REQUIRED: The number of ticket sales required to break even in the large venue.

DISCUSSION: In a breakeven analysis, set the operating income equal to 0 and equate it with revenues minus fixed costs minus variable costs. The breakeven point is therefore

$$\begin{aligned} \$0 &= \$14x - \$4{,}800 - \$2x \\ \$12x &= \$4{,}800 \\ x &= 400 \end{aligned}$$

Answer (A) is incorrect because 343 ticket sales is found by not subtracting the variable cost per person from the revenues per person. Answers (B) and (C) are incorrect because 400 ticket sales is the amount needed to break even.

49. Harper owes the local music store $1,000 for equipment. If he intends on using the profit from the concert to pay back his debt, how many tickets must he sell at the small and large venue, respectively?

A. 300 and 429

B. 375 and 429

C. 375 and 500

D. 300 and 500

The correct answer is (C). *(Publisher)*

REQUIRED: The amount of ticket sales required to earn $1,000.

DISCUSSION: The breakeven formula equates operating income with revenue minus fixed costs minus variable cost. In this problem, set operating income to $1,000 dollars and solve:

Small

$$\begin{aligned} \$1{,}000 &= \$10x - \$2{,}000 - \$2x \\ \$8x &= \$3{,}000 \\ x &= 375 \end{aligned}$$

Large

$$\begin{aligned} \$1{,}000 &= \$14x - \$5{,}000 - \$2x \\ \$12x &= \$6{,}000 \\ x &= 500 \end{aligned}$$

Answer (A) is incorrect because 300 ticket sales is found by not subtracting the variable cost from revenue in the small venue, and 429 ticket sales is found by not subtracting the variable cost from revenues in the large venue. Answer (B) is incorrect because 429 ticket sales is found by not subtracting the variable cost from revenues in the large venue. Answer (D) is incorrect because 300 ticket sales is found by not subtracting the variable cost from revenue in the small venue.

50. Two companies are expected to have annual sales of 1,000,000 decks of playing cards next year. Estimates for next year are presented below:

	Company 1	Company 2
Selling price per deck	$3.00	$3.00
Cost of paper per deck	.62	.65
Printing ink per deck	.13	.15
Labor per deck	.75	1.25
Variable overhead per deck	.30	.35
Fixed costs	$960,000	$252,000

Given these data, which of the following responses is correct?

	Breakeven Point in Units for Company 1	Breakeven Point in Units for Company 2	Volume in Units at Which Profits of Company 1 and Company 2 Are Equal
A.	800,000	420,000	1,180,000
B.	800,000	420,000	1,000,000
C.	533,334	105,000	1,000,000
D.	533,334	105,000	1,180,000

The correct answer is (A). *(CIA, adapted)*

REQUIRED: The breakeven points and the indifference point for the companies.

DISCUSSION: The BEP in units is found by dividing the fixed costs by the UCM.

$$\textit{Company 1} = \frac{\$960{,}000}{\$3 - .62 - .13 - .75 - .30} = 80{,}000$$

$$\textit{Company 2} = \frac{\$252{,}000}{\$3 - .65 - .15 - 1.25 - .35} = 420{,}000$$

The volume at which profits are equal occurs when the difference between total revenue and total cost is also the same. Both charge $3 per deck, so total revenue for the two companies at the indifference volume (U) will be equal. Consequently, total costs should also be equal.

$$U(\$.62 + .13 + .75 + .30) + \$960{,}000 = U(\$.65 + .15 + 1.25 + .35) + \$252{,}000$$

$$\$1.80U + \$960{,}000 = \$2.40U + \$252{,}000$$

$$\$.60U = \$708{,}000$$

$$U = 1{,}180{,}000$$

Answer (B) is incorrect because the volume at which Company 1 and Company 2 have equal profits is 1,180,000 units. Answer (C) is incorrect because the breakeven point is determined by dividing fixed costs by unit contribution margin, not unit variable costs. Furthermore, the volume at which there are equal profits between the companies is also higher. Answer (D) is incorrect because the breakeven point is determined by dividing fixed costs by unit contribution margin, not unit variable costs.

51. Lyman Company has the opportunity to increase annual sales $100,000 by selling to a new, riskier group of customers. The uncollectible expense is expected to be 15%, and collection costs will be 5%. The company's manufacturing and selling expenses are 70% of sales, and its effective tax rate is 40%. If Lyman should accept this opportunity, the company's after-tax profits would increase by

A. $6,000

B. $10,000

C. $10,200

D. $14,400

The correct answer is (A). *(CMA, adapted)*

REQUIRED: The after-tax increase in profit.

DISCUSSION: Sales increase by $100,000. Collection and bad debt expense is 20% of sales (15% + 5%). Total variable expenses for the new sales will be 90% (20% + 70%) of each sales dollar.

Sales	$100,000	
Minus costs	(90,000)	(90% × $100,000)
Additional profit	$ 10,000	
Minus taxes	(4,000)	(40% × $10,000)
After-tax profit	$ 6,000	

Answer (B) is incorrect because $10,000 is the pretax profits. Answer (C) is incorrect because the after-tax increase in profits is calculated by subtracting uncollectible expenses, collection costs, and manufacturing and selling expenses from sales. This pretax increase in profit is then multiplied by 60% (1 – 40% tax rate) to determine after-tax profit. Answer (D) is incorrect because the after-tax increase in profits is calculated by subtracting uncollectible expenses, collection costs, and manufacturing and selling expenses from sales. This pretax increase in profit is then multiplied by 60% (1 – 40% tax rate) to determine after-tax profit.

52. Tonykinn Company is contemplating marketing a new product. Fixed costs will be $800,000 for production of 75,000 units or less and $1,200,000 if production exceeds 75,000 units. The variable cost ratio is 60% for the first 75,000 units. Total fixed costs will increase to 50% for units in excess of 75,000. If the product is expected to sell for $25 per unit, how many units must Tonykinn sell to break even?

A. 120,000

B. 111,000

C. 96,000

D. 80,000

The correct answer is (B). *(C.J. Skender)*

REQUIRED: The breakeven point in units with changing fixed and variable cost behavior patterns.

DISCUSSION: The BEP in units is equal to fixed cost divided by the difference between unit selling price and unit variable cost (the unit contribution margin). At less than 75,000 units, fixed costs are $800,000 and UCM is $10 [$25 – (60% x $25)]. At this UCM, 80,000 units ($800,000 ÷ $10) must be sold, but this volume is not within the relevant range. At any production level greater than 75,000 units, total fixed costs are $1,200,000 but there are two UCM layers. The first 75,000 units sold will produce a contribution margin of $750,000 (75,000 x $10). Hence, another $450,000 ($1,200,000 – $750,000) must be contributed. The UCM is $12.50 [$25 – (50% x $25)] for units in excess of 75,000, and 36,000 ($450,000 ÷ $12.50) additional units must be sold. Total unit sales at the BEP are 111,000 (75,000 + 36,000).

Answer (A) is incorrect because only the first 75,000 units have a UCM of $10. Additional units in excess of 75,000 have a UCM of $12.50. Answer (C) is incorrect because the first 75,000 units have a UCM of only $10. Answer (D) is incorrect because production of units in excess of 75,000 units will increase fixed costs from $800,000 to $1,200,000.

53. The Childers Company sells widgets. The company breaks even at an annual sales volume of 75,000 units. Actual annual sales volume was 100,000 units, and the company reported a profit of $200,000. The annual fixed costs for the Childers Company are

A. $800,000

B. $600,000

C. $75,000

D. Insufficient information to determine amount of fixed costs.

The correct answer is (B). *(A. Wilson)*

REQUIRED: The annual fixed costs.

DISCUSSION: The profit increased by $200,000 and the sales volume increased by 25,000 units from the BEP. This increase in profit is a result of an increase in the contribution margin (sales – variable costs). Unit contribution margin is $8 ($200,000 ÷ 25,000 units). The BEP in units is equal to fixed costs divided by the UCM.

$$BEP = \frac{\textit{Fixed costs}}{\$8 \textit{ unit}} = 75{,}000 \textit{ units}$$

$$\textit{Fixed costs} = \$600{,}000$$

Answer (A) is incorrect because $800,000 is the contribution margin of sales of 100,000 units. Answer (C) is incorrect because the annual fixed costs are determined by finding the UCM ($200,000 profit ÷ 25,000 increase in units over breakeven point) and multiplying it by the number of units at the breakeven point. Answer (D) is incorrect because there is sufficient information to determine the amount of fixed costs. The annual fixed costs are determined by finding the UCM ($200,000 profit ÷ 25,000 increase in units over breakeven point) and multiplying it by the number of units at the breakeven point.

Questions 54 and 55 are based on the following information. Panyer Co. is a producer of a tank component. This product, J-5, has the following selling price and costs per unit:

Selling price	$300
Direct materials	125
Direct labor	25
Variable manufacturing overhead	50
Shipping and handling	5
Fixed manufacturing overhead	15
Fixed selling and administrative	10
Total costs	$230

54. Panyer has recently received a special, one-time order for 2,000 units of J-5. Panyer currently has enough excess capacity for this order. What should be the minimum price charged by Panyer?

A. $155

B. $205

C. $230

D. $300

The correct answer is (B). *(Publisher)*

REQUIRED: The minimum price to be charged on a special, one-time offer with excess capacity.

DISCUSSION: With the assumption that idle capacity is sufficient to manufacture 2,000 extra units, the goal is to find a minimum price that will not affect the price or quantity of other units sold. This minimum price is equal to the sum of the variable costs, $205.

Answer (A) is incorrect because $155 ignores the variable manufacturing overhead from total variable costs. Answer (C) is incorrect because $230 is the amount of the total costs, not the total variable costs. Answer (D) is incorrect because $300 is the normal selling price.

55. Panyer has again received a special, one-time offer for 2,000 units of J-5. Panyer is now operating at full capacity, 10,000 units, at a total cost of $2,300,000. To produce this order would cause a 20% increase in fixed costs. What is the minimum price that is acceptable for this one-time, special order?

A. $205

B. $230

C. $260

D. $300

The correct answer is (C). *(Publisher)*

REQUIRED: The differential cost for producing at 120% of capacity.

DISCUSSION: To determine the minimum price per unit, while at full capacity, the differential costs must be found. The total cost of producing 12,000 units is $2,820,000 {12,000 units x [$205 variable cost + 1.2($25 fixed cost)]}. Thus, the total differential cost is $520,000 ($2,820,000 – $2,300,000), and the unit differential cost is $260 ($520,000 ÷ 2,000 units).

Answer (A) is incorrect because $205 is the minimum price at less than full capacity. Answer (B) is incorrect because $230 is the unit cost at full capacity. Answer (D) is incorrect because $300 is the price for regular orders.

56. A manufacturer produces a product that sells for \$10 per unit. Variable costs per unit are \$6 and total fixed costs are \$12,000. At this selling price, the company earns a profit equal to 10% of total dollar sales. By reducing its selling price to \$9 per unit, the manufacturer can increase its unit sales volume by 25%. Assume that there are no taxes and that total fixed costs and variable costs per unit remain unchanged. If the selling price is reduced to \$9 per unit, the profit will be

A. \$3,000

B. \$4,000

C. \$5,000

D. \$6,000

The correct answer is (A). *(CIA, adapted)*

REQUIRED: The expected profit given the change in selling price.

DISCUSSION: To determine the profit under the new pricing policy, the sales volume under the old policy must be calculated.

$$\$10X - \$6X - \$12{,}000 = 0.1(\$10X)$$
$$X = 4{,}000 \text{ units}$$

Profit at the new price (\$9) is

$$\text{Expected volume at } \$9 = 125\% \times 4{,}000 = 5{,}000$$
$$\text{Profit} = (\$9 \times 5{,}000) - (\$6 \times 5{,}000) - \$12{,}000 = \$3{,}000$$

Answer (B) is incorrect because the sales volume under the old policy is 4,000 units [\$10X – \$6X – \$12,000 – .1(\$10X)]. The sales will increase 25% when the price is lowered to \$9; thus, the new volume is 5,000 units (4,000 x 125%) and the profit will be \$3,000. Answer (C) is incorrect because the sales volume under the old policy is 4,000 units [\$10X – \$6X – \$12,000 – .1(\$10X)]. The sales will increase 25% when the price is lowered to \$9; thus, the new volume is 5,000 units (4,000 x 125%), and the profit will be \$3,000. Answer (D) is incorrect because the sales volume under the old policy is 4,000 units [\$10X – \$6X – \$12,000 – .1(\$10X)]. The sales will increase 25% when the price is lowered to \$9; thus, the new volume is 5,000 units (4,000 x 125%) and the profit will be \$3,000.

57. A manufacturing concern sells its sole product for \$10 per unit, with a unit contribution margin of \$6. The fixed manufacturing cost rate per unit is \$2 based on a denominator capacity of 1 million units, and fixed marketing costs are \$1.5 million. If 900,000 units are produced, the absorption-costing breakeven point in units sold is

A. 425,000 units.

B. 583,333 units.

C. 900,000 units.

D. 1,000,000 units.

The correct answer is (A). *(Publisher)*

REQUIRED: The absorption-costing breakeven point in units sold.

DISCUSSION: The absorption-costing breakeven point in units sold equals the sum of (1) the total fixed costs and (2) the product of the fixed manufacturing cost application rate and the difference between the BEP in units sold (X) and units produced, with the sum divided by the UCM. Thus, the absorption-costing breakeven point in units sold is 425,000 units:

$$X = \frac{(\$2 \times 1{,}000{,}000 \textit{ denominator capacity}) + \$1{,}500{,}000 + \$2(X - 900{,}000)}{\$6}$$

$$X = \frac{(\$3{,}500{,}000 + \$2X - \$1{,}800{,}000}{\$6}$$

$$\$6X = \$1{,}700{,}000 + \$2X$$

$$X = 425{,}000$$

Answer (B) is incorrect because 583,333 units is the variable-costing BEP. Answer (C) is incorrect because 900,000 units is the actual production level. Answer (D) is incorrect because 1,000,000 units is the capacity used to calculate the fixed overhead application rate.

Questions 58 through 63 are based on the following information. The statement of income for Dimmell Co. presented below represents the operating results for the fiscal year just ended. Dimmell had sales of 1,800 tons of product during the current year. The manufacturing capacity of Dimmell's facilities is 3,000 tons of product.

Dimmell Co.
Statement of Income
For the Year Ended December 31, 2001

Sales	$900,000
Variable costs	
Manufacturing	$315,000
Selling costs	180,000
Total variable costs	$495,000
Contribution margin	$405,000
Fixed costs	
Manufacturing	$ 90,000
Selling	112,500
Administration	45,000
Total fixed costs	$247,500
Net income before income taxes	$157,500
Income taxes (40%)	(63,000)
Net income after income taxes	$ 94,500

58. The breakeven volume in tons of product for 2001 is

A. 400 tons.

B. 1,100 tons.

C. 900 tons.

D. 550 tons.

The correct answer is (B). *(Publisher)*

REQUIRED: The breakeven volume in tons.

DISCUSSION: The breakeven formula sets sales equal to the sum of fixed and variable costs. Thus, sales equal $247,500 plus 0.55S. The variable costs of $495,000 are 55% of the total sales of $900,000. Working through the formula, sales are equal to $550,000. Because 1,800 tons were sold for $900,000, the selling price is $500 per ton ($900,000 ÷ 1,800 tons). The breakeven point in tons is the breakeven point in dollars of $550,000 divided by $500 per ton, or 1,100 tons.

$$S = \$247,500 + 0.55S$$
$$0.45S = \$247,500$$
$$S = \$550,000$$

Answer (A) is incorrect because 400 tons used only fixed manufacturing costs rather than all fixed costs. Answer (C) is incorrect because 900 tons used variable costs in the denominator instead of contribution margin. Answer (D) is incorrect because 550 tons is half of the breakeven level.

59. Refer to the information preceding question 58 on page 101. If the sales volume is estimated to be 2,100 tons in the next year, and if the prices and costs stay at the same levels and amounts next year, the after-tax net income that Dimmell can expect for 2001 is

A. $135,000

B. $110,250

C. $283,500

D. $184,500

The correct answer is (A). *(Publisher)*

REQUIRED: The after-tax net income.

DISCUSSION: Because the selling price is $500 per ton, sales will be $1,050,000. The contribution margin ratio is 45% ($405,000 ÷ $900,000), resulting in a contribution margin of $472,500. Net income before taxes is the contribution margin minus fixed costs of $247,500, which equals $225,000. Net income is $135,000 [(1.0 – 40%) tax rate x $225,000].

S = 2,100 × $500 = $1,050,000
CM = 0.45 × $1,050,000 = $472,500
NIBT = $472,500 – $247,500 = $225,000
NI = 0.60 × $225,000 = $135,000

Answer (B) is incorrect because $110,250 inverses the variable cost and contribution margin. Answer (C) is incorrect because $283,500 fails to deduct non-manufacturing fixed costs. Answer (D) is incorrect because $184,500 fails to deduct some fixed costs.

60. Refer to the information preceding question 58 on page 101. Dimmell has a potential foreign customer that has offered to buy 1,500 tons at $450 per ton. Assume that all of Dimmell's costs would be at the same levels and rates as in 2000. What net income after taxes would Dimmell make if it took this order and rejected some business from regular customers so as not to exceed capacity?

A. $297,500

B. $252,000

C. $211,500

D. $256,500

The correct answer is (C). *(Publisher)*

REQUIRED: The after-tax income following a special order.

DISCUSSION: Total sales equal $1,425,000. The unit variable cost of sales is $275 (55% of $500), and total variable cost of sales is $825,000 ($275 x 3,000 units). The contribution margin is $1,425,000 of sales minus $825,000 cost of sales. Net income before taxes is $600,000 minus $247,500 fixed costs, or $352,500. Net income is 60% (1.0 – 40% tax rate) of the net income before taxes of $352,500, or $211,500.

S = (1,500 × $450) + (1,500 × $500) = $1,425,000
COGS = (3,000 × $275) = $825,000
CM = $1,425,000 – $825,000 = $600,000
NIBT = $600,000 – $247,500 = $352,500
NI = 0.60 × $352,500 = $211,500

Answer (A) is incorrect because $297,500 fails to deduct selling expenses. Answer (B) is incorrect because $252,000 ignores administrative expenses. Answer (D) is incorrect because $256,500 mixes variable costs and contribution margin.

61. Refer to the information preceding question 58 on page 101. Dimmell plans to market its product in a new territory. Dimmell estimates that an advertising and promotion program costing $61,500 annually would need to be undertaken for the next 2 or 3 years. In addition, a $25 per ton sales commission over and above the current commission to the sales force in the new territory would be required. How many tons would have to be sold in the new territory to maintain Dimmell's current after-tax income of $94,500?

A. 307.5 tons.

B. 1,095.0 tons.

C. 273.333 tons.

D. 1,545.0 tons.

The correct answer is (A). *(Publisher)*

REQUIRED: The tons to be sold in a new territory to maintain the current level of income.

DISCUSSION: Given that $61,500 is required for advertising and promotion, these are fixed costs that will have to be covered by the contribution margin for sales in the new market of $200 each. The unit contribution margin for regular sales was $225 (45% x $500 selling price), and there is a $25 additional variable commission expense for sales in the new territory. The 45% contribution margin is computed by dividing the $405,000 contribution margin by the $900,000 of sales. Thus, the new unit contribution of $200 is divided into $61,500 of incremental fixed costs, resulting in additional sales from the new program of 307.5 tons at the breakeven point.

Answer (B) is incorrect because 1,095.0 tons would result in a greater profit. Answer (C) is incorrect because 273.333 tons ignores the increased commissions in the new territory. Answer (D) is incorrect because 1,545.0 tons results in increased income.

62. Refer to the information preceding question 58 on page 101. Dimmell is considering replacing a highly labor intensive process with an automatic machine. This would result in an increase of $58,500 annually in manufacturing fixed costs. The variable manufacturing costs would decrease $25 per ton. The new breakeven volume in tons would be

A. 990 tons.

B. 1,224 tons.

C. 1,854 tons.

D. 612 tons.

The correct answer is (B). *(Publisher)*

REQUIRED: The new breakeven point.

DISCUSSION: The new manufacturing fixed cost would be $306,000. The new contribution margin would be $250 per unit, which is the old contribution margin of $225 plus the $25 of variable cost savings per ton. Thus, the new breakeven point would be $306,000 of fixed costs divided by the new $250 contribution margin per unit, or 1,224 tons.

FC = $247,500 + $58,500 = $306,000
UCM = $225 + $25 = $250
BE = $306,000 ÷ $250 = 1,224 tons

Answer (A) is incorrect because 990 tons ignores the increase in fixed costs. Answer (C) is incorrect because 1,854 tons would produce a profit. Answer (D) is incorrect because 612 tons is half the breakeven level.

63. Refer to the information preceding question 58 on page 101. Ignore the facts presented in the previous question and now assume Dimmell estimates that the per ton selling price would decline 10% next year. Variable costs would increase $40 per ton and the fixed costs would not change. What sales volume in dollars would be required to earn an after-tax net income of $94,500 next year?

A. $1,140,000

B. $825,000

C. $1,500,000

D. $1,350,000

The correct answer is (D). *(Publisher)*

REQUIRED: The new sales volume that would maintain the same level of income.

DISCUSSION: The selling price per unit will decrease from $500 to $450. The unit variable costs will increase from $275 to $315. Also, $157,500 of pretax profit is required to obtain $94,500 of after-tax profit [$94,500 ÷ (1.0 – 40% tax rate)]. Thus, the number of units to be sold equals 3,000 [($247,500 FC + $157,500 pretax profit) ÷ ($450 – $315) UCM] . Multiplying 3,000 by the selling price of $450 results in $1,350,000 of sales.

Answers (A) and (B) are incorrect because they result in an income of less than $94,500. Answer (C) is incorrect because $1,500,000 results in an income above $94,500.

64. A not-for-profit social agency provides home health care assistance to as many patients as possible. Its budgeted appropriation (X) for next year must cover fixed costs of $5 million, and the annual per-patient cost (Y) of its services. However, the agency is preparing for a possible 10% reduction in its appropriation that will lower the number of patients served from 5,000 to 4,000. The reduced appropriation and the annual per-patient cost equal

	Reduced Appropriation	Per-Patient Annual Cost
A.	$5,000,000	$4,000
B.	$8,333,333	$833
C.	$9,000,000	$1,000
D.	$10,000,000	$5,000

The correct answer is (C). *(Publisher)*

REQUIRED: The reduced appropriation and the annual per-patient cost.

DISCUSSION: This question applies CVP analysis in a not-for-profit context in which the agency wishes to assist as many people as possible. Thus, a breakeven point must be calculated. Total revenue (the appropriation) equals fixed cost plus the product of unit variable cost (per-patient annual cost) and the number of patients who can be assisted given the available resources. The following are simultaneous equations stated in the two unknowns:

$$X - 5{,}000Y = \$5{,}000{,}000$$
$$.9X - 4{,}000Y = \$5{,}000{,}000$$

Because X must equal 5,000Y + $5,000,000, the second equation may be solved as follows for the per-patient annual cost (Y):

$$.9(5{,}000Y + \$5{,}000{,}000) - 4{,}000Y = \$5{,}000{,}000$$
$$4{,}500Y + \$4{,}500{,}000 - 4{,}000Y = \$5{,}000{,}000$$
$$500Y = \$500{,}000$$
$$Y = \$1{,}000$$

Accordingly, the budgeted appropriation (X) must be $10,000,000 [(5,000 x $1,000) VC + $5,000,000 FC], and the reduced appropriation must be $9,000,000 (90% x $10,000,000).

Answer (A) is incorrect because $5,000,000 is the fixed cost and 4,000 is the number served given a reduced appropriation. Answer (B) is incorrect because $8,333,333 and $833 are the appropriation and per-patient annual cost, respectively, for the next year that result from a 10% increase in the appropriation of the current year instead of a 10% decrease in the next year's appropriation. Answer (D) is incorrect because $10,000,000 is the budgeted appropriation and 5,000 is the number of people it would serve.

65. The percentage change in earnings before interest and taxes associated with the percentage change in sales volume is the degree of

A. Operating leverage.

B. Financial leverage.

C. Breakeven leverage.

D. Combined leverage.

The correct answer is (A). *(CIA, adapted)*

REQUIRED: The term for the percentage change in earnings before interest and taxes associated with the percentage change in sales.

DISCUSSION: Operating leverage is based on the degree to which fixed costs are used in production. Firms may increase fixed costs, such as by automation, to reduce variable costs. The result is a greater degree of operating leverage (DOL), which is the percentage change in net operating income (earnings before interest and taxes) divided by the percentage change in unit sales. Thus, operating leverage is related to the price elasticity concept in economics. It can also be determined from dividing the total contribution margin by operating income as expressed in the following formula, given that Q is quantity of units sold, P is unit price, V is unit variable cost, and F is fixed cost:

$$\frac{Q(P - V)}{Q(P - V) - F}$$

Answer (B) is incorrect because the degree of financial leverage equals the percentage change in net income divided by the percentage change in operating income. Answer (C) is incorrect because the breakeven point is the sales volume at which total revenue equals total costs. Answer (D) is incorrect because the degree of total (combined) leverage equals the percentage change in net income divided by the percentage change in sales.

66. A higher degree of operating leverage compared with the industry average implies that the firm

A. Has higher variable costs.

B. Has profits that are more sensitive to changes in sales volume.

C. Is more profitable.

D. Is less risky.

The correct answer is (B). *(CMA, adapted)*

REQUIRED: The effect of a higher degree of operating leverage (DOL).

DISCUSSION: Operating leverage is a measure of the degree to which fixed costs are used in the production process. A company with a higher percentage of fixed costs (higher operating leverage) has greater risk than one in the same industry that relies more heavily on variable costs. The DOL equals the percentage change in net operating income divided by the percentage change in sales. Thus, profits become more sensitive to changes in sales volume as the DOL increases.

Answer (A) is incorrect because a firm with higher operating leverage has higher fixed costs and lower variable costs. Answer (C) is incorrect because a firm with higher leverage will be relatively more profitable than a firm with lower leverage when sales are high. The opposite is true when sales are low. Answer (D) is incorrect because a firm with higher leverage is more risky. Its reliance on fixed costs is greater.

67. Carlisle Company currently sells 400,000 bottles of perfume each year. Each bottle costs $.84 to produce and sells for $1.00. Fixed costs are $28,000 per year. The firm has annual interest expense of $6,000, preferred stock dividends of $2,000 per year, and a 40% tax rate. The degree of operating leverage for Carlisle Company is

A. 2.4

B. 1.78

C. 2.13

D. 1.2

The correct answer is (B). *(CMA, adapted)*

REQUIRED: The degree of operating leverage.

DISCUSSION: Operating leverage is the percentage change in operating income resulting from a percentage change in sales. It measures how a change in volume affects profits. An alternative formula for operating leverage divides contribution margin by the contribution margin minus fixed costs. Companies with larger investments and greater fixed costs ordinarily have higher contribution margins and more operating leverage. The degree of operating leverage measures the extent to which fixed assets are used in the production process. A company with a high percentage of fixed costs is more risky than a firm in the same industry that relies more on variable costs to produce. Based on a contribution margin of $.16 per unit ($1 – $.84 variable cost), the degree of operating leverage is

$$(400{,}000 \times \$.16) \div [(400{,}000 \times \$.16) - \$28{,}000] = 1.78$$

Answer (A) is incorrect because 2.4 is obtained by overstating the contribution margin or the fixed costs. Answer (C) is incorrect because 2.13 includes a nonoperating expense (interest) as a fixed cost. Answer (D) is incorrect because 1.2 is obtained by understating the $64,000 contribution margin or understating the $28,000 of fixed costs.

Questions 68 and 69 are based on the following information. A company has sales of one of its products of $400,000 per year and a contribution margin ratio of 20%. Its margin of safety is $40,000.

68. What is the company's breakeven point?

A. $360,000

B. $320,000

C. $288,000

D. $80,000

The correct answer is (A). *(Publisher)*

REQUIRED: The breakeven point in dollars.

DISCUSSION: The margin of safety equals sales above the breakeven point. Thus, if the margin of safety is $40,000, the breakeven point must be $360,000 ($400,000 sales – $40,000).

Answers (B), (C), and (D) are incorrect because the breakeven point in dollars is the difference between actual sales and the margin of safety.

69. What are the company's fixed costs?

A. $72,000

B. $80,000

C. $288,000

D. $320,000

The correct answer is (A). *(Publisher)*

REQUIRED: The amount of fixed costs.

DISCUSSION: The margin of safety equals sales above the breakeven point. Thus, if the margin of safety is $40,000, the breakeven point in sales dollars must be $360,000. If the contribution margin ratio is 20%, the contribution margin (sales – variable costs) at the breakeven point is $72,000 (20% x $360,000). This amount must equal fixed costs because sales equal the sum of fixed and variable costs at the breakeven point.

Answer (B) is incorrect because $80,000 results from not subtracting the margin of safety from sales. Answer (C) is incorrect because $288,000 equals total variable costs. Answer (D) is incorrect because $320,000 equals total variable costs if $400,000 is the breakeven point.

Questions 70 and 71 are based on the following information. For one of its divisions, Buona Fortuna Company has fixed costs of $300,000 and a variable-cost percentage equal to 60% of its $10 per unit selling price. It would like to earn a pre-tax income of $90,000 per year from the division.

70. What is the breakeven point in dollars?

A. $300,000

B. $500,000

C. $750,000

D. $1,050,000

The correct answer is (C). *(Publisher)*

REQUIRED: The breakeven point in dollars.

DISCUSSION: Breakeven point is calculated by dividing the fixed costs of $300,000 by the contribution margin percentage, which was 40%. Dividing $300,000 by .4 results in a breakeven point of $750,000.

Answer (A) is incorrect because $300,000 equals fixed costs. Answer (B) is incorrect because $500,000 is based on the variable-cost percentage. Answer (D) is incorrect because $1,050,000 added fixed costs to the breakeven level.

71. How many units will Buona Fortuna have to sell to earn a pre-tax income of $90,000 per year?

A. 65,000 units.

B. 75,000 units.

C. 77,250 units.

D. 97,500 units.

The correct answer is (D). *(Publisher)*

REQUIRED: The unit sales necessary to generate a given pre-tax income.

DISCUSSION: The breakeven point is calculated by dividing the fixed costs by the unit contribution margin in dollars. To determine how many units need to be sold to yield a specific income, that income should be treated as a fixed cost. Thus, adding the $90,000 of desired income to the $300,000 of fixed costs results in a numerator of $390,000. The necessary unit sales equals 97,500 units [$390,000 ÷ ($10 unit price – $6 unit variable cost)].

Answer (A) is incorrect because 65,000 units results from using the unit variable cost in the denominator. Answer (B) is incorrect because 75,000 units is the breakeven point. Answer (C) is incorrect because 77,250 units would yield a profit less than the desired $90,000.

Questions 72 through 74 are based on the following information. Catfur Company has fixed costs of $300,000. It produces two products, X and Y. Product X has a variable cost percentage equal to 60% of its $10 per unit selling price. Product Y has a variable cost percentage equal to 70% of its $30 selling price. For the past several years, sales of Product X have averaged 66 ⅔% of the sales of Product Y. That ratio is not expected to change.

72. What is Catfur's breakeven point in dollars?

A. $300,000

B. $750,000

C. $857,142

D. $942,857

The correct answer is (D). *(Publisher)*

REQUIRED: The breakeven point in dollars for a two-product company when the products have different unit contribution margins (UCMs).

DISCUSSION: A helpful approach in a multiproduct situation is to make calculations based on the composite unit, i.e., 2 units of Product X and 3 units of Product Y (a 66 ⅔% ratio). The selling price of this composite unit is $110 [(2 x $10) + (3 x $30)]. The UCM of the composite unit is $35 {[2 x ($10 – $6)] + [3 x ($30 – $21)]}. Consequently, the breakeven point in composite units is 8,571.43 ($300,000 FC ÷ $35 UCM), and the breakeven point in sales dollars is $942,857 (8,571.43 x $110).

Answer (A) is incorrect because $300,000 equals the fixed costs. Answer (B) is incorrect because $750,000 assumes a 40% contribution margin ratio. Answer (C) is incorrect because $857,142 assumes a 35% contribution margin ratio.

73. How many units of Product Y will Catfur sell at the breakeven point?

A. 8,571 units.

B. 20,454 units.

C. 23,377 units.

D. 25,714 units.

The correct answer is (D). *(Publisher)*

REQUIRED: The units of Product Y sold at the breakeven point.

DISCUSSION: In the preceding question, a composite unit was defined as consisting of 2 units of Product X and 3 units of Product Y. The preceding question also calculated the breakeven point in composite units to be 8,571.43. Accordingly, the units of Product Y sold at the breakeven point are equal to 25,714 (8,571.43 x 3).

Answer (A) is incorrect because 8,571 units equals the number of composite units sold at the breakeven point. Answer (B) is incorrect because 20,454 assumes a 40% contribution margin ratio. Answer (C) is incorrect because 23,377 assumes a 35% contribution margin ratio.

74. Assume that Catfur Company achieved its planned breakeven level of sales in dollars, but the mix of products sold was one-to-one. All actual costs and unit selling prices equaled budgeted amounts. What is the impact on profitability?

A. The company is operating at the breakeven point.

B. The company earned a profit.

C. The company sustained a loss.

D. Cannot be determined from the information given.

The correct answer is (B). *(Publisher)*

REQUIRED: The impact on profitability of selling more higher margin products.

DISCUSSION: The expected sales mix is 40% for Product X and 60% for Product Y. Given that Product X has a 40% contribution margin ratio and Product Y has a 30% contribution margin ratio, selling more of Product X and less of Product Y increases the average contribution margin ratio. The effect is to lower the breakeven point. Thus, if the new composite unit includes 2 units of Product X and 2 units of Product Y, the composite unit selling price is $80 [(2 x $10) + (2 x $30)], and the composite UCM is $26 {[2 x ($10 – $6)] + [2 x ($30 – $21)]}. The new breakeven point in composite units is therefore 11,538.46 ($300,000 FC ÷ $26 UCM), and the new breakeven point in sales dollars is $923,077 (11,538.46 x $80). Given that sales reached the budgeted breakeven point of $942,857 (see question 57), Catfur must have made a profit of $19,780 ($942,857 – $923,077).

Answers (A), (C), and (D) are incorrect because the company is profitable due to the higher average contribution margin.

75. A company wants to open a new store in one of two nearby shopping malls. In Mall A, the rent will be $250,000 per year. In Mall B, the rent will be 4% of gross revenues. Assuming that revenues and all other elements under consideration are the same for both malls, at what level of revenues will the company be indifferent between the two malls?

A. $1,000,000

B. $4,000,000

C. $6,250,000

D. $12,500,000

The correct answer is (C). *(Publisher)*

REQUIRED: The indifference level of revenues for two leasing options.

DISCUSSION: The level of indifference is calculated by setting the equation for the first mall (a flat $250,000) equal to the equation for the second mall. Solving for that equation involves dividing $250,000 by 4%, resulting in an indifference level of revenues of $6,250,000.

Answers (A) and (B) are incorrect because, at revenues below $6,250,000, Mall B is preferable. Answer (D) is incorrect because, at any level of revenues over $6,250,000, Mall A is preferable.

Questions 76 through 78 are based on the following information. A company wants to open a new store in one of three nearby shopping malls. In Mall A, the rent will be $300,000 per year. In Mall B, the rent will be 4% of gross revenues. In Mall C, the rent will be $150,000 per year plus 3% of gross revenues. Assume that revenues and all other elements under consideration are the same for all three malls.

76. Which mall should the company choose if revenues are expected to be $6,000,000 per year?

A. Mall A.

B. Mall B.

C. Mall C.

D. The company will be indifferent between two of the choices.

The correct answer is (B). *(Publisher)*

REQUIRED: The lowest cost facility when revenues are expected to be $600,000.

DISCUSSION: The answer depends on the expected level of revenues. If the company expects revenues to be $6,000,000 per year, the calculation is as follows:

Mall A: $300,000.
Mall B: $6,000,000 x 4% = $240,000.
Mall C: $6,000,000 x 3% = $180,000 + $150,000 = $330,000.

Thus, Mall B is preferable.

Answers (A), (C), and (D) are incorrect because Mall B's rent is $240,000, Mall A's rent is $300,000, and Mall C's rent is $330,000.

77. If the company expects revenues to be $10,000,000 per year, which mall should be chosen?

A. Mall A.

B. Mall B.

C. Mall C.

D. The company will be indifferent between two of the choices.

The correct answer is (A). *(Publisher)*

REQUIRED: The lowest cost facility when revenues are expected to be $10,000,000 per year.

DISCUSSION: If the company expects revenues to be $10,000,000 per year, the calculation is as follows:

Mall A: $300,000.
Mall B: $10,000,000 x 4% = $400,000.
Mall C: $10,000,000 x 3% = $300,000 + $150,000 = $450,000.

Thus, Mall A is preferable.

Answers (B), (C), and (D) are incorrect because Mall A's rent is $300,000, Mall B's rent is $400,000, and Mall C's rent is $450,000.

78. What is the maximum level of revenues at which Mall C will be the most desirable of the three options?

A. $149,999

B. $5,000,000

C. $15,000,000

D. Mall C will never be the most desirable choice.

The correct answer is (D). *(Publisher)*

REQUIRED: The level of revenues, if any, at which Mall C is the most desirable (lowest cost) option.

DISCUSSION: Mall C will never be the optimal choice because it will be less desirable than Mall B as long as 1% of gross revenue is less than $150,000, that is, until revenues reach $15,000,000 [$150,000 minimum ÷ (4% – 3%)]. However, at any level of revenues greater than $7,500,000 ($300,000 ÷ 4%), Mall A (a flat $300,000 rental) will be more desirable than either of the other choices. Thus, Mall C will never be the most desirable.

Answers (A), (B), and (C) are incorrect because Mall C will never be the lowest cost choice.

79. Based on potential sales of 1,000 units per year, a new product has estimated costs of $600,000. What is the target price to obtain a 20% return on sales?

A. $720

B. $750

C. $1,080

D. $3,000

The correct answer is (B). *(Publisher)*

REQUIRED: The target selling price that yields a specified return on sales.

DISCUSSION: If the company wants a 20% return on sales, the $600,000 of costs represent 80% of the selling price. Dividing $600,000 by .8 results in a total selling price of $750,000, or $750 per unit.

Answer (A) is incorrect because $720 is based on a 20% return on costs. Answer (C) is incorrect because $1,080 is based on an 80% return on costs. Answer (D) is incorrect because $3,000 is based on an 80% return on sales.

Questions 80 and 81 are based on the following information. Pontotoc Industries manufactures a product that is used as a subcomponent by other manufacturers. It has the following price and cost structure:

Selling price		$300
Costs		
Direct materials	$40	
Direct labor	30	
Variable manufacturing overhead	24	
Fixed manufacturing overhead	60	
Variable selling	6	
Fixed selling and administrative	20	(180)
Operating margin		$120

80. What will the contribution margin per unit be if the company sells 10,000 units?

A. $206

B. $200

C. $140

D. $120

The correct answer is (B). *(Publisher)*

REQUIRED: The contribution margin per unit.

DISCUSSION: Contribution margin is the excess of sales over variable costs. Sales will be at $300 per unit. Variable costs are $100, consisting of $40 of direct materials, $30 of direct labor, $24 of variable overhead, and $6 of variable selling costs. Thus, the contribution margin will be $200 per unit ($300 – $100).

Answer (A) is incorrect because $206 excludes variable selling costs. Answer (C) is incorrect because $140 includes fixed manufacturing overhead. Answer (D) is incorrect because $120 is the operating margin.

81. During the next year, sales are expected to be 10,000 units. All costs will remain the same except for fixed manufacturing overhead, which will increase 20%, and direct materials, which will increase 10%. The selling price per unit for next year will be $320. Based on this information, Pontotoc's contribution margin for next year will be

A. $1,240,000

B. $1,360,000

C. $2,160,000

D. $2,200,000

The correct answer is (C). *(Publisher)*

REQUIRED: The contribution margin for next year.

DISCUSSION: Contribution margin is the excess of sales over variable costs. Sales of 10,000 units at $320 each will produce total revenue of $3,200,000. Variable costs will be $104 per unit, consisting of $44 for direct materials, $30 for direct labor, $24 for variable overhead, and $6 for selling costs. At $104 per unit, the 10,000 units will have total variable costs of $1,040,000, resulting in a contribution margin of $2,160,000 ($3,200,000 – $1,040,000).

Answer (A) is incorrect because $1,240,000 is based on total costs. Answer (B) is incorrect because $1,360,000 includes $80 per unit of fixed costs. Answer (D) is incorrect because $2,200,000 does not consider the 10% increase in direct materials costs.

STUDY UNIT 3: CAPITAL BUDGETING

13 pages of outline
88 multiple-choice questions

This study unit is particularly important because it is the only one devoted to investment decision analysis, a major topic to which a relative weight range of 20% to 30% of Part 4 has been assigned. However, the subject is well-defined, so preparing for capital budgeting questions is straightforward. This study unit provides definitions and outlines (with examples) of the most important techniques. Risk is also discussed. Furthermore, the tax consequences of business decision making should be considered. Hence, the outlines in the appendix on income taxes should be reviewed.

A. Definitions

1. Capital budgeting is the process of planning expenditures for assets, the returns on which are expected to continue beyond 1 year.
   a. Capital budgeting decisions are long-term.
      1) Once made, they tend to be relatively inflexible because the commitments extend well into the future.
   b. Without proper timing, additional capacity generated by the acquisition of capital assets may not coincide with changes in demand for output, resulting in capacity excess or shortage.
      1) Accurate forecasting is needed to anticipate changes in product demand so that full economic benefits flow to the firm when the capital asset is available for use.
   c. A capital budget usually involves substantial expenditures. The sources of these funds are critical.
      1) Planning is important because of possible changes in capital markets, inflation, interest rates, and the money supply.
2. **Types of Costs Considered in Capital Budgeting Analysis**
   a. An **avoidable cost** may be eliminated by ceasing an activity or by improving efficiency.
   b. A **common cost** is shared by all options and is not clearly allocable to any one of them.
   c. The **cost of capital** is the weighted average of the interest cost of debt (net of tax) and the implicit cost of equity capital to be invested in long-term assets. It represents a required minimum return of a new investment to prevent dilution of shareholders' interests.
   d. A **deferrable cost** may be shifted to the future with little or no effect on current operations.

e. A **fixed cost** does not vary with the level of activity within the relevant range.

f. An **imputed cost** may not entail a specified dollar outlay formally recognized by the accounting system, but it is nevertheless relevant to establishing the economic reality analyzed in the decision-making process.

g. An **incremental cost** is the difference in cost resulting from selecting one option instead of another.

h. An **opportunity cost** is the benefit forgone, for example, the contribution to income, by not selecting the best alternative use of scarce resources.

i. **Relevant costs** vary with the action. Other costs are constant and therefore do not affect the decision.

j. A **sunk cost** cannot be avoided because an expenditure or an irrevocable decision to incur the cost has been made.

k. **Taxes**. As with every other decision, the tax consequences of a new investment (and possible disinvestment of a replaced asset) must be considered.

1) All capital budgeting decisions need to be evaluated on an after-tax basis, because taxes may affect different decisions differently. Companies that operate in numerous locales may find the decision process more complex because they are subject to multiple tax jurisdictions. Another possibility is that special tax concessions may be negotiated for locating an investment in one locale versus another.

3. Capital budgeting requires choosing among investment proposals. Thus, a ranking procedure for such decisions is needed. The following are steps in the ranking procedure:

a. Determine the asset cost or net investment.

1) The net investment is the net outlay, or gross cash requirement, minus cash recovered from the trade or sale of existing assets, with any necessary adjustments for applicable tax consequences. Cash outflows in subsequent periods must also be considered.

2) The investment required also includes funds to provide for increases in working capital, for example, the additional receivables and inventories resulting from the acquisition of a new manufacturing plant. This investment in working capital is treated as an initial cost of the investment (a cash outflow) that will be recovered at the end of the project (i.e., the salvage value is equal to the initial cost).

b. Calculate estimated cash flows, period by period, using the acquired assets.

1) Reliable estimates of cost savings or revenues are necessary.

2) Net cash flow is the economic benefit or cost, period by period, resulting from the investment.

3) **Economic life** is the time period over which the benefits of the investment proposal are expected to be obtained, as distinguished from the physical or technical life of the asset involved.

4) **Depreciable life** is the period used for accounting and tax purposes over which cost is to be systematically and rationally allocated. It is based upon permissible or standard guidelines and may have no particular relevance to economic life.

c. Relate the cash-flow benefits to their cost by using one of several methods to evaluate the advantage of purchasing the asset.

d. Rank the investments.

4. The rating methods can sometimes yield different solutions to an investment problem. Managers should therefore use all the methods and then apply their judgment in making the final decision. All of these techniques consider only the economic factors in a decision, but sometimes noneconomic factors can take precedence.

5. Stop and review! You have completed the outline for this subunit. Study multiple-choice questions 1 through 5 beginning on page 124.

B. Present Value and Future Value. The basic **compound interest** concept is that a quantity of money sometime in the future is worth less than the same amount of money today. The difference is measured in terms of interest calculated according to the appropriate discount rate. Interest is the payment received by holders of money to forgo current consumption. Conversely, the current consumer of money pays interest for its use.

1. Standard tables have been developed to facilitate the calculation of present and future values. Each entry in one of these tables represents the result of substituting in the pertinent present value or future value equation a payment of $1, the number of periods (n), and an interest rate (i).

2. The **present value (PV)** of an amount is the value today of some future payment.

   a. It equals the present value of $1 (a factor found in a standard table) for the given number of periods and interest rate times the future payment.

   b. $$PV\ of\ \$1 = \frac{1}{(1+i)^n}$$

3. The **future value (FV)** of an amount is the amount available at a specified time in the future based on a single investment (deposit) today. The FV is the amount to be computed if one knows the present value and the appropriate discount rate.

   a. It equals the future value of $1 (a factor found in a standard table) for the given number of periods and interest rate times the current payment.

   b. $$FV\ of\ \$1 = (1+i)^n$$

4. **Annuities**. An annuity is usually a series of equal payments at equal intervals of time, e.g., $1,000 at the end of every year for 10 years.

   a. An **ordinary annuity (annuity in arrears)** is a series of payments occurring at the end of the periods involved. An **annuity due (annuity in advance)** is an annuity in which the payments are made (or received) at the beginning of the periods.

      1) The difference for PV calculations is that no interest is computed on the first payment of an annuity due.

   b. The **PV of an annuity**. The same present value tables may be used for both kinds of annuities. Most tables are for ordinary annuities. The factor for an ordinary annuity of one less period (n – 1), increased by 1.0 to include the initial payment (for which no interest is computed), is the factor for an annuity due.

      1) $$PV\ of\ an\ ordinary\ annuity\ of\ \$1 = \frac{1 - [1 \div (1+i)^n]}{i}$$

c. The **FV of an annuity** is the value that a series of equal payments will have at a certain moment in the future if interest is earned at a given rate.

1) In an ordinary annuity, the payments are made (or received) at the end of the periods, and the future value is determined at the date of the last payment. Hence, no interest is earned during the first period or on the last payment.

2) In an annuity due, the payments occur at the beginning of the periods. The difference for FV calculations is that interest is computed on the first payment of an annuity due, and the future value is determined one period after the last payment. Hence, the same future value tables may be used for both kinds of annuities. The factor for the FV of an ordinary annuity of one more period (n + 1), decreased by 1.0 to exclude the last payment (for which no interest is computed), is the factor for an annuity due.

3) $$\textit{FV of an ordinary annuity of \$1} = \frac{(1 + i)^n}{i} - 1$$

5. CMA/CFM candidates should note that this outline section is short but very important. Work through all questions carefully and be sure to fully understand this fundamental topic.

6. Stop and review! You have completed the outline for this subunit. Study multiple-choice questions 6 through 13 beginning on page 125.

C. Capital Budgeting Methods

1. Techniques for Ranking Investment Proposals

a. **Payback** is the number of years required to complete the return of the original investment, i.e., the time it takes for a new asset to pay for itself. This measure is computed by dividing the initial net investment by the periodic constant expected after-tax cash flow to be generated.

1) $$\frac{\textit{Initial net investment}}{\textit{Periodic constant expected cash flow}}$$

2) If periodic cash flows are not constant, the calculation must be in cumulative form.

3) Payback is easy to calculate, but it ignores cash flows received after the payback period, and it does not account for the time value of money.

4) Its strength is its simplicity.

5) It is sometimes used for foreign investments if foreign expropriation of company assets is feared.

6) Even in these circumstances, the payback method is most often used in addition to another, more sophisticated, method.

7) To some extent, it does measure risk. The longer the payback period, the more risky the investment.

8) The payback reciprocal (1 ÷ payback) is sometimes used as an estimate of the internal rate of return.

9) The **bailout payback** incorporates the salvage value of the asset into the calculation. It measures the length of the payback period when the periodic cash inflows are combined with the salvage value.

10) **Breakeven time** is a more sophisticated version of the payback method that is growing in popularity. Breakeven time is defined as the period required for the discounted cumulative cash inflows on a project to equal the discounted cumulative cash outflows (usually the initial cost). Thus, it is the time necessary for the present value of the discounted cash flows to equal zero. This period begins at the outset of a project, not when the initial cash outflow occurs.

a) An alternative that results in a longer breakeven time is to consider the time required for the present value of the cumulative cash inflows to equal the present value of all the expected future cash outflows.

b. **Net present value (NPV)** is broadly defined as the difference between the present value of the estimated net cash inflows and the present value of the net cash outflows.

1) The NPV method is used when the discount rate (the cost of capital, required rate of return, hurdle rate) is specified.

2) $$NPV = \sum_{t=0}^{n} \frac{PCF_t}{(1 + k)^t}$$

If: n = number of periods of future cash flows
PCF_t = periodic cash flow for period t
k = discount rate
t = time period

3) The reinvestment rate often becomes critical when choosing between the NPV and IRR methods. NPV assumes the cash flows from the investment can be reinvested at the particular project's cost of capital.

4) Present value tables are used to reduce the future cash flows to current dollars.

5) If the NPV is positive, the project should be accepted. If NPV is negative, the project should be rejected.

6) NPV is the best method for **capital rationing** situations. Capital rationing exists when a firm sets a limit on the amount of funds to be invested during a given period. Under capital rationing, a firm will not undertake all projects that are viewed as profitable. Only those projects that will return the greatest NPV for the limited capital available can be undertaken.

c. **Internal rate of return (IRR)** is an interest rate (r) computed such that the net present value (NPV) of the investment is zero. Hence, the present value of the expected cash outflows equals the present value of the expected cash inflows.

1) $$\sum_{t=0}^{n} \frac{PCF_t}{(1 + IRR)^t} = 0$$

If: n = number of periods of future cash flows
PCF_t = periodic cash flow for period t
IRR = internal rate of return
t = time period

2) The IRR can be found by trial and error using arbitrarily selected interest rates. One may go from column to column on a present value table until an NPV of zero is obtained.

   a) As long as r is greater than k, NPV must be greater than zero.
   b) The IRR and NPV methods may rank projects differently if

      i) The cost of one project is greater than the cost of another.
      ii) The timing of cash flows differs among projects.

3) The IRR method assumes that the cash flows will be reinvested at the internal rate of return (r).

   a) If the project's funds are not reinvested at the IRR, the ranking calculations obtained may be in error.

   b) The NPV method gives a better grasp of the problem in many decision situations because the reinvestment is assumed to be the cost of capital (k).

4) Under some circumstances, the IRR is not totally reliable because it can give multiple answers for the same set of facts. This anomaly occurs when more than one change in the direction of net periodic cash flows occurs. For example, one direction change occurs when an initial net cash outflow is followed by a series of net inflows. However, if a negative net cash flow occurs in a later period, another direction change and a second solution result. The number of solutions to the IRR formula equals the number of changes in the direction of the net cash flows.

   a) The reason multiple solutions may arise is that the IRR equation has n roots (solutions). All of these roots except one are imaginary numbers when one change in the direction of cash flows occurs. If more than one direction change occurs, the number of real roots increases on a one-to-one basis.

5) An improved version of the IRR is the modified IRR (MIRR). It assumes that reinvestment is at the cost of capital (k) rather than the IRR. The formula for the MIRR equates the present value of the investment's cash outflows discounted at k with the present value of the terminal value. The terminal value is the future value of the cash inflows assuming reinvestment at k. The MIRR is the rate at which the present value of the terminal value equals the present value of the cash outflows. If $PCOF_t$ equals the periodic cash outflow (a negative number) for time t, and $PCIF_t$ equals the periodic cash inflow for time t, the formula is

$$\sum_{t=0}^{n} \frac{PCOF_t}{(1+k)^t} = \frac{\sum_{t=0}^{n} PCIF_t(1+k)^{n-t}}{(1+MIRR)^n}$$

6) For example, an investment in a nuclear reactor requires an initial outflow of funds at the time of acquisition. This outflow is followed by years of inflows. Then, at either periodic intervals or at the end of the asset's life, substantial outflows will be necessary for cleanup costs (perhaps greater than the initial purchase cost). This change in direction of cash flows causes the IRR multiple-solution problem.

d. The **profitability or excess present value index** is the ratio of the present value of the future net cash inflows to the present value of the initial net investment. These amounts are the same ones used in the calculation of the NPV.

1) $$\frac{\textit{PV of future net cash inflows}}{\textit{PV of initial net investment}}$$

2) This variation of the net present value method facilitates comparison of different-sized investments.

e. The **accounting rate of return** (also called unadjusted rate of return or book value rate of return) is the increase in accounting net income divided by the required investment.

1) Sometimes the denominator is the average investment rather than the initial investment.

2) The accounting rate of return ignores the effects of the time value of money and the level of cash flows.

2. EXAMPLE: Hazman Company plans to replace an old piece of equipment that is obsolete and expected to be unreliable under the stress of daily operations. The equipment is fully depreciated, and no salvage value can be realized upon its disposal. One piece of equipment being considered as a replacement would provide an annual cash savings of $7,000 before income taxes and without regard to the effect of depreciation. The equipment would cost $18,000 and has an estimated useful life of 5 years. No salvage value would be used for depreciation purposes because the equipment is expected to have no value at the end of 5 years.

Hazman uses the straight-line depreciation method on all equipment for both book and tax purposes. Hence, annual depreciation is $3,600. The company is subject to a 40% tax rate. Hazman has an after-tax cost of capital of 14%, so it would use the 14% column from a present value table.

Analysis of cash flows

		Annual Before-Tax Cash Flow	Annual Tax Savings (Tax)	Annual After-Tax Cash Flow	Annual After-Tax Net Income
Investment	Year 0	$(18,000)	-0-	$(18,000)	-0-
Annual cash savings	Years 1-5	$ 7,000	$(2,800)	$ 4,200	$ 4,200
Depreciation tax shield	Years 1-5		1,440	1,440	(2,160)
Totals				$ 5,640	$ 2,040

a. Payback period $= \dfrac{\textit{Investment}}{\textit{After-tax cash flow}}$

$= \dfrac{\$18{,}000}{\$5{,}640}$

$= 3.19$ *years*

b. Net present value = *(After-tax cash flows × Present value of annuity) - Initial investment*

= ($5,640 × 3.43) - $18,000

= $1,345.20

c. Internal rate of return

Net present value at 16%	$5,640 × 3.27	=	$18,443
Net present value at 18%	$5,640 × 3.13	=	17,653
Difference			$ 790

Net present value at 16%	$18,443
Initial investment	18,000
Difference	$ 443

Estimated increment ($443 ÷ $790) × 2%	=	1.1%
Rate used		16.0
Internal rate of return		17.1%

d. Profitability index (PI) $= \dfrac{\textit{Present value of after-tax cash flows}}{\textit{Initial investment}}$

$= \dfrac{\$5{,}640 \times 3.43}{\$18{,}000}$

$= 1.07$

e. Accounting rate of return $= \dfrac{\textit{Annual after-tax net income}}{\textit{Investment (initial or average)}}$

Initial investment	Average investment
$\dfrac{\$2{,}040}{\$18{,}000} = 11.3\%$	$\dfrac{\$2{,}040}{\$9{,}000} = 22.7\%$

3. EXAMPLE: The management of Flesher Farms is trying to decide whether to buy a new team of mules at a cost of $1,000 or a new tractor at a cost of $10,000. They will perform the same job. But because the mules require more laborers, the annual return is only $250 of net cash inflows. The tractor will return $2,000 of net cash inflows per year. The mules have a working life of 8 years and the tractor 10 years. Neither investment is expected to have a salvage value at the end of its useful life.

   a. Payback period

      Mules: $\frac{\$1{,}000}{\$250} = 4 \textit{ years}$

      Tractor: $\frac{\$10{,}000}{\$2{,}000} = 5 \textit{ years}$

   b. Accounting rate of return

      Mules: $\frac{\$250 - \$125 \textit{ depreciation}}{\$1{,}000} = 12.5\%$

      Tractor: $\frac{\$2{,}000 - \$1{,}000 \textit{ depreciation}}{\$10{,}000} = 10\%$

   c. Net present value assuming that the company's cost of capital is 6%

Mules:	$250 × 6.209 =	$1,552
		(1,000)
	Net present value	$ 552
Tractor:	$2,000 × 7.360 =	$14,720
		(10,000)
	Net present value	$ 4,720

   d. Profitability index

      Mules: $\frac{\$1{,}552}{\$1{,}000} = 1.552$

      Tractor: $\frac{\$14{,}720}{\$10{,}000} = 1.472$

   e. Internal rate of return

      Mules: $250 × *Factor* = $1,000
      Factor = 4

      On the 8-year line, the factor of 4 results in a rate of return of approximately 18.72%.

      Tractor: $2,000 × *Factor* = $10,000
      Factor = 5

      On the 10-year line, the factor of 5 results in a rate of return of approximately 15.2%.

f. The mule investment has the better payback and the higher IRR and accounting rate of return. Also, the mules have the better profitability index. The tractor has a better net present value, however. The various methods thus give different answers to the investment question. Either investment would be profitable for the company. Management may decide to let noneconomic factors influence the decision. For example, the mules would require the use of more laborers. If unemployment in the community is high, management might wish to achieve a social goal of providing more jobs. Alternatively, a labor shortage might convince management to buy the tractor to reduce labor worries.

4. NPV and IRR are the soundest investment rules from a shareholder wealth maximization perspective.

a. In some cases, NPV and IRR will rank projects differently.

1) EXAMPLE:

Project	Initial Cost	Year-End Cash Flow	IRR	NPV (k=10%)
A	$1,000	$1,200	20%	$91
B	$ 50	$ 100	100%	$41

IRR preference ordering: B,A
NPV preference ordering: A,B

b. If one of two or more mutually exclusive projects is accepted, the others must be rejected.

1) EXAMPLE: The decision to build a shopping mall on a piece of land eliminates placing an office building on the same land.

2) When choosing between mutually exclusive projects, the ranking differences between NPV and IRR become very important. In the example 4.a.1), a firm using IRR would accept B and reject A. A firm using NPV would make exactly the opposite choice.

c. The problem can be seen more clearly using a net present value (NPV) profile. The NPV profile is a plot of a project's NPV at different discount rates. The NPV is plotted on the vertical axis and the rate of return (k) on the horizontal axis.

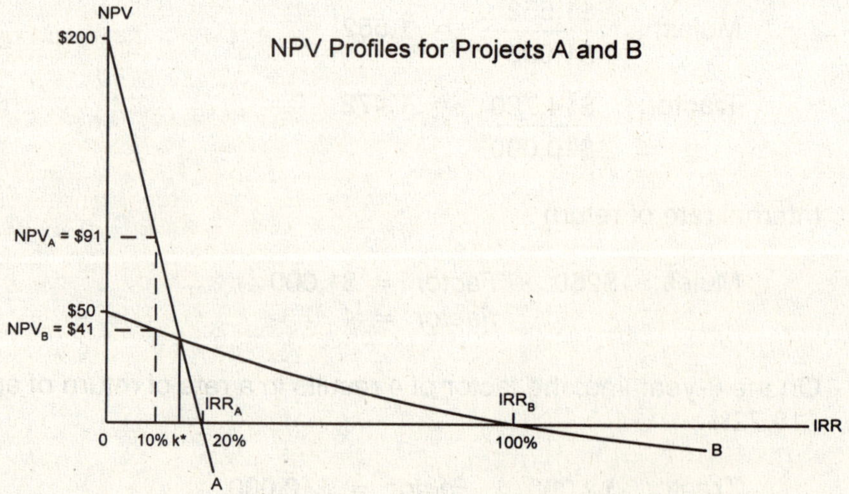

1) These profiles are downward sloping because a higher discount rate implies a lower NPV. The graph shows that, for all discount rates higher than k*, the firm should select project B over A because NPV_B is greater than NPV_A. This preference ordering also results from applying the IRR criterion. Below k*, however, NPV_A is greater than NPV_B, so A should be selected, even though IRR_B is greater than IRR_A.

2) These profiles show that IRR will always prefer B to A. NPV will prefer B to A only past some critical discount rate k*.

d. The manager concerned with shareholder wealth maximization should choose the project with the greatest NPV, not the largest IRR. IRR is a percentage measure of wealth, but NPV is an absolute measure. Shareholder well-being is also measured in absolute amounts.

1) The choice of NPV over IRR is easy to see with a simple example. Assume a choice between investing $1 and receiving $2 or investing $100,000 and receiving $150,000. The IRRs of the projects are 100% and 50%, respectively, which supports the first project. But assume instead that the interest rate is 10%. The NPVs of the projects are $.81 and $36,363, respectively. To select the first project because of the IRR criterion would lead to a return of $.81 instead of $36,363. Thus, the NPV is the better criterion when choosing between mutually exclusive projects.

e. The NPV profile can be of great practical use to managers trying to make investment decisions. It gives the manager a clear insight into the following questions:

1) At what interest rates is an investment project still a profitable opportunity?
2) How sensitive is a project's profitability to changes in the discount rate?

5. **Post-investment audits** should be conducted to serve as a control mechanism and to deter managers from proposing unprofitable investments.

a. Actual-to-expected cash flow comparisons should be made and unfavorable variances should be explained. The reason may be an inaccurate forecast or implementation problems.

b. Individuals who supplied unrealistic estimates should have to explain differences. Knowing that a post-investment audit will be conducted may cause managers to provide more realistic forecasts in the future.

6. Stop and review! You have completed the outline for this subunit. Study multiple-choice questions 14 through 85 beginning on page 128.

D. Risk

1. **Risk** is the possibility of an unfavorable event. **Investment risk** is therefore analyzed in terms of the probability that the actual return on an investment will be lower than the expected return. Accordingly, the concepts of probability distributions and expected value are basic to risk management. They are discussed in Study Unit 4.

a. The risk of an investment may be considered in isolation. It may also be viewed from the perspective of its inclusion in a portfolio of assets chosen to minimize the riskiness of the whole. For a discussion of portfolio management, see Part 2CFM, Study Unit 8.

2. **Specific Types of Risks**

   a. **Interest-rate risk** is the risk of fluctuations in the value of an asset due to changes in interest rates.

      1) In general, interest-rate risk is greater the longer the maturity of the asset.

      2) One component of interest-rate risk is **price risk**. Thus, the value of bonds declines when interest rates increase.

      3) A second component of interest-rate risk is **reinvestment-rate risk**. If interest rates decline, lower returns will be available for reinvestment of interest and principal payments received.

   b. **Purchasing-power risk** is the risk that a general rise in the price level will reduce what can be purchased with a fixed sum of money. Accordingly, required returns include an inflation premium.

   c. **Default risk** is the risk that a borrower will be unable to repay debt. Hence, the higher the default risk, the higher the return required by an investor.

   d. **Market risk** is the risk that changes in price will result from changes that affect all firms. Prices of all securities, even the values of portfolios, are correlated to some degree with broad swings in the economy caused by recession, inflation, high interest rates, etc. Market risk is also known as systematic risk or nondiversifiable risk.

   e. **Nonmarket risk** or **company-specific risk** is the risk that is influenced by an individual firm's policies and decisions. Nonmarket risk is diversifiable because it is specific to each firm. Thus, it is also known as diversifiable or unsystematic risk.

   f. **Portfolio risk** is the risk remaining after allowing for the risk-reducing effects of combining securities into a portfolio.

   g. **Total risk** is the risk of a single asset, whereas market risk is its risk if it is held in a large portfolio of diversified securities.

   h. **Liquidity risk** is the possibility that an asset cannot be sold on short notice for its market value. If an asset must be sold at a deep discount, it is said to have a substantial amount of liquidity risk.

   i. **Business risk** is the risk of fluctuations in earnings before interest and taxes or in operating income when the firm uses no debt. It is the risk inherent in its operations that excludes **financial risk**, which is the risk to the shareholders from the use of financial leverage. Business risk depends on factors such as

      1) Demand variability
      2) Sales price variability
      3) Input price variability
      4) Amount of operating leverage

   j. **Exchange-rate risk** is the risk that a foreign currency transaction will be negatively exposed to fluctuations in exchange rates.

3. **Risk analysis** attempts to measure the likelihood of the variability of future returns from the proposed investment. Risk cannot be ignored entirely, but mathematical approaches can be impossible because of a lack of critical information. The following approaches are frequently used to assess risk:
   a. **Informal method.** NPVs are calculated at the firm's k, and the possible projects are individually reviewed. If the NPVs are relatively close for two mutually exclusive projects, the apparently less risky project is chosen.
   b. **Risk-adjusted discount rates.** This technique adjusts k upward as the investment becomes riskier. By increasing the discount rate from 10% to 15%, for example, the expected flow from the investment must be relatively larger or the increased discount rate will generate a negative NPV, and the proposed acquisition/investment would be rejected. Although difficult to apply in extreme cases, this technique has much intuitive value.
   c. **Certainty equivalent adjustments.** This technique is directly drawn from the concept of utility theory. It forces the decision maker to specify at what point the firm is indifferent to the choice between a certain sum of money and the expected value of a risky sum. The technique is not frequently used because decision makers are not familiar with the concept.
   d. **Sensitivity analysis.** Forecasts of many calculated NPVs under various assumptions are compared to see how sensitive NPV is to changing conditions. Changing or relaxing the assumptions about a certain variable or group of variables may drastically alter the NPV. Thus, the asset may appear to be much riskier than was originally predicted.
   e. **Simulation analysis.** This method represents a refinement of standard profitability theory. The computer is used to generate many examples of results based upon various assumptions. Project simulation is frequently expensive. Unless a project is exceptionally large and expensive, full-scale simulation is usually not worthwhile.
   f. **The Capital Asset Pricing Model.** This method is derived from the use of portfolio theory. It assumes that all assets are held in a portfolio. Each asset has variability in its returns. Some of this variability is caused by movements in the market as a whole, and some is specific to each firm. In a portfolio, the firm's specific variability is eliminated through diversification, and the only relevant risk is the market component. The more sensitive an asset's rate of return is to changes in the market's rate of return, the riskier the asset.
   g. See also the section on Forecasting in Study Unit 6 in this book.

4. Stop and review! You have completed the outline for this subunit. Study multiple-choice questions 86 through 88 beginning on page 154.

MULTIPLE-CHOICE QUESTIONS

A. Definitions

1. The relevance of a particular cost to a decision is determined by

A. Riskiness of the decision.

B. Number of decision variables.

C. Amount of the cost.

D. Potential effect on the decision.

The correct answer is (D). *(CMA, adapted)*

REQUIRED: The determinant of relevance of a particular cost to a decision.

DISCUSSION: Relevance is the capacity of information to make a difference in a decision by helping users of that information to predict the outcomes of events or to confirm or correct prior expectations. Thus, relevant costs are those expected future costs that vary with the action taken. All other costs are constant and therefore have no effect on the decision.

Answers (A), (B), and (C) are incorrect because the ultimate determinant of relevance is the ability to influence the decision.

2. In equipment-replacement decisions, which one of the following does not affect the decision-making process?

A. Current disposal price of the old equipment.

B. Operating costs of the old equipment.

C. Original fair market value of the old equipment.

D. Cost of the new equipment.

The correct answer is (C). *(CMA, adapted)*

REQUIRED: The irrelevant factor when making an equipment replacement decision.

DISCUSSION: All relevant costs should be considered when evaluating an equipment replacement decision. These include the cost of the new equipment, the disposal price of the old equipment, and the operating costs of the old equipment versus the operating costs of the new equipment. The original cost or fair market value of the old equipment is a sunk cost and is irrelevant to future decisions.

Answers (A), (B), and (D) are incorrect because they should be considered when evaluating an equipment-replacement decision.

3. The term that refers to costs incurred in the past that are not relevant to a future decision is

A. Discretionary cost.

B. Full absorption cost.

C. Underallocated indirect cost.

D. Sunk cost.

The correct answer is (D). *(CMA, adapted)*

REQUIRED: The past costs not relevant to a future decision.

DISCUSSION: A sunk cost cannot be avoided because it represents an expenditure that has already been made or an irrevocable decision to incur the cost.

Answer (A) is incorrect because a discretionary cost is characterized by uncertainty about the input-output relationship; advertising and research are examples. Answer (B) is incorrect because full absorption costing includes in production costs materials, labor, and both fixed and variable overhead. Answer (C) is incorrect because underallocated indirect cost is a cost that has not yet been charged to production.

4. Which one of the following statements concerning cash flow determination for capital budgeting purposes is not correct?

A. Tax depreciation must be considered because it affects cash payments for taxes.

B. Book depreciation is relevant because it affects net income.

C. Sunk costs are not incremental flows and should not be included.

D. Net working capital changes should be included in cash flow forecasts.

The correct answer is (B). *(CMA, adapted)*

REQUIRED: The false statement about cash flow determination.

DISCUSSION: Tax depreciation is relevant to cash flow analysis because it affects the amount of income taxes that must be paid. However, book depreciation is not relevant because it does not affect the amount of cash generated by an investment.

Answer (A), (C), and (D) are incorrect because they are all true statements relating to capital budgeting.

5. A depreciation tax shield is

A. An after-tax cash outflow.

B. A reduction in income taxes.

C. The cash provided by recording depreciation.

D. The expense caused by depreciation.

The correct answer is (B). *(CMA, adapted)*

REQUIRED: The true statement about a depreciation tax shield.

DISCUSSION: A tax shield is something that will protect income against taxation. Thus, a depreciation tax shield is a reduction in income taxes due to a company's being allowed to deduct depreciation against otherwise taxable income.

Answer (A) is incorrect because a tax shield is not a cash flow but a means of reducing outflows for income taxes. Answer (C) is incorrect because cash is not provided by recording depreciation; the shield is a result of deducting depreciation from taxable revenues. Answer (D) is incorrect because depreciation is recognized as an expense even if it has no tax benefit.

B. Present Value and Future Value

6. Essex Corporation is evaluating a lease that takes effect on March 1. The company must make eight equal payments, with the first payment due on March 1. The concept most relevant to the evaluation of the lease is

A. The present value of an annuity due.

B. The present value of an ordinary annuity.

C. The future value of an annuity due.

D. The future value of an ordinary annuity.

The correct answer is (A). *(CMA, adapted)*

REQUIRED: The concept most relevant to evaluating a long-term lease given that the first payment is due at the inception of the lease.

DISCUSSION: An annuity is a series of cash flows or other economic benefits occurring at fixed intervals, ordinarily as a result of an investment. Present value is the value at a specified time of an amount or amounts to be paid or received later, discounted at some interest rate. In an annuity due, the payments occur at the beginning, rather than at the end, of the periods. Thus, the present value of an annuity due includes the initial payment at its undiscounted amount. Evaluation of an investment decision, e.g., a lease, that involves multiperiod cash payments (an annuity) requires an adjustment for the time value of money. Because of the interest factor, a dollar today is worth more than a dollar in the future. Consequently, comparison of different investments requires restating their future benefits and costs in terms of present values. This lease should therefore be evaluated using the present value of an annuity due.

Answer (B) is incorrect because an ordinary annuity assumes that each payment occurs at the end of a period. Answers (C) and (D) are incorrect because future value is the converse of a present value. It is an amount accumulated in the future. Evaluation of a lease, however, necessitates calculation of a present value.

7. Janet Taylor Casual Wear has $75,000 in a bank account as of December 31, 2001. If the company plans on depositing $4,000 in the account at the end of each of the next 3 years (2002, 2003, and 2004) and all amounts in the account earn 8% per year, what will the account balance be at December 31, 2004? Ignore the effect of income taxes.

	8% Interest Rate Factors	
Period	Future Value of an Amount of $1	Future Value of an Ordinary Annuity of $1
1	1.08	1.00
2	1.17	2.08
3	1.26	3.25
4	1.36	4.51

A. $87,000

B. $88,000

C. $96,070

D. $107,500

The correct answer is (D). *(CMA, adapted)*

REQUIRED: The future value of an investment.

DISCUSSION: Both future value tables will be used because the $75,000 already in the account will be multiplied times the future value factor of 1.26 to determine the amount 3 years hence, or $94,500. The three payments of $4,000 represent an ordinary annuity. Multiplying the three-period annuity factor (3.25) by the payment amount ($4,000) results in a future value of the annuity of $13,000. Adding the two elements together produces a total account balance of $107,500 ($94,500 + $13,000).

Answer (A) is incorrect because $87,000 is the amount that would be in the account if interest were zero. Answer (B) is incorrect because the $88,000 ignores the interest that would be earned on the $75,000 initial balance. Answer (C) is incorrect because $96,070 is the amount that would be available after 2 years (December 31, 2003).

Questions 8 through 11 are based on the following information. The following data pertain to a 4-year project being considered by Metro Industries:

- A depreciable asset that costs $1,200,000 will be acquired on January 1. The asset, which is expected to have a $200,000 salvage value at the end of 4 years, qualifies as 3-year property under the Modified Accelerated Cost Recovery System (MACRS).
- The new asset will replace an existing asset that has a tax basis of $150,000 and can be sold on the same January 1 for $180,000.
- The project is expected to provide added annual sales of 30,000 units at $20. Additional cash operating costs are variable, $12 per unit; fixed, $90,000 per year.
- A $50,000 working capital investment that is fully recoverable at the end of the fourth year is required.

Metro is subject to a 40% income tax rate and rounds all computations to the nearest dollar. Assume that any gain or loss affects the taxes paid at the end of the year in which it occurred. The company uses the net present value method to analyze investments and will employ the following factors and rates:

Period	Present Value of $1 at 12%	Present Value of $1 Annuity at 12%	MACRS
1	0.89	0.89	33%
2	0.80	1.69	45
3	0.71	2.40	15
4	0.64	3.04	7

8. The discounted cash flow for the fourth year MACRS depreciation on the new asset is

A. $0

B. $17,920

C. $21,504

D. $26,880

The correct answer is (C). *(CMA, adapted)*

REQUIRED: The discounted cash flow for the fourth year MACRS depreciation deduction on the new asset.

DISCUSSION: Tax law allows taxpayers to ignore salvage value when calculating depreciation under MACRS. Thus, the depreciation deduction is 7% of the initial $1,200,000 cost, or $84,000. At a 40% tax rate, the deduction will save the company $33,600 in taxes in the fourth year. The present value of this savings is $21,504 ($33,600 x 0.64 present value of $1 at 12% for four periods).

Answer (A) is incorrect because a tax savings will result in the fourth year from the MACRS deduction. Answer (B) is incorrect because $17,920 is based on a depreciation calculation in which salvage value is subtracted from the initial cost. Answer (D) is incorrect because the appropriate discount factor for the fourth period is 0.64, not 0.80.

9. The discounted, net-of-tax amount that relates to disposal of the existing asset is

A. $168,000

B. $169,320

C. $180,000

D. $190,680

The correct answer is (B). *(CMA, adapted)*

REQUIRED: The discounted, net-of-tax amount relating to the disposal of the existing asset.

DISCUSSION: The cash inflow from the existing asset is $180,000, but that amount is subject to tax on the $30,000 gain ($180,000 – $150,000 tax basis). The tax on the gain is $12,000 (40% x $30,000). Because the tax will not be paid until year-end, the discounted value is $10,680 ($12,000 x .89 PV of $1 at 12% for one period). Thus, the net-of-tax inflow is $169,320 ($180,000 – $10,680). NOTE: This asset was probably a Section 1231 asset, and any gain on sale qualifies for the special capital gain tax rates. Had the problem not stipulated a 40% tax rate, the capital gains rate would be used. An answer based on that rate is not among the options.

Answer (A) is incorrect because $168,000 fails to discount the outflow for taxes. Answer (C) is incorrect because $180,000 ignores the impact of income taxes. Answer (D) is incorrect because the discounted present value of the income taxes is an outflow and is deducted from the inflow from the sale of the asset.

10. The expected incremental sales will provide a discounted, net-of-tax contribution margin over 4 years of

A. $57,600

B. $92,160

C. $273,600

D. $437,760

The correct answer is (D). *(CMA, adapted)*

REQUIRED: The expected net-of-tax contribution margin over 4 years.

DISCUSSION: Additional annual sales are 30,000 units at $20 per unit. If variable costs are expected to be $12 per unit, the unit contribution margin is $8, and the total before-tax annual contribution margin is $240,000 ($8 x 30,000 units). The after-tax total annual contribution margin is $144,000 [(1.0 – .4) x $240,000]. This annual increase in the contribution margin should be treated as an annuity. Thus, its present value is $437,760 ($144,000 x 3.04 PV of an annuity of $1 at 12% for four periods).

Answers (A) and (B) are incorrect because $57,600 and $92,160 are based on only 1 year's results, not 4. Answer (C) is incorrect because $273,600 improperly includes fixed costs in the calculation of the contribution margin.

11. The overall discounted-cash-flow impact of the working capital investment on Metro's project is

A. $(2,800)

B. $(18,000)

C. $(50,000)

D. $(59,200)

The correct answer is (B). *(CMA, adapted)*

REQUIRED: The overall discounted-cash-flow impact of the working capital investment.

DISCUSSION: The working capital investment is treated as a $50,000 outflow at the beginning of the project and a $50,000 inflow at the end of 4 years. Accordingly, the present value of the inflow after 4 years should be subtracted from the initial $50,000 outlay. The overall discounted-cash-flow impact of the working capital investment is $18,000 [$50,000 – ($50,000 x .64 PV of $1 at 12% for four periods)].

Answer (A) is incorrect because the firm will have its working capital tied up for 4 years. Answer (C) is incorrect because the working capital investment is recovered at the end of the fourth year. Hence, the working capital cost of the project is the difference between $50,000 and the present value of $50,000 in 4 years. Answer (D) is incorrect because the answer cannot exceed $50,000, which is the amount of the cash outflow.

12. Kline Corporation is expanding its plant, which requires an investment of $8 million in new equipment. Kline's sales are expected to increase by $6 million per year as a result of the expansion. Cash investment in current assets averages 30% of sales, and accounts payable and other current liabilities are 10% of sales. What is the estimated total cash investment for this expansion?

A. $6.8 million.

B. $8.6 million.

C. $9.2 million.

D. $9.8 million.

The correct answer is (C). *(Publisher)*

REQUIRED: The estimated total cash investment for the expansion.

DISCUSSION: For capital budgeting purposes, the net investment is the net outlay or cash requirement. This amount includes the cost of the new equipment, minus any cash recovered from the trade or sale of existing assets. The investment required also includes funds to provide for increases in working capital, for example, the additional receivables and inventories resulting from the acquisition of a new manufacturing plant. The investment in working capital is treated as an initial cost of the investment, although it will be recovered at the end of the project (its salvage value equals its initial cost). For Kline, the additional current assets will be 30% of sales, but current liabilities can be used to fund assets to the extent of 10% of sales. Thus, the initial investment in working capital will equal 20% of the $6 million in sales, or $1,200,000. The total initial cash outlay will consist of the $8 million in new equipment plus $1,200,000 in working capital, a total of $9.2 million.

Answer (A) is incorrect because $6.8 million subtracted the net investment in working capital from the cost of the equipment. Answer (B) is incorrect because $8.6 million assumes current assets will increase by 10% of new sales but that current liabilities will not change. Answer (D) is incorrect because $9.8 million ignores the financing of incremental current assets with accounts payable.

13. Use the following 8% interest rate factors for this question.

Period	Future Value of $1	Future Value of Annuity of $1
1	1.08	1.00
2	1.17	2.08
3	1.26	3.25
4	1.36	4.51

The Suellen Company has $150,000 in a bank account as of December 31, 2001. If the company plans to deposit $8,000 in the account at the end of each of the next 3 years (2002, 2003, and 2004), and all amounts in the account earn 8% per year, what will the account balance be at December 31, 2004? Ignore the effect of income taxes.

A. $174,000

B. $176,000

C. $192,140

D. $215,000

The correct answer is (D). *(Publisher)*

REQUIRED: The balance in a savings account at the end of a specified period if a stated amount is deposited at the end of each year.

DISCUSSION: Both future value tables must be used. The future value of the beginning balance is found by multiplying it by the interest factor for the future value of $1 for three periods at 8%. This amount is $189,000 ($150,000 x 1.26). The $8,000 payments at the end of each of the next 3 years represent an ordinary annuity. The future value of this annuity is found by multiplying the periodic payment by the interest factor for the future value of an ordinary annuity for three periods at 8%. This amount is $26,000 ($8,000 x 3.25). Consequently, the balance at December 31, 2004 will be $215,000 ($189,000 + $26,000).

Answer (A) is incorrect because $174,000 is the amount that would be in the account if interest were $0. Answer (B) is incorrect because the $176,000 ignores the interest to be earned on the initial balance. Answer (C) is incorrect because $192,140 is the amount that will be available after 2 years (December 31, 2003).

C. Capital Budgeting Methods

14. Of the following decisions, capital budgeting techniques would least likely be used in evaluating the

A. Acquisition of new aircraft by a cargo company.

B. Design and implementation of a major advertising program.

C. Trade for a star quarterback by a football team.

D. Adoption of a new method of allocating nontraceable costs to product lines.

The correct answer is (D). *(CMA, adapted)*

REQUIRED: The decision least likely to be evaluated using capital budgeting techniques.

DISCUSSION: Capital budgeting is the process of planning expenditures for investments on which the returns are expected to occur over a period of more than one year. Thus, capital budgeting concerns the acquisition or disposal of long-term assets and the financing ramifications of such decisions. The adoption of a new method of allocating nontraceable costs to product lines has no effect on a company's cash flows, does not concern the acquisition of long-term assets, and is not concerned with financing. Hence, capital budgeting is irrelevant to such a decision.

Answer (A) is incorrect because a new aircraft represents a long-term investment in a capital good. Answer (B) is incorrect because a major advertising program is a high cost investment with long-term effects. Answer (C) is incorrect because a star quarterback is a costly asset who is expected to have a substantial effect on the team's long-term profitability.

15. Suzie owns a computer reselling business and is expanding her business. Suzie is presented with one proposal, Proposal A, such that the estimated investment for the expansion project is $85,000, and it is expected to produce cash flows after taxes of $25,000 for each of the next 6 years. An alternate proposal, Proposal B, involves an investment of $32,000 and after-tax cash flows of $10,000 for each of the next 6 years. The cost of capital that would make Suzie indifferent between these two proposals lies between

A. 10% and 12%

B. 14% and 16%

C. 16% and 18%

D. 18% and 20%

The correct answer is (C). *(Publisher)*

REQUIRED: The cost of capital that would lead to indifference between the two proposals.

DISCUSSION: The cost of capital at which the two projects will produce the same NPV can be found by calculating the IRR of the difference in cash flows between the two projects. Proposal A requires an additional investment of $53,000 and generates extra cash flows of $15,000 for 6 years. The IRR for this set of cash flows exceeds 16% but is less than 18%. This further means that, for any cost of capital below this range, Proposal A will have a higher NPV, and, for any cost of capital above this range, Proposal B will have a higher NPV.

Answers (A) and (B) are incorrect because Proposal A would be superior. Answer (D) is incorrect because Proposal B would be superior. Proposal A would have a negative NPV with regard to the incremental cash flows.

16. Jasper Company has a payback goal of 3 years on new equipment acquisitions. A new sorter is being evaluated that costs $450,000 and has a 5-year life. Straight-line depreciation will be used; no salvage is anticipated. Jasper is subject to a 40% income tax rate. To meet the company's payback goal, the sorter must generate reductions in annual cash operating costs of

A. $60,000

B. $100,000

C. $150,000

D. $190,000

The correct answer is (D). *(CMA, adapted)*

REQUIRED: The cash savings that must be generated to achieve a targeted payback period.

DISCUSSION: The payback period is calculated by dividing cost by the annual cash inflows, or cash savings. To achieve a payback period of 3 years, the annual increment in net cash inflow generated by the investment must be $150,000 ($450,000 ÷ 3-year targeted payback period). This amount equals the total reduction in cash operating costs minus related taxes. Depreciation is $90,000 ($450,000 ÷ 5 years). Because depreciation is a noncash deductible expense, it shields $90,000 of the cash savings from taxation. Accordingly, $60,000 ($150,000 – $90,000) of the additional net cash inflow must come from after-tax net income. At a 40% tax rate, $60,000 of after-tax income equals $100,000 ($60,000 ÷ 60%) of pretax income from cost savings, and the outflow for taxes is $40,000. Thus, the annual reduction in cash operating costs required is $190,000 ($150,000 additional net cash inflow required + $40,000 tax outflow).

Answer (A) is incorrect because $60,000 is after-tax net income from the cost savings. Answer (B) is incorrect because $100,000 is the pretax income from the cost savings. Answer (C) is incorrect because $150,000 ignores the impact of depreciation and income taxes.

17. The length of time required to recover the initial cash outlay of a capital project is determined by using the

A. Discounted cash flow method.

B. Payback method.

C. Weighted net present value method.

D. Net present value method.

The correct answer is (B). *(CMA, adapted)*

REQUIRED: The method of determining the time required to recover the initial cash outlay of a capital project.

DISCUSSION: The payback method measures the number of years required to complete the return of the original investment. This measure is computed by dividing the net investment by the average expected cash inflows to be generated, resulting in the number of years required to recover the original investment. The payback method gives no consideration to the time value of money, and there is no consideration of returns after the payback period.

Answer (A) is incorrect because the discounted cash flow method computes a rate of return. Answers (C) and (D) are incorrect because the net present value method is based on discounted cash flows; the length of time to recover an investment is not the result.

18. A characteristic of the payback method (before taxes) is that it

A. Incorporates the time value of money.

B. Neglects total project profitability.

C. Uses accrual accounting inflows in the numerator of the calculation.

D. Uses the estimated expected life of the asset in the denominator of the calculation.

The correct answer is (B). *(CMA, adapted)*

REQUIRED: The characteristic of the payback method.

DISCUSSION: The payback method calculates the number of years required to complete the return of the original investment. This measure is computed by dividing the net investment required by the average expected cash flow to be generated, resulting in the number of years required to recover the original investment. Payback is easy to calculate but has two principal problems: it ignores the time value of money, and it gives no consideration to returns after the payback period. Thus, it ignores total project profitability.

Answer (A) is incorrect because the payback method does not incorporate the time value of money. Answer (C) is incorrect because the payback method uses the net investment in the numerator of the calculation. Answer (D) is incorrect because payback uses the net annual cash inflows in the denominator of the calculation.

Questions 19 through 21 are based on the following information. Henderson Inc. has purchased a new fleet of trucks to deliver its merchandise. The trucks have a useful life of 8 years and cost a total of $500,000. Henderson expects its net increase in after-tax cash flow to be $150,000 in Year 1, $175,000 in Year 2, $125,000 in Year 3, and $100,000 in each of the remaining years.

19. Ignoring the time value of money, how long will it take Henderson to recover the amount of investment?

A. 3.5 years.

B. 4.0 years.

C. 4.2 years.

D. 5 years.

The correct answer is (A). *(Publisher)*

REQUIRED: The payback period for an investment, ignoring the time value of money.

DISCUSSION: The payback period for an investment, ignoring the time value of money, can be found by accumulating each year's net cash flows until the initial investment is recovered. The amount accumulated after 3 years is $450,000. Thus, 50% of year 4 cash flows is needed to recover the initial investment. The payback period is 3.5 years.

Answer (B) is incorrect because 4 years includes the additional $50,000 of Year 4. Answer (C) is incorrect because 4.2 years takes the average inflow of all 8 years and divides that into the $500,000 initial investment. Answer (D) is incorrect because 5 years uses only the cash flows from the remaining 5 years.

20. What is the payback reciprocal for the fleet of trucks?

A. 29%

B. 25%

C. 24%

D. 20%

The correct answer is (A). *(Publisher)*

REQUIRED: The payback reciprocal for an investment.

DISCUSSION: The payback reciprocal for an investment is found by dividing 1 by the payback time. The payback time for this investment is 3.5 years, and the payback reciprocal is 1 divided by 3.5, or 29%.

Answer (B) is incorrect because 25% includes the additional $50,000 of Year 4 in the payback time. Answer (C) is incorrect because 24% takes the average inflow of all 8 years and divides that into the $500,000 initial investment in the payback time. Answer (D) is incorrect because 20% uses only the cash flows from the remaining 5 years in the payback time.

21. Assume the net cash flow to be $130,000 a year. What is the payback time for the fleet of trucks?

A. 3 years.

B. 3.15 years.

C. 3.85 years.

D. 4 years.

The correct answer is (C). *(Publisher)*

REQUIRED: The payback period for an investment, ignoring the time value of money.

DISCUSSION: The payback period for an investment, ignoring the time value of money, can be found by accumulating each year's net cash flows until the initial investment is recovered. Therefore, dividing the $500,000 initial investment by the annual $130,000 inflow gives a payback time of 3.85 years.

Answers (A) and (B) are incorrect because 3.85 years is the length of the payback period. Answer (D) is incorrect because 4 years does not prorate the final $130,000.

22. Which one of the following statements about the payback method of investment analysis is correct? The payback method

A. Does not consider the time value of money.

B. Considers cash flows after the payback has been reached.

C. Uses discounted cash flow techniques.

D. Generally leads to the same decision as other methods for long-term projects.

The correct answer is (A). *(CMA, adapted)*

REQUIRED: The true statement about the payback method of investment analysis.

DISCUSSION: The payback method calculates the amount of time required to complete the return of the original investment, i.e., the time it takes for a new asset to pay for itself. Although the payback method is easy to calculate, it has inherent problems. The time value of money and returns after the payback period are not considered.

Answer (B) is incorrect because the payback method ignores cash flows after payback. Answer (C) is incorrect because the payback method does not use discounted cash flow techniques. Answer (D) is incorrect because the payback method may lead to different decisions.

23. The payback reciprocal can be used to approximate a project's

A. Profitability index.

B. Net present value.

C. Accounting rate of return if the cash flow pattern is relatively stable.

D. Internal rate of return if the cash flow pattern is relatively stable.

The correct answer is (D). *(CMA, adapted)*

REQUIRED: The item that can be approximated by a project's payback reciprocal.

DISCUSSION: The payback reciprocal (1 ÷ payback) has been shown to approximate the internal rate of return (IRR) when the periodic cash flows are equal and the life of the project is at least twice the payback period.

Answer (A) is incorrect because the payback reciprocal is not related to the profitability index. Answer (B) is incorrect because the payback reciprocal approximates the IRR, which is the rate at which the NPV is $0. Answer (C) is incorrect because the accounting rate of return is based on accrual-income-based figures, not on discounted cash flows.

24. The bailout payback method

A. Incorporates the time value of money.

B. Equals the recovery period from normal operations.

C. Eliminates the disposal value from the payback calculation.

D. Measures the risk if a project is terminated.

The correct answer is (D). *(CMA, adapted)*

REQUIRED: The true statement about the bailout payback method.

DISCUSSION: The payback period equals the net investment divided by the average expected cash flow, resulting in the number of years required to recover the original investment. The bailout payback incorporates the salvage value of the asset into the calculation. It determines the length of the payback period when the periodic cash inflows are combined with the salvage value. Hence, the method measures risk. The longer the payback period, the more risky the investment.

Answer (A) is incorrect because the bailout payback method does not consider the time value of money. Answer (B) is incorrect because the bailout payback includes salvage value as well as cash flow from operations. Answer (C) is incorrect because the bailout payback incorporates the disposal value in the payback calculation.

25. Whatney Co. is considering the acquisition of a new, more efficient press. The cost of the press is $360,000, and the press has an estimated 6-year life with zero salvage value. Whatney uses straight-line depreciation for both financial reporting and income tax reporting purposes and has a 40% corporate income tax rate. In evaluating equipment acquisitions of this type, Whatney uses a goal of a 4-year payback period. To meet Whatney's desired payback period, the press must produce a minimum annual before-tax operating cash savings of

A. $90,000

B. $110,000

C. $114,000

D. $150,000

The correct answer is (B). *(CMA, adapted)*

REQUIRED: The minimum annual before-tax operating cash savings yielding a specified payback period.

DISCUSSION: Payback is the number of years required to complete the return of the original investment. Given a periodic constant cash flow, the payback period equals net investment divided by the constant expected periodic after-tax cash flow. The desired payback period is 4 years, so the constant after-tax annual cash flow must be $90,000 ($360,000 ÷ 4). Assuming that the company has sufficient other income to permit realization of the full tax savings, depreciation of the machine will shield $60,000 ($360,000 ÷ 6) of income from taxation each year, an after-tax cash savings of $24,000 (40% x $60,000). Thus, the machine must generate an additional $66,000 ($90,000 – $24,000) of after-tax cash savings from operations. This amount is equivalent to $110,000 [$66,000 ÷ (1.0 – .4)] of before-tax operating cash savings.

Answer (A) is incorrect because $90,000 is the total desired annual after-tax cash savings. Answer (C) is incorrect because $114,000 results from adding, not subtracting, the $24,000 of tax depreciation savings to determine the minimum annual after-tax operating savings. Answer (D) is incorrect because $150,000 assumes that depreciation is not tax deductible.

26. When evaluating projects, breakeven time is best described as

A. Annual fixed costs ÷ monthly contribution margin.

B. Project investment ÷ annual net cash inflows.

C. The point where cumulative cash inflows on a project equal total cash outflows.

D. The point at which discounted cumulative cash inflows on a project equal discounted total cash outflows.

The correct answer is (D). *(CMA, adapted)*

REQUIRED: The definition of breakeven time.

DISCUSSION: Breakeven time is a capital budgeting tool that is widely used to evaluate the rapidity of new product development. It is the period required for the discounted cumulative cash inflows for a project to equal the discounted cumulative cash outflows. The concept is similar to the payback period, but it is more sophisticated because it incorporates the time value of money. It also differs from the payback method because the period covered begins at the outset of a project, not when the initial cash outflow occurs.

Answer (A) is incorrect because it is related to breakeven point, not breakeven time. Answer (B) is incorrect because the payback period equals investment divided by annual undiscounted net cash inflows. Answer (C) is incorrect because the payback period is the period required for total undiscounted cash inflows to equal total undiscounted cash outflows.

Questions 27 through 30 are based on the following information. In order to increase production capacity, Gunning Industries is considering replacing an existing production machine with a new technologically improved machine effective January 1. The following information is being considered by Gunning Industries:

- The new machine would be purchased for $160,000 in cash. Shipping, installation, and testing would cost an additional $30,000.
- The new machine is expected to increase annual sales by 20,000 units at a sales price of $40 per unit. Incremental operating costs include $30 per unit in variable costs and total fixed costs of $40,000 per year.
- The investment in the new machine will require an immediate increase in working capital of $35,000. This cash outflow will be recovered at the end of year 5.
- Gunning uses straight-line depreciation for financial reporting and tax reporting purposes. The new machine has an estimated useful life of 5 years and zero salvage value.
- Gunning is subject to a 40% corporate income tax rate.

Gunning uses the net present value method to analyze investments and will employ the following factors and rates:

Period	Present Value of $1 at 10%	Present Value of an Ordinary Annuity of $1 at 10%
1	.909	.909
2	.826	1.736
3	.751	2.487
4	.683	3.170
5	.621	3.791

---

27. Gunning Industries' net cash outflow in a capital budgeting decision is

A. $190,000

B. $195,000

C. $204,525

D. $225,000

The correct answer is (D). *(CMA, adapted)*

REQUIRED: The net cash outflow in a capital budgeting decision.

DISCUSSION: The machine costs $160,000 and will require $30,000 to install and test. In addition, the company will have to invest in $35,000 of working capital to support the production of the new machine. Thus, the total investment necessary is $225,000.

Answer (A) is incorrect because $190,000 ignores the investment in working capital. Answer (B) is incorrect because $195,000 ignores the $30,000 of shipping, installation, and testing costs. Answer (C) is incorrect because $204,525 is the present value of $225,000 at 10% for 1 year.

28. Gunning Industries' discounted annual depreciation tax shield for the year of replacement is

A. $13,817

B. $16,762

C. $20,725

D. $22,800

The correct answer is (A). *(CMA, adapted)*

REQUIRED: The discounted annual depreciation tax shield.

DISCUSSION: Gunning uses straight-line depreciation. Thus, the annual charge is $38,000 [($160,000 + $30,000) ÷ 5 years], and the tax savings is $15,200 (40% x $38,000). That benefit will be received in 1 year, so the present value is $13,817 ($15,200 tax savings x .909 present value of $1 for 1 year at 10%).

Answer (B) is incorrect because $16,762 is greater than the undiscounted tax savings. Answer (C) is incorrect because $20,725 assumes a 60% tax rate (the complement of the actual 40% rate). Answer (D) is incorrect because $22,800 assumes a 60% tax rate and no discounting.

29. The acquisition of the new production machine by Gunning Industries will contribute a discounted net-of-tax contribution margin of

A. $242,624

B. $303,280

C. $363,936

D. $454,920

The correct answer is (D). *(CMA, adapted)*

REQUIRED: The discounted net-of-tax contribution margin from the new machine.

DISCUSSION: The new machine will increase sales by 20,000 units a year. The increase in the pretax total contribution margin will be $200,000 per year [20,000 units x ($40 SP – $30 VC)], and the annual increase in the after-tax contribution margin will be $120,000 [(1.0 – .4) x $200,000]. The present value of the after-tax increase in the contribution margin over the 5-year useful life of the machine is $454,920 ($120,000 x 3.791 PV of an ordinary annuity for 5 years at 10%).

Answer (A) is incorrect because $242,624 deducts fixed costs from the pretax contribution margin and applies a 60% tax rate. Answer (B) is incorrect because $303,280 deducts fixed costs from the after-tax contribution margin before discounting. Answer (C) is incorrect because $363,936 deducts fixed costs from the contribution margin before calculating taxes and the present value.

30. The overall discounted cash flow impact of Gunning Industries' working capital investment for the new production machine would be

A. $(7,959)

B. $(10,080)

C. $(13,265)

D. $(35,000)

The correct answer is (C). *(CMA, adapted)*

REQUIRED: The overall discounted cash flow impact of the working capital investment.

DISCUSSION: The $35,000 of working capital requires an immediate outlay for that amount, but it will be recovered in 5 years. Thus, the net discounted cash outflow is $13,265 [$35,000 initial investment – ($35,000 future inflow x .621 PV of $1 for 5 years at 10%)].

Answer (A) is incorrect because $(7,959) assumes the initial investment and its return are reduced by applying a 40% tax rate. Answer (B) is incorrect because $(10,080) assumes the initial investment was discounted for 1 year. Answer (D) is incorrect because $(35,000) fails to consider that the $35,000 will be recovered (essentially in the form of a salvage value) after the fifth year.

31. The net present value (NPV) method of investment project analysis assumes that the project's cash flows are reinvested at the

A. Computed internal rate of return.

B. Risk-free interest rate.

C. Discount rate used in the NPV calculation.

D. Firm's accounting rate of return.

The correct answer is (C). *(CMA, adapted)*

REQUIRED: The rate at which the NPV method assumes early cash inflows are reinvested.

DISCUSSION: The NPV method is used when the discount rate (cost of capital) is specified. It assumes that cash flows from the investment can be reinvested at the particular project's cost of capital (the discount rate).

Answer (A) is incorrect because the internal rate of return method assumes that cash flows are reinvested at the internal rate of return. Answers (B) and (D) are incorrect because the NPV method assumes that cash flows are reinvested at the NPV discount rate.

32. Regal Industries is replacing a grinder purchased 5 years ago for $15,000 with a new one costing $25,000 cash. The original grinder is being depreciated on a straight-line basis over 15 years to a zero salvage value; Regal will sell this old equipment to a third party for $6,000 cash. The new equipment will be depreciated on a straight-line basis over 10 years to a zero salvage value. Assuming a 40% marginal tax rate, Regal's net cash investment at the time of purchase if the old grinder is sold and the new one purchased is

A. $19,000

B. $15,000

C. $17,400

D. $25,000

The correct answer is (C). *(CMA, adapted)*

REQUIRED: The net cash investment at the time of purchase of an old asset and the sale of a new one.

DISCUSSION: The old machine has a book value of $10,000 [$15,000 cost – 5($15,000 cost ÷ 15 years) depreciation]. The loss on the sale is $4,000 ($10,000 – $6,000 cash received), and the tax savings from the loss is $1,600 (40% x $4,000). Thus, total inflows are $7,600. The only outflow is the $25,000 purchase price of the new machine. The net cash investment is therefore $17,400 ($25,000 – $7,600).

Answer (A) is incorrect because $19,000 overlooks the tax savings from the loss on the old machine. Answer (B) is incorrect because $15,000 is obtained by deducting the old book value from the purchase price. Answer (D) is incorrect because the net investment is less than $25,000 given sales proceeds from the old machine and the tax savings.

33. Garfield, Inc. is considering a 10-year capital investment project with forecasted revenues of $40,000 per year and forecasted cash operating expenses of $29,000 per year. The initial cost of the equipment for the project is $23,000, and Garfield expects to sell the equipment for $9,000 at the end of the tenth year. The equipment will be depreciated over 7 years. The project requires a working capital investment of $7,000 at its inception and another $5,000 at the end of year 5. Assuming a 40% marginal tax rate, the expected net cash flow from the project in the tenth year is

A. $32,000

B. $24,000

C. $20,000

D. $11,000

The correct answer is (B). *(CMA, adapted)*

REQUIRED: The expected net cash flow from the project in the tenth year.

DISCUSSION: The project will have an $11,000 before-tax cash inflow from operations in the tenth year ($40,000 – $29,000). Also, $9,000 will be generated from the sale of the equipment. The entire $9,000 will be taxable because the basis of the asset was reduced to zero in the seventh year. Thus, taxable income will be $20,000 ($11,000 + $9,000), leaving a net after-tax cash inflow of $12,000 [(1.0 – .4) x $20,000]. To this $12,000 must be added the $12,000 tied up in working capital ($7,000 + $5,000). The total net cash flow in the tenth year will therefore be $24,000.

Answer (A) is incorrect because $32,000 omits the $8,000 outflow for income taxes. Answer (C) is incorrect because taxes will be $8,000, not $12,000. Answer (D) is incorrect because $11,000 is the net operating cash flow.

34. The rankings of mutually exclusive investments determined using the internal rate of return method (IRR) and the net present value method (NPV) may be different when

A. The lives of the multiple projects are equal and the size of the required investments are equal.

B. The required rate of return equals the IRR of each project.

C. The required rate of return is higher than the IRR of each project.

D. Multiple projects have unequal lives and the size of the investment for each project is different.

The correct answer is (D). *(CMA, adapted)*

REQUIRED: The circumstances in which IRR and NPV rankings of mutually exclusive projects may differ.

DISCUSSION: The two methods ordinarily yield the same results, but differences can occur when the duration of the projects and the initial investments differ. The reason is that the IRR method assumes cash inflows from the early years will be reinvested at the internal rate of return. The NPV method assumes that early cash inflows are reinvested at the cost of capital.

Answer (A) is incorrect because the two methods will give the same results if the lives and required investments are the same. Answer (B) is incorrect because, if the required rate of return equals the IRR (i.e., the cost of capital is equal to the IRR), the two methods would yield the same decision. Answer (C) is incorrect because, if the required rate of return is higher than the IRR, both methods would yield a decision not to acquire the investment.

35. The proper discount rate to use in calculating certainty equivalent net present value is the

A. Risk-adjusted discount rate.

B. Cost of capital.

C. Risk-free rate.

D. Cost of equity capital.

The correct answer is (C). *(CMA, adapted)*

REQUIRED: The proper discount rate to use in calculating certainty equivalent net present value.

DISCUSSION: Rational investors choose projects that yield the best return given some level of risk. If an investor desires no risk, that is, an absolutely certain rate of return, the risk-free rate is used in calculating net present value. The risk-free rate is the return on a risk-free investment such as government bonds. Certainty equivalent adjustments involve a technique directly drawn from utility theory. It forces the decision maker to specify at what point the firm is indifferent to the choice between a sum of money that is certain and the expected value of a risky sum.

Answer (A) is incorrect because a risk-adjusted discount rate does not represent an absolutely certain rate of return. A discount rate is adjusted upward as the investment becomes riskier. Answer (B) is incorrect because the cost of capital has nothing to do with certainty equivalence. Answer (D) is incorrect because the cost of equity capital does not equate to a certainty equivalent rate.

36. Amster Corporation has not yet decided on its hurdle rate for use in the evaluation of capital budgeting projects. This lack of information will prohibit Amster from calculating a project's

	Accounting Rate of Return	Net Present Value	Internal Rate of Return
A.	No	No	No
B.	Yes	Yes	Yes
C.	No	Yes	Yes
D.	No	Yes	No

The correct answer is (D). *(CMA, adapted)*

REQUIRED: The capital budgeting technique(s), if any, that require(s) determination of a hurdle rate.

DISCUSSION: A hurdle rate is not necessary in calculating the accounting rate of return. That return is calculated by dividing the net income from a project by the investment in the project. Similarly, a company can calculate the internal rate of return (IRR) without knowing its hurdle rate. The IRR is the discount rate at which the net present value is $0. However, the NPV cannot be calculated without knowing the company's hurdle rate. The NPV method requires that future cash flows be discounted using the hurdle rate.

Answers (A), (B), and (C) are incorrect because the accounting rate of return and the IRR but not the NPV can be calculated without knowing the hurdle rate.

37. When determining net present value in an inflationary environment, adjustments should be made to

A. Increase the discount rate only.

B. Increase the estimated cash inflows and increase the discount rate.

C. Increase the estimated cash inflows but not the discount rate.

D. Decrease the estimated cash inflows and increase the discount rate.

The correct answer is (B). *(CMA, adapted)*

REQUIRED: The adjustment in determining net present value in an inflationary environment.

DISCUSSION: In an inflationary environment, nominal future cash flows should increase to reflect the decrease in the value of the unit of measure. Also, the investor should increase the discount rate to reflect the increased inflation premium arising from the additional uncertainty. Lenders will require a higher interest rate in an inflationary environment.

Answer (A) is incorrect because future cash flows should also increase. Answer (C) is incorrect because the discount rate should be increased to take into consideration future uncertainty and the risk premium that lenders will require in an inflationary environment. Answer (D) is incorrect because cash flows should increase in an inflationary environment.

38. All of the following items are included in discounted cash flow analysis except

A. Future operating cash savings.

B. The current asset disposal price.

C. The future asset depreciation expense.

D. The tax effects of future asset depreciation.

The correct answer is (C). *(CMA, adapted)*

REQUIRED: The item not included in discounted cash flow analysis.

DISCUSSION: Discounted cash flow analysis, using either the internal rate of return (IRR) or the net present value (NPV) method, is based on the time value of cash inflows and outflows. All future operating cash savings are considered as well as the tax effects on cash flows of future depreciation charges. The cash proceeds of future asset disposals are likewise a necessary consideration. Depreciation expense is a consideration only to the extent that it affects the cash flows for taxes. Otherwise, depreciation is excluded from the analysis because it is a noncash expense.

Answers (A), (B), and (D) are incorrect because future operating cash savings, the current asset disposal price, and the tax effects of future asset depreciation are considerations in discounted cash flow analysis.

39. The capital budgeting model that is ordinarily considered the best model for long-range decision making is the

A. Payback model.

B. Accounting rate of return model.

C. Unadjusted rate of return model.

D. Discounted cash flow model.

The correct answer is (D). *(CMA, adapted)*

REQUIRED: The best capital budgeting model for long-range decision making.

DISCUSSION: The capital budgeting methods that are generally considered the best for long-range decision making are the internal rate of return and net present value methods. These are both discounted cash flow methods.

Answer (A) is incorrect because the payback method gives no consideration to the time value of money or to returns after the payback period. Answers (B) and (C) are incorrect because the accounting rate of return and the unadjusted rate of return do not consider the time value of money.

40. The use of an accelerated method instead of the straight-line method of depreciation in computing the net present value of a project has the effect of

A. Raising the hurdle rate necessary to justify the project.

B. Lowering the net present value of the project.

C. Increasing the present value of the depreciation tax shield.

D. Increasing the cash outflows at the initial point of the project.

The correct answer is (C). *(CMA, adapted)*

REQUIRED: The effect on NPV of using an accelerated depreciation method.

DISCUSSION: Accelerated depreciation results in greater depreciation in the early years of an asset's life compared with the straight-line method. Thus, accelerated depreciation results in lower income tax expense in the early years of a project and higher income tax expense in the later years. By effectively deferring taxes, the accelerated method increases the present value of the depreciation tax shield.

Answer (A) is incorrect because the hurdle rate can be reached more easily as a result of the increased present value of the depreciation tax shield. Answer (B) is incorrect because the greater depreciation tax shield increases the NPV. Answer (D) is incorrect because greater initial depreciation reduces the cash outflows for the taxes but has no effect on the initial cash outflows.

Questions 41 through 43 are based on the following information. Capital Invest Inc. uses a 12% hurdle rate for all capital expenditures and has done the following analysis for four projects for the upcoming year:

	Project 1	Project 2	Project 3	Project 4
Initial capital outlay	$200,000	$298,000	$248,000	$272,000
Annual net cash inflows				
Year 1	$ 65,000	$100,000	$ 80,000	$ 95,000
Year 2	70,000	135,000	95,000	125,000
Year 3	80,000	90,000	90,000	90,000
Year 4	40,000	65,000	80,000	60,000
Net present value	(3,798)	4,276	14,064	14,662
Profitability index	98%	101%	106%	105%
Internal rate of return	11%	13%	14%	15%

41. Which project(s) should Capital Invest Inc. undertake during the upcoming year assuming it has no budget restrictions?

A. All of the projects.

B. Projects 1, 2, and 3.

C. Projects 2, 3, and 4.

D. Projects 1, 3, and 4.

The correct answer is (C). *(CMA, adapted)*

REQUIRED: The project(s) undertaken assuming no budget restrictions.

DISCUSSION: A company using the NPV method should undertake all projects with a positive NPV, unless some of those projects are mutually exclusive. Given that Projects 2, 3, and 4 have positive NPVs, they should be undertaken. Project 1 has a negative NPV.

Answers (A), (B), and (D) are incorrect because Project 1 has a negative NPV and should not be undertaken, but Projects 2, 3, and 4 have positive NPVs and therefore should be included.

42. Which project(s) should Capital Invest Inc. undertake during the upcoming year if it has only $600,000 of funds available?

A. Projects 1 and 3.

B. Projects 2, 3, and 4.

C. Projects 2 and 3.

D. Projects 3 and 4.

The correct answer is (D). *(CMA, adapted)*

REQUIRED: The project(s) undertaken given capital rationing.

DISCUSSION: Given that only $600,000 is available and that each project costs $200,000 or more, no more than two projects can be undertaken. Because Projects 3 and 4 have the greatest NPVs, profitability indexes, and IRRs, they are the projects in which the company should invest.

Answer (A) is incorrect because Project 1 has a negative NPV. Answer (B) is incorrect because this answer violates the $600,000 limitation. Answer (C) is incorrect because the combined NPV of Projects 2 and 3 is less than the combined NPV of Projects 3 and 4.

43. Which project(s) should Capital Invest Inc. undertake during the upcoming year if it has only $300,000 of capital funds available?

A. Project 1.

B. Projects 2, 3, and 4.

C. Projects 3 and 4.

D. Project 3.

The correct answer is (D). *(CMA, adapted)*

REQUIRED: The project(s) undertaken given capital rationing.

DISCUSSION: Given that $300,000 is available and that each project costs $200,000 or more, only one project can be undertaken. Because Project 3 has a positive NPV and the highest profitability index, it is the best investment. The high profitability index means that the company will achieve the highest NPV per dollar of investment with Project 3. The profitability index facilitates comparison of different-sized investments.

Answer (A) is incorrect because Project 1 has a negative NPV. Answers (B) and (C) are incorrect because choosing more than one project violates the $300,000 limitation.

44. The NPV of a project has been calculated to be $215,000. Which one of the following changes in assumptions would decrease the NPV?

A. Decrease the estimated effective income tax rate.

B. Decrease the initial investment amount.

C. Extend the project life and associated cash inflows.

D. Increase the discount rate.

The correct answer is (D). *(CMA, adapted)*

REQUIRED: The change in assumption that would decrease the net present value (NPV).

DISCUSSION: An increase in the discount rate would lower the net present value, as would a decrease in cash flows or an increase in the initial investment.

Answer (A) is incorrect because a decrease in the tax rate would decrease tax expense, thus increasing cash flows and the NPV. Answer (B) is incorrect because a decrease in the initial investment amount would increase the NPV. Answer (C) is incorrect because an extension of the project life and associated cash inflows would increase the NPV.

45. Lawson Inc. is expanding its manufacturing plant, which requires an investment of $4 million in new equipment and plant modifications. Lawson's sales are expected to increase by $3 million per year as a result of the expansion. Cash investment in current assets averages 30% of sales; accounts payable and other current liabilities are 10% of sales. What is the estimated total investment for this expansion?

A. $3.4 million.

B. $4.3 million.

C. $4.6 million.

D. $4.9 million.

The correct answer is (C). *(CMA, adapted)*

REQUIRED: The estimated total cash investment.

DISCUSSION: The investment required includes increases in working capital (e.g., additional receivables and inventories resulting from the acquisition of a new manufacturing plant). The additional working capital is an initial cost of the investment but one that will be recovered (i.e., it has a salvage value equal to its initial cost). Lawson can use current liabilities to fund assets to the extent of 10% of sales. Thus, the total initial cash outlay will be $4.6 million {$4 million + [(30% – 10%) x $3 million sales]}.

Answer (A) is incorrect because $3.4 million deducts the investment in working capital from the cost of equipment. Answer (B) is incorrect because $4.3 million equals $4 million plus 10% of $3 million. Answer (D) is incorrect because $4.9 million fails to consider the financing of 33⅓% of the incremental current assets with accounts payable.

46. A disadvantage of the net present value method of capital expenditure evaluation is that it

A. Is calculated using sensitivity analysis.

B. Computes the true interest rate.

C. Does not provide the true rate of return on investment.

D. Is difficult to apply because it uses a trial-and-error approach.

The correct answer is (C). *(CMA, adapted)*

REQUIRED: The disadvantage of the NPV method.

DISCUSSION: The NPV is broadly defined as the excess of the present value of the estimated net cash inflows over the net cost of the investment. A discount rate has to be stipulated by the person conducting the analysis. A disadvantage is that it does not provide the true rate of return for an investment, only that the rate of return is higher than a stipulated discount rate (which may be the cost of capital).

Answer (A) is incorrect because the ability to perform sensitivity analysis is an advantage of the NPV method. Answer (B) is incorrect because the NPV method does not compute the true interest rate. Answer (D) is incorrect because the IRR method, not the NPV method, uses a trial-and-error approach when cash flows are not identical from year to year.

47. Barker Inc. has no capital rationing constraint and is analyzing many independent investment alternatives. Barker should accept all investment proposals

A. If debt financing is available for them.

B. That have positive cash flows.

C. That provide returns greater than the before-tax cost of debt.

D. That have a positive net present value.

The correct answer is (D). *(CMA, adapted)*

REQUIRED: The investment proposals that should be accepted by a company with no capital rationing constraints.

DISCUSSION: A company should accept any investment proposal, unless some are mutually exclusive, that has a positive net present value or an internal rate of return greater than the company's cost of capital.

Answer (A) is incorrect because the mere availability of financing is not the only consideration; more important is the cost of the financing, which must be less than the rate of return on the proposed investment. Answer (B) is incorrect because an investment with positive cash flows may be a bad investment due to the time value of money; cash flows in later years are not as valuable as those in earlier years. Answer (C) is incorrect because returns should exceed the weighted-average cost of capital, which includes the cost of equity capital as well as the cost of debt capital.

48. Jackson Corporation uses net present value techniques in evaluating its capital investment projects. The company is considering a new equipment acquisition that will cost $100,000, fully installed, and have a zero salvage value at the end of its five-year productive life. Jackson will depreciate the equipment on a straight-line basis for both financial and tax purposes. Jackson estimates $70,000 in annual recurring operating cash income and $20,000 in annual recurring operating cash expenses. Jackson's cost of capital is 12% and its effective income tax rate is 40%. What is the net present value of this investment on an after-tax basis?

A. $28,840

B. $8,150

C. $36,990

D. $80,250

The correct answer is (C). *(CMA, adapted)*

REQUIRED: The net present value on an after-tax basis.

DISCUSSION: Annual cash outflow for taxes is $12,000 {[$70,000 inflows – $20,000 cash operating expenses – ($100,000 ÷ 5) depreciation] x 40%}. The annual net cash inflow is therefore $38,000 ($70,000 – $20,000 – $12,000). The present value of these net inflows for a 5-year period is $136,990 ($38,000 x 3.605 present value of an ordinary annuity for 5 years at 12%), and the NPV of the investment is $36,990 ($136,990 – $100,000 investment).

Answer (A) is incorrect because $28,840 is the present value of the depreciation tax savings. Answer (B) is incorrect because $8,150 ignores the depreciation tax savings. Answer (D) is incorrect because $80,250 ignores taxes.

Question 49 is based on the following information. Willis Inc. has a cost of capital of 15% and is considering the acquisition of a new machine which costs $400,000 and has a useful life of 5 years. Willis projects that earnings and cash flow will increase as follows:

Year	Net Earnings	After-Tax Cash Flow
1	$100,000	$160,000
2	100,000	140,000
3	100,000	100,000
4	100,000	100,000
5	200,000	100,000

	15% Interest Rate Factors	
Period	Present Value of $1	Present Value of an Annuity of $1
1	0.87	0.87
2	0.76	1.63
3	0.66	2.29
4	0.57	2.86
5	0.50	3.36

49. The net present value of this investment is

A. Negative, $64,000.

B. Negative, $14,000.

C. Positive, $18,600.

D. Positive, $200,000.

The correct answer is (C). *(CMA, adapted)*

REQUIRED: The NPV of the investment.

DISCUSSION: The NPV is calculated by discounting the after-tax cash flows by the cost of capital. The cash flows of this company constitute an annuity of $100,000 per year plus additional flows of $60,000 in year 1 and $40,000 in year 2. Thus, the present value of these flows is $418,600 [($100,000 x 3.36 PV of an annuity of $1 for 5 periods at 15%) + ($60,000 x .87 PV of $1 for 1 period at 15%) + ($40,000 x .76 PV of $1 for 2 periods at 15%)]. The NPV is therefore $18,600 ($418,600 – $400,000).

Answer (A) is incorrect because a negative NPV of $64,000 assumes net cash flows of $100,000 per year. Answer (B) is incorrect because a negative NPV of $14,000 results from discounting the net earnings. Answer (D) is incorrect because $200,000 is the undiscounted excess of the cash flows over the investment.

50. A weakness of the internal rate of return (IRR) approach for determining the acceptability of investments is that it

A. Does not consider the time value of money.

B. Is not a straightforward decision criterion.

C. Implicitly assumes that the firm is able to reinvest project cash flows at the firm's cost of capital.

D. Implicitly assumes that the firm is able to reinvest project cash flows at the project's internal rate of return.

The correct answer is (D). *(CMA, adapted)*

REQUIRED: The weakness of the internal rate of return approach.

DISCUSSION: The IRR is the rate at which the discounted future cash flows equal the net investment (NPV = 0). One disadvantage of the method is that inflows from the early years are assumed to be reinvested at the IRR. This assumption may not be sound. Investments in the future may not earn as high a rate as is currently available.

Answer (A) is incorrect because the IRR method considers the time value of money. Answer (B) is incorrect because the IRR provides a straightforward decision criterion. Any project with an IRR greater than the cost of capital is acceptable. Answer (C) is incorrect because the IRR method implicitly assumes reinvestment at the IRR; the NPV method implicitly assumes reinvestment at the cost of capital.

51. The internal rate of return (IRR) is the

A. Hurdle rate.

B. Rate of interest for which the net present value is greater than 1.0.

C. Rate of interest for which the net present value is equal to zero.

D. Rate of return generated from the operational cash flows.

The correct answer is (C). *(CMA, adapted)*

REQUIRED: The true statement about the internal rate of return (IRR).

DISCUSSION: The IRR is the interest rate at which the present value of the expected future cash inflows is equal to the present value of the cash outflows for a project. Thus, the IRR is the interest rate that will produce a net present value (NPV) equal to zero. The IRR method assumes that the cash flows will be reinvested at the internal rate of return.

Answer (A) is incorrect because the hurdle rate is a concept used to calculate the NPV of a project; it is determined by management prior to the analysis. Answer (B) is incorrect because the IRR is the rate of interest at which the NPV is zero. Answer (D) is incorrect because the IRR is a means of evaluating potential investment projects.

52. A project requires an initial cash investment at its inception of $10,000, and no other cash outflows are necessary. Cash inflows from the project over its 3-year life are $6,000 at the end of the first year, $5,000 at the end of the second year, and $2,000 at the end of the third year. The future value interest factors for an amount of $1 at the cost of capital of 8% are

Period			
1	2	3	4
1.080	1.166	1.260	1.360

The present value interest factors for an amount of $1 for three periods are as follows:

Interest Rate				
8%	9%	10%	12%	14%
.794	.772	.751	.712	.675

The modified IRR (MIRR) for the project is closest to

A. 8%

B. 9%

C. 10%

D. 12%

The correct answer is (D). *(Publisher)*

REQUIRED: The approximate MIRR.

DISCUSSION: The MIRR is the interest rate at which the present value of the cash outflows discounted at the cost of capital equals the present value of the terminal value. The terminal value is the future value of the cash inflows assuming they are reinvested at the cost of capital. The present value of the outflows is $10,000 because no future outflows occur. The terminal value is $14,396 [(1.166 x $6,000) + (1.08 x $5,000) + (1.00 x $2,000)]. Accordingly, the present value of $1 interest factor for the rate at which the present value of the outflows equals the present value of the terminal value is .695 ($10,000 ÷ $14,396). This factor is closest to that for 12%.

Answers (A), (B), and (C) are incorrect because the MIRR must be greater than 12%. The present value of $1 interest factor for this project that equates the present values of the costs and the terminal value is lower than the factor for 12%. The lower the present value factor, the higher the interest rate.

Questions 53 and 54 are based on the following information. A firm with an 18% cost of capital is considering the following projects (on January 1, year 1):

	January 1, Year 1 Cash Outflow (000's Omitted)	December 31, Year 5 Cash Inflow (000's Omitted)	Project Internal Rate of Return
Project A	$3,500	$7,400	16%
Project B	4,000	9,950	?

Present Value of $1 Due at the End of "N" Periods

N	12%	14%	15%	16%	18%	20%	22%
4	.6355	.5921	.5718	.5523	.5158	.4823	.4230
5	.5674	.5194	.4972	.4761	.4371	.4019	.3411
6	.5066	.4556	.4323	.4104	.3704	.3349	.2751

53. Using the net-present-value (NPV) method, project A's net present value is

A. $316,920

B. $23,140

C. $(265,460)

D. $(316,920)

The correct answer is (C). *(CIA, adapted)*

REQUIRED: The net present value of project A.

DISCUSSION: The cash inflow occurs 5 years after the cash outflow, and the NPV method uses the firm's cost of capital of 18%. The present value of $1 due at the end of 5 years discounted at 18% is .4371. Thus, the NPV of project A is $(265,460) [(.4371 x $7,400,000 cash inflow) – $3,500,000 cash outflow].

Answer (A) is incorrect because $316,920 discounts the cash inflow over a 4-year period. Answer (B) is incorrect because $23,140 assumes a 16% discount rate. Answer (D) is incorrect because $(316,920) discounts the cash inflow over a 4-year period and also subtracts the present value of the cash inflow from the cash outflow.

54. Project B's internal rate of return is closest to

A. 15%

B. 16%

C. 18%

D. 20%

The correct answer is (D). *(CIA, adapted)*

REQUIRED: The percentage closest to project B's internal rate of return.

DISCUSSION: The internal rate of return is the discount rate at which the NPV is zero. Consequently, the cash outflow equals the present value of the inflow at the internal rate of return. The present value of $1 factor for project B's internal rate of return is therefore .4020 ($4,000,000 cash outflow ÷ $9,950,000 cash inflow). This factor is closest to the present value of $1 for 5 periods at 20%.

Answer (A) is incorrect because 15% results in a positive NPV for project B. Answer (B) is incorrect because 16% is the approximate internal rate of return for project A. Answer (C) is incorrect because 18% is the company's cost of capital.

55. The profitability index approach to investment analysis

A. Fails to consider the timing of project cash flows.

B. Considers only the project's contribution to net income and does not consider cash flow effects.

C. Always yields the same accept/reject decisions for independent projects as the net present value method.

D. Always yields the same accept/reject decisions for mutually exclusive projects as the net present value method.

The correct answer is (C). *(CMA, adapted)*

REQUIRED: The true statement about the profitability index.

DISCUSSION: The profitability index is the ratio of the present value of future net cash inflows to the initial net cash investment. It is a variation of the net present value (NPV) method and facilitates the comparison of different-sized investments. Because it is based on the NPV method, the profitability index will yield the same decision as the NPV for independent projects. However, decisions may differ for mutually exclusive projects of different sizes.

Answer (A) is incorrect because the profitability index, like the NPV method, discounts cash flows based on the cost of capital. Answer (B) is incorrect because the profitability index is cash based. Answer (D) is incorrect because the NPV and the profitability index may yield different decisions if projects are mutually exclusive and of different sizes.

56. If an investment project has a profitability index of 1.15, the

A. Project's internal rate of return is 15%.

B. Project's cost of capital is greater than its internal rate of return.

C. Project's internal rate of return exceeds its net present value.

D. Net present value of the project is positive.

The correct answer is (D). *(CMA, adapted)*

REQUIRED: The meaning of a profitability index in excess of 1.0.

DISCUSSION: The profitability index is the ratio of the present value of future net cash inflows to the initial net cash investment. It is a variation of the NPV method that facilitates comparison of different-sized investments. A profitability index greater than 1.0 indicates a profitable investment, or one that has a positive net present value.

Answer (A) is incorrect because the IRR is the discount rate at which the NPV is $0, which is also the rate at which the profitability index is 1.0. The IRR cannot be determined solely from the index. Answer (B) is incorrect because, if the index is 1.15 and the discount rate is the cost of capital, the NPV is positive, and the IRR must be higher than the cost of capital. Answer (C) is incorrect because the IRR is a discount rate, whereas the NPV is an amount.

57. Flex Corporation is studying a capital acquisition proposal in which newly acquired assets will be depreciated using the straight-line method. Which one of the following statements about the proposal would be incorrect if a switch is made to the Modified Accelerated Cost Recovery System (MACRS)?

A. The net present value will increase.

B. The internal rate of return will increase.

C. The payback period will be shortened.

D. The profitability index will decrease.

The correct answer is (D). *(CMA, adapted)*

REQUIRED: The false statement about the effect of a switch from straight-line depreciation to MACRS.

DISCUSSION: MACRS is an accelerated method of depreciation under which depreciation expense will be greater during the early years of an asset's life. Thus, the outflows for income taxes will be less in the early years, but greater in the later years, and the NPV (present value of net cash inflows – investment) will be increased. The profitability index (present value of net cash inflows ÷ the investment) must increase if the NPV increases.

Answer (A) is incorrect because the NPV will increase. The present value of the net inflows will increase with no change in the investment. Answer (B) is incorrect because the IRR will increase. Deferring expenses to later years increases the discount rate needed to reduce the NPV to $0. Answer (C) is incorrect because the payback period will be shortened. Switching to MACRS defers expenses and increases cash flows early in the project's life.

58. The technique used to evaluate all possible capital projects of different dollar amounts and then rank them according to their desirability is the

A. Profitability index method.

B. Net present value method.

C. Payback method.

D. Discounted cash flow method.

The correct answer is (A). *(CMA, adapted)*

REQUIRED: The technique used to evaluate and rank all possible capital projects of varying dollar amounts.

DISCUSSION: The profitability index is the ratio of the present value of future net cash inflows to the initial cash investment; that is, the figures are those used to calculate the net present value (NPV), but the numbers are divided rather than subtracted. This variation of the NPV method facilitates comparison of different-sized investments. It provides an optimal ranking in the absence of capital rationing.

Answers (B) and (D) are incorrect because the profitability index method and the NPV method are discounted cash flow methods. However, the profitability index method is the variant that purports to calculate a return per dollar of investment. Answer (C) is incorrect because the payback method gives no consideration to the time value of money or to returns after the payback period.

59. The profitability index (present value index)

A. Represents the ratio of the discounted net cash outflows to cash inflows.

B. Is the relationship between the net discounted cash inflows less the discounted cash outflows divided by the discounted cash outflows.

C. Is calculated by dividing the discounted profits by the cash outflows.

D. Is the ratio of the discounted net cash inflows to discounted cash outflows.

The correct answer is (D). *(CMA, adapted)*

REQUIRED: The true statement about the profitability index.

DISCUSSION: The profitability index, also known as the excess present value index, is the ratio of the present value of future net cash inflows to the initial net cash investment (discounted cash outflows). This tool is a variation of the NPV method that facilitates comparison of different-sized investments.

Answer (A) is incorrect because the cash inflows are also discounted in the profitability index. Answer (B) is incorrect because the numerator is the discounted net cash inflows. Answer (C) is incorrect because the profitability index is based on cash flows, not profits.

60. If income tax considerations are ignored, how is depreciation handled by the following capital budgeting techniques?

	Internal Rate of Return	Accounting Rate of Return	Payback
A.	Excluded	Included	Excluded
B.	Included	Excluded	Included
C.	Excluded	Excluded	Included
D.	Included	Included	Included

The correct answer is (A). *(CMA, adapted)*

REQUIRED: The manner in which depreciation is handled by each capital budgeting technique.

DISCUSSION: If taxes are ignored, depreciation is not a consideration in any of the methods based on cash flows because it is a noncash expense. Thus, the internal rate of return, net present value, and payback methods would not consider depreciation because these methods are based on cash flows. However, the accounting rate of return is based on net income as calculated on an income statement. Because depreciation is included in the determination of accrual accounting net income, it would affect the calculation of the accounting rate of return.

Answers (B), (C), and (D) are incorrect because the IRR and the payback period are based on cash flows. Depreciation is not needed in their calculation. However, the accounting rate of return cannot be calculated without first deducting depreciation.

61. The method that divides a project's annual after-tax net income by the average investment cost to measure the estimated performance of a capital investment is the

A. Internal rate of return method.

B. Accounting rate of return method.

C. Payback method.

D. Net present value (NPV) method.

The correct answer is (B). *(CMA, adapted)*

REQUIRED: The capital budgeting method that divides annual after-tax net income by the average investment cost.

DISCUSSION: The accounting rate of return uses undiscounted net income (not cash flows) to determine a rate of profitability. Annual after-tax net income is divided by the average book value (or the initial value) of the investment in assets.

Answer (A) is incorrect because the internal rate of return is the rate at which NPV is zero. The minimum desired rate of return is not used in the discounting. Answer (C) is incorrect because the payback period is the time required to complete the return of the original investment. This method gives no consideration to the time value of money or to returns after the payback period. Answer (D) is incorrect because the NPV method computes the discounted present value of future cash inflows to determine whether it is greater than the initial cash outflow.

Questions 62 through 64 are based on the following information. Jorelle Company's financial staff has been requested to review a proposed investment in new capital equipment. Applicable financial data is presented below. There will be no salvage value at the end of the investment's life and, due to realistic depreciation practices, it is estimated that the salvage value and net book value are equal at the end of each year. All cash flows are assumed to take place at the end of each year. For investment proposals, Jorelle uses a 12% after-tax target rate of return.

Investment Proposal

Year	Purchase Cost and Book Value	Annual Net After-Tax Cash Flows	Annual Net Income
0	$250,000	$ 0	$ 0
1	168,000	120,000	35,000
2	100,000	108,000	39,000
3	50,000	96,000	43,000
4	18,000	84,000	47,000
5	0	72,000	51,000

Discounted Factors for a 12% Rate of Return

Year	Present Value of $1.00 Received at the End of Each Period	Present Value of an Annuity of $1.00 Received at the End of Each Period
1	.89	.89
2	.80	1.69
3	.71	2.40
4	.64	3.04
5	.57	3.61
6	.51	4.12

62. The accounting rate of return on the average investment proposal is

A. 12.0%

B. 17.2%

C. 28.0%

D. 34.4%

The correct answer is (D). *(CMA, adapted)*

REQUIRED: The accounting rate of return.

DISCUSSION: The accounting rate of return (unadjusted rate of return or book value rate of return) equals accounting net income divided by the required average investment. The accounting rate of return ignores the time value of money. The average income over 5 years is $43,000 per year [($35,000 + $39,000 + $43,000 + $47,000 + $51,000) ÷ 5]. Hence, the accounting rate of return is 34.4% [$43,000 ÷ ($250,000 ÷ 2)].

Answer (A) is incorrect because 12.0% is the discount rate. Answer (B) is incorrect because 17.2% equals $43,000 divided by $250,000. Answer (C) is incorrect because 28.0% equals $35,000 divided by $125,000.

63. The net present value for the investment proposal is

A. $106,160

B. $(97,970)

C. $356,160

D. $96,560

The correct answer is (A). *(CMA, adapted)*

REQUIRED: The net present value (NPV).

DISCUSSION: The NPV is the sum of the present values of all cash inflows and outflows associated with the proposal. If the NPV is positive, the proposal should be accepted. The NPV is determined by discounting each expected cash flow using the appropriate 12% interest factor for the present value of $1. Thus, the NPV is $106,160 [(.89 x $120,000) + (.80 x $108,000) + (.71 x $96,000) + (.64 x $84,000) + (.57 x $72,000) – (1.00 x $250,000)].

Answer (B) is incorrect because $(97,970) is based on net income instead of cash flows. Answer (C) is incorrect because $356,160 excludes the purchase cost. Answer (D) is incorrect because $96,560 equals average after-tax cash inflow times the interest factor for the present value of a 5-year annuity, minus $250,000.

64. The traditional payback period for the investment proposal is

A. Over 5 years.

B. 2.23 years.

C. 1.65 years.

D. 2.83 years.

The correct answer is (B). *(CMA, adapted)*

REQUIRED: The traditional payback period.

DISCUSSION: The traditional payback period is the number of periods required for the undiscounted expected cash flows to equal the original investment. When periodic cash flows are not expected to be uniform, a cumulative calculation is necessary. The first year's cash inflow is $120,000. Adding another $108,000 in Year 2 brings the total payback after 2 years to $228,000. Accordingly, the traditional payback period is about 2.23 years {2 + [($250,000 – $228,000) ÷ $96,000 cash inflow in year 3]}.

Answer (A) is incorrect because the 5-year total expected cash inflow is $480,000. Answer (C) is incorrect because the total expected cash flow for 1.65 years is $190,200. Answer (D) is incorrect because the total expected cash flow for 2.83 years is $307,680.

65. Which one of the following capital investment evaluation methods does not take the time value of money into consideration?

A. Net present value.

B. Discounted payback.

C. Internal rate of return.

D. Accounting rate of return.

The correct answer is (D). *(CMA, adapted)*

REQUIRED: The capital investment evaluation method that does not discount cash flows.

DISCUSSION: The accounting rate of return (unadjusted rate of return or book value rate of return) equals accounting net income divided by the required initial or average investment. The accounting rate of return ignores the time value of money.

Answer (A) is incorrect because the net present value is the sum of the present values of all the cash inflows and outflows associated with an investment. Answer (B) is incorrect because the discounted payback method calculates the payback period by determining the present values of the future cash flows. Answer (C) is incorrect because the internal rate of return is the discount rate at which the NPV is zero.

Questions 66 through 69 are based on the following information. MS Trucking is considering the purchase of a new piece of equipment that has a net initial investment with a present value of $300,000. The equipment has an estimated useful life of 3 years. For tax purposes, the equipment will be fully depreciated at rates of 30%, 40%, and 30% in years one, two, and three, respectively. The new machine is expected to have a $20,000 salvage value. The machine is expected to save the company $170,000 per year in operating expenses. MS Trucking has a 40% marginal income tax rate and a 16% cost of capital. Discount rates for a 16% rate are:

	Present Value of an Ordinary Annuity of $1	Present Value of $1
Year 1	.862	.862
Year 2	1.605	.743
Year 3	2.246	.641

66. What is the net present value of this project?

A. $31,684

B. $26,556

C. $94,640

D. $18,864

The correct answer is (B). *(Publisher)*

REQUIRED: The net present value (NPV) of the new machine.

DISCUSSION: The NPV method discounts the expected cash flows from a project using the required rate of return. A project is acceptable if its NPV is positive. The future cash inflows consist of $170,000 of saved expenses per year minus income taxes after deducting depreciation. In the first year, the after-tax cash inflow is $170,000 minus taxes of $32,000 {[$170,000 – (30% x $300,000) depreciation] x 40%}, or $138,000. In the second year, the after-tax cash inflow is $170,000 minus taxes of $20,000 {[$170,000 – (40% x $300,000) depreciation] x 40%}, or $150,000. In the third year, the after-tax cash inflow (excluding salvage value) is again $138,000. Also in the third year, the after-tax cash inflow from the salvage value is $12,000 [$20,000 x (1.0 – .40)]. Accordingly, the total for the third year is $150,000 ($138,000 + $12,000). The sum of these cash flows discounted using the factors for the present value of $1 at a rate of 16% is $326,556.

$138,000 × .862	=	$118,956
$150,000 × .743	=	111,450
$150,000 × .641	=	96,150
Discounted cash inflows		$326,556

Thus, the NPV is $26,556 ($326,556 – $300,000 initial outflow).

Answer (A) is incorrect because $31,684 does not consider the income taxes on the salvage value. The asset was fully depreciated, so any sales receipts would be taxable income. Answer (C) is incorrect because $94,640 does not consider income taxes. Answer (D) is incorrect because $18,864 does not consider the salvage value.

67. What is the profitability index for the project?

A. 1.089

B. 1.106

C. 1.315

D. 1.063

The correct answer is (A). *(Publisher)*

REQUIRED: The profitability index.

DISCUSSION: The profitability index is the present value of the future net cash inflows divided by the present value of the net initial investment. The present value of the future net cash inflows is $326,556 (see the preceding question). Hence, the profitability index is 1.089 ($326,556 ÷ $300,000).

Answer (B) is incorrect because 1.106 does not consider the income taxes on the salvage value. The asset was fully depreciated, so any sales receipts would be taxable income. Answer (C) is incorrect because 1.315 ignores all cash flows for income taxes. Answer (D) is incorrect because 1.063 does not consider the salvage value.

68. The payback period for this investment is

A. 2.84 years.

B. 1.76 years.

C. 2.08 years.

D. 3.00 years.

The correct answer is (C). *(Publisher)*

REQUIRED: The payback period.

DISCUSSION: The payback period is the time required to recover the original investment. As determined in question 66, the annual net after-tax cash inflows for years 1 through 3 are $138,000, $150,000, and $150,000, respectively. After 2 years, $288,000 ($138,000 + $150,000) will have been recovered. Consequently, the first $12,000 received in year 3 will recoup the initial investment. Because $12,000 represents .08 of the third year's net after-tax cash inflows ($12,000 ÷ $150,000), the payback period is 2.08 years.

Answer (A) is incorrect because 2.84 years does not consider depreciation. Answer (B) is incorrect because 1.76 years does not consider taxes. Answer (D) is incorrect because 3.00 years is the time required to depreciate the asset fully, not the payback period.

69. Assume the same facts as on page 146, except that the salvage value at the end of the investment's useful life is zero. What is the new payback period?

A. 2.84 years.

B. 1.76 years.

C. 2.08 years.

D. 2.09 years.

The correct answer is (D). *(Publisher)*

REQUIRED: The payback period assuming zero salvage value.

DISCUSSION: The payback period is the time required to recover the original investment. As determined in question 66, the annual net after-tax cash inflows for years 1 through 3 are $138,000, $150,000, and $138,000 (excluding salvage value), respectively. After 2 years, $288,000 ($138,000 + $150,000) will have been recovered. Consequently, the first $12,000 received in year 3 will recoup the initial investment. Because $12,000 represents .09 (rounded) of the third year's net after-tax cash inflows ($12,000 ÷ $138,000), the payback period is 2.09 years.

Answer (A) is incorrect because 2.84 years includes salvage value and does not consider depreciation. Answer (B) is incorrect because 1.76 years includes salvage value and does not consider taxes. Answer (C) is incorrect because 2.08 years includes salvage value.

Questions 70 through 72 are based on the following information. Maloney Company uses a 12% hurdle rate for all capital expenditures and has done the following analysis for four projects for the upcoming year:

	Project 1	Project 2	Project 3	Project 4
Initial outlay	$4,960,000	$5,440,000	$4,000,000	$5,960,000
Annual net cash inflows				
Year 1	1,600,000	1,900,000	1,300,000	2,000,000
Year 2	1,900,000	2,500,000	1,400,000	2,700,000
Year 3	1,800,000	1,800,000	1,600,000	1,800,000
Year 4	1,600,000	1,200,000	800,000	1,300,000
Net present value	281,280	293,240	(75,960)	85,520
Profitability Index	106%	105%	98%	101%
Internal rate of return	14%	15%	11%	13%

70. Which project(s) should Maloney undertake during the upcoming year assuming it has no budget restrictions?

A. All of the projects.

B. Projects 1, 2, and 3.

C. Projects 1, 2, and 4.

D. Projects 1 and 2.

The correct answer is (C). *(Publisher)*

REQUIRED: The project(s) that should be undertaken given no capital rationing.

DISCUSSION: A company using the net present value (NPV) method should undertake all projects with positive NPVs that are not mutually exclusive. Given that Projects 1, 2, and 4 have positive NPVs, those projects should be undertaken. Furthermore, a company using the internal rate of return (IRR) as a decision rule ordinarily chooses projects with a return greater than the cost of capital. Given a 12% cost of capital, Projects 1, 2, and 4 should be chosen using an IRR criterion if they are not mutually exclusive. Use of the profitability index yields a similar decision because a project with an index greater than 100% should be undertaken.

Answers (A) and (B) are incorrect because Project 3 has a negative NPV. Answer (D) is incorrect because Project 4 has a positive NPV and should be undertaken.

71. Which projects should Maloney undertake during the upcoming year if it has only $12,000,000 of investment funds available?

A. Projects 1 and 3.

B. Projects 1, 2, and 4.

C. Projects 1 and 4.

D. Projects 1 and 2.

The correct answer is (D). *(Publisher)*

REQUIRED: The project(s) that should be undertaken given a capital rationing limitation of $12,000,000.

DISCUSSION: With only $12,000,000 available, and each project costing $4,000,000 or more, no more than two projects can be undertaken. Accordingly, Projects 1 and 2 should be selected because they have the greatest NPVs and profitability indexes.

Answer (A) is incorrect because Project 3 has a negative NPV. Answer (B) is incorrect because choosing three projects violates the $12,000,000 limitation. Answer (C) is incorrect because the combined NPV of Projects 1 and 4 is less than the combined NPV of Projects 1 and 2.

72. Which project(s) should Maloney undertake during the upcoming year if it has only $6,000,000 of funds available?

A. Project 3.

B. Projects 1 and 2.

C. Project 1.

D. Project 2.

The correct answer is (C). *(Publisher)*

REQUIRED: The project(s) that should be undertaken given a capital rationing limitation of $6,000,000.

DISCUSSION: With only $6,000,000 available, and each project costing $4,000,000 or more, no more than one project can be undertaken. Project 1 should be chosen because it has a positive NPV and the highest profitability index. The high profitability index means that the company will achieve the highest NPV per dollar of investment with Project 1. The profitability index facilitates comparison of different-sized investments.

Answer (A) is incorrect because Project 3 has a negative NPV and should not be selected regardless of the capital available. Answer (B) is incorrect because selecting two projects violates the $6,000,000 limitation on funds. Answer (D) is incorrect because, despite having the highest NPV, Project 2 has a lower profitability index than Project 1. Consequently, Project 1 offers the greater return per dollar of investment.

Questions 73 through 77 are based on the following information. A proposed investment is not expected to have any salvage value at the end of its 5-year life. For present value purposes, cash flows are assumed to occur at the end of each year. The company uses a 12% after-tax target rate of return.

Year	Purchase Cost and Book Value	Annual Net After-Tax Cash Flows	Annual Net Income
0	$500,000	$ 0	$0
1	336,000	240,000	70,000
2	200,000	216,000	78,000
3	100,000	192,000	86,000
4	36,000	168,000	94,000
5	0	144,000	102,000

Discount Factors for a 12% Rate of Return

Year	Present Value of $1 at the End of Each Period	Present Value of an Annuity of $1 at the End of Each Period
1	.89	.89
2	.80	1.69
3	.71	2.40
4	.64	3.04
5	.57	3.61
6	.51	4.12

73. The accounting rate of return based on the average investment is

A. 84.9%

B. 34.4%

C. 40.8%

D. 12%

The correct answer is (B). *(Publisher)*

REQUIRED: The accounting rate of return.

DISCUSSION: The accounting rate of return (the unadjusted rate of return or book value rate of return) equals the increase in accounting net income divided by either the initial or the average investment. It ignores the time value of money. The average income over 5 years is $86,000 [($70,000 + $78,000 + $86,000 + $94,000 + $102,000) ÷ 5]. Dividing the $86,000 average net income by the $250,000 average investment ($500,000 cost ÷ 2) produces an accounting rate of return of 34.4%.

Answer (A) is incorrect because 84.9% equals the NPV divided by the average investment. Answer (C) is incorrect because 40.8% equals year 5 net income divided by the average investment. Answer (D) is incorrect because 12% equals the after-tax target rate of return.

74. The net present value is

A. $304,060

B. $212,320

C. $(70,000)

D. $712,320

The correct answer is (B). *(Publisher)*

REQUIRED: The net present value (NPV).

DISCUSSION: The NPV method discounts the expected cash flows from a project using the required rate of return. A project is acceptable if its NPV is positive. Based on the interest factors for the present value of $1 at 12% and the annual after-tax cash flows, the NPV of the project over its 5-year life is

$240,000 ×	.89	=	$213,600
216,000 ×	.80	=	172,800
192,000 ×	.71	=	136,320
168,000 ×	.64	=	107,520
144,000 ×	.57	=	82,080
Total present value			$712,320
Purchase cost			(500,000)
NPV			$212,320

Answer (A) is incorrect because $304,060 is the present value of the net income amounts. Answer (C) is incorrect because $(70,000) is the excess of the net initial investment over the sum of the undiscounted net income amounts. Answer (D) is incorrect because $712,320 is the present value of the cash inflows.

Questions 75 through 77 are based on the information preceding question 73 on page 149.

75. The traditional payback period is

A. Over 5 years.

B. 2.23 years.

C. 1.65 years.

D. 2.83 years.

The correct answer is (B). *(Publisher)*

REQUIRED: The traditional payback period.

DISCUSSION: The payback period is the number of years required to complete the return of the original investment. The cash flows are not time adjusted. When the annual cash flows are not uniform, a cumulative computation is necessary. Thus, the total payback after 2 years is $456,000 ($240,000 + $216,000), and another $44,000 ($500,000 – $456,000) must be recovered in the third year. The third year fraction is found by assuming that cash flows occur evenly throughout the period. Dividing $44,000 by the $192,000 of third year inflows yields a ratio of .23. Hence, the payback period is 2.23 years.

Answer (A) is incorrect because the payback period would exceed 5 years if the calculation were based on net income instead of cash flows. Answer (C) is incorrect because 1.65 years is based on the sum of cash flows and net income. Answer (D) is incorrect because it is based on discounted cash flows--not actual cash flows.

76. The profitability index is

A. .61

B. .42

C. .86

D. 1.425

The correct answer is (D). *(Publisher)*

REQUIRED: The profitability index.

DISCUSSION: The profitability index is the present value of the estimated net cash inflows over the investment's life divided by the present value of the net initial investment. If the profitability index is greater than 1.0, the investment project should be accepted. The present value of the cash inflows is $712,320 (see Q. 25). Thus, the profitability index is 1.425 ($712,320 ÷ $500,000 net initial investment).

Answer (A) is incorrect because .61 is the present value of the net income amounts divided by the net initial investment. Answer (B) is incorrect because .42 is the NPV divided by the net initial investment. Answer (C) is incorrect because .86 is the sum of the undiscounted net income amounts divided by the net initial investment.

77. Which statement about the internal rate of return of the investment is true?

A. The IRR is exactly 12%.

B. The IRR is over 12%.

C. The IRR is under 12%.

D. No information about the IRR can be determined.

The correct answer is (B). *(Publisher)*

REQUIRED: The internal rate of return (IRR).

DISCUSSION: The NPV at an after-tax target rate of return of 12% is $212,320 (see Q. 25). Given that the NPV is positive, the investment project should be accepted assuming no capital rationing. Furthermore, the IRR (the discount rate that reduces the NPV to $0) must be greater than the 12% hurdle rate that produced a positive NPV. The higher the discount rate, the lower the NPV.

Answer (A) is incorrect because the IRR would be 12% if the NPV were $0. Answer (C) is incorrect because the IRR would be under 12% if the NPV were negative. Answer (D) is incorrect because whether the IRR is equal to, less than, or greater than the after-tax target rate of return can be determined from the amount of the NPV.

Questions 78 through 81 are based on the following information. The Dickins Corporation is considering the acquisition of a new machine at a cost of $180,000. Transporting the machine to Dickins' plant will cost $12,000. Installing the machine will cost an additional $18,000. It has a 10-year life and is expected to have a salvage value of $10,000. Furthermore, the machine is expected to produce 4,000 units per year with a selling price of $500 and combined direct materials and direct labor costs of $450 per unit. Federal tax regulations permit machines of this type to be depreciated using the straight-line method over 5 years with no estimated salvage value. Dickins has a marginal tax rate of 40%.

78. What is the net cash outflow at the beginning of the first year that Dickins should use in a capital budgeting analysis?

A. $(170,000)

B. $(180,000)

C. $(192,000)

D. $(210,000)

The correct answer is (D). *(Publisher)*

REQUIRED: The net cash outflow at the beginning of the first year that should be used in a capital budgeting analysis.

DISCUSSION: Delivery and installation costs are essential to preparing the machine for its intended use. Thus, the company must initially pay $210,000 for the machine, consisting of the invoice price of $180,000, the delivery costs of $12,000, and the $18,000 of installation costs.

Answer (A) is incorrect because $(170,000) includes salvage value and ignores delivery and installation costs. Answer (B) is incorrect because $(180,000) ignores the outlays needed for delivery and installation. Answer (C) is incorrect because $(192,000) excludes installation costs.

79. What is the net cash flow for the third year that Dickins Corporation should use in a capital budgeting analysis?

A. $136,800

B. $136,000

C. $128,400

D. $107,400

The correct answer is (A). *(Publisher)*

REQUIRED: The net cash flow for the third year.

DISCUSSION: The company will receive net cash inflows of $50 per unit ($500 selling price – $450 variable costs), a total of $200,000 per year for 4,000 units. This amount will be subject to taxation. However, for the first 5 years, a depreciation deduction of $42,000 per year ($210,000 cost ÷ 5 years) will be available. Thus, annual taxable income will be $158,000 ($200,000 – $42,000). At a 40% tax rate, income tax expense will be $63,200, and the net cash inflow will be $136,800 ($200,000 – $63,200).

Answer (B) is incorrect because $136,000 results from subtracting salvage value when calculating depreciation expense. Answer (C) is incorrect because $128,400 assumes depreciation is recognized over 10 years. Answer (D) is incorrect because $107,400 assumes that depreciation is recognized over 10 years and that it requires a cash outlay.

Questions 80 and 81 are based on the information preceding question 78 on page 151.

80. What is the net cash flow for the tenth year of the project that Dickins should use in a capital budgeting analysis?

A. $200,000

B. $158,000

C. $136,800

D. $126,000

The correct answer is (D). *(Publisher)*

REQUIRED: The net cash flow for the tenth year of the project.

DISCUSSION: The company will receive net cash inflows of $50 per unit ($500 selling price – $450 of variable costs), a total of $200,000 per year for 4,000 units. This amount will be subject to taxation, as will the $10,000 gain on sale of the investment, resulting in taxable income of $210,000. No depreciation will be deducted in the tenth year because the asset was fully depreciated after 5 years. Because the asset was fully depreciated (book value was $0), the $10,000 received as salvage value is fully taxable. At 40%, the tax on $210,000 is $84,000. After subtracting $84,000 of tax expense from the $210,000 of inflows, the net inflows amount to $126,000.

Answer (A) is incorrect because $200,000 overlooks the salvage proceeds and the taxes to be paid. Answer (B) is incorrect because $158,000 equals annual taxable income for each of the first 5 years. Answer (C) is incorrect because $136,800 is the annual net cash inflow in the second through the fifth years.

81. What is the approximate payback period on the new machine?

A. 1.05 years.

B 1.54 years.

C. 1.33 years.

D. 2.22 years.

The correct answer is (B). *(Publisher)*

REQUIRED: The payback period.

DISCUSSION: When annual cash inflows are uniform, the payback period is calculated by dividing the initial investment ($210,000, as determined in question 41) by the annual net cash inflows ($136,800, as determined in question 42). Dividing $210,000 by $136,800 produces a payback period of 1.54 years.

Answer (A) is incorrect because 1.05 years fails to subtract income taxes. Answer (C) is incorrect because 1.33 years includes taxable income in the denominator instead of cash flows. Answer (D) is incorrect because 2.22 years subtracts depreciation from cash flows.

Questions 82 through 84 are based on the following information. Tonya Inc. has a cost of capital of 15% and is considering the acquisition of a new machine that costs $800,000 and has a useful life of 5 years. Tonya projects that earnings and cash flow will increase as follows:

Year	Net Earnings	After-Tax Cash Flow
1	$200,000	$320,000
2	200,000	280,000
3	200,000	200,000
4	200,000	200,000
5	200,000	200,000

Interest rate factors at 15% are as follows:

Period	Present Value of $1	Present Value of an Annuity
1	.87	0.87
2	.76	1.63
3	.66	2.29
4	.57	2.86
5	.50	3.36

82. The net present value of this investment is

A. $(128,000)

B. $200,000

C. $37,200

D. $400,000

The correct answer is (C). *(Publisher)*

REQUIRED: The net present value (NPV).

DISCUSSION: The NPV is calculated by discounting the after-tax cash flows from an investment by the cost of capital:

Year 1	$320,000	×	.87	=	$278,400
Year 2	280,000	×	.76	=	212,800
Year 3	200,000	×	.66	=	132,000
Year 4	200,000	×	.57	=	114,000
Year 5	200,000	×	.50	=	100,000
Present value of inflows					$837,200

Subtracting the $800,000 initial cost from the $837,200 present value of the future inflows produces an NPV of $37,200.

Answer (A) is incorrect because $(128,000) is based on a discounting of the net earnings rather than on cash flows. Answer (B) is incorrect because $200,000 equals total undiscounted net earnings minus the initial investment. Answer (D) is incorrect because $400,000 equals the undiscounted after-tax cash flows minus the initial cost.

83. What is the profitability index for the investment?

A. 0.05

B. 0.96

C. 1.05

D. 1.25

The correct answer is (C). *(Publisher)*

REQUIRED: The profitability index.

DISCUSSION: The profitability index is the present value of the future net cash flows divided by the present value of the net initial investment. The present value of the future net cash flows is $837,200 (see preceding question). Dividing the $800,000 initial cost of the investment into the $837,200 present value of the future net cash inflows produces a profitability index of approximately 1.05. Any profitability index greater than one is considered acceptable.

Answer (A) is incorrect because 0.05 results from using the NPV in the numerator. Answer (B) is incorrect because 0.96 reverses the numerator and denominator of the calculation. Answer (D) is incorrect because 1.25 uses undiscounted earnings in the numerator instead of the present value of future net cash flows.

84. What is the payback period of this investment?

A. 1.5 years.

B. 3.0 years.

C. 3.3 years.

D. 4.0 years.

The correct answer is (B). *(Publisher)*

REQUIRED: The payback period for the investment.

DISCUSSION: The payback method is simply the time required to recover the investment. It does not consider the time value of money or returns after the payback period. When cash flows are not uniform, a cumulative calculation is necessary. Thus, 3.0 years is the payback period ($320,000 in the first year, $280,000 in the second year, and $200,000 in the third year).

Answer (A) is incorrect because only a little over half of the initial $800,000 investment would have been recovered by the end of 1.5 years. Answer (C) is incorrect because $800,000 will have been recovered by the end of 3 years. Answer (D) is incorrect because 4.0 years is based on net earnings.

85. Metrejean Industries is analyzing a capital investment proposal for new equipment to produce a product over the next 8 years. At the end of 8 years, the equipment must be removed from the plant and will have a net book value of $0, a tax basis of $150,000, a cost to remove of $80,000, and scrap salvage value of $20,000. Metrejean's effective tax rate is 40%. What is the appropriate "end-of-life" cash flow related to these items that should be used in the analysis?

A. $90,000

B. $54,000

C. $24,000

D. $(36,000)

The correct answer is (C). *(Publisher)*

REQUIRED: The appropriate end-of-life cash flow related to the investment.

DISCUSSION: The tax basis of $150,000 and the $80,000 cost to remove are deductible expenses, but the $20,000 scrap value is an offsetting cash inflow. Thus, the taxable loss is $210,000 ($150,000 + $80,000 – $20,000). At a 40% tax rate, the $210,000 loss will produce a tax savings (inflow) of $84,000. Accordingly, the final cash flows will consist of an outflow of $80,000 (cost to remove) and inflows of $20,000 (scrap) and $84,000 (tax savings), a net inflow of $24,000.

Answer (A) is incorrect because $90,000 assumes that the loss on disposal is a cash inflow. It also ignores income taxes. Answer (B) is incorrect because $54,000 assumes that the loss on disposal involves a cash inflow. Answer (D) is incorrect because $(36,000) assumes that the tax basis is $0.

D. Risk

86. For capital budgeting purposes, management would select a high hurdle rate of return for certain projects because management

A. Wants to use equity funding exclusively.

B. Believes too many proposals are being rejected.

C. Believes bank loans are riskier than capital investments.

D. Wants to factor risk into its consideration of projects.

The correct answer is (D). *(CMA, adapted)*

REQUIRED: The reason for selecting a high hurdle rate for certain projects.

DISCUSSION: Risk analysis attempts to measure the likelihood of the variability of future returns from the proposed investment. Risk can be incorporated into capital budgeting decisions in a number of ways, one of which is to use a hurdle rate higher than the firm's cost of capital, that is, a risk-adjusted discount rate. This technique adjusts the interest rate used for discounting upward as an investment becomes riskier. The expected flow from the investment must be relatively larger or the increased discount rate will generate a negative net present value, and the proposed acquisition will be rejected.

Answer (A) is incorrect because the nature of the funding may not be a sufficient reason to use a risk-adjusted rate. The type of funding is just one factor affecting the risk of a project. Answer (B) is incorrect because a higher hurdle will result in rejection of more projects. Answer (C) is incorrect because a risk-adjusted high hurdle rate is used for capital investments with greater risk.

87. Mega Inc., a large conglomerate with operating divisions in many industries, uses risk-adjusted discount rates in evaluating capital investment decisions. Consider the following statements concerning Mega's use of risk-adjusted discount rates.

I. Mega may accept some investments with internal rates of return less than Mega's overall average cost of capital.

II. Discount rates vary depending on the type of investment.

III. Mega may reject some investments with internal rates of return greater than the cost of capital.

IV. Discount rates may vary depending on the division.

Which of the above statements are correct?

A. I and III only.

B. II and IV only.

C. II, III, and IV only.

D. I, II, III, and IV.

The correct answer is (D). *(CMA, adapted)*

REQUIRED: The true statement about use of risk-adjusted discount rates.

DISCUSSION: Risk analysis attempts to measure the likelihood of the variability of future returns from the proposed investment. Risk can be incorporated into capital budgeting decisions in a number of ways, one of which is to use a hurdle rate higher than the firm's cost of capital, that is, a risk-adjusted discount rate. This technique adjusts the interest rate used for discounting upward as an investment becomes riskier. The expected flow from the investment must be relatively larger, or the increased discount rate will generate a negative net present value, and the proposed acquisition will be rejected. Accordingly, the IRR (the rate at which the NPV is zero) for a rejected investment may exceed the cost of capital when the risk-adjusted rate is higher than the IRR. Conversely, the IRR for an accepted investment may be less than the cost of capital when the risk-adjusted rate is less than the IRR. In this case, the investment presumably has very little risk. Furthermore, risk-adjusted rates may also reflect the differing degrees of risk, not only among investments, but by the same investments undertaken by different organizational subunits.

Answer (A) is incorrect because discount rates may vary with the project or with the subunit of the organization. Answers (B) and (C) are incorrect because the company may accept some projects with IRRs less than the cost of capital, or reject some project with IRRs greater than the cost of capital.

88. Sensitivity analysis, if used with capital projects,

A. Is used extensively when cash flows are known with certainty.

B. Measures the change in the discounted cash flows when using the discounted payback method rather than the net present value method.

C. Is a "what-if" technique that asks how a given outcome will change if the original estimates of the capital budgeting model are changed.

D. Is a technique used to rank capital expenditure requests.

The correct answer is (C). *(CMA, adapted)*

REQUIRED: The true statement about sensitivity analysis.

DISCUSSION: After a problem has been formulated into any mathematical model, it may be subjected to sensitivity analysis, which is a trial-and-error method used to determine the sensitivity of the estimates used. For example, forecasts of many calculated NPVs under various assumptions may be compared to determine how sensitive the NPV is to changing conditions. Changing the assumptions about a certain variable or group of variables may drastically alter the NPV, suggesting that the risk of the investment may be excessive.

Answer (A) is incorrect because sensitivity analysis is useful when cash flows, or other assumptions, are uncertain. Answer (B) is incorrect because sensitivity analysis can be used with any of the capital budgeting methods. Answer (D) is incorrect because sensitivity analysis is not a ranking technique; it calculates results under varying assumptions.

89. An analysis of a company's planned equity financing using the Capital Asset Pricing Model (or Security Market Line) incorporates only the

A. Expected market earnings, the current U.S. treasury bond yield, and the beta coefficient.

B. Expected market earnings and the price-earnings ratio.

C. Current U.S. treasury bond yield, the price-earnings ratio, and the beta coefficient.

D. Current U.S. treasury bond yield and the dividend payout ratio.

The correct answer is (A). *(CMA, adapted)*

REQUIRED: The components of the capital asset pricing model.

DISCUSSION: The capital asset pricing model adds the risk-free rate to the product of the market risk premium and the beta coefficient. The market risk premium is the amount above the risk-free rate (approximated by the U.S. treasury bond yield) that must be paid to induce investment in the market. The beta coefficient of an individual stock is the correlation between the price volatility of the stock market as a whole and the price volatility of the individual stock.

Answers (B) and (C) are incorrect because the price-earnings ratio is not a component of the model. Answer (D) is incorrect because the dividend payout ratio is not a component of the model.

90. A company uses portfolio theory to develop its investment portfolio. If the company wishes to obtain optimal risk reduction through the portfolio effect, it should make its next investment in

A. An investment that correlates negatively to the current portfolio holdings.

B. An investment that is uncorrelated to the current portfolio holdings.

C. An investment that is highly correlated to the current portfolio holdings.

D. An investment that is perfectly correlated to the current portfolio holdings.

The correct answer is (A). *(CIA, adapted)*

REQUIRED: The relation of the next investment to the existing portfolio that results in optimal risk reduction.

DISCUSSION: A common general definition is that risk is an investment with an unknown outcome but a known probability distribution of returns (a known mean and standard deviation). An increase in the standard deviation (variability) of returns is synonymous with an increase in the riskiness of a project. Risk is also increased when the project's returns are positively (directly) correlated with other investments in the company's portfolio; that is, risk increases when returns on all projects rise or fall together. Consequently, the overall risk is decreased when projects have low variability and are negatively correlated (the diversification effect).

Answer (B) is incorrect because uncorrelated investments are more risky than negatively correlated investments. Answers (C) and (D) are incorrect because correlated investments are very risky.

STUDY UNIT 4: DECISION MAKING UNDER UNCERTAINTY

21 pages of outline
62 multiple-choice questions

A. Probability
B. Expected Value
C. Statistics
D. Hypothesis Testing

This study unit is the first of three that concern quantitative methods, a major topic that has been assigned a relative weight range of 10% to 15% of Part 4. Such a presentation makes the quantitative methods material more manageable for CMA candidates. The next two study units explain various managerial optimization and planning techniques. This study unit is important because it outlines probability and statistical material that underlies many of these models.

Remember to consider the tax consequences of business decision making. For a review, see the outlines in Appendix B, Taxes as Tested on the CMA/CFM Exams.

A. Probability

1. Probability is important to management decision making because of the unpredictability of future events. Probability estimation techniques assist in making the best decisions given doubt concerning outcomes.
   a. Decision making under conditions of **risk** occurs when the probability distribution of the possible future states of nature is known.
   b. Decision making under conditions of **uncertainty** occurs when the probability distribution of possible future states of nature is not known and must be subjectively determined.
2. Probability provides a method for mathematically expressing doubt or assurance about the occurrence of a chance event. The probability of an event varies from 0 to 1.
   a. A probability of 0 means the event cannot occur, whereas a probability of 1 means the event is certain to occur.
   b. A probability between 0 and 1 indicates the likelihood of the event's occurrence; e.g., the probability that a fair coin will yield heads is 0.5 on any single toss.
3. The two types of probability are objective and subjective. They differ in how they are calculated.
   a. **Objective probabilities** are calculated from either logic or actual experience. For example, in rolling dice one would logically expect each face on a single die to be equally likely to turn up at a probability of 1/6. Alternatively, the die could be rolled a great many times, and the fraction of times each face turned up could then be used as the frequency or probability of occurrence.
   b. **Subjective probabilities** are estimates, based on judgment and past experience, of the likelihood of future events. Weather forecasts often include the subjective probability of rain. In business, subjective probability can indicate the degree of confidence a person has that a certain outcome will occur, e.g., future performance of a new employee.

4. **Basic Terms Used with Probability**
   a. Two events are **mutually exclusive** if they cannot occur simultaneously (e.g., heads and tails cannot both occur on a single toss of a coin).
   b. The **joint probability** for two events is the probability that **both** will occur.
   c. The **conditional probability** of two events is the probability that one will occur given that the other has already occurred.
   d. Two events are **independent** if the occurrence of one has no effect on the probability of the other (e.g., rolling two dice).
      1) If one event has an effect on the other event, they are **dependent**.
      2) Two events are **independent** if their joint probability equals the product of their individual probabilities.
      3) Two events are **independent** if the conditional probability of each event equals its unconditional probability.
5. Probabilities can be combined using the following rules:
   a. The joint probability for two events equals the probability (Pr) of the first event multiplied by the conditional probability of the second event, given that the first has already occurred.
      1) EXAMPLE: If 60% of the students at a university are male, Pr(male) is 6/10. If 1/6 of the male students have a B average, Pr(B average given male) is 1/6. Therefore, the probability that any given student (male or female) selected at random, is **both** male **and** has a B average is

$$\begin{array}{ccccc} Pr(male) & \times & Pr(B|male) & = & Pr(male \cap B) \\ 6/10 & \times & 1/6 & = & 1/10 \end{array}$$

         $Pr(male \cap B)$ is .10; that is, the probability that the student is male **and** has a B average is 10%.
   b. The probability that either one or both of two events will occur equals the sum of their separate probabilities minus their joint probability.
      1) EXAMPLE: If two fair coins are thrown, the probability that at least one will come up heads is Pr(coin #1 is heads) plus Pr(coin #2 is heads) minus Pr(coin #1 and coin #2 are both heads), or

$$(.5) + (.5) - (.5 \times .5) = .75$$

      2) EXAMPLE: If in the earlier example 1/3 of all students, male or female, have a B average [Pr(B average) is 1/3], the probability that any given student either is male or has a B average is

$$\begin{array}{ccccccc} Pr(male) & + & Pr(B\ avg.) & - & Pr(B \cap male) & = & Pr(male\ or\ has\ B\ avg.) \\ 6/10 & + & 1/3 & - & 1/10 & = & 25/30 \end{array}$$

         The term $Pr(B \cap male)$ must be subtracted to avoid double counting those students who belong to both groups.
   c. The probabilities for all possible mutually exclusive outcomes of a single experiment must add up to one.

1) EXAMPLE: Flipping two coins (H = heads, T = tails)

If Coin #1 is	If Coin #2 is	Probability of This Combination
H	H	.25
H	T	.25
T	H	.25
T	T	.25
Probability that one of the four possible combinations will occur		1.00 (certainty)

6. A probability distribution specifies the values of the variables and their respective probabilities. Certain standard distributions seem to occur frequently in nature and have proven useful in business.

   a. Discrete distributions include the following:

      1) **Uniform distribution**. All outcomes are equally likely, such as the flipping of one coin, or even of two coins, as in the example above.

      2) **Binomial distribution**. Each trial has only two possible outcomes, e.g., accept or reject, heads or tails. This distribution shows the likelihood of each of the possible combinations of trial results. It is used in quality control.

         a) EXAMPLE: The social director of a cruise ship is concerned that the occupants at each dining room table be balanced evenly between men and women. The tables have only 6, 10, or 16 seats. If the population of the ship is exactly 50% male and 50% female [Pr(male) = .5 and Pr(female) = .5], what is the random chance for an equal balance of males and females at each table?

         The binomial formula is

$$\frac{n!}{r!(n-r)!} \times p^{r}(1-p)^{n-r}$$

         If: p is the probability of the given condition.
         n is the sample size.
         r is the number of occurrences of the condition within the sample.
         ! is factorial, i.e., 1 × 2 x 3 × ... n, or 1 × 2 x 3 × ... r.

         Given a .5 probability that any person in the population is male, the probability that exactly three males (and therefore three females) will be seated randomly at a table for six persons is

$$\frac{6!}{3!(6-3)!} \times .5^{3}(1-.5)^{6-3} = .3125$$

         The probability is only 31% that an equally balanced group of males and females will be randomly chosen. For the tables with 10 and 16 seats, the probabilities are .2461 and .1964, respectively. The social director will have to assign seats.

3) The **Bernoulli distribution** involves only one trial, whereas the binomial distribution deals with as many as necessary. Thus, the binomial distribution reduces to the Bernoulli distribution when n is 1.

4) The **hypergeometric distribution** is similar to the binomial distribution. It is used for sampling without replacement.

   a) For finite populations, sampling without replacement removes each item sampled from the population, thus changing the composition of the population from trial to trial.

      i) EXAMPLE: For a standard, well-shuffled deck of cards, the probability of picking the 2 of spades on the first draw is 1/52. If the card drawn is not replaced, the probability is 1/51 on the second draw, 1/50 on the third draw, etc. (unless, of course, the 2 of spades is drawn, in which case subsequent trial probabilities are 0). Sampling with replacement returns the card to the deck. The probability that any given card will be drawn is then 1/52 on every trial when sampling with replacement.

   b) For large populations and small samples, the binomial distribution approximates the hypergeometric distribution and is computationally more convenient.

5) The **Poisson distribution** is useful when the event being studied may happen more than once with random frequency during a given period.

   a) The Poisson distribution is defined as

$$f(k) = \frac{\lambda^k e^{-\lambda}}{k!} \qquad \textit{mean and variance} = \lambda$$

   If: k is the number of occurrences.
   e is the natural logarithm (2.71828...).

   b) When sample size is large and λ (lambda) is small (preferably less than 7), the Poisson distribution approaches the binomial distribution. In that case, λ is assumed to equal np.

   If: n = number of items sampled
   p = probability of a binomial event's occurrence

   c) EXAMPLE: A trucking company has established that, on average, two of its trucks are involved in an accident each month. It wishes to calculate the probability of a month with no (0) crashes and the probability of a month with four crashes. The mean of the Poisson distribution is λ, and the average monthly crash rate is two, so λ = 2. The probability of zero crashes in a given month is

$$f(0) = \frac{\lambda^0 e^{-\lambda}}{0!} = \frac{1e^{-2}}{1} = e^{-2} = .135 \qquad (\textit{Note}:\ 0! = 1)$$

   The probability of four crashes in a given month is

$$f(4) = \frac{\lambda^4 e^{-\lambda}}{4!} = \frac{2^4 e^{-2}}{4!} = .09$$

b. Continuous distributions include the following:

1) **Normal distribution**. The most important and useful of all probability distributions, it describes many physical phenomena. In sampling, it describes the distribution of the sample mean regardless of the distribution of the population. It has a symmetrical, bell-shaped curve centered about the mean (see the diagram below). For the normal distribution, about 68% of the area (or probability) lies within plus or minus 1 standard deviation of the mean, 95.5% lies within 2 standard deviations, and 99% lies within 3 standard deviations of the mean.

a) A special type of normal distribution is called the standard normal distribution. It has a mean of 0 and variance of 1. All normal distribution problems are first converted to the standard normal distribution to permit use of standard normal distribution tables.

b) Normal distributions have the following fixed relationships concerning the area under the curve and the distance from the mean.

Distance in Standard Deviations (confidence coefficient)	Area under the Curve (confidence level)
1.0	68%
1.64	90%
1.96	95%
2.0	95.5%
2.57	99%

EXAMPLE: If standard deviation (σ) = 10

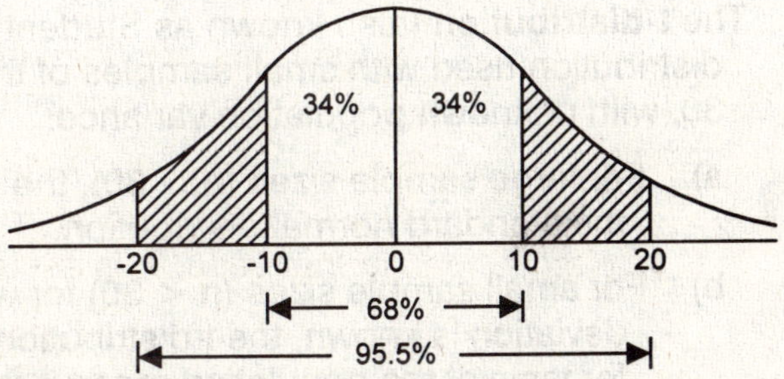

c) The standard deviation (sigma) is explained under the Statistics sideheading beginning on page 167.

2) The **exponential distribution** is related to the Poisson distribution. It is the probability of zero occurrences in a time period T.

a) The Poisson distribution has been defined as

$$f(k) = \frac{\lambda^k e^{-\lambda}}{k!}$$

The example in 6.a.5)c) stated that, when no events have occurred in the time period (k = 0),

$$f(0) = e^{-\lambda}$$

b) For the exponential distribution, M is used instead of λ.

$P = e^{-M}$
M = LT = frequency of the event's occurrence in this period
L = frequency or rate of the event's occurrence per unit of time
T = number of time units in this period
P = probability under exponential distribution

c) EXAMPLE: The trucking company mentioned in 6.a.5)c) on page 160 may wish to calculate the probability that a month will elapse without a single crash (the probability of a 1-month period between crashes).

L = 2 crashes per month
T = 1 month
$\therefore$ M = 2 × 1 = 2

The probability of a 1-month interval between crashes is then

$$P = e^{-M} = 2.71828^{-2} = .135$$

The probability of a 2-month period between crashes is

L = 2
T = 2
M = 2 × 2 = 4
$P = e^{-M} = 2.71828^{-4} = .0183$, or much less likely

3) The **t-distribution** (also known as Student's distribution) is a special distribution used with small samples of the population, usually less than 30, with unknown population variance.

a) For large sample sizes ($n > 30$), the t-distribution is almost identical to the standard normal distribution.

b) For small sample sizes ($n < 30$) for which only the sample standard deviation is known, the t-distribution provides a reasonable estimate for tests of the population mean if the population is normally distributed.

c) The t-distribution is useful in business because large samples are often too expensive. For a small sample, the t-statistic (from a t-table) provides a better estimate of the variance than that from a table for the normal distribution.

d) Also see section 5. under the sideheading on Hypothesis Testing.

4) The **chi-square distribution** is used in testing the goodness of fit between actual data and the theoretical distribution. In other words, it tests whether the sample is likely to be from the population, based on a comparison of the sample variance and the population variance.

a) The chi-square statistic (χ^2) is the sample variance (s^2), multiplied by its degree of freedom (n – 1), and divided by the hypothesized population variance (σ^2), if n is the number of items sampled.

b) A calculated value of the chi-square statistic greater than the critical value in the χ^2 table indicates that the sample chosen comes from a population with greater variance than the hypothesized population variance.

c) The chi-square test is useful in business for testing hypotheses concerning populations. If the variance of a process is known and a sample is tested to determine if it has the same variance, the chi-square statistic may be calculated.

d) EXAMPLE: A canning machine fills cans with a product and has exhibited a long-term standard deviation of .4 ounces, (σ = .4). A new machine is tested, but because the tests are expensive, only 15 cans are examined. The following is the result:

Sample standard deviation (s) = .311

The chi-square statistic is calculated as follows:

$$\chi^2 = \frac{(n - 1)s^2}{\sigma^2} = \frac{(15 - 1).311^2}{.4^2} = 8.463$$

Assume the null hypothesis (H_o) is that the new machine has a variance lower than or equal to the variance of the old machine, and that a probability of error (α) of .05 is acceptable. The χ^2 statistic for an α of .05 and 14 degrees of freedom is 23.68 in the χ^2 table. This critical value is much greater than the sample statistic of 8.463, so the null hypothesis cannot be rejected. Alpha (α) error is the error of incorrectly rejecting the true hypothesis. (See Hypothesis Testing on page 174.)

7. Stop and review! You have completed the outline for this subunit. Study multiple-choice questions 1 through 15 beginning on page 178.

B. Expected Value

1. For decisions involving risk, the concept of expected value provides a rational means for selecting the best alternative. The expected value of an action is found by multiplying the probability of each outcome by its payoff and summing the products. It represents the long-term average payoff for repeated trials. The best alternative is the one having the highest expected value.

   a. EXAMPLE: An investor is considering the purchase of two different pieces of property. The value of the property will change if a road, currently planned by the state, is built. The following are estimates that road construction will occur:

Future State of Nature (SN)	Probability
SN 1: No road is ever built.	.1
SN 2: A road is built this year.	.2
SN 3: A road is built more than 1 year from now.	.7

   The following are estimates of the values of the properties under each of the three possible future states of nature.

Property	Value if SN 1	Value if SN 2	Value if SN 3
Property 1	$10,000	$40,000	$35,000
Property 2	$20,000	$50,000	$30,000

   The expected value of each property is determined by multiplying the probability of each state of nature by the value under that state of nature and adding all of the products. Thus, Property 1 is the better investment if the two properties cost the same.

   P1: (.1 × $10,000) + (.2 × $40,000) + (.7 × $35,000) = $33,500 expected value
   P2: (.1 × $20,000) + (.2 × $50,000) + (.7 × $30,000) = $33,000 expected value

2. Some managers are reluctant to use subjectively derived numbers, arguing that they are not measurable. Certainly, there is no way to prove the accuracy of the estimates. Their use, however, makes explicit what would otherwise be decided subconsciously or intuitively. Once stated, the reasonableness of subjectively derived numbers can be examined.

   a. Another criticism of expected value is that it is based on repetitive trials, whereas many business decisions involve only one trial.

      1) EXAMPLE: A company wishes to launch a communications satellite. The probability of launch failure is .2, and the value of the satellite if the launch fails is $0. The probability of a successful launch is .8, and the value of the satellite would then be $25,000,000. The expected value is thus

      (.2 × $0) + (.8 × $25,000,000) = $20,000,000

      But $20,000,000 is not a possible value for a single satellite; either it flies for $25,000,000 or it crashes for $0.

3. **Decision trees** and **payoff tables** may be used to describe decision situations.

   a. EXAMPLE: A dealer in luxury yachts may order 0, 1, or 2 yachts for this season's inventory. The cost of carrying each excess yacht is $50,000, and the gain for each yacht sold is $200,000. The situation may be described by a payoff table as follows:

Season's Actual Demand	Order 0	Order 1	Order 2
0 yachts	0	$ (50,000)	$(100,000)
1 yacht	0	200,000	150,000
2 yachts	0	200,000	400,000

   Given the probabilities of the season's demand,

Pr	Demand
.10	0
.50	1
.40	2

   the dealer may calculate the expected value of each course of action as follows:

Order 0	Order 1	Order 2
.1 × 0	.1 × $ (50,000)	.1 × $(100,000)
.5 × 0	+ .5 × 200,000	+ .5 × 150,000
.4 × 0	+ .4 × 200,000	+ .4 × 400,000
EV(0) = 0	EV(1) = $175,000	EV(2) = $ 225,000

   The decision with the greatest expected value is to order two yachts, so, in the absence of additional information, the dealer should order two. The decision tree representation of this situation is

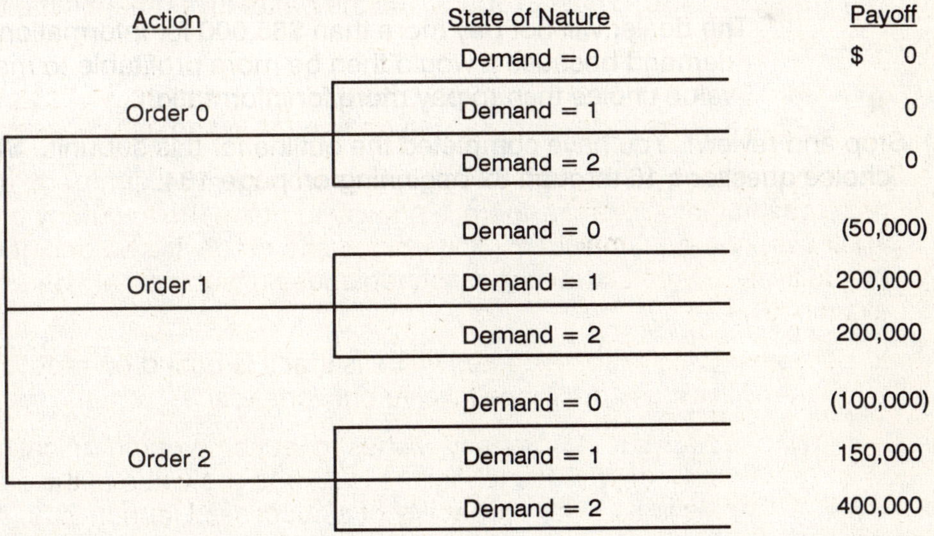

4. **Value of Perfect Information**

   a. **Perfect information** is the knowledge that a future state of nature will occur with certainty, i.e., being sure of what will occur in the future.

      1) The **expected value of perfect information (EVPI)** is the difference between the expected value without perfect information and the return if the best action is taken given perfect information.

      2) EXAMPLE (from the yacht dealer problem above): If the yacht dealer were able to poll all potential customers and they truthfully stated whether they would purchase a yacht this year (i.e., if perfect information about this

year's yacht sales could be purchased), what is the greatest amount of money the dealer should pay for this information? What is EVPI?

If the dealer had perfect knowledge of demand, (s)he would make the best decision for each state of nature. The cost of the other decisions is the conditional cost of making other than the best choice. This cost may be calculated by subtracting the expected value from the expected value given perfect information. This difference measures how much better off the decision maker would be with perfect information. From the payoff table on page 165, we find the expected value of the best choice under each state of nature.

Pr	State of Nature	Best Action	Best Action Payoff	Expected Value (Pr × Payoff)
.1	Demand = 0	Buy 0	$ 0	$ 0
.5	Demand = 1	Buy 1	200,000	100,000
.4	Demand = 2	Buy 2	400,000	160,000
				$260,000

The dealer expects to make $260,000 with perfect information about future demand, and $225,000 if the choice with the best expected value is made. The expected value of perfect information (EVPI) is then

Expected value with perfect information	$260,000
Expected value of the best choice	(225,000)
EVPI =	$ 35,000

The dealer will not pay more than $35,000 for information about future demand because it would then be more profitable to make the expected value choice than to pay more for information.

5. Stop and review! You have completed the outline for this subunit. Study multiple-choice questions 16 through 32 beginning on page 184.

C. Statistics

1. The field of statistics concerns information calculated from sample data. The field is divided into two categories: descriptive statistics and inferential statistics. Both are widely used in business.

   a. **Descriptive statistics** includes ways to summarize large amounts of raw data.

   b. **Inferential statistics** draws conclusions about a population based on a sample of the population.

   c. A **parameter** is a numerical characteristic of a population computed using every element in the population. For example, the mean and the mode are parameters of a population.

   d. A **statistic** is a numerical characteristic of a sample (taken from a population) computed using only the elements of the sample of the population. For example, the mean and the mode are statistics of the sample.

   e. Nonparametric, or distribution-free, statistics is applied to problems for which rank order is known, but the specific distribution is not. Thus, various metals may be ranked in order of hardness without having any measure of hardness.

2. **Descriptive statistics** summarizes large amounts of data. Measures of central tendency and measures of dispersion are such summaries.

   a. **Measures of central tendency** are values typical of a set of data.

      1) The **mean** is the arithmetic average of a set of numbers.

         a) The mean of a sample is often represented with a bar over the letter for the variable ($\bar{x}$).

         b) The mean of a population is often represented by the Greek letter μ (mu).

      2) The **median** is the halfway value if raw data are arranged in numerical order from lowest to highest. Thus, half the values are smaller than the median and half are larger. It is the 50th percentile.

      3) The **mode** is the most frequently occurring value. If all values are unique, no mode exists.

      4) **Asymmetrical distributions**

         a) The following is a frequency distribution that is asymmetrical to the right (positively skewed); that is, the mean is greater than the mode.

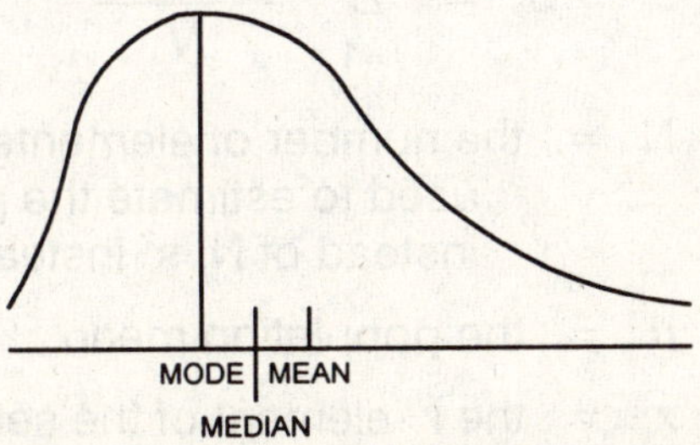

         b) Accounting distributions tend to be asymmetrical to the right. Recorded amounts are zero or greater, many low-value items are included, but a few high-value items may also be recognized.

c) The following is a distribution that is asymmetrical to the left; that is, the mode is greater than the mean.

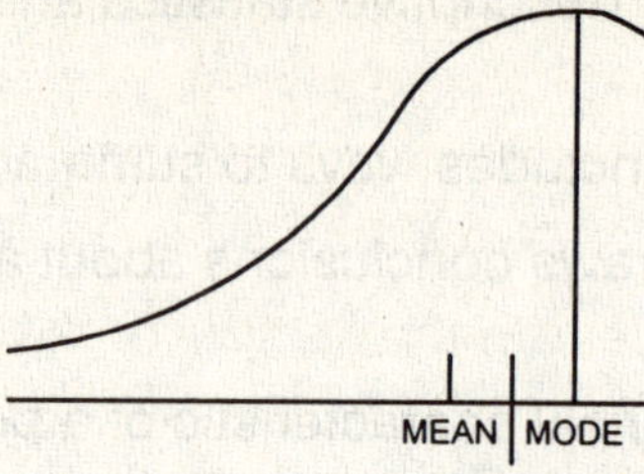

5) For symmetrical distributions, the mean, median, and mode are the same, and the tails are identical. Hence, there is no skew. The normal and t-distributions are symmetrical distributions.

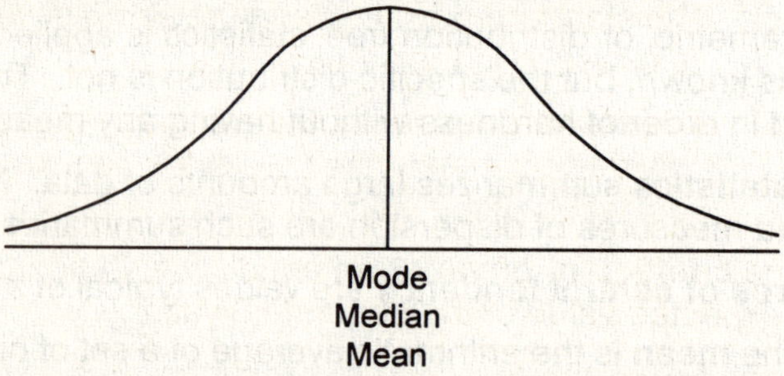

b. **Measures of dispersion** indicate the variation within a set of numbers.

1) An important operation involved is summation, represented by the capital Greek letter Σ (sigma). The summation sign means to perform the required procedure on every member of the set (every item of the sample) and then add all of the results.

2) The **variance** is the average of the squared deviations from the mean. It is found by subtracting the mean from each value, squaring each difference, adding the squared differences, and then dividing the sum by the number of data points. The variance of a population is represented by σ^2 (the lowercase Greek letter sigma squared).

a) The formula for the variance of a set is

$$\sigma^2 = \sum_{i=1}^{N} \frac{(x_i - \mu)^2}{N}$$

If: N = the number of elements in the population. If a sample is used to estimate the population variance, use n – 1 instead of N, s^2 instead of σ^2, and $\bar{x}$ instead of μ.)

μ = the population mean

x_i = the ith element of the set

3) The **standard deviation** is the square root of the variance.

a) The formula is the square root of the preceding one.

$$\sigma = \sqrt{\sum_{i=1}^{N} \frac{(x_i - \mu)^2}{N}}$$

b) For a sample with the sample mean $\bar{x}$, the population standard deviation (sigma) may be estimated from the **sample standard deviation**, s.

$$s = \sqrt{\sum_{i=1}^{n} \frac{(x_i - \bar{x})^2}{n - 1}}$$

c) The standard deviation and the sample standard deviation are always expressed in the same units as the data.

d) The **standard error of the mean** is the population standard deviation divided by the square root of the sample size. It is the standard deviation of the distribution of sample means.

4) The **coefficient of variation** equals the standard deviation divided by the expected value of the dependent variable.

a) For example, assume that a stock has a 10% expected rate of return with a standard deviation of 5%. The coefficient of variation is .5 (5% ÷ 10%).

b) Converting the standard deviation to a percentage permits comparison of numbers of different sizes. In the example above, the riskiness of the stock is apparently greater than that of a second stock with an expected return of 20% and a standard deviation of 8% (8% ÷ 20% = .4).

5) The **range** is the difference between the largest and smallest values of any group.

6) **Percentiles and quartiles** are other types of location parameters (the mean and median are special cases of these parameters). A percentile is a value of X such that p% of the observations are less and (100 – p)% are greater. Quartiles are the 25th, 50th, and 75th percentiles. For example, the 50th percentile (second quartile) is the median.

c. A **frequency distribution** is used to summarize raw data by segmenting the range of possible values into equal intervals and showing the number of data points within each interval.

1) EXAMPLE: A **histogram** is a bar graph of a frequency distribution with a vertical bar to represent the count of each interval. A histogram of the number of occurrences of various net cash flows looks like the following diagram:

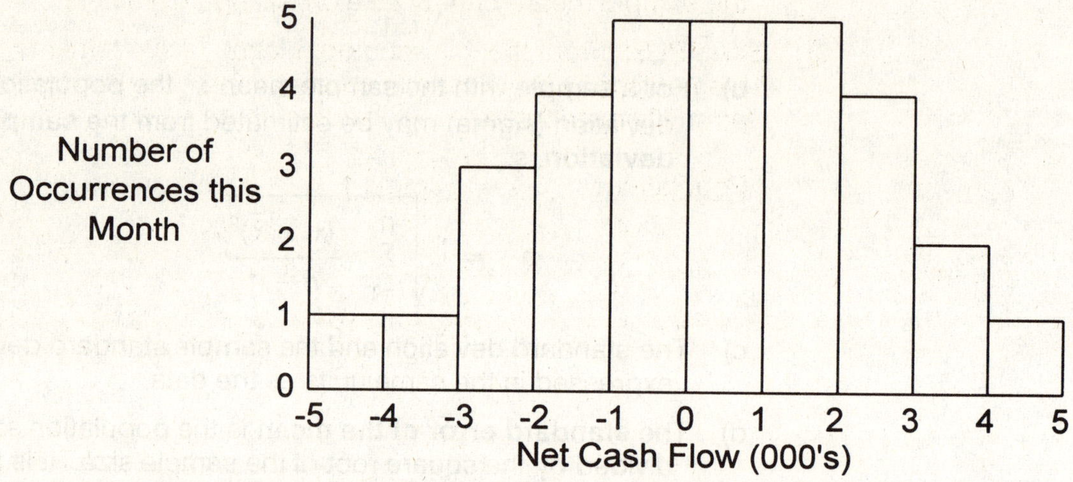

If: 5 days had cash flows between 0 and \$1,000.
5 days had cash flows between \$1,000 and \$2,000.
3 days had cash flows between –\$2,000 and –\$3,000, etc.

3. **Inferential statistics** provides methods for drawing conclusions about populations based on sample information.

   a. Inferential statistics applies to

      1) Estimating population parameters
      2) Testing hypotheses
      3) Examining the degree of relationship between two or more random variables

   b. **Sampling** is important in business because a complete census, i.e., measuring the entire population, is usually too costly, too time-consuming, impossible (as in the case of destructive testing), and error-prone. Sampling is used extensively in auditing, quality control, market research, and analytical studies of business operations.

   c. The **central limit theorem** states that, regardless of the distribution of the population from which random samples are drawn, the shape of the sampling distribution of $\bar{x}$ (the mean) approaches the normal distribution as the sample size is increased.

      1) Given simple random samples of size n, the mean of the sampling distribution of $\bar{x}$ will be μ (the population mean), its variance will be $\sigma^2 \div n$, and its standard deviation will be $\sigma \div \sqrt{n}$ (the standard error of the mean).

      2) Thus, whenever a process includes the average of independent samples of the same sample size from the same distribution, the normal distribution can be used as an approximation of that process even if the underlying population is not normally distributed. The central limit theorem explains why the normal distribution is so useful.

d. Population parameters may be estimated from sample statistics.

1) Every statistic has a sampling distribution that gives every possible value of the statistic and the probability of each of those values.

2) Hence, the point estimate calculated for a population parameter (such as the sample mean, $\overline{x}$) may take on a range of values.

3) EXAMPLE: From the following population of 10 elements, samples of three elements may be chosen in several ways. Assume that the population is normally distributed.

Population	Sample 1	Sample 2	Sample 3
4	4	7	6
7	5	6	9
9	3	5	5
5	Σx_i = 12	Σx_i = 18	Σx_i = 20
6	n = 3	n = 3	n = 3
5			
3	$\overline{x}$ = 12÷3	$\overline{x}$ = 18÷3	$\overline{x}$ = 20÷3
5	= 4	= 6	= 6.67
6			
6			
Σx_i = 56	μ = 56 ÷ 10 = 5.6		
N = 10			

σ (*standard deviation*) = 1.562 *based on* $\sqrt{\Sigma(x_i - \mu)^2 \div N}$

NOTE: This sample population was chosen for computational convenience only. The population in this example is so small that inference is not required, and the samples are so small that the t-distribution would be more appropriate than the normal distribution.

e. The quality of the estimates of population parameters depends on two things: the sample size and the variance of the population

f. The **precision (interval estimator, confidence interval**, or **prediction interval)** incorporates both the sample size and the population standard deviation along with a probability that the interval includes the true population parameter.

1) For the population mean, this interval is

$$\overline{x} \pm z\ (\sigma \div \sqrt{n})$$

If: $\overline{x}$ = the sample mean, which is a point estimate of the population mean

z = the number of standard deviations needed to ensure a specified level of confidence

σ = the standard deviation of the population

n = the sample size

$\sigma \div \sqrt{n}$ = the standard error of the mean (square root of the variance of the sampling distribution of $\overline{x}$)

a) This representation of precision assumes that the variance (σ^2) of the population is known, the sample means are normally distributed with a mean equal to the true population mean (μ), and the variance of the sampling distribution is $\sigma^2 \div n$.

b) In the more realistic case in which the population variance is not known, and a sample is being evaluated, the distribution is described as a t-distribution with mean equal to μ and variance equal to $s^2 \div n$, when s^2 is the sample variance.

c) To compute the precision for the mean of the population given in 3.d. based on a sample, the mean of the sample must be determined. For example, the mean ($\bar{x}$) of Sample 2 is 6. Thus, the sample standard deviation is

$$s = \sqrt{\frac{(7-6)^2 + (6-6)^2 + (5-6)^2}{3-1}} = 1.0$$

d) To compute the precision, the z-value is found in a table for the standard normal distribution. If a two-tailed test is desired and the confidence level is set at 95%, 2.5% of the area under the normal curve will lie in each tail. Thus, the entries in the body of the table will be .9750 and .0250. These entries correspond to z-values of 1.96 and −1.96, respectively. Accordingly, 95% of the area under the curve for the standard normal distribution lies within 1.96 standard deviations of the mean. Hence, the 95% precision interval is $6 \pm 1.96(\sigma \div \sqrt{n})$. Because the population standard deviation is not known, the sample standard deviation ($s = 1.0$) is used (see C.2.b.3) on page 169). Precision then becomes

$$6 \pm 1.96\ (1.0 \div \sqrt{3}) = 6 \pm 1.13 = 4.87 \textit{ to } 7.13$$

Consequently, the probability is 95% that this interval contains the population mean.

4. **Statistical Quality Control**

a. Statistical quality control is a method of determining whether the shipment or production run of units lies within acceptable limits. It is also used to determine whether production processes are out of control.

1) Items are either good or bad, i.e., inside or outside of control limits.
2) Statistical quality control is based on the binomial distribution.

b. Acceptance sampling is a method of determining the probability that the defective rate in a batch is less than a certain level.

1) EXAMPLE: Assume a sample is taken and the probability that the sample was from a population of 500 with a specified deviation rate is calculated. According to standard acceptance sampling tables, if the sample consists of 25 items and none is defective, there is a 93% probability that the population deviation rate is less than 10%. If 60 items are examined and no defectives are found, the probability is 99% that the deviation rate is less than 10%. If two defectives in 60 units are observed, there is a 96% probability that the deviation rate is less than 10%.

c. Statistical control charts are graphic aids for monitoring the status of any process subject to random variations. Originally developed to control the quality of production processes, they also have applications of direct interest to the management accountant, for example, unit cost of production, direct labor hours used, ratio of actual expenses to budgeted expenses, number of calls by sales personnel, or total accounts receivable.

d. The chart consists of three horizontal lines plotted on a horizontal time scale. The center line represents the average or mean value for the process being controlled. The other two lines are the upper control limit (UCL) and the lower control limit (LCL). The processes are measured periodically, and the values are plotted on the chart (X). If the value falls within the control limits, no action is taken. If the value falls outside the limits, the process is considered out of control, and an investigation is made for possible corrective action. Another advantage of the chart is that it makes trends visible.

1) **P charts** are based on an attribute (acceptable/not acceptable) rather than a measure of a variable. Specifically, it shows the percentage of defects in a sample.

2) **C charts** are also attribute control charts. They show defects per item.

3) An **R chart** shows the range of dispersion of a variable, such as size or weight.

4) An **X-bar chart** shows the sample mean for a variable.

5) EXAMPLE:

```
Unit Cost ($)                 X  Out of control
1.05   -------------------------------------------------UCL
1.00   ______________X__________________________________
0.95   -----X-------------------------------------------LCL
          March     April     May
```

e. Variations in the value of some process parameter may have several causes.

1) **Random variations** occur by chance. Present in virtually all processes, they are not correctable because they will not repeat themselves in the same manner. Excessively narrow control limits will result in many investigations of what are simply random fluctuations.

2) **Implementation deviations** occur because of human or mechanical failure to achieve target results.

3) **Measurement variations** result from errors in the measurements of actual results.

4) **Model fluctuations** can be caused by errors in the formulation of a decision model.

5) **Prediction variances** result from errors in forecasting data used in a decision model.

f. Establishing control limits based on benchmarks is a common method. A more objective method is to use the concept of expected value. The limits are important because they are the decision criteria for determining whether a deviation will be investigated.

g. Cost-benefit analysis using expected value provides a more objective basis for setting control limits. The limits of controls should be set so that the cost of an investigation is less than or equal to the benefits derived.

1) The expected costs include investigation cost and the cost of corrective action.

(*Probability of being out of control* × *Cost of corrective action*)
\+ (*Probability of being in control* × *Investigation cost*)
Total expected cost

2) The benefit of an investigation is the avoidance of the costs of continuing to operate an out-of-control process. The expected value of benefits is the probability of being out of control multiplied by the cost of not being corrected.

5. Stop and review! You have completed the outline for this subunit. Study multiple-choice questions 33 through 53 beginning on page 189.

D. Hypothesis Testing

1. A hypothesis is a supposition about the true state of nature. Hypothesis testing calculates the conditional probability that both the hypothesis is true **and** the sample results observed have occurred. Four steps are

a. A hypothesis is formulated to be tested.

b. Sample evidence is obtained.

c. The probability that the hypothesis is true, given the observed evidence, is computed.

d. If that probability is too low, the hypothesis is rejected. Whether a probability is too low is a subjective measure dependent on the situation. A probability of .6 that a team will win may be sufficient to place a small bet on the next game. A probability of .95 that a parachute will open is too low to justify skydiving.

2. The hypothesis to be tested is the **null hypothesis** or H_o; the alternative hypothesis is denoted H_a.

a. The null hypothesis, H_o, may state an equality (=) or indicate that the parameter is equal to or greater (less) than (≥ or ≤) some value.

b. The alternative hypothesis, H_a, contains every other possibility.

1) It may be stated as not equal to (≠), greater than (>), or less than (<) some value, depending on the null hypothesis.

c. The probability of error in hypothesis testing is usually labeled as

State of Nature	Decision: Do not reject H_o	Decision: Reject H_o
H_o is true	Correct	Type I Error $P(I) = \alpha$
H_o is false	Type II Error $P(II) = \beta$	Correct

1) These are the same α (alpha) and β (beta) errors familiar to auditors.

d. EXAMPLE: The hypothesis is that a component fails at a pressure of 80 or more lb. on the average; i.e., the average component will not fail at a pressure below 80 lb. For a group of 36 components, the average failure pressure was found to be 77.48 lb. Given that n is 36, $\bar{x}$ is 77.48 lb., and σ is 13.32 lb., the following are the hypotheses:

H_o: The average failure pressure of the population of components is ≥ 80 lb.

H_a: The average failure pressure is < 80 lb.

1) If a 5% chance of being wrong is acceptable, α (Type I error or the chance of incorrect rejection of the null hypothesis) is set equal to .05 and the confidence level at .95. In effect, 5% of the area under the curve of the standard normal distribution will constitute a rejection region. For this one-tailed test, the 5% rejection region will fall entirely in the left-hand tail of the distribution because the null hypothesis will not be rejected for any values of the test statistic that fall in the right-hand tail. According to standard tables, 5% of the area under the standard normal curve lies to the left of the z-value of –1.645.

The following is the formula for the z-statistic:

$$z = \frac{\bar{x} - \mu_o}{\sigma \div \sqrt{n}}$$

If:
- σ = given population standard deviation
- μ_o = hypothesized true population mean
- n = sample size
- z = number of standard deviations needed to ensure the specified level of confidence
- $\bar{x}$ = the sample mean

2) Substituting the hypothesized value of the population mean failure pressure (μ_o = 80 lb.) determines the z-statistic.

$$z = \frac{77.48 - 80}{13.32 \div \sqrt{36}} = -1.135$$

3) Because the calculated z-value corresponding to the sample mean of 77.48 is greater than the critical value of –1.645, the null hypothesis cannot be rejected.

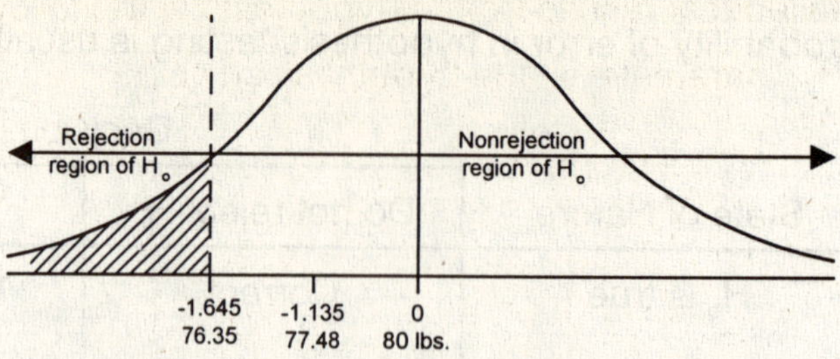

a) The lower limit of the 95% nonrejection area under the curve is

80 *lb.* – 1.645 ($\sigma \div \sqrt{n}$) =
80 *lb.* – 1.645 ($13.32 \div \sqrt{36}$) =
76.35 *lb.*

b) Because a sample average of 77.48 lb. (a z of –1.135) falls within the nonrejection region (i.e., > 76.35 lb.), the null hypothesis that the average failure pressure of the population is ≥ 80 lb. cannot be rejected. The null hypothesis is rejected only if the sample average is equal to or less than the critical value (76.35 lb.).

3. Hypothesis tests may be one-tailed or two-tailed.

   a. A one-tailed test results from a hypothesis of the following form:

      H_o: parameter ≤ or ≥ the hypothesized value
      H_a: parameter > or < the hypothesized value

      1) One-tailed test, upper tail

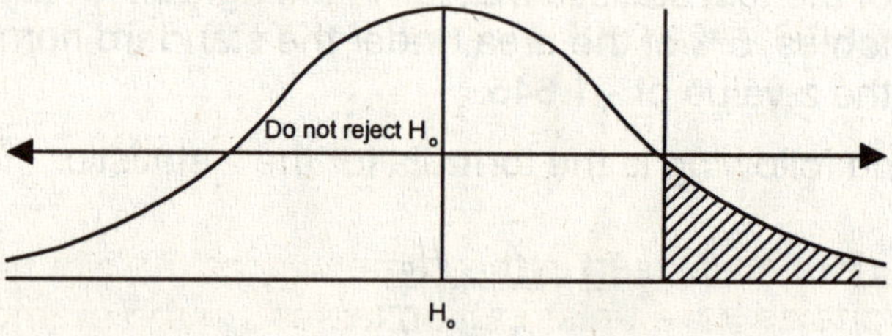

      H_o: parameter ≤ the hypothesized value
      H_a: parameter > the hypothesized value

      2) One-tailed test, lower tail

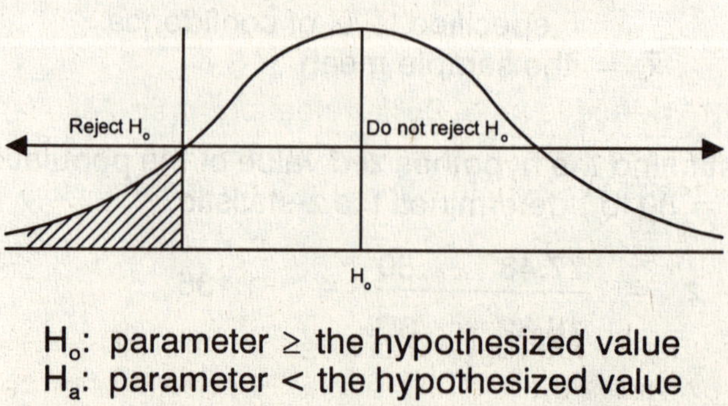

      H_o: parameter ≥ the hypothesized value
      H_a: parameter < the hypothesized value

b. A two-tailed test results from a hypothesis of the following form:

H_o: parameter = the hypothesized value
H_a: parameter ≠ the hypothesized value

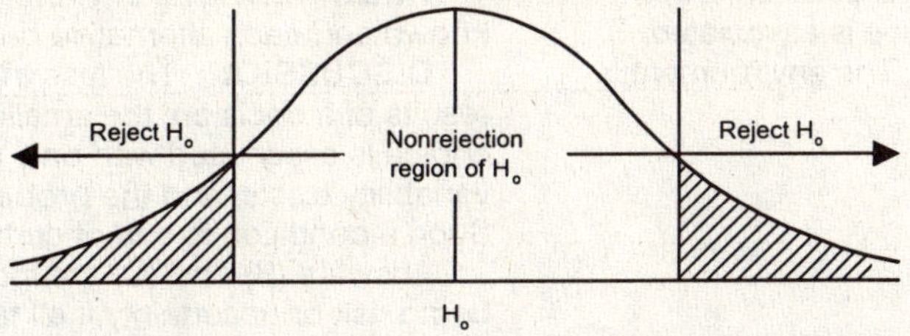

4. A failure to prove H_o is false does not prove that it is true. This failure simply means that H_o is not a rejectable hypothesis. In practice, however, auditors often use acceptance as a synonym for nonrejection.

5. Given a small sample (less than 30), the t-statistic (t-distribution) must be used.

   a. The **t-distribution** requires a number called **the degrees of freedom**, which is (n – k) for k parameters. When one parameter (such as the mean) is estimated, the number of degrees of freedom is (n – 1). The degrees of freedom is a correction factor that is necessary because, given k parameters and n elements, only (n – k) elements are free to vary. After (n – k) elements are chosen, the remaining k elements' values are already determined.

      1) EXAMPLE: Two numbers have an average of 5.

$$\bar{x} = \frac{(x_1 + x_2)}{2} = 5$$

If x_1 is allowed to vary but the average remains the same, x_1 determines x_2 because only 1 degree of freedom (n – 1) or (2 – 1) is available.

If: $x_1 = 2, x_2 = 8$
$x_1 = 3, x_2 = 7$

   b. The t-distribution is used in the same way as the z or normal distribution. Standard texts have t-distribution tables. In the example in D.2.d., if the sample size had been 25 and the sample standard deviation had been given instead of the population value, the t-statistic would have been

$$t = \frac{\bar{x} - \mu_o}{s \div \sqrt{25}} = \frac{77.48 - 80}{13.32 \div \sqrt{25}} = -.946$$

      1) At a confidence level of 95% (rejection region of 5%) and 24 degrees of freedom (sample of 25 – 1 parameter estimated), the t-distribution table indicates that 5% of the area under the curve is to the left of a t-value of –1.711. Because the computed value is greater than –1.711, the null hypothesis cannot be rejected in this one-tailed test.

      2) As the number of degrees of freedom increases, the t-distribution approximates the z-distribution. For degrees of freedom > 30, the z-distribution may be used.

6. Stop and review! You have completed the outline for this subunit. Study multiple-choice questions 54 through 62 beginning on page 198.

MULTIPLE-CHOICE QUESTIONS

A. Probability

1. A decision maker is operating in an environment in which all the facts surrounding a decision are known exactly, and each alternative is associated with only one possible outcome. The environment is known as

A. Certainty.

B. Risk.

C. Uncertainty.

D. Conflict.

The correct answer is (A). *(CMA, adapted)*

REQUIRED: The environment in which all facts are known and each alternative has only one possible outcome.

DISCUSSION: The less the variability of the future results of a decision, the smaller the risk involved. If each choice is associated with only one possible outcome, no variability exists, and the probability of the outcome is 1.0. Such a condition is one of certainty (zero variance).

Answers (B) and (C) are incorrect because there would be no risk or uncertainty if all facts were known and there was only one possible outcome for each alternative. Answer (D) is incorrect because conflict between alternatives may exist even when all facts are known and each choice has only one outcome.

2. A company uses two major material inputs in its production. To prepare its manufacturing operations budget, the company has to project the cost changes of these material inputs. The cost changes are independent of one another. The purchasing department provides the following probabilities associated with projected cost changes:

Cost Change	Material 1	Material 2
3% increase	.3	.5
5% increase	.5	.4
10% increase	.2	.1

The probability of a 3% increase in the cost of both Material 1 and Material 2 is

A. 15%

B. 40%

C. 80%

D. 20%

The correct answer is (A). *(CIA, adapted)*

REQUIRED: The probability of joint increases.

DISCUSSION: The joint probability of occurrence of two independent events equals the product of their individual probabilities. The probability that the cost of Material 1 will increase by 3% is .3. The probability that the cost of Material 2 will increase by 3% is .5. The probability that both will occur is .15 (.3 × .5).

Answer (B) is incorrect because 40% is the average of the probabilities of a 3% increase in the costs of Material 1 and Material 2. Answer (C) is incorrect because 80% is the sum of the probabilities of a 3% increase in the costs of Material 1 and Material 2. Answer (D) is incorrect because 20% is the difference between the probabilities of an increase in the costs of Material 1 and Material 2.

Questions 3 and 4 are based on the following information.

A simple two-state cost investigation model calls for a decision to investigate if the following inequality holds:

$$L(1 - P) > I + C(1 - P)$$

If: L = excess operating cost if out of control
P = probability that the process is in control
I = cost of investigation
C = cost of correction if out of control

3. The value (1 – P) must be

A. The probability that the process is out of control.

B. A measure of the seriousness of the consequences.

C. The inverse of the probability matrix.

D. The transpose of the probability matrix.

The correct answer is (A). *(CMA, adapted)*

REQUIRED: The true statement about the given value.

DISCUSSION: The sum of the probabilities that the process is in or out of control is 1.0. Thus, if P is the probability that a process is in control, (1 – P) is its complement, or the probability that the process is not in control.

Answers (B), (C), and (D) are incorrect because each is a nonsense answer.

4. The critical probability that the process is in control is

A. $I/(L - C) - 1$.

B. $(I + C)/L$.

C. $1 - I/(L - C)$.

D. Greater than I/C.

The correct answer is (C). *(CMA, adapted)*

REQUIRED: The item describing the critical probability that the process is in control.

DISCUSSION: The critical probability that the process is in control is defined as P. Transforming the inequality into an equality and solving for P determines the value of the critical probability. For any lesser value of P, given constant values of L, I, and C, investigation will be justified.

$$L(1 - P) = I + C(1 - P)$$
$$(L - C)(1 - P) = I$$
$$1 - P = I \div (L - C)$$
$$P = 1 - I \div (L - C)$$

For example, if I equals \$18, L equals \$100, and C equals \$10, then the lowest (critical) probability at which the process is considered to be in control is 0.8 [1 – \$18 ÷ (\$100 – \$10)]. At this value, the expected value of the operating cost savings from investigation equals the sum of the cost of investigation and the expected value of the cost of correction.

Answers (A), (B), and (D) are incorrect because transforming the inequality to an equality and solving for P gives the expression 1 – [I ÷ (L – C)].

Questions 5 and 6 are based on the following information. A computer store sells four computer models designated as P104, X104, A104, and S104. The store manager has made random number assignments to represent customer choices based on past sales data. The assignments are shown below.

Model	Random Numbers
P104	0-1
X104	2-6
A104	7-8
S104	9

5. The probability that a customer will select model P104 is

A. 10%

B. 20%

C. 50%

D. Some percentage other than those given.

The correct answer is (B). *(CMA, adapted)*

REQUIRED: The probability that a customer will select model P104.

DISCUSSION: Ten random numbers have been assigned. Of these, two (0 and 1) have been assigned to model P104. Thus, there are two chances out of ten, or 20%, that a customer will select that model.

Answer (A) is incorrect because 10% is the probability of selecting S104. Answer (C) is incorrect because 50% is the probability of selecting X104. Answer (D) is incorrect because 20% is among the responses given.

6. In running a simulation of the computer demand, the following numbers are drawn in sequence: 2, 8, and 6. The simulation indicates that the third customer will purchase.

A. Model P104.

B. Model X104.

C. Model A104.

D. Model S104.

The correct answer is (B). *(CMA, adapted)*

REQUIRED: The model that will be purchased by the third customer when the numbers drawn are 2, 8, and 6.

DISCUSSION: The third customer is simulated by the third number drawn. Hence, the third customer's purchase is represented by the number 6. The numbers 2 through 6 correspond to model X104. Thus, the third customer is expected to purchase model X104.

Answer (A) is incorrect because Model P104 corresponds to numbers 0 and 1. Answer (C) is incorrect because Model A104 corresponds to numbers 7 and 8. Answer (D) is incorrect because Model S104 corresponds to number 9.

7. Ron Bagley is contemplating whether to investigate a labor efficiency variance in the Assembly Department. It will cost $6,000 to undertake the investigation and another $18,000 to correct operations if the department is found to be operating improperly. If the department is operating improperly and Bagley fails to investigate, operating costs from the various inefficiencies are expected to amount to $33,000. Bagley will be indifferent between investigating and not investigating the variance if the probability of improper operation is

A. 0.29

B. 0.40

C. 0.60

D. 0.71

The correct answer is (B). *(CMA, adapted)*

REQUIRED: The probability of improper operation at which the manager is indifferent about investigating the variance.

DISCUSSION: Costs will be either $6,000 or $24,000 if the variance is investigated and either $0 or $33,000 if not. The probabilities at which the expected costs of investigation are equal to the expected costs of not investigating can be determined from the following equation (X = the probability that operations are improper):

$$\begin{aligned} \$24{,}000X + \$6{,}000\,(1 - X) &= \$33{,}000X - \$0\,(1 - X) \\ \$24{,}000X + \$6{,}000 - 6{,}000X &= \$33{,}000X \\ X &= .4 \\ 1 - X &= .6 \end{aligned}$$

Answer (A) is incorrect because, if X is less than 0.40, the investigation is not cost beneficial. Answer (C) is incorrect because 0.60 is the probability that the department is operating properly at which the manager is indifferent about investigating the variance. Answer (D) is incorrect because the investigation should be made if X is greater than 0.40.

Questions 8 and 9 are based on the following information. The probabilities shown in the table represent the estimate of sales for a new product.

Sales (Units)	Probability
0-200	15%
201-400	45%
401-600	25%
601-800	15%

8. What is the probability of selling between 201 and 600 units of the product?

A. 0%

B. 11.25%

C. 70%

D. 25%

The correct answer is (C). *(Publisher)*

REQUIRED: The probability of selling between 201 and 600 units of the new product.

DISCUSSION: The probability of selling between 201 and 400 units is 45%, and the probability of selling between 401 and 600 units is 25%. Hence, the probability of selling between 201 and 600 units is the sum of these probabilities, or 70%.

Answers (A) and (B) are incorrect because there is a 70% probability of selling between 201 and 600 units. Answer (D) is incorrect because 25% is the probability of selling between 401 and 600 units.

9. What is the best estimate of the expected sales of the new product?

A. 480

B. 380

C. 400

D. 800

The correct answer is (B). *(Publisher)*

REQUIRED: The best estimate of the expected sales of the new product.

DISCUSSION: The expected sales levels should be weighted by the individual probabilities of their occurrence. The midpoint of each sales level is used as the estimate for that level. Thus, sales are expected to be 380 units.

$$\begin{aligned} 100 \times 15\% &= 15 \\ 300 \times 45\% &= 135 \\ 500 \times 25\% &= 125 \\ 700 \times 15\% &= \underline{105} \\ & \quad \underline{380} \end{aligned}$$

Answer (A) is incorrect because 480 is based on the maximum value in each range. Answer (C) is incorrect because 400 is the maximum value in the modal range. Answer (D) is incorrect because 800 is the highest estimate.

10. Ryerson Company has three sales departments, each contributing the following percentages of total sales: clothing, 50%; hardware, 30%; and household sundries, 20%. Each department has had the following average annual damaged goods rates: clothing, 2%; hardware, 5%; and household sundries, 2.5%. A random corporate audit has found a weekly damaged goods rate of sufficient magnitude to alarm Ryerson's management. The probability (rounded) that this rate occurred in the Clothing Department is

A. 50%

B. 1%

C. 25%

D. 33⅓%

The correct answer is (D). *(CMA, adapted)*

REQUIRED: The probability that a quality control problem has occurred in the Clothing Department.

DISCUSSION: Weighting the percentage of sales by the percentage of quality control problems for each department determines the annual company rate of damaged goods.

Department	Sales	Damages	Weighted
Clothing	50%	2%	1%
Hardware	30%	5%	1.5%
Household	20%	2.5%	0.5%
			3%

The Clothing Department accounts for 33⅓% (1% ÷ 3%) of the total company rate. Thus, the probability that an observed problem is in the Clothing Department is 33⅓%.

Answer (A) is incorrect because the rate is 50% (1.5% ÷ 3%) for the Hardware Department. Answer (B) is incorrect because damages in the Clothing Department contribute to 1% of the company rate, but the question asks for the probability that a problem occurred in the Clothing Department. Answer (C) is incorrect because the Clothing Department quality problems make up 1/3 of the overall company problems.

11. A quantitative technique useful in projecting a firm's sales and profits is

A. Probability distribution theory.

B. Gantt charting.

C. Learning curves.

D. Queuing theory.

The correct answer is (A). *(CMA, adapted)*

REQUIRED: The quantitative technique useful in projecting sales and profits.

DISCUSSION: Probability distribution theory can be used to project sales. It is a mathematical method for making decisions about the likelihood of future events (such as sales) in the face of uncertainty. Various estimates of sales (generated from the sales force) can be weighted with different probabilities.

Answer (B) is incorrect because a Gantt chart is a bar chart used to measure progress toward a goal. Answer (C) is incorrect because a learning curve measures the benefit of experience in the early stages of a new task. Answer (D) is incorrect because queuing (waiting-line) theory is used to determine the optimum balance between the cost of providing service to reduce waiting lines and the cost of allowing waiting lines to exist when items in the queue arrive at random.

Questions 12 and 13 are based on the following information. Brown Maintenance Services has been performing technical maintenance service on a wide variety of business machines for small businesses in a large metropolitan area for many years. Brown has been considering a change in its fee structure. The analysis requires a probability distribution for each week for the number of anticipated service calls related to each of several different types of equipment.

12. Phil Brown suggested establishing the distribution using the relative frequency of various numbers of weekly service calls over the last 4 years. The following tabulation was prepared:

Number of Calls	Number of Occurrences
801-850	4
851-900	10
901-950	80
951-1,000	40
1,001-1,050	20
1,051-1,100	12
1,101-1,150	12
1,151-1,200	10
1,201-1,250	8
1,251-1,300	4

Based on this probability distribution, the probability of more than 1,150 calls for service during a given week is

A. .06

B. .89

C. .17

D. .11

The correct answer is (D). *(Publisher)*

REQUIRED: The probability that more than 1,150 calls will be received in a week.

DISCUSSION: Of the 200 recorded occurrences, only 22 were over 1,150 per week: 10 in the 1,151-1,200 range, 8 in the 1,201-1,250 range, and 4 in the 1,251-1,300 range. Thus, the probability of a number of calls greater than 1,150 is simply (10 + 8 + 4) ÷ 200 = .11.

Answer (A) is incorrect because .06 is the probability that over 1,200 calls will be received. Answer (B) is incorrect because .89 is the probability that fewer than 1,151 calls will be received. Answer (C) is incorrect because .17 is the probability that more than 1,100 calls will be received.

13. Billy Brown disagreed with his brother, arguing that the number and size of the companies with which they held maintenance contracts had changed recently. He concluded that this past experience was not relevant, and the firm would be better off recognizing their ignorance of the future in the rate restructuring. Billy proposed the assumption of a uniform distribution with every possible number of calls from 801 to 1,300 being equally likely. Based on this uniform distribution, the probability of more than 1,150 calls during a week is

A. .40

B. .30

C. .70

D. .20

The correct answer is (B). *(Publisher)*

REQUIRED: The probability that more than 1,150 calls will be received in a week.

DISCUSSION: If each of 500 outcomes (1,300 – 800) is equally likely, the probability of more than 1,150 calls in a week is 30% [(1,300 – 1,150) ÷ 500].

Answers (A), (C), and (D) are incorrect because the probability is determined by dividing the number of outcomes greater than 1,150 by the total number of outcomes.

14. Two firms share customers in the same market. Firm A sampled its customers' buying habits and found that about 70% of its customers were repeat customers each week, while 30% went to Firm B. Firm B found that 80% of its customers remained loyal each week, while 20% switched to Firm A. If this retention and loss of customers continues for a long period, the percentage of customers Firm A will have is

A. 70%

B. 80%

C. 60%

D. 40%

The correct answer is (D). *(CMA, adapted)*

REQUIRED: The percentage of customers Firm A will have in the long run.

DISCUSSION: Each firm's retention and loss of customers should be defined in equation form, and the results should be solved simultaneously. Thus, 100% of A's customers minus B's customers equals 70% repeat business of A plus 20% of new business from B's former customers. In equation form, A's customers are .7A + .2B = 100% – B. Similarly, the equation for Firm B's customers is .8B + .3A = 100% – A. Simplifying the equations gives

$$A = \frac{100\% - 1.2B}{.7} \qquad B = \frac{100\% - 1.3A}{.8}$$

Substituting the B equation into the equation for A,

$$A = \frac{100\% - 1.5\ (100\% - 1.3A)}{.7}$$

$$A = \frac{100\% - 150\% + 1.95A}{.7} = \frac{-50\% + 1.95A}{.7}$$

$$.7A = -50\% + 1.95A$$

$$1.25A = 50\%$$

$$A = 40\%$$

Answer (A) is incorrect because 70% equals Firm A's percentage of repeat customers. Answer (B) is incorrect because 80% equals Firm B's percentage of repeat customers. Answer (C) is incorrect because 60% is the percentage of customers Firm B will have in the long run.

15. Two projects are going to be implemented. The probability for positive cash flows in each one is 1/2, and the probability for negative cash flows in each one is 1/2. What is the probability that at least one of the projects will have positive cash flows?

A. 0.5

B. 0.67

C. 0.75

D. 0.95

The correct answer is (C). *(Publisher)*

REQUIRED: The probability that at least one of two projects will have positive cash flows.

DISCUSSION: The probability that either one or both of two events will occur equals the sum of their separate probabilities minus their joint probability. Therefore, the probability that at least one of the two projects will have positive cash flows is the probability that project 1 will have positive cash flows plus the probability that project 2 will have positive cash flows, minus the probability that both project 1 and project 2 will have positive cash flows. Thus, (0.5) + (0.5) – [(0.5)(0.5)] = 0.75.

Answers (A) and (B) are incorrect because the probability of at least one positive project would be greater. Answer (D) is incorrect because there is a .25 chance that both projects will have negative cash flows.

B. Expected Value

16. Pongo Company's managers are attempting to value a piece of land that they own. One potential occurrence is that the old road that borders the land gets paved. Another possibility is that the road does not get paved. A third outcome is that the road might be destroyed and completely replaced by a new road. Based on the following future states of nature, their probabilities, and subsequent values of the land, what is the expected value of the land?

Future States of Nature (SN)	Probability
SN 1: Current road gets paved	.5
SN 2: Road does not get paved	.4
SN 3: Current road destroyed and replaced with new road	.1

Estimates of land value under each possible future state of nature:

Value if SN 1: $200,000
Value if SN 2: $100,000
Value if SN 3: $550,000

A. $133,333

B. $195,000

C. $225,000

D. $283,333

The correct answer is (B). *(Publisher)*

REQUIRED: The expected value of the land.

DISCUSSION: The expected value of the land is determined by multiplying the probability of each state of nature by the value under that particular state of nature, and adding all of the products. Thus, the land's expected value is (0.5)($200,000) + (0.4)($100,000) + (0.1)($550,000) = $195,000.

Answer (A) is incorrect because $133,333 places too much weighting on the second option. Answer (C) is incorrect because $225,000 gives too much weight to the third option. Answer (D) is incorrect because $283,333 uses a simple unweighted average of the returns from the three options.

17. Sweivel Company is preparing its budget and, taking into consideration the recent pace of economic recovery, has developed several sales forecasts and the estimated probability associated with each sales forecast. To determine the sales forecast to be used for budgeting purposes, which one of the following techniques should Sweivel use?

A. Expected value analysis.

B. Continuous probability simulation.

C. Exponential distribution analysis.

D. Sensitivity analysis.

The correct answer is (A). *(CMA, adapted)*

REQUIRED: The technique for determining a sales forecast given different forecasts and their probabilities.

DISCUSSION: Expected value analysis provides a rational means for selecting the best alternative in decisions involving risk. The expected value of an alternative is found by multiplying the probability of each outcome by its payoff, and summing the products. It represents the long-term average payoff for repeated trials.

Answer (B) is incorrect because simulation is not necessary. Several estimates are known. Answer (C) is incorrect because the exponential distribution is the probability of zero occurrences in a specified time period. Answer (D) is incorrect because sensitivity analysis involves making several estimates of key variables and recalculating results based on the alternative estimates; it determines how sensitive a solution is to changes in estimates.

18. Under favorable weather conditions, the management of Flesher Farms expects its raspberry crop to have a $120,000 market value. An unprotected crop subject to frost has an expected market value of $80,000. If Flesher protects the raspberries against frost, the market value of the crop is still expected to be $120,000 under frost-free conditions and $180,000 if a frost occurs. What must be the probability of a frost for Flesher to be indifferent to spending $20,000 for tents to provide frost protection?

A. .167

B. .200

C. .250

D. .333

The correct answer is (B). *(Publisher)*

REQUIRED: The probability of frost at which management is indifferent between providing and not providing frost protection.

DISCUSSION: Without protection, the company will have sales of either $80,000 or $120,000. With protection, the company will have sales of either $120,000 or $180,000. The $20,000 cost of tents must be subtracted from these amounts to compute the net contribution. Thus, the contribution if the berries are protected will be either $100,000 or $160,000. The problem can be solved algebraically with the following formula (X = the probability of frost):

$$X(\$80{,}000) + (1 - X)(\$120{,}000) = X(\$160{,}000) + (1 - X)(\$100{,}000)$$
$$\$80{,}000X + \$120{,}000 - \$120{,}000X = \$160{,}000X + \$100{,}000 - \$100{,}000X$$
$$\$120{,}000 - \$40{,}000X = \$100{,}000 + \$60{,}000X$$
$$\$120{,}000 = \$100{,}000 + \$100{,}000X,$$
$$\$20{,}000 = \$100{,}000X;$$
$$X = .200$$

In other words, if the probability of frost is 20%, management will be indifferent between providing and not providing protection. At a frost expectancy of greater than 20%, management should provide protection.

Answer (A) is incorrect because, if frost has a probability of .167, the expected value of not providing protection exceeds that of providing protection. Answers (C) and (D) are incorrect because, if the probability of frost is greater than .200, an expected value analysis indicates the company should provide protection.

19. During the past few years, Wilder Company has experienced the following average number of power outages:

Number per month	Number of Months
0	3
1	2
2	4
3	3
	12

Each power outage results in out-of-pocket costs of $800. For $1,000 per month, Wilder can lease a generator to provide power during outages. If Wilder leases a generator in the coming year, the estimated savings (or additional expense) for the year will be

A. ($15,200)

B. ($1,267)

C. $3,200

D. $7,200

The correct answer is (C). *(Publisher)*

REQUIRED: The estimated savings or additional expense from leasing the generator, given expected levels of occurrence.

DISCUSSION: Each outage costs $800, but this expense can be avoided by paying $1,000 per month ($12,000 for the year). The expected-value approach uses the probability distribution derived from past experience to determine the average expected outages per month.

$$3 \div 12 \times 0 = 0.$$
$$2 \div 12 \times 1 = .16667$$
$$4 \div 12 \times 2 = .66667$$
$$3 \div 12 \times 3 = .75000$$
$$\underline{1.58334}$$

The company can expect to have, on average, 1.58334 outages per month. At $800 per outage, the expected cost is $1,266.67. Thus, paying $1,000 to avoid an expense of $1,266.67 saves $266.67 per month, or $3,200 per year.

Answer (A) is incorrect because $(15,200) is the annual amount the company will lose without a generator. Answer (B) is incorrect because ($1,267) is the monthly amount the company will lose without a generator. Answer (D) is incorrect because $7,200 is the amount the company will save if two outages occur per month.

20. Philip Enterprises, distributor of compact disks (CDs), is developing its budgeted cost of goods sold for 1998. Philip has developed the following range of sales estimates and associated probabilities for the year:

Sales Estimate	Probability
$ 60,000	25%
85,000	40
100,000	35

Philip's cost of goods sold averages 80% of sales. What is the expected value of Philip's 1998 budgeted cost of goods sold?

A. $85,000

B. $84,000

C. $68,000

D. $67,200

The correct answer is (D). *(CMA, adapted)*

REQUIRED: The expected value of cost of goods sold.

DISCUSSION: The expected value is calculated by weighting each sales estimate by the probability of its occurrence. Consequently, the expected value of sales is $84,000 [$60,000 x .25) + ($85,000 x .40) + ($100,000 x .35)]. Cost of goods sold is therefore $67,200 (80% x $84,000).

Answer (A) is incorrect because $85,000 is the sales estimate with the highest probability. Answer (B) is incorrect because $84,000 is the expected value of sales. Answer (C) is incorrect because $68,000 is 80% of the sales estimate with the highest probability.

Questions 21 through 24 are based on the following information.

The College Honor Society sells hot pretzels at the home football games. The pretzels are sold for $1.00 each, and the cost per pretzel is $.30. Any unsold pretzels are discarded because they will be stale before the next home game.

The frequency distribution of the demand for pretzels per game is presented below.

Unit Sales Volume	Probability
2,000 pretzels	.10
3,000 pretzels	.15
4,000 pretzels	.20
5,000 pretzels	.35
6,000 pretzels	.20

21. The estimated demand for pretzels at the next home football game using an expected value approach is

A. 4,000 pretzels.

B. 4,400 pretzels.

C. 5,000 pretzels.

D. Some amount other than those given.

The correct answer is (B). *(CMA, adapted)*

REQUIRED: The estimated demand using an expected value approach.

DISCUSSION: The calculation using an expected value approach weights each possible sales volume by its probability. Thus, the estimated demand is 4,400 pretzels.

Volume		Probability		Result
2,000	×	.10	=	200
3,000	×	.15	=	450
4,000	×	.20	=	800
5,000	×	.35	=	1,750
6,000	×	.20	=	1,200
Expected value				4,400

Answer (A) is incorrect because 4,000 pretzels assumes each outcome is equally likely. Answer (C) is incorrect because 5,000 pretzels is the estimated demand using a deterministic approach based on the most likely outcome. Answer (D) is incorrect because 4,400 pretzels is among the responses given.

22. The estimated demand for pretzels at the next home football game using a deterministic approach based on the most likely outcome is

A. 4,000 pretzels.

B. 4,400 pretzels.

C. 5,000 pretzels.

D. 6,000 pretzels.

The correct answer is (C). *(CMA, adapted)*

REQUIRED: The estimated demand using a deterministic approach based on the most likely outcome.

DISCUSSION: A deterministic approach assumes that a value is known with certainty. If that value is deemed to be the most likely outcome, assumed demand will be 5,000 pretzels, the volume with the highest probability (35%).

Answer (A) is incorrect because 4,000 pretzels assumes each outcome is equally likely. Answer (B) is incorrect because 4,400 pretzels is the estimated demand using an expected value approach. Answer (D) is incorrect because 6,000 pretzels is merely the greatest demand.

23. The conditional profit per game of having 4,000 pretzels available but only selling 3,000 pretzels is

A. $1,800

B. $2,100

C. $2,800

D. Some amount other than those given.

The correct answer is (A). *(CMA, adapted)*

REQUIRED: The conditional profit for specified inventory and sales levels.

DISCUSSION: Each pretzel costs $.30. Thus, the cost of 4,000 pretzels is $1,200 (4,000 x $.30). Selling 3,000 pretzels at $1 each produces revenue of $3,000. Subtracting the $1,200 of costs from the $3,000 of revenue results in a conditional profit of $1,800.

Answer (B) is incorrect because $2,100 assumes 3,000 are available and 3,000 are sold. Answer (C) is incorrect because $2,800 is the conditional profit given that 4,000 are sold. Answer (D) is incorrect because $1,800 is among the responses given.

24. The conditional profit per game of having 4,000 pretzels available and selling all 4,000 pretzels is

A. $1,200

B. $2,100

C. $2,800

D. $800

The correct answer is (C). *(CMA, adapted)*

REQUIRED: The conditional profit for specified inventory and sales levels.

DISCUSSION: Each pretzel costs $.30. Thus, the cost of 4,000 pretzels is $1,200 (4,000 x $.30). Selling 4,000 pretzels at $1 each produces revenue of $4,000. Subtracting the $1,200 of costs from the $4,000 of revenue results in a conditional profit of $2,800.

Answer (A) is incorrect because $1,200 is the cost of 4,000 pretzels. Answer (B) is incorrect because $2,100 assumes 3,000 are available and 3,000 are sold. Answer (D) is incorrect because $800 assumes 2,000 are sold.

25. The modeling technique to be used for situations involving a sequence of events with several possible outcomes associated with each event is

A. Queuing theory.

B. Dynamic programming.

C. The critical path method.

D. Decision tree analysis.

The correct answer is (D). *(CMA, adapted)*

REQUIRED: The technique used to model a sequence of events each with several possible outcomes.

DISCUSSION: Decision tree analysis is useful when the most beneficial series of decisions is to be chosen. The possible decisions for each decision point, the events that might follow from each decision, the probabilities of these events, and the quantified outcomes of the events should be known.

Answer (A) is incorrect because queuing (waiting-line) theory is a group of mathematical models of systems involving waiting lines. Answer (B) is incorrect because dynamic programming is an approach to solving problems, not a particular algorithm. It divides a large mathematical model into smaller, more manageable pieces in such a way that, once the smaller problems have been solved, the result is the optimal solution to the overall model. Answer (C) is incorrect because the critical path method is a network model used to help control large-scale, complex projects.

26. Which one of the following statements does not apply to decision tree analysis?

A. The sum of the probabilities of the events is less than one.

B. All of the events are mutually exclusive.

C. All of the events are included in the decision.

D. The branches emanate from a node from left to right.

The correct answer is (A). *(CMA, adapted)*

REQUIRED: The statement not applicable to decision tree analysis.

DISCUSSION: In a decision tree, the events following from a decision are mutually exclusive. Also, all possible events are included. Thus, the sum of the probabilities of the events is 1.0.

Answers (B), (C), and (D) are incorrect because they are true statements with respect to decision trees.

27. The modeling technique to be employed in a situation involving a sequence of events with several possible outcomes associated with each event is

A. Network analysis.

B. Decision tree analysis.

C. Monte Carlo simulation.

D. Linear programming.

The correct answer is (B). *(CMA, adapted)*

REQUIRED: The modeling technique used given a sequence of events with several possible outcomes associated with each event.

DISCUSSION: Decision trees may be used to describe complex decision situations. Each branch of the tree represents a different decision, and each twig extending from each branch represents several possible outcomes of the decision.

Answer (A) is incorrect because network analysis is a means of scheduling long-term projects; PERT and the critical path method (CPM) are examples. Answer (C) is incorrect because Monte Carlo simulation is a technique for experimenting with mathematical models using a computer, which generates random values for each variable. Answer (D) is incorrect because linear programming is a means of allocating scarce resources given a specified objective and a variety of constraints.

28. The use of a decision tree is appropriate for decision making under conditions of

A. Uncertainty and risk.

B. Uncertainty and subjective likelihoods.

C. Certainty.

D. Risk.

The correct answer is (D). *(CMA, adapted)*

REQUIRED: The conditions under which decision tree analysis is appropriate.

DISCUSSION: Decision tree analysis is used when various options, their consequences, and the probabilities of those consequences can be ascertained with a high degree of confidence. Under conditions of uncertainty, however, probabilities are unknown.

Answers (A) and (B) are incorrect because each includes uncertainty as part of the answer. Answer (C) is incorrect because, under a condition of certainty, a decision tree would not be needed.

Questions 29 through 31 are based on the following information. A beverage stand can sell either soft drinks or coffee on any given day. If the stand sells soft drinks and the weather is hot, it will make $2,500; if the weather is cold, the profit will be $1,000. If the stand sells coffee and the weather is hot, it will make $1,900; if the weather is cold, the profit will be $2,000. The probability of cold weather on a given day at this time is 60%.

29. The expected payoff for selling coffee is

A. $1,360

B. $2,200

C. $3,900

D. $1,960

The correct answer is (D). *(CMA, adapted)*

REQUIRED: The expected payoff for selling coffee.

DISCUSSION: The expected payoff calculation for coffee is

$$.4(\$1,900) + .6(\$2,000) = \$1,960$$

Answer (A) is incorrect because the least the company can make by selling coffee is $1,900. Answers (B) and (C) are incorrect because the most the company can make by selling coffee is $2,000.

30. The expected payoff if the vendor has perfect information is

A. $3,900

B. $2,200

C. $1,360

D. $1,960

The correct answer is (B). *(CMA, adapted)*

REQUIRED: The expected payoff if the vendor has perfect information.

DISCUSSION: The vendor would like to sell coffee on cold days ($2,000) and soft drinks on hot days ($2,500). Hot days are expected 40% of the time. Hence, the probability is 40% of making $2,500 by selling soft drinks. The chance of making $2,000 by selling coffee is 60%. The payoff equation is

$$.4(\$2,500) + .6(\$2,000) = \$2,200$$

Answer (A) is incorrect because the most the vendor can make is $2,500 per day. Answers (C) and (D) are incorrect because the least the vendor could make by having perfect information is $2,000 on cold days.

31. Considering only the information given in the fact pattern, if the probability of hot weather given a hot weather forecast is 50%, how much would the vendor be willing to pay for the forecast?

A. $600

B. $300

C. $1,000

D. $500

The correct answer is (B). *(CMA, adapted)*

REQUIRED: The amount the vendor should be willing to pay for a forecast.

DISCUSSION: If the weather is hot and coffee is served, the vendor earns $1,900. If the vendor knows the weather will be hot, (s)he sells soft drinks and makes $2,500, a $600 increase. Thus, the vendor should be willing to pay up to $600 for perfect information regarding hot weather. However, if the forecasts are only 50% accurate, the information is not perfect. Accordingly, the vendor should be willing to pay only $300 (50% x the $600 potential increase in profits) for the sometimes accurate forecasts.

Answer (A) is incorrect because the vendor would pay $600 for perfect information, but the forecasts are only 50% accurate. Answer (C) is incorrect because the most the vendor could profit from perfect information on hot days would be $600 ($2,500 – $1,900). Answer (D) is incorrect because perfect information is worth $600, but information that is 50% accurate warrants a $300 payment.

32. The expected value of perfect information is the

A. Same as the expected profit under certainty.

B. Sum of the conditional profit (loss) for the best event of each act times the probability of each event occurring.

C. Difference between the expected profit under certainty and the expected opportunity loss.

D. Difference between the expected profit under certainty and the expected monetary value of the best act under uncertainty.

The correct answer is (D). *(CMA, adapted)*

REQUIRED: The true statement about the expected value of perfect information.

DISCUSSION: Perfect information permits certainty that a future state of nature will occur. The expected value of perfect information determines the maximum amount a decision maker is willing to pay for information. It is the difference between the expected value without perfect information, that is, the expected value of the best action under uncertainty, and the expected value under certainty. Under certainty, a decision maker knows in each case which state of nature will occur and can act accordingly.

Answer (A) is incorrect because the expected value of perfect information is the difference between the expected profit under certainty and the profit from the best decision under uncertainty. Answer (B) is incorrect because the expected value of perfect information is the excess of the total conditional profits under certainty over the profit from the best decision under uncertainty. Answer (C) is incorrect because there is no expected opportunity loss under conditions of certainty.

C. Statistics

33. Bobby, a market analyst for Investing Magazine, is analyzing the stock of Vandelay Industries. Bobby decides to use the market data for the year to date to approximate the standard deviation on monthly returns for Vandelay. Given the following, what should Bobby estimate the standard deviation to be?

Month	Return
January	10%
February	2%
March	19%
April	25%
May	13%
June	9%

A. 4.79

B. 8.07

C. 14.53

D. 22.91

The correct answer is (B). *(Publisher)*

REQUIRED: The standard deviation.

DISCUSSION: The population standard deviation may be estimated from the sample standard deviation. This can be found by first determining the variance of the sample set, which equals the summation of the values obtained by subtracting each individual monthly return from the average return for the 6 months given (13%), squaring this difference, and then dividing that difference by (n – 1), if n is the number of elements in the sample. Thus, (n – 1) equals 5. Accordingly, the sample variance is 65.2. The standard deviation is the square root of the variance, or 8.07.

Answers (A), (C), and (D) are incorrect because the standard deviation is the square root of the variance.

Questions 34 through 36 are based on the following information. Pace Company's accountant has gathered a sample of sales volume in units for the past 11 working days.

Day	Sales Volume in Units
1	100
2	150
3	200
4	100
5	100
6	300
7	135
8	180
9	160
10	190
11	310
	1,925

34. The mean value for daily sales for this period is

A. 1,925

B. 100

C. 160

D. 175

The correct answer is (D). *(CMA, adapted)*

REQUIRED: The mean value for daily sales.

DISCUSSION: The mean is a simple average. It is computed by dividing the total sales of 1,925 units by the 11 days. Thus, the mean is 175 units.

Answer (A) is incorrect because 1,925 is the unit sales volume. Answer (B) is incorrect because 100 is the mode. Answer (C) is incorrect because 160 is the median.

35. The modal value for daily sales for this period is

A. 300

B. 100

C. 160

D. Some amount other than those given.

The correct answer is (B). *(CMA, adapted)*

REQUIRED: The modal value for daily sales.

DISCUSSION: The mode is the number that occurs most often in a set. Because 100 units is the only recurring number, 100 is the mode.

Answer (A) is incorrect because 300 is the volume on the day representing the midpoint of the 11-day period. Answer (C) is incorrect because 160 is the median. Answer (D) is incorrect because 100 is among the responses given.

36. The median value for daily sales for this period is

A. 300

B. 100

C. 160

D. 175

The correct answer is (C). *(CMA, adapted)*

REQUIRED: The median value for daily sales.

DISCUSSION: When data are arranged in numerical order, one-half of the values are smaller than the median and one-half of the values are greater. Accordingly, this value occurs at the 50th percentile. With 11 days of sales, the middle value is the sixth item. The following is the ranking of sales in order of volume: 100, 100, 100, 135, 150, 160, 180, 190, 200, 300, 310. The middle item (median) is 160.

Answer (A) is incorrect because 300 is the volume on the day representing the midpoint of the 11-day period. Answer (B) is incorrect because 100 is the mode. Answer (D) is incorrect because 175 is the mean.

37. A sales supervisor is concerned with what appear to be unusually high travel expenses of one salesperson. The measure by which the sales supervisor can best evaluate the difference between the expenses of one salesperson and those of other salespersons is the

A. Coefficient of correlation.

B. Mode.

C. Median.

D. Standard deviation.

The correct answer is (D). *(CMA, adapted)*

REQUIRED: The measure used to evaluate the difference between the expenses incurred.

DISCUSSION: The standard deviation is a measure of the degree of compactness of the values in a population. This measure is used to help determine appropriate sample sizes. If σ is the population standard deviation, μ is the population mean, and N is the population, the formula is

$$\sigma = \sqrt{\frac{\Sigma(x_i - \bar{\mu})^2}{N}}$$

The denominator is n – 1 when calculating the standard deviation of a sample.

Answer (A) is incorrect because a coefficient of correlation indicates the strength of the linear relationship among two or more variables. Answer (B) is incorrect because the mode is the most common variable in a set. Answer (C) is incorrect because the median is the middle value in a set.

38. In a large computer manufacturer, there has been much concern about the consistency across departments in adhering to new and unpopular purchasing guidelines. An auditor has a list that rank-orders all departments according to the percentage of purchases that are consistent with the guidelines and indicates which division the department is from. The auditor performs a t-test for differences in means on the average rank of departments in divisions A and B to see if there is any difference in compliance with the policy and finds that division A (which has more departments) has a significantly higher (i.e., better) average rank than division B. Which one of the following conclusions should be drawn from this analysis?

A. Division A is complying better with the new policy.

B. A random sample of departments should be drawn and the analysis recalculated.

C. A t-test is not valid when the tested groups differ in size.

D. A t-test is inappropriate for this data, and another type of analysis should be used.

The correct answer is (D). *(CIA, adapted)*

REQUIRED: The true statement about application of a t-test to rank-ordered data.

DISCUSSION: The t-distribution is used for small samples but in the same way as the normal distribution. A t-test is not valid when used with ordinal-level data. A t-test by definition is an application of parametric statistics. Nonparametric (distribution-free) statistics is applied to problems for which rank order, but not a specific distribution, is known.

Answer (A) is incorrect because a t-test is not valid in this case. Answer (B) is incorrect because the auditor already has a list of the entire population, and no sampling is needed. Answer (C) is incorrect because a t-test can be used with groups that differ in size.

39. Dick Rixard is a quality control specialist for the Lasser Company. Lasser makes ball bearings that require close tolerances. In a recent production run, Rixard found that the bearings had an average diameter of .755 inches with a standard deviation of .01 inches for a sample of 100 bearings. In his quality control testing, he uses a 95% confidence interval. If the Z score is 1.96, the lower boundary of the confidence interval for this production run is

A. .755 inches.

B. .7354 inches.

C. .7746 inches.

D. Some amount other than those given.

The correct answer is (D). *(CMA, adapted)*

REQUIRED: The lower boundary of the confidence interval given standard deviation and desired confidence interval.

DISCUSSION: Because a 95% confidence interval equals 1.96 standard deviations in either direction from the mean, the desired lower boundary is found using the following formula:

$$\bar{x} - z\ (s \div \sqrt{n})$$
$$.755 - 1.96\ (.01 \div \sqrt{100})$$
$$.755 - .00196 = .75304$$

Answer (A) is incorrect because .755 inches is the mean diameter. Answer (B) is incorrect because .7354 inches equals the mean minus 1.96 times the standard deviation. Answer (C) is incorrect because .7746 inches equals the mean plus 1.96 times the standard deviation.

Questions 40 through 42 are based on the following information. The management at Altron Corporation is evaluating two new products to undertake during the next 1-year period, digital clocks and calculators. For each product, the Marketing Department has estimated the possible net cash flows and the probability of each cash flow.

Digital Clocks		Calculators	
Probability	Cash Flow	Probability	Cash Flow
25%	$4,000	10%	$ 0
50	5,000	50	3,000
25	8,000	40	10,000

Since both products have equal expected values of $5,500, management is still undecided. John Wills, a financial analyst, has suggested that the calculation of the standard deviation for each product's cash flows may aid in the decision. He used the following data:

$$S = \sqrt{\sum^{n} (X_e - \bar{X})^2 P_e}$$

If: e = 1
X_e = individual cash flows
$\bar{X}$ = expected value of all cash flows
P_e = probabilities assigned to each cash flow

40. John Wills is using the standard deviation to assist in the decision process because he is

A. Measuring the size of the constraints.

B. Determining the frequency of an event.

C. Measuring the dispersion of the probability for each event.

D. Measuring the range of each event.

The correct answer is (D). *(CMA, adapted)*

REQUIRED: The reason the standard deviation is used in the decision process.

DISCUSSION: The standard deviation measures the range (dispersion) of observations around the mean. For example, about 95.5% of all observations fall within two standard deviations of the mean of a normally distributed variable.

Answer (A) is incorrect because the standard deviation measures dispersion (range), not size of constraints. Answer (B) is incorrect because frequency is given. Answer (C) is incorrect because the standard deviation measures the dispersion of observations, not probabilities.

41. The standard deviation of the cash flows of the digital clocks is

A. $2,250,000.00

B. $1,500.00

C. $750.00

D. $353.55

The correct answer is (B). *(CMA, adapted)*

REQUIRED: The standard deviation of the cash flows of the clocks.

DISCUSSION: Substituting the cash flow figures into the formula produces the following:

X_e	$\bar{X}$	$(X_e - \bar{X})^2$		P_e		$(X_e - \bar{X})^2 P_e$
$4,000	$5,500	$2,250,000	×	.25	=	$ 562,500
5,000	5,500	250,000	×	.50	=	125,000
8,000	5,500	6,250,000	×	.25	=	1,562,500
Variance						$2,250,000

The square root of $2,250,000 is $1,500.

Answer (A) is incorrect because $2,250,000 is the variance. Answer (C) is incorrect because $750 is the square root of $562,500. Answer (D) is incorrect because $353.55 is the square root of $125,000.

42. If the standard deviation for the cash flows of the calculators is $3,775, the management at Altron would

A. Choose to introduce digital clocks because this product has the lesser risk.

B. Choose to introduce calculators because this product has a 40% probability for the highest cash flow.

C. Not choose either product.

D. Choose to introduce calculators because the product has the lesser risk.

The correct answer is (A). *(CMA, adapted)*

REQUIRED: The decision management should make based on the given standard deviation for calculators.

DISCUSSION: As calculated in the discussion of the preceding question, the standard deviation for clocks ($1,500) is lower than the standard deviation for calculators. Thus, the range of outcomes is less dispersed, and producing clocks is less risky.

Answer (B) is incorrect because the greatest probability is for a lower rate of return. Answer (C) is incorrect because both products are profitable. Answer (D) is incorrect because calculators have the higher standard deviation and the greater risk.

43. Statistical quality control often involves the use of control charts whose basic purpose is to

A. Determine when accounting control procedures are not working.

B. Control labor costs in production operations.

C. Detect performance trends away from normal operations.

D. Monitor internal control applications in computerized operations.

The correct answer is (C). *(CMA, adapted)*

REQUIRED: The purpose of statistical quality control charts.

DISCUSSION: Statistical control charts are graphic aids for monitoring the status of any process subject to random variations. The chart consists of three horizontal lines plotted on a horizontal time scale. The vertical scale represents the appropriate quantitative measure. The center line represents the average or mean value for the process being controlled. The other two lines are the upper control limit and the lower control limit. The processes are measured periodically, and the values are plotted on the chart. If the value falls within the control limits, no action is taken. If the value falls outside the limits, the process is considered "out of control," and an investigation is made for possible corrective action. Another advantage of the chart is that it makes trends visible.

Answers (A), (B), and (D) are incorrect because quality control concerns product quality, not controls over accounting procedures, costs, or computerized operations.

44. The statistical quality control department prepares a control chart showing the percentages of defective production. Simple statistical calculations provide control limits that indicate whether assignable causes of variation are explainable on chance grounds. The chart is particularly valuable in determining whether the quality of materials received from outside vendors is consistent from month to month. What is the best term for this chart?

A. C chart.

B. P chart.

C. R chart.

D. X-bar chart.

The correct answer is (B). *(CIA, adapted)*

REQUIRED: The statistical quality control chart described.

DISCUSSION: A P chart is based on an attribute (acceptable/not acceptable) rather than a measure of a variable, specifically, the percentage of defects in a sample.

Answer (A) is incorrect because a C chart is also an attribute control chart. It shows defects per item. Answer (C) is incorrect because an R chart displays the range of dispersion of a variable, such as size or weight. Answer (D) is incorrect because an X-bar chart plots the sample mean for a variable.

45. A health insurance company uses a computer application to monitor physician bill amounts for various surgical procedures. This program allows the company to better control reimbursement rates. The X-bar chart below is an example of the output from this application.

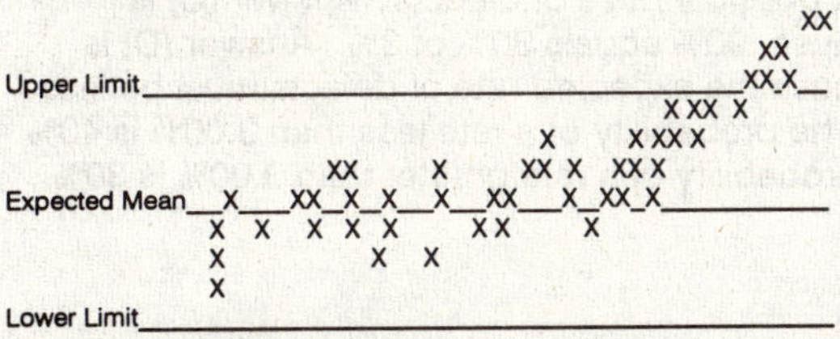

Select the interpretation that best explains the data plotted on the chart.

A. Random variation.

B. Abnormal variation.

C. Normal variation.

D. Cyclic variation.

The correct answer is (B). *(CIA, adapted)*

REQUIRED: The interpretation that best explains the data plotted on the chart.

DISCUSSION: Statistical quality control charts are graphic aids for monitoring the status of any process subject to random variations. For example, an X-bar chart depicts the sample means for a variable. If the values fall within the upper and lower control limits, no action is taken. Accordingly, values outside these limits are abnormal and should be investigated for possible corrective action.

Answers (A) and (C) are incorrect because random or normal variations should fall within realistically determined control limits. Answer (D) is incorrect because, in time series analysis, cyclic variation is the fluctuation in the value of a variable caused by change in the level of general business activity.

46. Many firms are beginning to use the statistical processing control techniques as part of their total quality management approach. Which of the following would not constitute a part of statistical processing control techniques?

A. Acceptance sampling.

B. Dollar-unit sampling.

C. Quality control charts.

D. Continuous monitoring and feedback.

The correct answer is (B). *(CIA, adapted)*

REQUIRED: The item not part of statistical processing control.

DISCUSSION: Dollar-unit sampling has been uniquely applied to auditing account balances. It is not used in statistical processing control.

Answer (A) is incorrect because acceptance sampling is a standard statistical process control technique. Answer (C) is incorrect because quality control charts are an integral part of TQM approaches. Answer (D) is incorrect because continuous monitoring and frequent feedback are two of the important elements of statistical quality control.

Questions 47 through 49 are based on the following information. The table below shows the estimated probabilities of the proportions of defective units resulting from a production run. Each defective unit costs the company $100, and each production run consists of 100 units.

Percentage Defective	Probability
1%	20%
2%	20%
3%	30%
4%	20%
5%	10%

47. The expected percentage of defective units is

A. 2.80%

B. 3.00%

C. 0.90%

D. 3.45%

The correct answer is (A). *(Publisher)*

REQUIRED: The expected percentage of defective units resulting from a production run.

DISCUSSION: To determine the expected value, each outcome is multiplied by its probability as follows:

1% × .20	=	.20%
2% × .20	=	.40%
3% × .30	=	.90%
4% × .20	=	.80%
5% × .10	=	.50%
Total		2.80%

The expected rate of defective units is 2.80%.

Answer (B) is incorrect because 3.00% is the unweighted average of the possible rates of defects. Answer (C) is incorrect because .90% equals 30% of 3%. Answer (D) is incorrect because the expected rate of defects must be less than 3.00%. The probability of a rate less than 3.00% is 40%, whereas the probability of a rate greater than 3.00% is 30%.

48. The variance of the percentage of defective units is

A. 0.00%

B. 3.16%

C. 10.20%

D. 1.56%

The correct answer is (D). *(Publisher)*

REQUIRED: The variance of the percentage of defective units.

DISCUSSION: The variance is the average of the squared deviations from the mean. It is found by subtracting the mean from each value, squaring each difference, adding the squared differences, and dividing the sum by the number of data points. The mean, or expected value, of the percentage of defective units was determined in the preceding question to be 2.80%. Moreover, there are effectively 100 data points in this problem because the probabilities of possible outcomes are given instead of a set of actual outcomes. For example, a 20% probability of a 1% rate is the equivalent of 20 occurrences of a 1% rate in a sample of 100 production runs. Accordingly, instead of dividing by the sum of the data points in the variance formula, each squared difference can be multiplied by the related probability as shown below:

$$(1.0\% - 2.80\%)^2 \times .20 = .648$$
$$(2.0\% - 2.80\%)^2 \times .20 = .128$$
$$(3.0\% - 2.80\%)^2 \times .30 = .012$$
$$(4.0\% - 2.80\%)^2 \times .20 = .288$$
$$(5.0\% - 2.80\%)^2 \times .10 = \underline{.484}$$
$$\text{Variance} \quad \underline{\underline{1.56\ \%}}$$

Answer (A) is incorrect because a variance occurs unless all data points are identical. Answer (B) is incorrect because this variance is not possible. None of the data points are more than 2.2% from the mean. Answer (C) is incorrect because 10.2% is the simple sum of the squares assuming an unweighted list of only five data points.

49. The company hires a worker to oversee the machinery and assure that no defective units are produced. What is the breakeven amount per production run to be paid to the worker?

A. $100

B. $200

C. $280

D. $295

The correct answer is (C). *(Publisher)*

REQUIRED: The breakeven amount paid to a worker to eliminate defective units.

DISCUSSION: As shown in the solution to question 47, the expected rate of defective units is 2.80% per production run, or 2.8 units per 100. If each defective unit costs the company $100, avoiding 2.8 defective units saves the company $280. Thus, any labor cost less than $280 will increase profits.

Answer (A) is incorrect because $100 is the amount that a defective unit costs the company. Answer (B) is incorrect because more than $200 can be paid and the company's profits will increase. Answer (D) is incorrect because the company's profits will decline if $295 is paid to the worker.

50. SuperTrucks Corporation is undertaking a new division of transport truck production for the upcoming calendar year. In the division's first year, management for SuperTrucks has predicted that there is a 20% chance that cash inflows from the new division will be $8,000, a 30% chance that cash inflows will be $12,000, and a 50% chance that cash inflows will be $10,000. What is the standard deviation of the expected cash flows for SuperTrucks Corporation's new division?

A. $2,720.00

B. $1,400.00

C. $767.00

D. $366.67

The correct answer is (B). *(Publisher)*

REQUIRED: The standard deviation of expected cash flows.

DISCUSSION: Substituting the cash flow figures into the probability-based formula for the population variance produces the following:

X_e	$\overline{X}$	$(X_e - \overline{X})^2$		P_e		$(X_e - \overline{X})^2 P_e$
$ 8,000	$10,200	$4,840,000	x	0.20	=	$ 968,000
$12,000	$10,200	$3,240,000	x	0.30	=	$ 972,000
$10,000	$10,200	$ 40,000	x	0.50	=	$ 20,000
				(Variance) σ^2	=	$1,960,000
				(Standard deviation) σ	=	$ 1,400

$\overline{X}$ = (0.20)($8,000) + (0.30)($12,000) + (0.50)($10,000)
= $1,600 + $3,600 + $5,000
= $10,200

Answers (A), (C), and (D) are incorrect because the standard deviation is the square root of the variance.

Questions 51 and 52 are based on the following information. Stock A has the following possibilities for earnings in the 2000 year:

Earnings	Probability
$1,000,000	18%
$1,200,000	15%
$1,700,000	24%
$2,600,000	22%
$3,300,000	21%

51. What are the expected earnings for Stock A in 2001?

A. $843,000

B. $1,333,333

C. $1,960,000

D. $2,033,000

The correct answer is (D). *(Publisher)*

REQUIRED: The expected earnings for the year.

DISCUSSION: To determine the expected value, each potential outcome is multiplied by its probability, and then these values are summed to give the expected value.

$1,000,000 x .18	=	$ 180,000
1,200,000 x .15	=	180,000
1,700,000 x .24	=	408,000
2,600,000 x .22	=	572,000
3,300,000 x .21	=	693,000
		$2,033,000

Answers (A) and (B) are incorrect because they are lower than possible given any of the probabilities given.
Answer (C) is incorrect because it represents the arithmetic average of the five possibilities, without regard to probabilities.

52. What is the variance of potential earnings of Stock A?

A. $854,758

B. $223,111,000,000

C. $576,272,000,000

D. $730,611,000,000

The correct answer is (D). *(Publisher)*

REQUIRED: The variance of potential earnings.

DISCUSSION: The population variance is the average of the squared deviations from the mean. It is found by subtracting the mean from each value, squaring each difference, adding these squared differences, and dividing the sum by the number of data points. The mean, or expected value, was determined to be $2,033,000 in the previous question. In this case, probabilities are given, so instead of dividing by the sum of the data points in the variance formula, each squared difference can be multiplied by the appropriately related probability, as shown below.

$(\$1,000,000 - \$2,033,000)^2 \times 0.18 =$	$192,076,020,000
$(\$1,200,000 - \$2,033,000)^2 \times 0.15 =$	104,083,350,000
$(\$1,700,000 - \$2,033,000)^2 \times 0.24 =$	26,613,360,000
$(\$2,600,000 - \$2,033,000)^2 \times 0.22 =$	70,727,580,000
$(\$3,300,000 - \$2,033,000)^2 \times 0.21 =$	337,110,690,000
Variance	$730,611,000,000

The variance is an intermediate step in the calculation of the standard deviation. The standard deviation is the square root of the variance.

Answer (A) is incorrect because it is the standard deviation. Answers (B) and (C) are incorrect because they were not properly weighted for the probabilities of occurrence.

53. A U.S. company currently has domestic operations only. It is considering an equal-size investment in either Canada or Britain. The data on expected rate of return and the risk associated with each of these proposed investments are given below.

Proposed Investment	Mean Return	Standard Deviation
British Investment	22%	10%
Canadian Investment	28%	15%

The mean return on the company's current (domestic only) business is 20%, with a standard deviation of 15%. Using the above data and the correlation coefficients, the company calculated the following portfolio risk and return (based on a ratio of 50% U.S. domestic operations and 50% international operations).

Investments	Mean Return	Standard Deviation
U.S. and Britain	21%	3%
U.S. and Canada	24%	15%

The company plans to select the optimal combination of countries based on risk and return for the domestic and international investments taken together. Based on the above data, which one of the following alternatives provides the best risk adjusted return to the firm?

A. Undertake the British investment.

B. Undertake the Canadian investment.

C. Do not undertake either investment.

D. Unable to determine based on data given.

The correct answer is (A). *(CMA, adapted)*

REQUIRED: The option that provides the best risk adjusted return to the firm.

DISCUSSION: The coefficient of variation is useful when the rates of return and standard deviations of two investments differ. It measures the risk per unit of return because it divides the standard deviation (σ) by the expected return ($\hat{k}$).

$$\textit{Coefficient of variation} = \frac{\sigma}{\hat{k}}$$

The investment in the U.S. and Britain has a coefficient of variation of .143 (3% ÷ 21%), whereas the investment in the U.S. and Canada has a coefficient of variation of .625 (15% ÷ 24%). Accordingly, the company should undertake the British investment because it has substantially less risk per unit of return.

Answer (B) is incorrect because the Canadian investment is significantly riskier per unit of return. Answers (C) and (D) are incorrect because the company should undertake the British investment. As a result, its return will increase and its risk will decrease.

D. Hypothesis Testing

54. An audit of accounts payable was made to determine if the error rate was within the stated policy of 0.5%. One hundred of the 10,000 accounts payable transactions were randomly selected using a 95% confidence level. No errors were found. With 95% certainty, one can conclude that the sample results

A. Indicate another sample is needed.

B. Prove there are no errors in accounts payable.

C. Indicate the null hypothesis is false.

D. Fail to prove the error rate is above 0.5%.

The correct answer is (D). *(CIA, adapted)*

REQUIRED: The sound conclusion if a sample contained no errors.

DISCUSSION: The null hypothesis is that the error rate is equal to or less than 0.5%. Given that no errors were found, no basis for disproving the null hypothesis is presented.

Answer (A) is incorrect because the sample is adequate. Answer (B) is incorrect because even an examination of all 10,000 transactions would not necessarily prove that there are no errors in accounts payable. Answer (C) is incorrect because the null hypothesis has not been disproved.

55. A t-test is a statistical test used to compare the

A. Variances of two distributions.

B. Independence of two variables.

C. Correlation of two variables.

D. Means of two distributions.

The correct answer is (D). *(CMA, adapted)*

REQUIRED: The use of a t-test.

DISCUSSION: Student's "t" distribution is a special distribution to be used when only small samples are available and the population variance is not given. It is mound-shaped and symmetric like a normal distribution, only typically flatter with more variation. A small sample is usually deemed to be less than 30. For samples larger than 30, the "t" distribution gives results similar to the standard normal distribution. It is used to compare the means of two distributions to determine whether there is any significant difference between the two groups from which the means were taken.

Answer (A) is incorrect because the t-test compares means. Answers (B) and (C) are incorrect because correlation analysis measures the relationship among variables.

Questions 56 through 62 are based on the following information. The ABC Company has specified that the mean number of calories in a can of its diet soda is 1 or less. A consumer testing service examined nine cans with the following amounts of calories: .9, .95, 1.0, 1.05, .85, 1.0, .95, .95, and .9. The mean of these observations is .95. The sum of the squared deviations from the mean is .03. Assume the underlying population is approximately normal.

---

56. What is the sample standard deviation?

A. .0577

B. .0612

C. .0316

D. .00375

The correct answer is (B). *(Publisher)*

REQUIRED: The sample standard deviation.

DISCUSSION: The sample mean is the sum of the observations divided by the sample size. The sample mean is typically denoted as $\bar{x}$. The following is the sample variance (s^2): $\Sigma(x_i - \bar{x})^2 \div (n - 1)$. The sample standard deviation (s) is the square root of the variance. Thus, the sample variance is .00375 [.03 ÷ (9 – 1)], and the sample standard deviation is .0612 ($\sqrt{.00375}$).

Answer (A) is incorrect because .0577 results from using n, not n–1, in the denominator. Answer (C) is incorrect because .0316 equals .03 divided by .95. Answer (D) is incorrect because .00375 is the variance.

57. Let μ denote the true mean calories of all diet sodas produced by ABC Company. What hypothesis should be tested to determine whether ABC's claim is valid?

A. $H_o: \mu = 1$
   $H_a: \mu \neq 1$

B. $H_o: \mu \leq 1$
   $H_a: \mu > 1$

C. $H_o: \mu = 1$
   $H_a: \mu > 1$

D. $H_o: \mu = 0$
   $H_a: \mu < 1$

The correct answer is (B). *(Publisher)*

REQUIRED: The hypothesis to be tested to determine whether the advertising claim is valid.

DISCUSSION: ABC asserts that its diet soda has, on the average, 1 calorie or less. Thus, a possible null hypothesis is $H_o: \mu \leq 1$. The alternative hypothesis is that the mean is greater than 1, that is, $H_a: \mu > 1$. Because the null hypothesis cannot be rejected if the test statistic falls in the left-hand tail of the distribution, the test is one-tailed.

Answers (A), (C), and (D) are incorrect because the null hypothesis is that the population mean is equal to or less than 1 calorie. The alternate hypothesis is that it is greater than 1.

58. The appropriate means for testing this hypothesis is

A. A z-statistic.

B. A t-statistic.

C. An F-statistic.

D. A Q-statistic.

The correct answer is (B). *(Publisher)*

REQUIRED: The appropriate statistic to test the hypothesis.

DISCUSSION: The t-statistic is appropriate for tests of hypotheses based on small samples. It measures how the sample mean differs from the hypothesized true mean in terms of standard deviations. The formula is

$$t = \frac{\bar{x} - \mu}{s \div \sqrt{n}}$$

If: $\bar{x}$ = sample mean,
μ = hypothesized true mean
s = sample standard deviation
n = sample size.

Answer (A) is incorrect because the z-statistic is appropriate when the standard deviation of the population is known or a large sample ($n > 30$) permits a reasonable approximation of the population standard deviation. Answer (C) is incorrect because the F-statistic tests differences in variances. Answer (D) is incorrect because Q-statistic is a nonsense term.

59. The value of the t-statistic is

A. −2.45

B. −.05

C. 2.45

D. 4.65

The correct answer is (A). *(Publisher)*

REQUIRED: The value of the t-statistic.

DISCUSSION: Given that $\bar{x}$ equals .95, the hypothesized value of the true mean is 1, the sample standard deviation equals .0612 (see Q. 56), and the sample size is 9, the value of the t-statistic is

$$t = \frac{.95 - 1}{.0612 \div \sqrt{9}} = \frac{-.05}{.0612 \div 3} = \frac{-.05}{.0204} = -2.45$$

Answer (B) is incorrect because −.05 is the difference between the sample mean and the hypothesized true mean. Answers (C) and (D) are incorrect because the value is negative.

60. The appropriate number of degrees of freedom for this t-statistic is

A. 8

B. 9

C. 0

D. 1

The correct answer is (A). *(Publisher)*

REQUIRED: The appropriate number of degrees of freedom for the t-statistic.

DISCUSSION: The degrees of freedom associated with the test statistic equals the sample size minus the number of parameters being tested. Measurements of calorie content were made for nine cans of diet soda; i.e., nine distinct observations were included in the sample. The only parameter tested is μ, the mean calorie content of the diet sodas, so the number of degrees of freedom is 8 (9 − 1).

Answer (B) is incorrect because 9 is the sample size. Answers (C) and (D) are incorrect because one parameter is tested.

61. Refer to the information preceding question 56 on page 198. The following data are from a table of critical values of t:

d.f.	$t_{.10}$	$t_{.05}$	$t_{.025}$
5	1.476	2.015	2.571
6	1.440	1.943	2.447
7	1.415	1.895	2.365
8	1.397	1.860	2.306
9	1.383	1.833	2.262

The value of t defining the rejection region for testing the hypothesis that ABC's soda has 1 calorie or less per can, assuming a 95% confidence level, is

A. 2.306

B. 1.86

C. 1.833

D. 2.262

The correct answer is (B). *(Publisher)*

REQUIRED: The value of t defining the rejection region for testing the hypothesis at a 95% confidence level.

DISCUSSION: The rejection region is determined by the value of t at the appropriate degrees of freedom and the specific level of confidence. Choosing a 95% level of confidence means that the value of t will restrict the probability of a Type 1 error (rejecting the null hypothesis when the null hypothesis is true) to 5%. The rejection region is bounded by this value. Hence, the null hypothesis will be rejected if the t-statistic calculated from the formula is equal to or greater than the t-value from a table of t-values. The t-value correlates the appropriate degrees of freedom (n – 1) and the appropriate probability of making a Type 1 error (.05). In this case, with 8 degrees of freedom, the appropriate rejection region is defined by a t value of 1.86. Thus, a calculated t-statistic equal to or greater than 1.86 permits rejection of the null hypothesis H_o: $\mu \leq 1$.

Answer (A) is incorrect because 2.306 is for a confidence level of 97.5%. Answer (C) is incorrect because 1.833 is for 9 degrees of freedom. Answer (D) is incorrect because 2.262 is for 9 degrees of freedom and a confidence level of 97.5%.

62. Refer to the information preceding question 56 on page 198. What conclusion can be drawn with 95% confidence?

A. The manufacturer's claim can be rejected.

B. The manufacturer's claim cannot be rejected.

C. No decision is possible based on current information.

D. All sodas have at most 1 calorie.

The correct answer is (B). *(Publisher)*

REQUIRED: The conclusion with 95% confidence.

DISCUSSION: The critical value of t is 1.86 (see preceding question), and the calculated t-statistic is –2.45 (see Q. 50). Accordingly, the 5% rejection region includes, and lies to the right of, the t-value of 1.86. Because the computed value is less than (to the left of) 1.86 in this one-tailed test, the null hypothesis cannot be rejected.

Answers (A) and (C) are incorrect because the null hypothesis cannot be rejected. Answer (D) is incorrect because some sodas have more than 1 calorie.

Study Unit 5: QUANTITATIVE METHODS I

22 pages of outline
42 multiple-choice questions

A. Basic Mathematical Techniques
B. Linear Programming

This study unit is the second of three that concern quantitative methods, a major topic that has been assigned a relative weight range of 10% to 15% of Part 4.

A. Basic Mathematical Techniques

1. **Quantitative methods**, also referred to as operations research, is a broad term used to describe various applications of mathematics in business or any complex system. They are used to find the best estimates of expected results. **Management science** is a broader term that includes computerized information systems. All three terms refer to the scientific approach to problem solving for managers. The construction of mathematical, econometric, or statistical models of business situations allows the rational treatment of complex business problems that often include uncertainty.
   a. **Operations research** is, more specifically, the discipline of applying quantitative methods oriented to planning.
2. Typical quantitative methods applications include the following:
   a. Inventory control determines how much and when to order.
   b. Forecasting projects future sales, costs, etc.
   c. Statistical control aids in the detection of variances in production quality, costs, etc.
   d. Allocating resources finds the most economical production plan, assignment of sales territories, locations of warehouses, etc.
   e. Corporate planning models help managers explore the consequences of alternative policies.
3. Quantitative methods are important to the management accountant because
   a. Familiarity with these methods and their uses permits the management accountant to recognize possible applications within the firm and to seek the assistance of an expert.
   b. Quantitative methods often make use of accounting data. Knowledge of quantitative methods permits the management accountant to judge whether the accounting data are being used correctly.
   c. Quantitative methods can significantly reduce the cost of obtaining information concerning the operation of a business.
4. **Advantages of Using Mathematics**
   a. To make explicit (i.e., to quantify) most of the decision-making factors
   b. To make rational decisions that can be repeated, possibly using a computer
   c. To uncover solutions that might appear contrary to intuition, or to support intuitive conclusions
   d. To provide better insight into complex problems that could not otherwise be understood

5. **Limitations of Quantitative Methods**
   a. It is seldom possible to include all relevant variables in a mathematical model. Thus, managers should not rely solely on a model. They should review the model's results in light of other considerations.
   b. Mathematical models are supposed to be representations of the real world. But errors in the data used, the variables, or their relationships may never be eliminated. Also, models are usually oversimplifications.
   c. The techniques used can become complex, cumbersome, and costly. Their use should be justified on a cost-benefit basis.
6. Computers have allowed rapid advances in the use of quantitative methods.
   a. Computers have led to the widespread use of techniques such as linear programming.
   b. Managerial decision making in many fields can be automated by incorporating mathematical models into the data-processing programs.
   c. Although the first applications of computers merely replaced human clerical effort, most modern applications involve the use of methods impossible to implement in manual form.
   d. Study Units 7 and 8 concern information systems and controls.
7. **Graphs, Charts, and Diagrams**
   a. Graphs ordinarily depict functional relationships.
      1) Independent variables are shown on the horizontal or x-axis.
      2) Dependent variables are shown on the vertical or y-axis.
      3) EXAMPLE: Overhead cost (dependent variable) applied on the basis of labor hours is graphed on the vertical axis, with labor cost (independent variable) plotted on the horizontal axis.
   b. Natural numbers, e.g., 1, 2, 3, 4, 5, usually calibrate both horizontal and vertical axes of graphs.
      1) If the vertical axis is calibrated in logarithms, the growth rate of the independent variable is expressed, and the graph is a semilog graph.
      2) If both axes are calibrated in logarithms, the rate of growth of the dependent variable is expressed in terms of the rate of growth of the independent variable (a log-log graph).
   c. **Bar charts** (bar graphs) are a two-dimensional (variables plotted on x and y axes) graphic means of quantitative comparison by using bars or rectangles with lengths proportional to the measure of the data or things being compared.
      1) Bar charts may be vertical (column charts) or horizontal.
      2) **Pareto analysis** is based on the concept that about 80% of results or effects are caused by about 20% of people or events. Thus, managers should concentrate on the relatively few people or events that have the most significant effects. **Pareto diagrams** are bar charts in which the frequencies of the various types of adverse conditions that may occur in a given process are depicted.
         a) For example, an auditor might construct a Pareto diagram of the types and numbers of control deviations found in the receiving department. The tallest (or longest) bar signifies the most common type of problem.

3) A **histogram** is a bar chart used to depict a frequency distribution. It is used to create a graphic representation of the results of repeated measurements of a single variable, with the class intervals on the horizontal axis and the frequency (or relative frequency) of occurrence on the vertical axis.

a) Whether the histogram approximates or deviates from a normal distribution is of interest for control purposes.

b) EXAMPLE: A histogram of the number of occurrences of various net cash flows is given below. Cash flows between $0 and $1,000 occurred on 5 days; cash flows between $1,000 and $2,000 also occurred on 5 days; cash flows between –$2,000 and –$3,000 occurred on 3 days, etc.

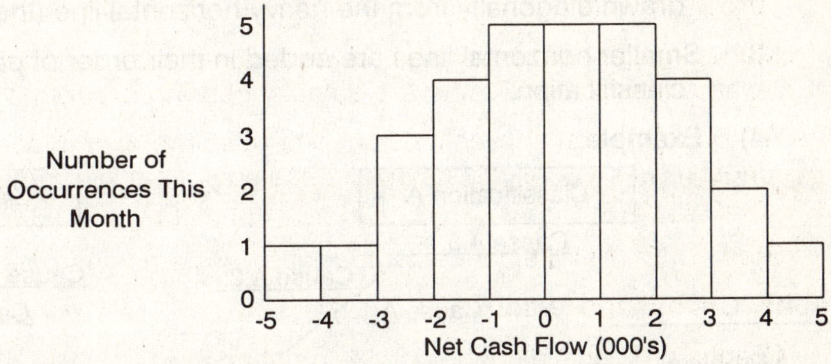

d. **Line charts** (also known as run, time series, or trend charts) show the frequency or amount of a variable (vertical axis) over time (horizontal axis). For example, sales might be plotted over a period of years on a simple line chart.

1) A multiple-line chart compares items, such as sales of several products over a period of years.

e. **Flowcharting** is a pictorial method of analyzing and understanding the processes and procedures involved in operations, whether manual or computerized.

1) Flowcharting is therefore useful in a preliminary survey of operations and in obtaining an understanding of internal control.

2) It is also helpful in systems development.

3) Flowcharting is based on standardized symbols. However, different systems of symbols have been devised, so care must be taken that the usage is consistent.

f. **PERT** (Program Evaluation and Review Technique) and **CPM** (Critical Path Method) are similar techniques that use network diagrams to plan and control complex projects in which the proper sequencing of activities is crucial.

1) A **Gantt chart** is a simpler project control device. It does not acknowledge the interrelationships of activities and omits calculations of activity times and costs.

g. **Scatter diagrams** may be used to detect misstated data, relationships among variables, and unusual operating results.

h. **Statistical control charts** are graphic aids for monitoring the status of any process subject to random variations. Statistical control charts are discussed at greater length in Study Unit 4.

i. A **fishbone diagram** (also called a cause-and-effect diagram or an Ishikawa diagram) is a process improvement technique that is useful in studying causation (why the actual and desired situations differ). This format organizes the analysis of causation and helps to identify possible interactions among causes.

   1) The head of the skeleton represents the statement of the problem.
   2) The principal classifications of causes are represented by lines (bones) drawn diagonally from the heavy horizontal line (the spine).
   3) Smaller horizontal lines are added in their order of probability in each classification.
   4) Example:

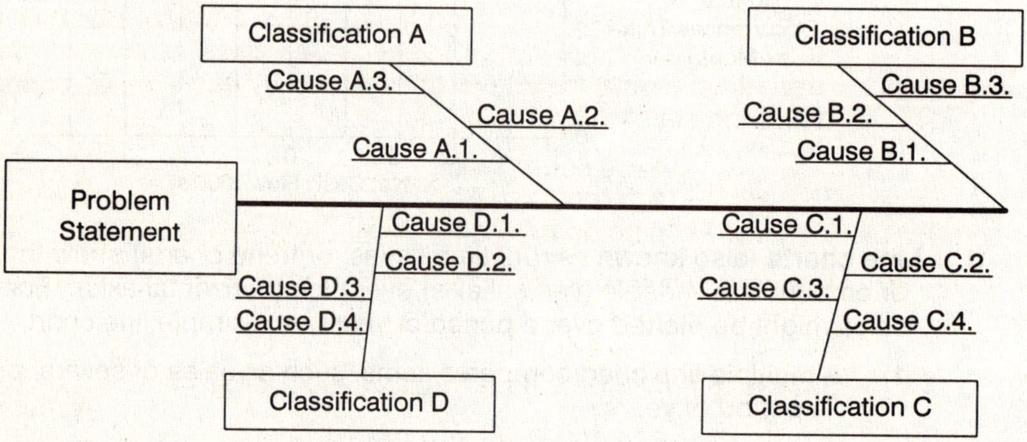

8. **Algebra**

a. Algebra is a generalization of arithmetic in which letters may represent numbers and be combined using the laws of arithmetic. Thus, it is a means of expressing relationships among variables (e.g., $2X = 4Y$). Through basic arithmetic operations (addition, subtraction, multiplication, and division) on both sides of an equation, solutions are obtainable (e.g., $X = 2Y$). Given the value of one variable, the value of the other can be obtained (e.g., if $Y = 4$, $X = 8$).

b. Linear equations are those with variables only to the first order (not squared, cubed, etc.). For example, $X = 10$ or $X + Y + 2 = 77.7$. These equations, when graphed, are depicted by straight lines (linear relationships).

   1) Many economic relationships are expressed in linear terms to facilitate analysis.
      a) Cost-volume-profit analysis
      b) Linear programming
      c) Method of least squares
   2) Procedures for implementing curvilinear analysis by adapting these and similar methodologies either are not known or are more cumbersome.

c. Nonlinear (curvilinear) functions are higher-order functions; their variables are squared, cubed, or of a higher order. Their graphs depict curves or other shapes, but not a straight line.

1) Cost and revenue functions, such as the EOQ model, are frequently nonlinear.

2) Managers are interested in the point at which revenue is maximized or cost minimized.

3) Differential calculus is used to calculate the maxima or minima of curvilinear functions.

d. Multiple unknowns (variables) require as many equations as there are unknowns for a finite solution. Thus, for two variables (i.e., a two-dimensional space), two lines and two equations are needed to define a point. In a three-dimensional space, three lines (three variables and three equations) are necessary to define a point.

1) The number of equations must be equal to the number of variables.

2) EXAMPLE: Three service departments (K, L, and M) provide some services to each other as well as to user departments. The relationships among the departments and the total costs incurred by each can be expressed in the following set of three equations:

$$K = \$100{,}000 + .10L$$
$$L = \$\ 70{,}000 + .15K$$
$$M = \$\ 50{,}000 + .05K + .09L$$

3) These simultaneous equations can be solved algebraically by using substitution.

Substitute K into the L equation, and solve for L.

$$L = \$70{,}000 + .15(\$100{,}000 + .10L)$$
$$L = \$70{,}000 + \$15{,}000 + .015L$$
$$.985L = \$85{,}000$$
$$L = \$86{,}294$$

Substitute this value for L into the K equation, and solve for K.

$$K = \$100{,}000 + .10(\$86{,}294)$$
$$K = \$108{,}629$$

Finally, solve the M equation by plugging in these values for K and L.

$$M = \$50{,}000 + .05(\$108{,}629) + .09(\$86{,}294)$$
$$M = \$50{,}000 + \$5{,}431 + \$7{,}766$$
$$M = \$63{,}197$$

9. **Linear, Matrix, and Boolean Algebra**

a. **Linear algebra** uses the mathematical theory of straight lines and vectors.

1) It is not used to solve managerial problems directly.

2) It is the basis for more applied mathematical techniques such as matrix algebra.

b. **Matrix algebra** is an efficient method of manipulating multiple linear equations.

1) A matrix is a group of numbers ordered into rows and columns.

$$\begin{bmatrix} 6 & 4 \\ 3 & 7 \end{bmatrix}$$

6 4 is the first row.

3 7 is the second row.

6 is the first column. 4 is the second column.
3 7

2) Each number in a matrix is called an element. Elements are numbered by row and column.

$$\begin{bmatrix} X_{11} & X_{12} \\ X_{21} & X_{22} \end{bmatrix}$$

X_{11} is in the first row and first column.
X_{12} is in the first row and second column.
X_{21} is in the second row and first column.
X_{22} is in the second row and second column.

3) Matrices can be added, subtracted, and multiplied in a fashion very similar to that of variables in single equations. Matrices must be of compatible sizes before they can be manipulated, however.

a) Addition and subtraction require matrices of the same dimensions.

i) For example, a matrix of 2 rows and 2 columns can only be added to or subtracted from another 2-row-2-column matrix.

b) Adding and subtracting matrices is accomplished by adding (subtracting) each element in the first matrix to (from) its corresponding element in the second matrix.

i) EXAMPLE:

$$\begin{bmatrix} 2 & 6 \\ 4 & 7 \end{bmatrix} + \begin{bmatrix} 3 & 9 \\ 2 & 3 \end{bmatrix} = \begin{bmatrix} 5 & 15 \\ 6 & 10 \end{bmatrix}$$

$$\begin{bmatrix} 2 & 6 \\ 4 & 7 \end{bmatrix} - \begin{bmatrix} 3 & 9 \\ 2 & 3 \end{bmatrix} = \begin{bmatrix} -1 & -3 \\ 2 & 4 \end{bmatrix}$$

c) Multiplication of matrices requires that the matrices to be multiplied conform.

i) If matrices are identified as

A (m × n) and B (n × p)

the two can be multiplied; thus, if the number of columns in the first equals the rows in the second, the matrices conform.

$$\underset{A}{\begin{bmatrix} x & x & x \\ x & x & x \end{bmatrix}} \quad \underset{B}{\begin{bmatrix} x & x \\ x & x \\ x & x \end{bmatrix}}$$

ii) Furthermore, the resulting matrix will have the number of rows of the first and columns of the second.

iii) In other words,

$$A\ (m \times \underline{n})\quad B\ (\underline{n} \times p)$$

A can be multiplied by B, and AB = (m × p).

d) Matrices are actually large groups of numbers that are being manipulated.

4) Matrix division is accomplished by multiplying by an inverse matrix.

a) Multiplying an inverse matrix by its original matrix yields an identity matrix.

b) An identity matrix is a matrix in which all elements are zeros except the primary diagonal elements, which are ones. The primary diagonal consists of the elements that make up the line from the top left corner to the bottom right corner of the matrix. Thus, only a square matrix (rows = columns) has an inverse.

5) Matrix algebra is the basis for solving complicated linear problems.

a) These are problems with a large number of variables (which require a large number of equations).

b) A good example is the simplex tableau. See Linear Programming on page 211.

c. **Boolean algebra** is the algebra of logic. It is sometimes called symbolic logic.

1) Boolean algebra is a method of expressing logic in a mathematical context.

a) It is primarily concerned with binary operations.

b) Boolean algebra is based on the commutative, associative, and distributive laws of binary operations.

2) Boolean algebra provides the theoretical concepts for computer design.

a) It is not used to solve managerial problems directly.

10. Calculus

a. The primary business application of **differential calculus** is to identify the maxima or minima of curvilinear functions.

1) In business and economics, these are the points of revenue or profit maximization (maximum) and cost minimization (minimum).

2) Maxima or minima occur if the slope is 0. When a line becomes horizontal, its slope is 0. Thus, the first step is to find the slope of the equation. Slope is the change in y over the change in x.

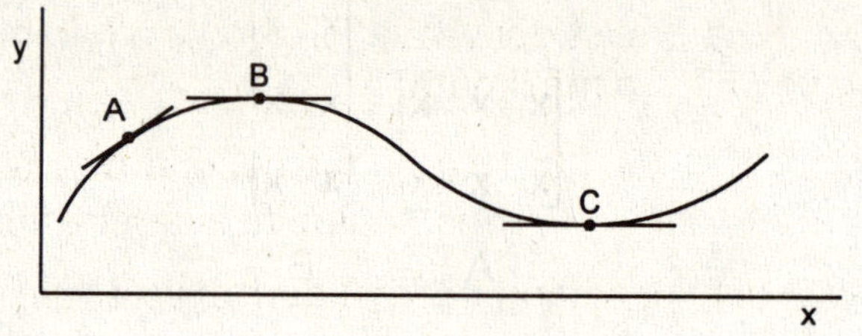

The straight lines tangent to the curve at points B and C have a slope of 0. Point B is a maximum; point C is a minimum.

3) The derivative of a function (equation) is its slope.

4) Derivatives are frequently expressed as f', $\frac{dy}{dx}$, or $\frac{df(x)}{dx}$, which means differentiate the function (equation) of y in terms of x.

a) A function of y in terms of x might be an expression such as $y = f(x) = x^2 + 12x + 7$, if x is an independent variable determining the value of y.

b) The derivative $= f' = \frac{df(x)}{dx} = \frac{d(x^2 + 12x + 7)}{dx}$

c) Alternatively, it can be expressed $f'(x^2 + 12x + 7)$.

5) The basic formula for finding the derivative is

$$f'(Ax^n) = nAx^{n-1}$$

If: A = the constant, i.e., the number of independent variables or x's

x = the variable being differentiated

n = the exponent of the independent variable

6) The derivative formula is applied to each term in the equation. The sum of the results is the derivative of the original equation.

7) EXAMPLE: $f'(x^2 + 12x + 7) = ?$

$f'(x^2) = 2x$, $f'(12x) = 12$, $f'(7) = 0$

Thus, $f'(x^2 + 12x + 7) = 2x + 12$

8) EXAMPLES: $f'(12) = 0$

$f'(12x) = 12$

$f'(12x^2) = 24x$

$f'(12x^3) = 36x^2$

$f'(12x^4) = 48x^3$

$f'(12x^{-2}) = -24x^{-3}$

9) The point(s) with zero slope on the graph of a function can be identified by

a) Calculating the first derivative

b) Setting the first derivative equal to zero and solving the equation

c) Calculating the second derivative and determining whether it is positive or negative

i) The second derivative of a function is the derivative of the first derivative.

ii) If the second derivative is positive, a point found by solving the first derivative for zero is a minimum; if it is negative, the point is a maximum.

- In 10.a.7), the second derivative is positive $[f'(2x+12)=2]$. Thus, the point is a minimum.

iii) If setting the first derivative equal to zero yielded two solutions, each is substituted into the second derivative.

iv) EXAMPLE:

- $f'(\frac{x^3}{3} - 4x^2 + 7x + 10) = x^2 - 8x + 7$

$$(x - 7)(x - 1) = 0$$
$$x = 7$$
$$x = 1$$

- $f'(x^2 - 8x + 7) = 2x - 8$

$$2(7) - 8 = 6$$
$$2(1) - 8 = -6$$

- Accordingly, x=7 is a minimum; x=1 is a maximum.

10) EXAMPLE: The formula for the economic order quantity (EOQ) (see Study Unit 1) results from differentiating the total cost with regard to the order quantity (x) and solving for x. This expression must be a minimum because the second derivative must be positive.

$$c = \frac{Da}{x} + \frac{xk}{2} + F$$

$$f' = \frac{-Da}{x^2} + \frac{k}{2}$$

If:
c = total cost
x = EOQ
a = cost per purchase order
D = annual demand in units
k = unit annual carrying cost
F = fixed cost of ordering

The second derivative ($2Da \div x^3$) must be positive. Accordingly, setting the first derivative equal to zero gives the minimum total cost.

$$\frac{-Da}{x^2} + \frac{k}{2} = 0$$

$$\frac{k}{2} = \frac{Da}{x^2}$$

$$x^2 = \frac{2Da}{k}$$

$$x = \sqrt{\frac{2Da}{k}}$$

b. **Integral calculus** also has a very important business application. It permits computation of the area under a curve.

1) The area under a probability curve equals the probability. Thus, integral calculus is used to evaluate the probability that events will occur.

2) Integral calculus finds an antiderivative, i.e., the function that, if differentiated, would result in the function being integrated.

3) The general form of the integral is $\int_a^b f(x)]dx$, which is the integral of f(x) with respect to x.

4) Some formulas for integration are

a) $\int_a^b [kf(x)]dx = k \int_a^b f(x)dx$

i) This formula indicates that constants (k) can be brought out of the integral.

b) $\int_a^b x^n \, dx = \left[\frac{x^{n+1}}{n+1}\right]_a^b$

i) The integral of x^n is x^{n+1} divided by n + 1. The degenerate case is n = 1, which is the same as x^1, i.e., $x = x^1$.

c) EXAMPLE: For the function y = 4x,

$$\int_a^b ydx = \int_a^b (4x)dx = 4 \int_a^b (x)dx = 4(x^2 \div 2) = 2x^2$$

11. Stop and review! You have completed the outline for this subunit. Study multiple-choice questions 1 through 23 beginning on page 223.

B. Linear Programming

1. Linear programming is a technique used to maximize a revenue or profit function, or minimize a cost function, subject to constraints, e.g., limited (scarce) resources or minimum/maximum levels of production, performance, etc. In business, linear programming is used for planning resource allocations. Managers are often faced with problems of selecting the most profitable or least costly way to use available resources.

   a. EXAMPLE: A manufacturer should minimize production costs while satisfying production requirements, maintaining required inventory levels, staying within production capacities, and using available employees. The objective function is the production cost to be minimized; the constraints are production requirements, inventory levels, production capacity, and available employees.

   b. Other business applications include

      1) Selecting a product mix
      2) Blending chemical products
      3) Scheduling flight crews
      4) Assigning jobs to machines
      5) Determining transportation routes

   c. The conditions that restrict the optimal value of the **objective function** are the **constraints**.

      1) A **shadow price** is the amount by which the value of the optimal solution of the objective function in a linear programming problem will change if a one-unit change is made in a binding constraint.

         a) A nonbinding constraint is one that has excess capacity; i.e., the optimal solution does not use all of the given resource. The shadow price for a nonbinding constraint is zero because a one-unit change will not affect the optimal solution when excess capacity exists.

         b) The calculation of shadow prices is a simple example of **sensitivity analysis**, which is any procedure to test the responsiveness of the solution indicated by a model to changes in variables, alternative decisions, or errors.

   d. The unknowns used to construct the objective function and the constraints are the decision **variables**. The values of the variables are limited by the constraints.

   e. Values that are fixed for purposes of solving the model (but are not fixed forever) are cost/profit coefficients for the objective function, technical coefficients for the constraints, and the constants on the right-hand side of the constraints.

   f. When the number of constraint equations equals the number of variables, a unique solution exists. When the number of variables exceeds the number of constraint equations, the number of possible solutions is usually infinite.

      1) The values of the variables are the outputs from the linear programming process.

      2) The constraint equations usually reflect the types of input or types of resources being allocated, e.g., available machine hours or raw materials.

      3) The objective of linear programming is to choose the best solution from a potentially infinite number of possibilities.

2. Management accountants should recognize situations in which linear programming may be applicable. When an application is attempted, qualified operations research experts should assist in providing guidance in formulating and solving the problem. Linear programming is a powerful planning tool for management, but it is complex and usually requires computer software.

3. Several methods are available to solve linear programming problems.

   a. The **graphical method**, although the easiest technique, is limited to problems with two variables.

   b. The **algebraic method** is a trial-and-error technique. Pairs of constraints are solved algebraically to find their intersection. The values of the decision variables are then substituted into the objective function and compared to find the best combination.

      1) The basic rule is that the optimal solution will be at the intersection of two or more constraint equations.

      2) Thus, all intersections can be computed and each solution evaluated in the objective function to determine which solution is optimal.

   c. The **simplex method** is the technique most commonly used to solve linear programming problems. It is an algorithm to move from one corner solution to a better corner solution. When a better solution cannot be found, the optimal solution has been reached.

      1) The simplex method relies on an area of mathematics called matrix algebra. The equations that form the constraints are arranged in a matrix of coefficients and manipulated as a group with matrix algebra.

      2) Almost all practical applications of linear programming require the use of computers. Most computer facilities have a linear programming package that uses the simplex algorithm to find the optimal solution.

4. EXAMPLE: A company produces products G and J. Product G contributes profits of \$5,000 per unit sold, and product J contributes \$4,000 per unit.

   a. The company seeks to maximize profits, so the objective function is

      Maximize 5G + 4J (the maximum profit in thousands of dollars)

   b. The objective function is subject to the following constraints:

$$
\begin{array}{llll}
G & + \; J \geq 5 & & \text{(minimal production requirement)} \\
G & \leq 3J & & \text{(market balance requirement)} \\
10G & + \; 15J \leq 150 & & \text{(production capacity constraint)} \\
G, J & \geq 0 & & \text{(nonnegativity constraint)}
\end{array}
$$

c. **Graphical solution**. To plot this set of constraints,

1) Change inequalities to equalities.
2) Plot the equalities.
3) Identify the correct side of the line for the original inequalities.

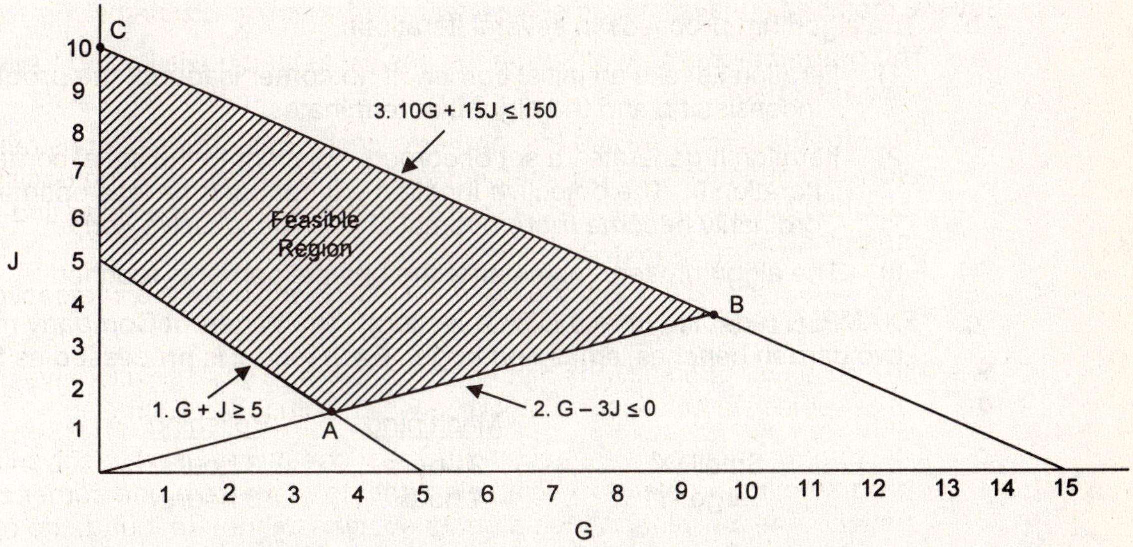

d. **Algebraic solution**. The combination of G and J that maximizes the objective function will occur at one of the extreme points or corners of the feasible region.

1) Simultaneously solving the constraints intersecting at those corners and substituting into the objective function,

a) At point A:

$$\begin{array}{r} G + J = 5 \\ -G + 3J = 0 \\ \hline 4J = 5 \end{array}$$

$J = 1.25$ and $G = 3.75$

Profit $= \$5,000(3.75) + \$4,000(1.25) =$ <u>$23,750</u>

b) At point B:

$$\begin{array}{r} G - 3J = 0 \\ \underline{10G + 15J = 150} \end{array} \longrightarrow \begin{array}{r} 5G - 15J = 0 \\ + \underline{10G + 15J = 150} \\ 15G \quad = 150 \end{array}$$

$G = 10$ and $J = 3.333$

Profit $= \$5,000(10) + \$4,000(3.333) =$ <u>$63,333</u>

c) At point C:

$$\begin{array}{r} \underline{G \qquad = 0} \\ 10G + 15J = 150 \end{array}$$ (from nonnegativity constraint)

$J = 10$ and $G = 0$

Profit $= \$5,000(0) + \$4,000(10) =$ <u>$40,000</u>

d) Point D does not lie in the feasible region.

2) The firm should choose to produce at point B (10 of product G and 3.333 of product J). This solution assumes that partial units of J may be produced. If product J is produced in single units, the answer may be interpreted to be to produce at a rate of 3⅓ units in a given period.

5. **Simplex Algorithm**

a. A simplex algorithm is a systematic method of examining the corners (also called vertices or extreme points) of the feasible region while in search of the optimal solution. The algorithm always moves in the direction of improving the solution and terminates at the optimal solution.

b. The algorithm proceeds in several iterations.

1) Iteration I seeks an initial corner. If no corner is found, the problem is inconsistent, and the algorithm terminates.

2) Iteration II generates a set of corners adjacent to the initial corner (found in Iteration I). The objective function will not become less optimal and will ordinarily become more optimal at each successive corner.

3) The algorithm terminates after the optimal corner is reached.

c. EXAMPLE (adapted from a CPA question): The Wineroot Company manufactures two garden benches, small and large. Each model is processed as follows:

	Machining	Polishing
Small (X)	2 hours	3 hours
Large (Y)	4 hours	3 hours

The available time for processing the two models is 100 hours a week in the Machining Department and 90 hours a week in the Polishing Department. The contribution margin expected is $5 for X and $7 for Y.

1) The objective function is the formula for the maximum total contribution margin. Because each X produces a $5 CM and each Y produces a $7 CM, the total CM will be 5X + 7Y. This is what Wineroot wants to maximize.

2) A constraint equation can be set up for the Machining Department, which has a total of 100 hours available each week. Each X requires 2 hours of time from that department, and each Y requires 4 hours. Accordingly, the total Machining Department time taken by production is 2X + 4Y, which must be equal to or less than the 100 hours available.

$$2X + 4Y \leq 100$$

3) The constraint equation for the Polishing Department is found in a similar manner. The Polishing Department has a total of 90 hours available each week. Each unit of X or Y requires 3 hours. Thus, the constraint for this department is that 3X + 3Y must be equal to or less than 90.

$$3X + 3Y \leq 90$$

4) Both constraints are inequality expressions of the form less than or equal. To transform them to an equality form, a positive variable called the slack variable must be added to the left-hand side of each constraint.

$$2X + 4Y \leq 100$$
$$2X + 4Y + S_1 = 100$$

$$3X + 3Y \leq 90$$
$$3X + 3Y + S_2 = 90$$

a) In general, each slack variable makes zero contribution to the value of the solution.

b) Before the tableau is reconstructed, the linear programming problem may be restated as follows:

Maximize $5X + 7Y + 0S_1 + 0S_2$ subject to

$$2X + 4Y + 1S_1 + 0S_2 = 100$$
$$3X + 3Y + 0S_1 + 1S_2 = 90$$

If: $X \geq 0$
$Y \geq 0$
$S_1, S_2 \geq 0$

d. The initial simplex tableau (in its most common format) can be constructed from the Wineroot Company's LP problem. Rows run horizontally and columns run vertically.

C_j	5	7	0	0	RHS
Variables / CB	X	Y	S_1	S_2	
0	2	4	1	0	100
0	3	3	0	1	90
Z_j	0	0	0	0	0
$C_j - Z_j$	5	7	0	0	

1) After the constraint equations are converted to equalities, construction of the tableau begins with the variables row, which consists first of the constraint variables (X and Y) and then the slack variables (S_1 and S_2). It is essentially a row of labels.

2) The C_j row lists the payoff coefficients for a profit-maximizing problem (or the cost coefficients for a cost-minimizing problem) given by the coefficients of the objective function.

a) The slack variables have payoff/cost coefficients of 0.

b) Thus, in constructing this tableau, the C_j row consists of the coefficients from the objective function ($5X + 7Y + 0S_1 + 0S_2$).

c) The C_j row is 5, 7, 0, 0.

3) The machining and polishing constraint rows in the simplex tableau are formed from the constraint equalities, which were

$$2X + 4Y + 1S_1 + 0S_2 = 100$$
$$3X + 3Y + 0S_1 + 1S_2 = 90$$

a) Using these coefficients, the constraint rows (resource coefficient rows) are

2	4	1	0	100
3	3	0	1	90

b) The resource coefficient rows in the tableau must equal the constraints.

4) There are as many basic variables as constraints.

a) Basic variables exist in the current tableau as a possible solution (not necessarily optimal).

b) In this example, the initial basic solution sets S_1 equal to 100 and S_2 equal to 90.

c) Nonbasic variables are the variables that are equal to zero in the current solution, e.g., X and Y.

5) The basic variables are those that form an identity matrix. Their corresponding variable columns consist of zeros and a single 1. The variable columns are those beneath each variable, i.e., beneath X, Y, S_1, and S_2.

a) In the sample tableau, S_1 has 1, 0 in its column, and S_2 has 0, 1. Consequently, they form an identity matrix.

b) In this tableau, the basic variables are also the slack variables.

i) But they may not necessarily be next to each other in other problems.

6) The values for the basic variables are given in the right-hand side (RHS) column of the tableau.

a) Because the entry 1 for the S_1 variable is on the first row of the identity matrix, the value of S is given in the first row of the RHS column as 100.

b) The entry of 1 for S_2 is in the second row, and its value is found in the second row under RHS to be 90.

c) The values 100 and 90 come from the right-hand side of the constraint equations. Thus, the RHS values are also the amounts of available resources.

7) The CB column consists of payoff (cost) coefficients of the basic variables in profit-maximization (cost-minimization) problems.

a) Hence, the column consists of as many numbers as there are basic variables (i.e., two for this tableau).

b) The numbers entered are the same as for the basic variables found in the C_j row (0 and 0).

c) Z_j and $C_j - Z_j$ are not part of the CB column.

8) The first element of the Z_j row is the sum of the products of multiplying each element in the CB column by each element in the first column [(0 x 2) + (0 x 3) = 0].

a) The subsequent elements are obtained by multiplying each element in the CB column by each element in the next column and adding the products.

b) For example, the third element in the Z_j row is 0 [(0 x 1) + (0 x 0)].

c) The elements in the Z_j row are opportunity costs.

9) The index row is the last (bottom) row. It provides a measure of the direction of improvement of the objective function value for a corresponding change in the tableau.

a) It is obtained by subtracting each element in the Z_j row from each element in the C_j row.

$$C_j - Z_j = 5\ 7\ 0\ 0$$

b) Positive elements in the index row for a maximization problem or negative elements for a minimization problem indicate that the solution could be improved.

10) The current (but not the optimal) solution to this simplex problem is obtained by multiplying the CB column times the RHS column and adding the products.

a) The result is found in the far right column (RHS) in the Z_j row, which here is zero.

b) The current solution does not always equal zero, although it does in this case because CB = 0.

e. The following is the graphical solution to the foregoing problem:

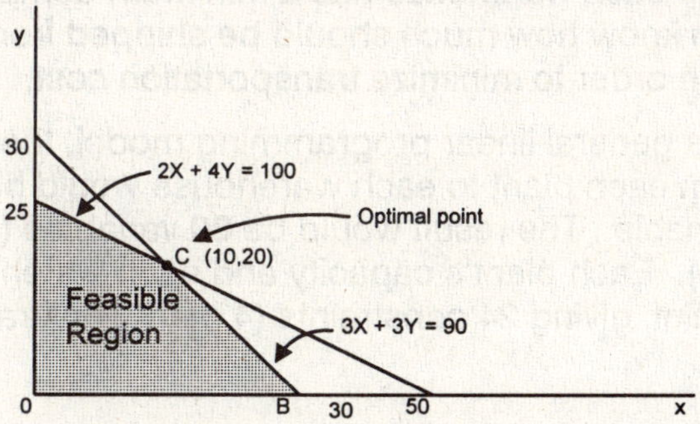

1) Given the objective function (OF) 5X + 7Y, the extreme points of the feasible region yield the following values:

At point	OF
(0, 0)	0
(0, 25)	(0 × 5) + (25 × 7) = 175
(30, 0)	(30 × 5) + (0 × 7) = 150
(10, 20)	(10 × 5) + (20 × 7) = 190

f. Computer software packages have been developed to manipulate linear programming problems with numerous constraints. Recognizing the basic method and its implications is usually sufficient for the management accountant.

6. **Transportation Model**

a. The transportation model is a special type of LP problem. It involves physical movement (transportation) of goods from sources of supply (such as factories) to destinations (warehouses).

1) EXAMPLE: A fleet of trucks may be available for daily deliveries in a city. Each day the transportation manager must determine the best route for each truck. Many possible combinations of trucks and destinations satisfy the constraints of meeting all deliveries with the available trucks, but the manager would like to have the minimum cost combination. In effect, the manager must allocate the available trucks (scarce resources) to the required destinations in a way that minimizes costs.

b. The objective function includes the transportation cost of each item from each source to each destination.

1) The constraints are the output for each supply point and the demand by each destination.

2) EXAMPLE: A manufacturer has four plants. Assume that each plant could ship products to any of 20 warehouses. Each plant has a maximum capacity, and each warehouse has a minimum demand. Management would like to know how much should be shipped from each plant to each warehouse in order to minimize transportation cost.

3) To formulate a general linear programming model, the amount to be shipped from each plant to each warehouse would have to be stated as a separate variable. The result would be 80 variables (4 plants x 20 warehouses). Each plant's capacity and each warehouse's demand would be a constraint, giving 24 constraints (4 plant constraints + 20 warehouse constraints).

c. The **assignment method** is a special case of the transportation method. It is used to assign employees to jobs, jobs to machines, sales associates to territories, bidders to contracts, etc.

1) The difference from the transportation method is that each agent (employee, machine, etc.) can be assigned to only one task.

2) The purpose of the assignment is to optimize a given objective, such as minimizing cost, minimizing time, or maximizing profits.

3) The assignment model can be modified to accommodate certain variations. They include the following:

a) Tasks exceed the available number of agents.
b) Certain assignments are unacceptable.
c) More than one agent may be required for some tasks.

d. **Vogel's Approximation**, or the **penalty method**, solves transportation problems easily. The name comes from the extra amount that must be paid (the penalty) if the lowest cost alternative is not selected in each row and each column.

1) EXAMPLE: Assume that a company has two factories that produce 18 and 14 units of product, respectively, during a period of time. Three distribution warehouses demand 15, 10, and 7 units, respectively, during the same period. The table below depicts this information along with the shipping costs per unit (e.g., $4 to ship one unit from factory 1 to warehouse C). The objective is to minimize total shipping costs. The penalties (in dollars) are shown in parentheses to the right of, and below, the table.

from \ to	A	B	C	Supply	
Row 1	$8	$5	$4	18	(1)
Row 2	$5	$6	$7	14	(1)
Demand	15	10	7	32 / 32	
	(3)	(1)	(3)		

a) For row 1: If the lowest cost alternative is not used ($4 for C), the shipment must go to B, whose cost of shipping is $1 higher than shipping to C. The firm would suffer a $1 penalty for not selecting the lowest cost box. For column A, the penalty would be $3 ($8 – $5) if the lowest cost box is not used.

b) The objective under the penalty method is to avoid the highest penalties; those columns or rows where the highest penalties might occur are filled first. Because columns A and C have the highest penalties, one of these (it does not matter which) should be filled first. Choose column C. The lowest cost box in the selected column should be filled with the maximum possible number of units. The maximum that can be placed into box 1C is 7 units because warehouse C needs only 7 units; this step eliminates column C from further consideration. Draw a line through column C and recompute the penalties for the remaining columns and rows (if necessary) as shown below.

from \ to	A	B	C	Supply		
1	$8	$5	$4 7	18	~~(1)~~	(3)
2	$5	$6	$7	14	(1)	
Demand	15	10	7	32 32		
	(3)	(1)	(3)			

c) Because column A still shares the highest penalty, fill its lowest cost box to the maximum extent possible; i.e., assign 14 units to box 2A. The row 2 total supply is 14, thus limiting the number of units that can be assigned to box 2A. This assignment of 14 units equals the row 2 constraint, so row 2 is crossed out (it no longer needs to be considered).

d) The final units are then assigned to row 1, columns A and B, as shown in the following table.

from \ to	A	B	C	Supply		
1	$8 1	$5 10	$4 7	18	~~(1)~~	(3)
~~2~~	$5 ~~14~~	$6	$7	~~14~~	~~(1)~~	
Demand	15	10	7	32 32		
	(3)	(1)	(3)			

e) The total shipping costs based on the assignments can be calculated by multiplying the number of units shipped times the shipping cost per unit, as follows:

$$1(\$8) + 10(\$5) + 7(\$4) + 14(\$5) = \$156$$

f) The use of the penalty method resulted in the optimal solution in this example, which is not always the case. The penalty method will occasionally result in a very good solution, but not the optimum. There is a quantitative technique to test for the optimality of a solution, but it is quite complicated and beyond the scope of this discussion.

7. **Goal Programming**

a. Goal programming was first introduced as a specialized type of linear programming to solve problems that had no feasible solution under regular linear programming.

b. Goal programming permits an ordinal specification (ranking) and solution to problems with multiple goals.

1) With its ability to incorporate multiple goals, the method has the potential for solving problems with both financial and nonfinancial goals.

8. **Integer Programming**

a. Integer programming (IP) concerns problems in which some variables are not continuous. Linear programming, on the other hand, assumes that all variables are continuous and all functions are linear (i.e., graphed with straight lines). Variables are thus allowed to take fractional values; 9, 9.367, and 10 are all possible values. In integer programming, fractions are not possible.

1) EXAMPLE: It might be possible for an oil company to build 12 or 13 gas stations on a stretch of highway, but it is not clear what an optimal solution of 12.72 gas stations means.

2) Integer programming problems are also known as **discrete** models because the variables take on discrete, noncontinuous values.

b. One way to solve IP problems is to solve them as LP problems and round to satisfy the integrality condition.

1) In the example, we would decide to build 13 gas stations by rounding 12.72 to 13.0.

2) Rounding raises several problems.

a) The rounded solution may violate constraints.

b) The rounded solution may be farther from the optimal solution than some other integer combination.

c. A certain class of IP problems concerns variables that are constrained to take on values of 1 or 0. These are yes-and-no or go, no-go decisions. This class of problems may be solved only with algorithms designed to handle these constraints. Rounding is never acceptable because the error induced would be too great.

9. **Dynamic programming** is an approach to solving problems, not a particular algorithm. It divides a large mathematical model into smaller, more manageable pieces in such a way that, once the smaller problems have been solved, the result is the optimal solution to the overall model.

   a. This type of model is useful for problems that involve multistage decisions, that is, decisions affected by previous decisions.

      1) Typically, they are applied to much smaller problems than LP, in terms of number of variables and constraints.

      2) Dynamic programming is useful for problems such as

         a) Inventory reordering rules
         b) Production scheduling in the face of fluctuating demand
         c) Spare parts level determination
         d) Capital budgeting for new ventures
         e) Systematic search for scarce resources
         f) Finding the shortest route through a network

   b. The key features of dynamic programming problems are

      1) The probabilistic (stochastic) nature of the variables. However, deterministic variations exist.

      2) The iterative formulation of the problems. The optimal solution is reached by a repetitive process.

   c. A decision tree with probabilities of occurrence for each branch is a good example of a multistage decision process that could be evaluated with dynamic programming.

10. Stop and review! You have completed the outline for this subunit. Study multiple-choice questions 24 through 42 beginning on page 231.

MULTIPLE-CHOICE QUESTIONS

A. Basic Mathematical Techniques

Questions 1 and 2 are based on the following information. An organization has collected data on the complaints made by personal computer users and has categorized the complaints.

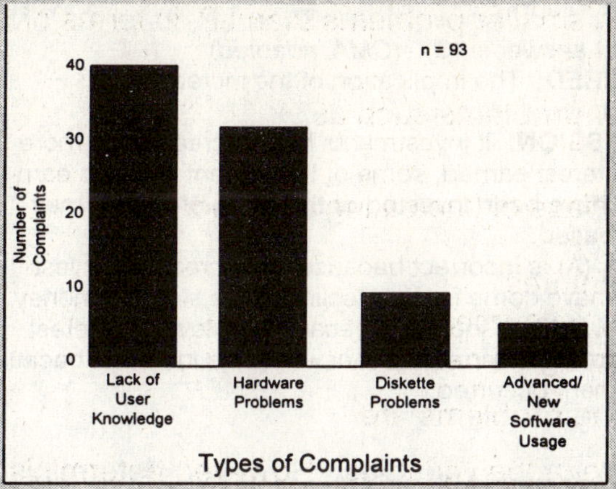

1. Using the information collected, the organization should focus on

A. The total number of personal computer complaints that occurred.

B. The number of computer complaints associated with diskette problems and new software usage.

C. The number of computer complaints associated with the lack of user knowledge and hardware problems.

D. The cost to alleviate all computer complaints.

The correct answer is (C). *(CIA, adapted)*

REQUIRED: The organization's focus based on the data.

DISCUSSION: Complaints based on lack of user knowledge and hardware problems are by far the most frequent according to this chart. Consequently, the company should devote its resources primarily to these issues.

Answer (A) is incorrect because more detailed information is not available. The Pareto diagram does not focus on the total quantity of computer complaints. Answer (B) is incorrect because complaints about diskettes and software are infrequent. Answer (D) is incorrect because cost information is not provided.

2. The chart displays

A. The arithmetic mean of each computer complaint.

B. The relative frequency of each computer complaint.

C. The median of each computer complaint.

D. The absolute frequency of each computer complaint.

The correct answer is (D). *(CIA, adapted)*

REQUIRED: The information provided by the chart.

DISCUSSION: This Pareto diagram depicts the frequencies of complaints in absolute terms. It displays the actual number of each type of complaint.

Answers (A), (B), and (C) are incorrect because the chart does not display arithmetic means, relative frequencies, or medians of each type of complaint.

3. Quality control programs employ many tools for problem definition and analysis. A scatter diagram is one of these tools. The objective of a scatter diagram is to

A. Display a population of items for analysis.

B. Show frequency distribution in graphic form.

C. Divide a universe of data into homogeneous groups.

D. Show the vital trend and separate trivial items.

The correct answer is (A). *(CIA, adapted)*

REQUIRED: The objective of a scatter diagram.

DISCUSSION: The objective of a scatter diagram is to demonstrate correlations. Each observation is represented by a dot on a graph corresponding to a particular value of X (the independent variable) and Y (the dependent variable).

Answer (B) is incorrect because the objective of a histogram is to show frequency distribution in graphic form. Answer (C) is incorrect because the objective of stratification is to divide a universe of data into homogeneous groups. Answer (D) is incorrect because regression analysis is used to find trend lines.

Questions 4 through 6 are based on the following information. The equality between a household's total sources and uses of funds can be represented in the form of the budget constraint model presented in the next column.

$$Py_t + b_{t-1}(1 + R) + m_{t-1} = Pc_t + b_t + m_t$$

If: Py_t = earned income during time period t
Pc_t = consumption spending during time period t
b_t = financial investments held at time period t
R = interest rate
m_t = stock of money held at time period t

4. If $b_t > b_{t-1}(1 + R)$,

A. Earned income must exceed consumption spending.

B. Interest rates must be very high.

C. Net investment took place.

D. Net disinvestment took place.

The correct answer is (C). *(CMA, adapted)*

REQUIRED: The implication of the increase in investment.

DISCUSSION: If investments have increased by more than the interest earned, some of the current period's earnings must have been invested or the stock of money may have decreased.

Answer (A) is incorrect because the increase in investment may have come from a decline in the stock of money held. Answer (B) is incorrect because the level of interest rates cannot be determined. Answer (D) is incorrect because net investment occurred.

5. If $Pc_t - Py_t > b_{t-1}(R)$,

A. Saving would occur.

B. Earned income would exceed interest income.

C. m_t would be negative.

D. Investments and/or stock of money would decline.

The correct answer is (D). *(CMA, adapted)*

REQUIRED: The implication of the excess of consumption over the sum of earned income and return of investment.

DISCUSSION: This model recognizes three sources of funds: current-period earned income, prior-period return on investment, and prior-period money held. The three uses of funds in the current period are consumption, investment, and money held. If the excess of consumption spending over earned income in period t is greater than the return on investment in period (t – 1), investment or the stock of money must have declined because spending is greater than income and earnings on investments.

Answer (A) is incorrect because, assuming saving is defined as investment, the effect on saving is indeterminate. The stock of money may or may not have been reduced in favor of greater investment. If saving is defined as money held plus investment, however, no saving could occur. Answer (B) is incorrect because the relationship between earned income and interest income cannot be determined. Answer (C) is incorrect because the stock of money cannot be negative.

6. If a household's need to hold money declines,

A. $b_t > b_{t-1}$.

B. $m_t < m_{t-1}$.

C. $Pc_t < Py_t$.

D. Negative saving occurs.

The correct answer is (B). *(CMA, adapted)*

REQUIRED: The effect of a decline in a household's need to hold money.

DISCUSSION: If a household's need to hold money declines, the stock of money in the current period will be less than the stock held in the preceding period.

Answer (A) is incorrect because the decline in money held does not necessarily result in greater investment. Answer (C) is incorrect because consumption may be more or less than earnings since the additional funds could have been used for consumption or investment. Answer (D) is incorrect because the reduction in the money stock may have been consumed or invested.

7. A company has two production and two service departments. Each service department provides service to the two production departments and to the other service department. In allocating service department costs, the company would like to account for the service provided by each service department to the other. To do so, it could use

A. Matrix algebra.

B. Regression analysis.

C. Game theory.

D. Sensitivity analysis.

The correct answer is (A). *(CIA, adapted)*

REQUIRED: The quantitative method useful for applying the reciprocal method of service cost allocation.

DISCUSSION: Matrix algebra is a mathematical technique for solving a set of simultaneous equations. Matrices consisting of the coefficients of the variables of the equations may be manipulated by addition, subtraction, multiplication, and inversion. Because the two service departments simultaneously provide services to each other, recognition of these reciprocal services requires solution of a set of simultaneous equations.

Answer (B) is incorrect because regression analysis is a statistical procedure for estimation of the relation between variables. Answer (C) is incorrect because game theory is a mathematical approach to decision making in which each decision maker takes into account the actions of competitors. Answer (D) is incorrect because sensitivity analysis is a method of studying the effects of changes in one or more variables on the results of a decision model.

8. A set of simultaneous equations reads:

$$6X_1 + 3X_2 = 49$$

$$12X_1 + 2X_2 = 17$$

If the equations were placed in a matrix format it would be

A. $\begin{bmatrix} X_1 \\ X_2 \end{bmatrix} \begin{bmatrix} 6 & 3 & 49 \\ 12 & 2 & 17 \end{bmatrix}$

B. $\begin{bmatrix} 6 & 3 \\ 12 & 2 \end{bmatrix} \begin{bmatrix} X_1 \\ X_2 \end{bmatrix} \begin{bmatrix} 49 \\ 17 \end{bmatrix}$

C. $\begin{bmatrix} 6 & 3 \\ 12 & 2 \end{bmatrix} \begin{bmatrix} 49 \\ 17 \end{bmatrix} \begin{bmatrix} X_1 \\ X_2 \end{bmatrix}$

D. $\begin{bmatrix} X_1 \\ X_2 \end{bmatrix} \begin{bmatrix} 6 & 3 \\ 12 & 2 \end{bmatrix} \begin{bmatrix} 49 \\ 17 \end{bmatrix}$

The correct answer is (B). *(CIA, adapted)*

REQUIRED: The proper matrix format for the equations.

DISCUSSION: Systems of linear equations can often be more conveniently manipulated by using matrix notation and matrix algebra techniques. In this format, the ordered numerical coefficients of the unknowns are presented in a matrix (in this case, two rows by two columns). The variables are described by a row or, in this problem, a column vector. Finally, the right side of the system is described by a vector (here, a column) of constants.

Answer (A) is incorrect because 49 and 17 are constants, not coefficients. Answer (C) is incorrect because the variables vector and the constants vector are misplaced. Answer (D) is incorrect because the coefficients matrix and the variables vector are misplaced.

9. The product of a matrix and its inverse matrix is

A. The matrix.

B. The inverse matrix.

C. An identity matrix.

D. A nonconforming matrix.

The correct answer is (C). *(Publisher)*

REQUIRED: The product of multiplying a matrix by its inverse matrix.

DISCUSSION: In matrix operations, ordinary division is not possible. In lieu of division, one multiplies by an inverse matrix. By definition, a matrix times its inverse will give an identity matrix. This operation is similar to dividing one number or one variable by itself and obtaining one. An identity matrix is a matrix with zeros in all elements except the principal diagonal, which contains ones.

Answers (A), (B), and (D) are incorrect because multiplying any matrix by a different matrix (whether inverse or not) will not yield the same result as either of the multiplicands.

10. The Apex Fertilizer Company is planning a new formulation to appeal to the increasing market of herb growers. Each unit of the product will require 3 pounds of chemical A, 1 pound of chemical B, and 4 pounds of chemical C. The per-pound costs of chemical A are $7.95; chemical B, $3.28; and chemical C, $6.14. Which of the following matrix algebra formulations will lead to the cost of one unit of the new fertilizer?

A. $[7.95,\ 3.28,\ 6.14] \begin{bmatrix} 3 & 0 & 0 \\ 0 & 1 & 0 \\ 0 & 0 & 4 \end{bmatrix}$

B. $\begin{bmatrix} 7.95 \\ 3.28 \\ 6.14 \end{bmatrix} [3,\ 1,\ 4]$

C. $\begin{bmatrix} 7.95 \\ 3.28 \\ 6.14 \end{bmatrix} \begin{bmatrix} 3 \\ 1 \\ 4 \end{bmatrix}$

D. $[3\ \ 1\ \ 4] \begin{bmatrix} 7.95 \\ 3.28 \\ 6.14 \end{bmatrix}$

The correct answer is (D). *(CIA, adapted)*

REQUIRED: The matrix algebra formulation to determine unit cost of fertilizer.

DISCUSSION: The objective is to multiply the number of pounds of each material times the cost of each material and sum the answers. This matrix appears in answer (D).

$$(3 \times 7.95) + (1 \times 3.28) + (4 \times 6.14)$$

When multiplying two matrices together, multiply the items in each row (horizontal) of the first matrix times each item in the second matrix's column (vertical), and then add the products. Here, a 1-by-3 matrix is multiplied by a 3-by-1 matrix, resulting in a 1-by-1 matrix representing the cost. Thus, the dimensions of the matrix resulting from multiplying 2 matrices together are equal to the number of rows in the first and the number of columns in the second.

Answer (A) is incorrect because a 1-by-3 matrix times a 3-by-3 matrix results in a 1-by-3 matrix, i.e., 3 separate numbers. Answer (B) is incorrect because a 3-by-1 matrix times a 1-by-3 matrix results in a 3-by-3 matrix, i.e., 9 numbers. Answer (C) is incorrect because a 3-by-1 matrix times a 3-by-1 matrix results in a 3-by-1 matrix, i.e., 3 numbers.

11. What process is used when calculus is employed to determine a firm's maximum profit for a given revenue function?

A. Integration.

B. Differentiation.

C. Operations research.

D. Regression analysis.

The correct answer is (B). *(CIA, adapted)*

REQUIRED: The process used when calculus is employed to determine a firm's maximum profit.

DISCUSSION: Differential calculus is used to identify the maxima or minima of nonlinear functions. The derivative of a function measures the slope or rate of change of that function. Maxima or minima occur where the slope is equal to zero. A maximum exists when the derivative of the first derivative of the function is negative. A minimum exists when the second derivative is positive.

Answer (A) is incorrect because integral calculus is used to compute the area under a curve. It finds antiderivatives; that is, the first derivative of the integral is the function that was integrated. Answer (C) is incorrect because operations research is a broad term for the application of scientific and quantitative methods to the understanding of systems, especially human-machine systems used in business. Answer (D) is incorrect because regression analysis creates an equation to explain the variation in a dependent variable caused by changes in one or more independent variables.

12. An internal auditing department developed the formula, Total audit cost (TC) = $a + bX + cX^2$, with X equaling internal audit resources. The director wanted to minimize TC with respect to X. The appropriate technique to use is

A. Linear programming.

B. Least squares.

C. Differential calculus.

D. Integral calculus.

The correct answer is (C). *(CIA, adapted)*

REQUIRED: The appropriate technique to solve the equation.

DISCUSSION: Only differential calculus can be used to derive the normal equations for regression (least squares) analysis. This form of calculus identifies the maxima or minima of curvilinear functions. Hence, it is applicable to the problem of finding the function that minimizes the squared deviations of the observed values from the regression line.

Answer (A) is incorrect because the formula is nonlinear and there are no stated constraints. Answer (B) is incorrect because least squares is related to regression analysis. Answer (D) is incorrect because integral calculus concerns areas and volume.

Questions 13 through 15 are based on the following information. Polar Bear Inc. has three service departments. The relationship between the three departments, maintenance, cleaning, and food prep, and the total costs incurred by each are expressed below:

Maintenance = $40,000 + .25C
Cleaning = $60,000 + .2M
Food Service = $50,000 + .1M + .45C

In addition, there are two production departments: Assembly and Finishing. Maintenance workers spend about 35% of their time in each of the two production departments. Cleaning employees spend about 15% of their time in each production department. Food service devotes 90% of its efforts to Assembly and 10% to Finishing.

13. Using the direct method, how much maintenance cost would be allocated to Assembly?

A. $14,000
B. $20,000
C. $20,264
D. $28,948

The correct answer is (B). *(Publisher)*

REQUIRED: The allocation of maintenance costs using the direct method.

DISCUSSION: Under the direct method, the costs incurred in a service department are allocated directly to production without consideration of services performed for other service departments. Thus, $40,000 of maintenance costs would be allocated to production. Since the same amount of work is performed for each production department, they would each be allocated $20,000.

Answer (A) is incorrect because it is based on a 35% allocation rather than a 50% allocation. Answers (C) and (D) are incorrect because they are not based on the direct method in that they include some interdepartmental allocations.

14. Using the reciprocal method, what is the total cleaning department cost to be allocated to all other departments?

A. $57,895
B. $71,579
C. $68,000
D. $60,000

The correct answer is (B). *(Publisher)*

REQUIRED: The total costs to be allocated for cleaning.

DISCUSSION: The three simultaneous equations can be solved using substitution. Substitute the maintenance department values into the cleaning department equation.

C = $60,000 + .2($40,000 + .25C)
C = $60,000 + $8,000 + .05C
.95C = $68,000
C = $71,579

Answer (A) is incorrect because $57,895 [$40,000 + (.25 x $71,579) cleaning costs] is the cost allocated to the maintenance department. Answer (C) is incorrect because $68,000 does not include the 25% of the cleaning department that maintenance is dependent upon. Answer (D) is incorrect because $60,000 does not include the 20% of the maintenance department that the cleaning department is dependent upon.

15. Using the reciprocal method, what is the total food service department cost to be allocated to all other departments?

A. $88,001
B. $55,790
C. $82,211
D. $83,211

The correct answer is (A). *(Publisher)*

REQUIRED: The total costs to be allocated for food service.

DISCUSSION: The three simultaneous equations can be solved using substitution. Plug in the values previously found for the maintenance and cleaning departments into the food service equation.

F = $50,000 + .1($57,895) + .45($71,579)
F = $50,000 + $5,790 + $32,211
F = $88,001

Answer (B) is incorrect because $55,790 does not include the 45% of the cleaning department that the food service department is dependent upon. Answer (C) is incorrect because $82,211 does not include the 10% of the maintenance department that the food service department is dependent upon. Answer (D) is incorrect because $83,211 reverses the 10% of maintenance in the equation with the 45% of cleaning in the equation.

16. Which of the following statements about marginal costs is correct?

A. Marginal cost equals the derivative of the total cost function.

B. Marginal cost increases as output quantity increases.

C. Marginal cost decreases as output quantity increases.

D. Marginal cost increases when total cost increases.

The correct answer is (A). *(CIA, adapted)*

REQUIRED: The true statement about marginal costs.

DISCUSSION: Marginal cost is the incremental cost incurred to produce one additional unit of output. The derivative of the total cost function is the change in total cost per unit change in output quantity. Accordingly, it is the same as marginal cost. Marginal cost increases or decreases as the derivative (slope) of the total cost function increases or decreases.

Answer (B) is incorrect because marginal cost may increase or decrease as total cost increases. Whether it increases or decreases depends on whether the derivative of the total cost function increases or decreases. Answer (C) is incorrect because marginal cost may increase or decrease as total cost increases. Answer (D) is incorrect because marginal cost may increase or decrease as total cost increases.

Questions 17 through 20 are based on the following information. Redler Company manufactures and sells an industrial strength cleaning fluid. The equations presented below represent the revenue (R) and cost (C) functions for the company if x is stated in terms of thousands of gallons of fluid.

$$R(x) = -\$100 + \$26x - \$0.05x^2$$
$$C(x) = \$50 + \$8x + \$0.01x^2$$

---

17. The function that represents the marginal cost to manufacture 1 gallon of cleaning fluid is

A. $\$8x + \$0.01x^2$

B. $\$8 + \$0.01x$

C. $\$8 + \$0.02x$

D. $\$8 \div \$0.02x$

The correct answer is (C). *(CMA, adapted)*

REQUIRED: The function representing the marginal cost to manufacture 1 gallon of product.

DISCUSSION: The marginal cost function is the derivative of the total cost function. The basic formula for the derivative is

$$f'(Ax^n) = nAx^{n-1}$$

If: A = the constant, or number of x's
x = the variable being differentiated
n = the exponent of the independent variable

Thus, the derivative of the cost function is

$$\begin{aligned} f'(\$50 + \$8x + \$0.01x^2) &= f'(\$50) + f'(\$8x) + f'(\$0.01x^2) \\ &= 0 + \$8 + \$0.02x \\ &= \$8 + \$0.02x \end{aligned}$$

Answer (A) is incorrect because $\$8x + \$0.01x^2$ is the cost function without the constant. Answer (B) is incorrect because the first derivative of $\$0.01x^2$ is $\$0.02x$. Answer (D) is incorrect because the derivative of the entire equation is the sum of the derivatives of the individual elements.

18. The quantity of cleaning fluid that should be produced to maximize revenue is

A. 520,000 gallons.

B. 260,000 gallons.

C. 160,000 gallons.

D. 150,000 gallons.

The correct answer is (B). *(CMA, adapted)*

REQUIRED: The quantity of product to be sold to maximize revenue.

DISCUSSION: A maximization problem is solved by setting the first derivative of the revenue function equal to zero and solving for x. The first derivative of the revenue function is 26 – .1x. Thus,

$$.1x = 26$$
$$x = 260$$

Selling 260,000 gallons of cleaning fluid would maximize revenues.

Answer (A) is incorrect because 520,000 gallons assumes the first derivative is 26 – .05x. Answer (C) is incorrect because 260,000 gallons maximizes revenue. Answer (D) is incorrect because 150,000 gallons is the profit-maximizing amount.

19. The quantity of cleaning fluid that should be produced to maximize profit is

A. 520,000 gallons.

B. 260,000 gallons.

C. 160,000 gallons.

D. 150,000 gallons.

The correct answer is (D). *(CMA, adapted)*

REQUIRED: The production level at which profit will be maximized.

DISCUSSION: Profit is equal to the revenue function minus the cost function [(–\$100 + \$26x – \$0.05x^2) – (\$50 + 8x + \$0.01x^2)]. It is not necessary to combine terms. The profit maximum is the first derivative of this profit function (26 – .1x) – (8 + .02x), or 18 – .12x. Set this derivative equal to zero and solve for x, or 150. Thus, producing 150,000 gallons of cleaning fluid will maximize profit.

Answer (A) is incorrect because 520,000 gallons is the quantity that maximizes revenue if the first derivative of the revenue function is 26 – .05x. Answer (B) is incorrect because 260,000 gallons is the quantity of product to be sold to maximize revenue. Answer (C) is incorrect because 150,000 gallons is the profit-maximizing production.

20. What would the profit (loss) be if the company manufactured and sold 300,000 gallons of cleaning fluid?

A. \$3,200,000

B. \$(50,000)

C. \$(3,350,000)

D. \$(150,000)

The correct answer is (D). *(CMA, adapted)*

REQUIRED: The amount of profit (loss) generated given productivity and sales.

DISCUSSION: This question is easily solved by substituting 300 for x in the previous two equations.

$$\text{Revenue} = -\$100 + \$26(300) - \$.05(90{,}000)$$
$$\text{Cost} = \$50 + \$8(300) + \$.01(90{,}000)$$

The cost equation is subtracted from the revenue equation to determine profit or loss per gallon of x.

$$[-\$100 + \$7{,}800 - \$4{,}500 - (\$50 + \$2{,}400 + \$900)]$$
$$\text{Profit} = \$3{,}200 \text{ (revenue)} - \$3{,}350 \text{ (costs)}$$
$$= \$(150)$$

Because x is stated in thousands of gallons, the total loss is \$150,000.

Answer (A) is incorrect because \$3,200,000 equals total revenues. Answer (B) is incorrect because \$(50,000) equals fixed costs. Answer (C) is incorrect because \$(3,350,000) equals total costs.

Questions 21 through 23 are based on the following information. A financial analyst has been asked to forecast net income over the next 3 years using the computer spreadsheet below. The actual results for year 1 are shown in column B. Net sales and cost of goods sold are projected to increase at the rate of 20% each year. Selling and administrative expenses are projected to increase at an annual rate of 10%. However, several mistakes have been made in the spreadsheet below.

	A	B	C	D	E
1		Year 1	Year 2	Year 3	Year 4
2	Net sales	$400,000	$480,000	$480,000	$576,000
3	Cost of goods sold	310,000	372,000	372,000	446,400
4	Gross profit	90,000	108,000	108,000	129,600
5	Selling expenses	30,000	33,000	36,300	39,930
6	Administrative expenses	40,000	44,000	48,400	58,080
7	Income before taxes	20,000	31,000	23,300	31,590
8	Income taxes at 40%	8,000	12,400	9,320	12,636
9	Net income	12,000	18,600	13,980	18,954

21. A correct formula for cell D2 is

A. +C2 × 1.2

B. +B2 × 1.2

C. +C2 + (.2 × B2)

D. +B2 + (.2 × B2)

The correct answer is (A). *(CMA, adapted)*

REQUIRED: The correct formula for cell D2.

DISCUSSION: Cell D2 is sales for year 3. Given that sales are expected to increase 20% each year, the amount in that cell should be 120% of the preceding year's sales. The formula is C2 x 1.2.

Answer (B) is incorrect because +B2 x 1.2 is the formula for cell C2. Answer (C) is incorrect because +C2 + (.2 x B2) bases the increase in sales on 20% of the year 1 amount instead of the year 2 amount. Answer (D) is incorrect because +B2 + (.2 x B2) is an alternative formula for cell C2.

22. Which one of the cells listed below contains an incorrect formula?

A. D4

B. D8

C. E6

D. C1

The correct answer is (C). *(CMA, adapted)*

REQUIRED: The cell that contains an incorrect formula.

DISCUSSION: Cell E6 contains the amount of administrative expenses for year 4. These expenses are expected to increase at the rate of 10% annually. This rate of increase is reflected in cells C6 and D6. However, the amount in cell E6 is 120% of the amount in cell D6 (1.2 x $48,400 = $58,080). The proper amount is $53,240 (1.1 x $48,400).

Answers (A) and (B) are incorrect because cell D4 (D2 – D3) and cell D8 (40% x D7) reflect proper applications of formulas, although the amounts are incorrect as a result of errors in other cells. Answer (D) is incorrect because C1 is the date.

23. If all of the errors in the formulas in the spreadsheet were corrected, the correct projected net income for year 4 would be

A. $21,858

B. $37,410

C. $90,978

D. $88,074

The correct answer is (B). *(CMA, adapted)*

REQUIRED: The projected net income for year 4 if all errors are corrected.

DISCUSSION: If the error in cell D2 is corrected, the sales for year 3 become $576,000 (1.2 x $480,000). Cell E2 should therefore be $691,200 (1.2 x $576,000). Cell C3 is correct, but cell D3 is not. Thus, cell E3 should be $535,680 (1.2 x 1.2 x $372,000), and the gross profit in cell E4 should be $155,520 ($691, 200 – $535,680). Selling expenses in cell E5 are correct at $39,930. Administrative expenses in cell E6 should be $53,240 (1.1 x $48,400). Hence, gross profit in cell E7 is $62,350 ($155,520 – $39,930 – $53,240). Income taxes (cell E8) become $24,940 (.4 x $62,350). Finally, net income in cell E9 is $37,410 ($62,350 – $24,940).

Answer (A) is incorrect because $21,858 results from not considering the changes in sales and cost of goods sold in year 3 and the impact those changes had in year 4. Answer (C) is incorrect because $90,978 results from not considering the change in cost of goods sold in year 3 and the impact it had on year 4. Answer (D) is incorrect because $88,074 results from not considering the change in cost of goods sold in year 3 or the change in administrative expenses in year 4.

B. Linear Programming

24. One limitation of the linear programming technique is that it is effective only for

A. Two-product situations.

B. Manufacturing resource constraint situations.

C. Straight-line relationship situations.

D. Income maximization situations.

The correct answer is (C). *(CMA, adapted)*

REQUIRED: The item that is a limitation of the linear programming technique.

DISCUSSION: Linear programming assumes straight-line relationships over a relevant range.

Answer (A) is incorrect because, although a linear programming problem can only be solved graphically in two-product situations, it can incorporate additional alternatives with the aid of a computer. Answer (B) is incorrect because linear programming can be used for product mix, raw material mix, staff assignment, and a variety of other problems. Answer (D) is incorrect because linear programming can be used to solve maximization problems.

25. In order to solve a linear programming problem, slack, surplus, and artificial variables must be employed. A slack variable represents

A. Opportunity costs.

B. Unused capacity.

C. Outside variables with high cost.

D. The variable with the most negative value.

The correct answer is (B). *(CMA, adapted)*

REQUIRED: The definition of a slack variable.

DISCUSSION: A slack variable represents unused capacity. Because linear programming formulations are often stated as inequalities, there is a possibility of unused capacity even at the optimal production level. To convert inequalities to equalities, slack variables are introduced to account for this unused capacity.

Answers (A), (C), and (D) are incorrect because a slack variable represents unused capacity.

26. The procedure employed to solve linear programming problems is

A. Differential calculus.

B. Integral calculus.

C. Simulation.

D. Matrix algebra.

The correct answer is (D). *(CMA, adapted)*

REQUIRED: The procedure used to solve linear programming problems.

DISCUSSION: Matrix algebra is a method of manipulating multiple linear equations. It is used to solve linear programming problems. The use of matrix algebra to solve such problems is called the simplex method.

Answer (A) is incorrect because differential calculus identifies the maxima and minima of curvilinear functions. Answer (B) is incorrect because integral calculus permits computation of the area under a curve. Answer (C) is incorrect because simulation is a means of experimenting with logical and mathematical models using a computer.

Questions 27 through 29 are based on the following information. Manders Manufacturing Corporation uses the following model to determine its product mix for metal (M) and scrap metal (S).

$$\text{Max } Z = \$30M + \$70S$$

$$\text{If: } 3M + 2S \leq 15$$
$$2M + 4S \leq 18$$

27. These mathematical functions are an example of a(n)

A. Simulation model.

B. Linear programming model.

C. Economic order quantity model.

D. Present value model.

The correct answer is (B). *(CMA, adapted)*

REQUIRED: The type of model typified by the maximization and constraint functions illustrated.

DISCUSSION: Linear programming is a technique used to maximize a revenue or profit function, or minimize a cost function, subject to constraints such as limited resources. Linear programming is often used for planning scarce resource allocations. The optimization, or objective, function (Max Z = \$30M + \$70S) indicates that a company is attempting to maximize its contribution margin on two products. The two constraint functions suggest that either the scarcity of resources or a lack of demand limits the amount of each product that can be produced.

Answer (A) is incorrect because a simulation model is a technique for experimenting with models using a computer. It does not necessarily use maximization and constraint functions. Answer (C) is incorrect because the EOQ is based on a single equation that includes a square root. Answer (D) is incorrect because the present value model is based on a single equation that incorporates a rate of return.

28. The two inequality functions are

A. Contributions.

B. Shadow points.

C. Conditions.

D. Constraints.

The correct answer is (D). *(CMA, adapted)*

REQUIRED: The term for the two inequality functions.

DISCUSSION: The restrictions on the optimal value of the objective function in linear programming are known as constraints. They are depicted by means of inequalities.

Answer (A) is incorrect because contributions to revenue are included in the objective (maximization) equation. Answer (B) is incorrect because a shadow point, or shadow price, is the amount by which the value of the optimal solution of the objective function will change if a one-unit change is made in a binding constraint. Answer (C) is incorrect because conditions are usually not shown through inequalities.

29. The point at which $M = 2$ and $S = 3$ would

A. Minimize cost.

B. Lie in a corner.

C. Be a feasible point.

D. Be an infeasible point.

The correct answer is (C). *(CMA, adapted)*

REQUIRED: The true statement about a given point in a linear programming model.

DISCUSSION: The point at which $M = 2$ and $S = 3$ is a feasible point because it does not violate either of the constraint functions. A graphic depiction of the constraints will show whether the point lies in a corner, which is not the case. Because it does not lie in a corner, it cannot be the optimal solution point. In a linear programming problem, the optimal solution lies at a corner in the feasible area. Thus, all that can be said for certain about the point is that it lies within the feasible area.

Answer (A) is incorrect because nothing can be said with respect to costs. Costs are not mentioned in any of the equations. Answer (B) is incorrect because the point does not lie in a corner. Answer (D) is incorrect because the point does lie within the feasible area.

30. In linear programming, the shadow price refers to the

A. Measurement of the value of relaxing a constraint in a problem with dual variables.

B. Marginal change in profit associated with a change in the contribution margin of one of the variables.

C. Unused capacity available once the optimal solution is obtained.

D. Variables that are included in the final solution of the linear programming model.

The correct answer is (A). *(CMA, adapted)*

REQUIRED: The definition of a shadow price.

DISCUSSION: A shadow price is the amount by which the value of the optimal solution of the objective function will change if a one-unit change is made in a binding constraint. The calculation of shadow prices is an example of sensitivity analysis, which is any procedure that tests the responsiveness of a solution to changes in variables.

Answer (B) is incorrect because the change is in the limited resource, not in the contribution margin of one of the variables. Answer (C) is incorrect because shadow prices are concerned with binding constraints, not nonbinding constraints; the shadow price of unused capacity, a nonbinding constraint, is zero. Answer (D) is incorrect because shadow prices are not used in the final solution of a linear programming model.

Questions 31 and 32 are based on the following information. A company produces two products, X and Y, which use material and labor as inputs. Fixed amounts of labor and material are available for production each month. In addition, the demand for product Y each month is limited; product X has no constraint on the number of units that can be sold.

A graphical depiction of these production and demand constraints is presented in the opposite column.

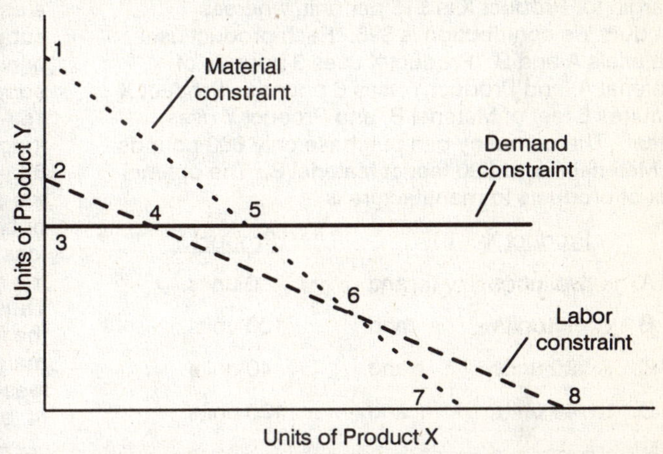

31. The feasible solution region is bounded by the lines connecting points

A. 3, 4, 6, and 7.

B. 1, 5, 6, and 8.

C. 2, 4, 5, 6, and 8.

D. 3, 5, 6, and 7.

The correct answer is (A). *(CIA, adapted)*

REQUIRED: The points bounding the feasible solution region.

DISCUSSION: A model consisting of a system of functions may be used to optimize an objective function. If the functions in the model are all linear, the model is a linear programming model. Linear programming is a technique to determine optimal resource allocation. Several solution methods are available to solve linear programming problems. The graphical method, the easiest technique, is limited to simple problems. Here, the graph consists of three lines, each representing a production constraint. The lines connecting points 3, 4, 6, and 7 bound the feasible solution region. Product mixes of X and Y that lie outside this boundary cannot be produced and/or sold because the demand constraint (line 3,4), the labor constraint (line 4,6), and the material constraint (line 6,7) are binding.

Answer (B) is incorrect because points 1, 5, and 8 are outside the boundary. Answer (C) is incorrect because points 2, 5, and 8 are outside the boundary. Answer (D) is incorrect because point 5 is outside the boundary.

32. Refer to the information preceding question 31 on page 233. If a series of profit lines for X and Y are drawn on the graph, the mix of X and Y that will result in the maximum profit can be determined from

A. The last point in the feasible solution region touched by a profit line.

B. Any point on the boundary of the feasible solution region touched by a profit line.

C. The first point on the feasible solution region boundary that intersects a profit line.

D. Any point on the demand constraint that intersects a profit line.

The correct answer is (A). *(CIA, adapted)*

REQUIRED: The means of determining the mix of X and Y that will result in the maximum profit.

DISCUSSION: A profit line has negative slope because the profit from sales of one product increases as the profit from sales of the other product declines. Moving the profit line rightward (while maintaining its slope) to the last point in the feasible region determines the solution.

Answers (B), (C), and (D) are incorrect because the last point in the feasible solution region touched by a profit line determines the solution.

33. The Mix and Match Company has two products, Product X and Product Y, that it manufactures through its production facilities. The contribution margin for Product X is $15 per unit, whereas Product Y's contribution is $25. Each product uses Materials A and B. Product X uses 3 pounds of Material A, and Product Y uses 6 pounds. Product X requires 6 feet of Material B, and Product Y uses 4 feet. The company can purchase only 600 pounds of Material A and 880 feet of Material B. The optimal mix of products to manufacture is

	Product X		Product Y
A.	146 units	and	0 units.
B.	0 units	and	100 units.
C.	120 units	and	40 units.
D.	40 units	and	120 units.

The correct answer is (C). *(CMA, adapted)*

REQUIRED: The optimal mix of products.

DISCUSSION: Linear programming is used to maximize a contribution margin function, or minimize a cost function, subject to constraints, such as scarce resources or minimum/maximum levels of production. In this problem, the equation to be maximized (the objective function, or $15X + 25Y$) is to be maximized subject to the two constraint functions (Material A: $3X + 6Y \leq 600$ and Material B: $6X + 4Y \leq 880$). One way to solve this problem is to graph the constraint lines and determine the feasible area. The optimal production level is at an extreme point within the feasible area. However, a simple alternative is to calculate the contribution margin for each feasible answer choice. Thus, a production level of 120 units of X and 40 units of Y is the feasible level that yields the maximum contribution margin of $2,800 [(120 x $15) + (40 x $25)]. This mix is feasible because it requires 600 pounds of A[(3 lbs. x 120) + (4 lbs. x 40)] and 880 feet of B[(6 ft. x 120) + (4ft. x 40)].

Answer (A) is incorrect because 146 units of X and 0 units of Y yield a contribution margin of $2,190. Answer (B) is incorrect because 0 units of X and 100 units of Y yield a contribution margin of $2,500. Answer (D) is incorrect because producing 120 units of Y violates the constraint for Material A.

Questions 34 and 35 are based on the following information. Keego Enterprises manufactures two products, boat wax and car wax, in two departments, the Mixing Department and the Packaging Department. The Mixing Department has 800 hours per month available, and the Packaging Department has 1,200 hours per month available. Production of the two products cannot exceed 36,000 pounds. Data on the two products follow:

	Contribution Margin (per 100 pounds)	Hours per 100 Pounds Mixing (M)	Hours per 100 Pounds Packaging (P)
Boat wax (B)	$200	5.0	3.6
Car wax (C)	150	2.4	6.0

34. The objective function for the linear program Keego would use to determine the optimal monthly production of each wax would be

A. Z = 150B + 200C.

B. 2B + 1.5C ≥ 36,000.

C. 2B + 1.5C ≤ 36,000.

D. Z = 200B + 150C.

The correct answer is (D). *(CMA, adapted)*

REQUIRED: The objective function for the linear program used to determine the optimal monthly production.

DISCUSSION: The objective function is the equation to maximize the contribution margin. Given that each 100 pounds of boat wax contributes $200 and each 100 pounds of car wax contributes $150, the company's goal is to maximize the total of these amounts. Consequently, if B is the number of 100-pound units of boat wax, C is the number of 100-pound units of car wax, and Z is the total contribution margin, the objective function is Z = $200B + $150C.

Answers (A), (B), and (C) are incorrect because the objective function is Z = 200B + 150C.

35. The mixing constraint for the Keego linear program would be

A. 2.4M + 6P ≥ 36,000.

B. 5B + 2.4C ≥ 800.

C. 5B + 2.4C ≤ 800.

D. 5B + 2.4C = 800.

The correct answer is (C). *(CMA, adapted)*

REQUIRED: The constraint equation for the Mixing Department.

DISCUSSION: The Mixing Department has only 800 hours available per month, which is a limitation (constraint) on production. Thus, the total mixing time is less than or equal to 800 hours. Every 100-pound batch of boat wax (B) requires 5 hours of mixing time, and every 100-pound batch of car wax (C) requires 2.4 hours of mixing time. Accordingly, the total mixing time available cannot exceed 5B plus 2.4C, and the constraint equation is 5B + 2.4C ≤ 800.

Answers (A), (B), and (D) are incorrect because the mixing constraint is 5B + 2.4C ≤ 800.

36. A small company makes only two products with the following production constraints representing two machines and their maximum availability:

$$2X + 3Y \leq 18$$
$$2X + Y \leq 10$$
$$X \geq 0;\ Y \geq 0$$

If: X = the units of the first product.
Y = the units of the second product.

If the profit equation is Z = \$4X + \$2Y, the maximum possible profit is

A. \$20

B. \$21

C. \$18

D. \$24

The correct answer is (A). *(CMA, adapted)*

REQUIRED: The maximum possible profit given the two machine constraints.

DISCUSSION: A simple means of solving this two-equation, two-variable linear programming problem is to graph the equations. The optimal solution is at one of the corners of the feasible space defined by the lines representing the equations. This space is the area that lies under both equations. The corners of the constraint area can be determined algebraically. If X equals 0, Y would equal 6 in the first constraint equation (assuming it is stated as an equality) and 10 in the second. If Y equals 0, X would equal 9 in the first equation and 5 in the second. Thus, the maximum units of X and Y that can be produced are 5 and 6, respectively. Substituting these corner solutions into the objective functions yields these results:

$$Z = \$4(5) + \$2(0) = \$20, \text{ or}$$
$$Z = \$4(0) + \$2(6) = \$12$$

Thus, the better solution is to produce 5 units of product X, which results in a profit of \$20.

A third feasible solution (corner of the feasible space) is to produce nothing (X = 0, Y = 0). This solution obviously yields a profit of \$0. The fourth corner is the intersection of the two simultaneous constraint equations (stated as equalities), which can be determined by subtracting the second equation from the first:

$$\begin{array}{r} 2X + 3Y = 18 \\ \underline{-2X + -Y = -10} \\ 2Y = 8 \\ Y = 4 \text{ and } X = 3 \end{array}$$

This solution also results in a profit of \$20 [(\$4 x 3) + (\$2 x 4)].

Answers (B), (C), and (D) are incorrect because the maximum profit is \$20.

37. Variables that are important to the decision-making process but are out of the control of the decision maker, e.g., economic conditions, are considered to be

A. Exogenous variables.

B. Decision variables.

C. Performance criteria.

D. Constraints.

The correct answer is (A). *(CMA, adapted)*

REQUIRED: The term for variables that are outside the control of the decision maker.

DISCUSSION: Exogenous or input variables are outside the control of the decision maker. Exogenous means "originating externally." These influence the decision model (system) but are not influenced by it.

Answer (B) is incorrect because at least one of the decision variables in a model must be under the decision maker's control; that is, at least one variable cannot be exogenous. Answer (C) is incorrect because performance criteria are the means of measuring the results of a decision after the fact. Answer (D) is incorrect because constraints are limitations (constants, not variables) that must be considered as part of the decision process.

38. A large retail chain has 10 warehouses and 50 retail locations. The quantitative technique it would likely use to determine the minimum cost of moving its goods from the warehouses to the retail stores is

A. A transportation model.

B. Markov analysis.

C. Calculus-based optimization.

D. A queuing model.

The correct answer is (A). *(CMA, adapted)*

REQUIRED: The quantitative technique to use to determine the minimum cost of moving goods from warehouses to retail stores.

DISCUSSION: The transportation model is a special application of linear programming to a type of network flow problem. It determines the minimum cost of, or maximum profit or revenue derived from, physically moving goods or delivering services from sources of supply (warehouses) to other destinations (retail stores).

Answer (B) is incorrect because Markov analysis is useful in decision problems in which the probability of the occurrence of a future state depends only on the current state of a model. Answer (C) is incorrect because calculus-based optimization does not incorporate the constraints of a transportation problem. Answer (D) is incorrect because a queuing model is a group of mathematical functions for systems involving waiting lines.

39. A transportation model is a special case of the

A. Economic order quantity model.

B. Markov model.

C. Linear programming model.

D. Dynamic programming model.

The correct answer is (C). *(CMA, adapted)*

REQUIRED: The model of which the transportation model is a special case.

DISCUSSION: The transportation model is a special type of linear programming model. A transportation model involves physical movement of goods from sources of supply to destinations. The objective function includes the transportation cost of each item from each source to each destination. The constraints are the output for each supply point and the demand by each destination.

Answer (A) is incorrect because the EOQ model is used to determine inventory order quantity and is not related to the transportation model. Answer (B) is incorrect because a Markov model is useful in decision problems in which the probability of the occurrence of a future state depends only on the current state of nature. Answer (D) is incorrect because dynamic programming is a type of model useful for problems that involve time-staged decisions such as decisions affected by previous decisions; typically, dynamic programming is used for much smaller problems than linear programming.

40. The personnel manager has just completed a survey on employee attitudes, working conditions, and productivity. He would like to analyze data subject to the constraints of his company (i.e., salary structure, capital equipment, etc.). The stated objectives of this company are to (1) maximize profits, (2) maximize safety, and (3) promote positive employee attitudes. The decision analysis technique that could accomplish this analysis is

A. Linear programming.

B. Statistical analysis.

C. Goal programming.

D. Dynamic programming.

The correct answer is (C). *(CMA, adapted)*

REQUIRED: The decision analysis technique to analyze the given information.

DISCUSSION: Goal programming permits an ordinal specification and solution to a problem with multiple goals. With its ability to incorporate multiple goals, goal programming has the potential to be applied to problems with financial, social, and environmental objectives. It was first introduced as a means of solving linear programming problems that did not have a feasible solution.

Answer (A) is incorrect because linear programming cannot deal with multiple objectives. Answer (B) is incorrect because statistical analysis is a mathematical technique. Answer (D) is incorrect because dynamic programming cannot deal with multiple objectives, although it can handle changes in constraints.

41. United Industries manufactures three products at its highly automated factory. The products are very popular, with demand far exceeding the company's ability to supply the marketplace. To maximize profit, management should focus on each product's

A. Gross margin.

B. Segment margin.

C. Contribution margin ratio.

D. Contribution margin per machine hour.

The correct answer is (D). *(CMA, adapted)*

REQUIRED: The measure used to determine the profit-maximizing output.

DISCUSSION: When demand far exceeds a company's ability to supply the marketplace, management will want to maximize its profits per unit of scarce resource. If the scarce resource is raw materials, the products that provide the greatest contribution margin per unit of raw materials should be emphasized. If machine hours are the constraint, profits are maximized by emphasizing the contribution margin per machine hour.

Answers (A) and (B) are incorrect because the company can sell as much of each product as it can produce. Thus, sales are limited by production constraints, e.g., machine hours. The company should therefore seek to maximize its return per unit of the constraint. Answer (C) is incorrect because the contribution margin ratio is only important as it translates to dollars. A high margin on a low sales volume will not be profitable.

42. Heniser Pet Foods manufactures two products, X and Y. The unit contribution margins for Products X and Y are $30 and $50, respectively. Each product uses Materials A and B. Product X uses 6 pounds of Material A and 12 pounds of Material B. Product Y uses 12 pounds of Material A and 8 pounds of Material B. The company can purchase only 1,200 pounds of Material A and 1,760 pounds of Material B. The optimal mix of products to manufacture is

A. 146 units of X and 0 units of Y.

B. 0 units of X and 100 units of Y.

C. 120 units of X and 40 units of Y.

D. 40 units of X and 120 units of Y.

The correct answer is (C). *(Publisher)*

REQUIRED: The optimal mix of products.

DISCUSSION: Linear programming is a technique used to maximize a contribution margin function or to minimize a cost function, subject to constraints such as scarce resources or minimum/maximum levels of production. Thus, linear programming is often used for planning resource allocations. In this problem, the equation to be maximized, called the objective function, is: $30X + $50Y. This equation is to be maximized subject to the constraints on materials. The two constraint functions are:

Material A: $6X + 12Y \leq 1{,}200$
Material B: $12X + 8Y \leq 1{,}760$

One way to solve this problem is to graph the constraint lines and determine the feasible area. The optimal production level is at an extreme point within the feasible area. The graph shows that a production level of 120 units of X and 40 units of Y is a feasible production level that maximizes the contribution margin.

Answer (A) is incorrect because the $4,380 of contribution margin yielded by 146 units of X and 0 units of Y is less than the $5,600 yielded by 120 units of X and 40 units of Y. Answer (B) is incorrect because the $5,000 of contribution margin yielded by 0 units of X and 100 units of Y is less than the $5,600 yielded by 120 units of X and 40 units of Y. Answer (D) is incorrect because 40 units of X and 120 units of Y is not within the feasible production area. It violates the constraint on Material A.

STUDY UNIT 6: QUANTITATIVE METHODS II

18 pages of outline
70 multiple-choice questions

This study unit is the third of three that concern quantitative methods, a major topic that has been assigned a relative weight range of 10% to 15% of Part 4.

A. Project Scheduling

1. Project scheduling techniques are designed to aid the planning and control of large-scale projects having many interrelated activities. Three of the more common techniques are Gantt or bar charts, PERT, and CPM. These techniques are suitable for any project having a target completion date and single start.
   a. Example applications
      1) Building construction
      2) Research and development projects
      3) New product planning
      4) Feasibility studies
      5) Audit studies
      6) Book publishing
2. **Gantt charts** or **bar charts** are simple to construct and use. To develop a Gantt chart, divide the project into logical subprojects called activities or tasks. Estimate the start and completion times for each activity. Prepare a bar chart showing each activity as a horizontal bar along a time scale.

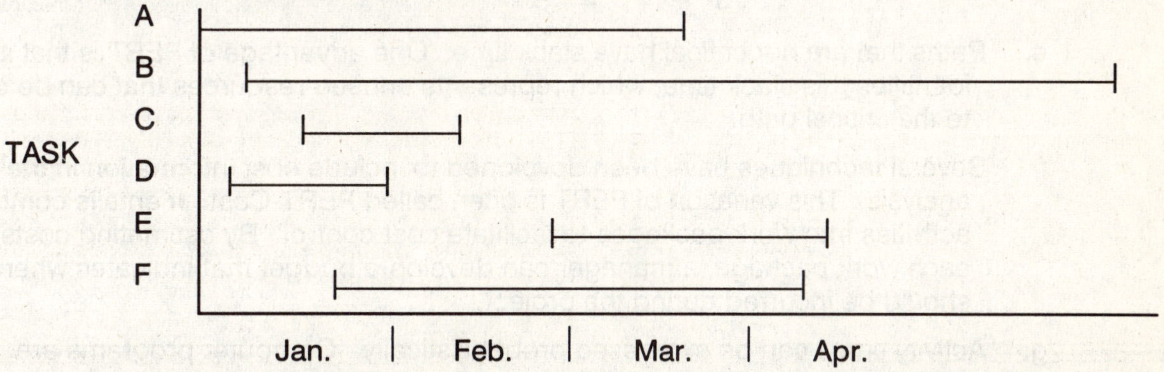

   a. The major advantage of the Gantt chart is its simplicity. It forces the planner to think ahead and define logical activities. As the project progresses, actual completion times can be compared with planned times. Furthermore, the technique requires no special tools or mathematics and can be used on small projects as well as large ones.
   b. The major disadvantage is that interrelationships among activities are not shown. Several special methods have been developed to show these on a Gantt chart, but they are feasible only for simple relationships.

3. **Program Evaluation and Review Technique (PERT)** was developed to aid managers in controlling large-scale, complex projects. PERT diagrams are free-form networks showing each activity as a line between events. A sequence of lines shows interrelationships among activities. PERT diagrams are more complex than Gantt charts, but they have the advantages of incorporating probabilistic time estimates and identifying the critical path.

   a. Events are discrete moments in time representing the start or finish of an activity. They consume no resources.

   b. Activities are tasks to be accomplished. They consume resources (including time) and have a duration over time.

   c. The network diagram is formed by

      1) The lines (activities) connected from left to right in the necessary sequence of their accomplishment. They can be marked with time lengths.

      2) Circles representing events and numbered for identification.

   d. The critical path is the longest path in time through the network. It is critical because, if any activity on the critical path takes longer than expected, the entire project will be delayed. Every network has at least one critical path. Some have more than one.

      1) The mean completion time for the critical path is the sum of the means of the activity times.

      2) The standard deviation of the completion time for the critical path is the square root of the sum of the variances (squares of the standard deviations) of the activity times.

         a) EXAMPLE: If the critical path has two activities, and the standard deviations of the completion times are 3 and 4, the standard deviation for the critical path is

$$\sqrt{3^2 + 4^2} = 5$$

   e. Paths that are not critical have slack time. One advantage of PERT is that it identifies this slack time, which represents unused resources that can be diverted to the critical path.

   f. Several techniques have been developed to include cost information in the analysis. This variation of PERT is often called PERT-Cost. It entails combining activities into work packages to facilitate cost control. By estimating costs for each work package, a manager can develop a budget that indicates when costs should be incurred during the project.

   g. Activity times can be expressed probabilistically. Computer programs are available to make the calculations and find critical paths.

   h. PERT analysis includes probabilistic estimates of activity completion times. Three time estimates are made -- optimistic, most likely, and pessimistic.

      1) The time estimates for an activity are assumed to approximate a beta probability distribution. In contrast with the normal distribution, this distribution has finite endpoints (the optimistic and pessimistic estimates) and is unimodal; that is, it has only one mode (the most likely time).

      2) PERT approximates the mean of the beta distribution by dividing the sum of the optimistic time, the pessimistic time, and four times the most likely time (the mode) by six.

3) The standard deviation is approximated by dividing the difference between the pessimistic and optimistic times by six. The basis for the latter approximation is that various probability distributions have tails that lie about plus or minus three standard deviations from the mean. For example, 99.9% of observations in the normal distribution are expected to lie within this range.

i. EXAMPLE: If an activity can be completed in 6 days (optimistic time), 10 days (most likely time), or 20 days (pessimistic time), the expected duration is 11 days {[6 + (4 x 10) + 20] ÷ 6}.

1) Thus, the most likely time is weighted the most heavily.
2) The standard deviation is 2.33 [(20 – 6) ÷ 6].

j. EXAMPLE:

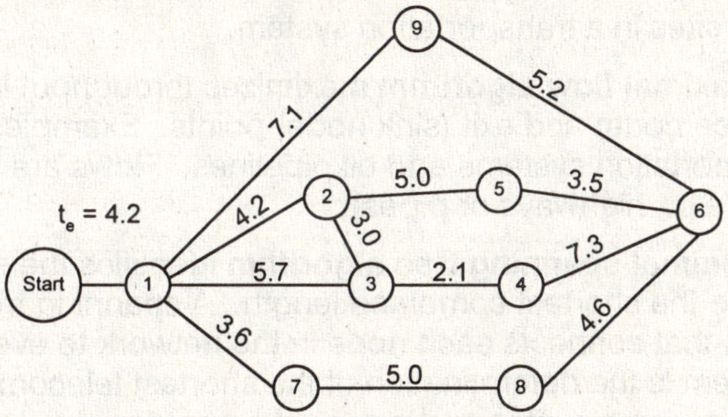

1) For the network above, the following are the paths and path times:

Path	Time (hours)
Start-1-9-6	16.5
Start-1-2-5-6	16.9
Start-1-2-3-4-6	20.8
Start-1-3-4-6	19.3
Start-1-7-8-6	17.4

2) Path Start-1-2-3-4-6 is the critical path because it has the longest time.

k. In the second example above, path 1-3 takes only 5.7 hours, whereas the critical path events (1-2-3) take 7.2 hours. The slack time represented by path 1-3 is thus 7.2 – 5.7, or 1.5. People assigned to path 1-3 have an extra 1.5 hours to help elsewhere.

4. The **critical path method (CPM)** was developed independently of PERT and is widely used in the construction industry. CPM may be thought of as a subset of PERT. Like PERT, it is a network technique, but, unlike PERT, it uses deterministic time and cost estimates. Its advantages include cost estimates plus the concept of "crash" efforts and costs.

a. Activity times are estimated for normal effort and crash effort. Crash time is the time to complete an activity assuming that all available resources were devoted to the task (overtime, extra crew, etc.).

b. Activity costs are also estimated for normal and crash efforts.

c. These estimates allow the project manager to estimate the costs of completing the project if some of the activities are completed on a crash basis.

d. The network diagram is constructed in the same manner as PERT diagrams. Once the diagram is constructed, the critical paths are found for normal and crash times. More than one critical path may exist for each diagram.

e. Crashing the network means finding the minimum cost for completing the project in minimum time.

f. CPM computer programs allow updating of the solution as work proceeds.

5. Network models are used to solve managerial problems pertaining to project scheduling, information systems design, and transportation systems design. Networks consisting of nodes and arcs may be created to represent in graphic form problems related to transportation, assignment, and transshipment. The shortest-route, minimal spanning tree, and maximal flow problems are other applications of network models.

   a. A **shortest-route algorithm** minimizes total travel time from one site to each of the other sites in a transportation system.

   b. The **maximal flow algorithm** maximizes throughput in networks with distinct entry (source node) and exit (sink node) points. Examples of applications are highway transportation systems and oil pipelines. Flows are limited by capacities of the arcs (e.g., highways or pipes).

   c. The **minimal spanning tree algorithm** identifies the set of connecting branches having the shortest combined length. A spanning tree is a group of branches (arcs) that connects each node in the network to every other node. An example problem is the determination of the shortest telecommunications linkage among users at remote sites and a central computer.

6. Controlling **job projects** requires attention to their scope, scheduling, quality, and costs. Thus, **variance analysis** is a tool for project management.

   a. The difference between the actual cost of work performed (CWP) and the budgeted cost of work scheduled (CWS) may be analyzed in terms of two variances:

      1) The **job project-performance cost variance** is the difference between the actual CWP and the budgeted CWP. This variance may be further divided into price and efficiency variances. It measures cost overruns or underruns.

      2) The **job project-schedule cost variance** is the difference between the budgeted CWP and the budgeted CWS. It measures the extent to which the project is ahead of or behind schedule.

7. Stop and review! You have completed the outline for this subunit. Study multiple-choice questions 1 through 19 beginning on page 257.

B. Queuing Theory

1. Queuing (waiting-line) theory is a group of mathematical models for systems involving waiting lines. In general, a queuing system consists of a queue, or waiting line, and a service facility.

   a. Examples of queuing systems

      1) Bank teller windows
      2) Grocery checkout counters
      3) Highway toll booths
      4) Docks for ships
      5) Airport holding patterns

   b. Two basic costs are involved.

      1) The cost of providing service includes facility costs and operating costs.

      2) The waiting cost is the cost of idle resources waiting in line. It may be a direct cost if paid employees are waiting or an opportunity cost in the case of waiting customers. In either case, there is always a cost of waiting.

   c. The objective of queuing theory is to minimize the total cost of the system, including both service and waiting costs, for a given rate of arrivals.

   d. The structure of queuing systems depends on the number of lines and service facilities and how they are coupled. Grocery stores usually have multiple-line, multiple-server systems. Some banks have single-line, multiple-teller systems. Job shops can be conceived of as multistage systems in which each machine is a server with its own queue.

   e. Mathematical solutions are available for simple systems having unscheduled random arrivals. For other systems, simulation must be used to find a solution.

   f. The arrivals and services in a queuing model occur in accordance with a Poisson process. Hence, the probability of occurrence of an event is constant, the occurrence of an event is independent of any other, the probability of an occurrence is proportional to the length of the interval, and, if the interval is small enough, the probability of more than one occurrence approaches zero.

      1) The Poisson probability distribution is used to predict the probability of a specified number of occurrences of an event in a specified time interval, given the expected number of occurrences per time unit.

      2) The related exponential distribution is used to approximate service times. This distribution gives the probability of zero events in a given interval. Accordingly, it gives the probability that service time will not exceed a given length of time.

2. **Formulas for a Single Service Facility and Random Arrivals of Work Units**

If: B = average number of work units arriving in one unit of time
T = average number of work units serviced in one unit of time (if no shortage of work units), and

$\frac{B}{T} < 1$ (otherwise the queue will grow to infinite length)

a. The average number of work units waiting in line or being serviced is

$$N = \frac{B}{(T - B)}$$

b. The average number in the waiting line (Nq) is $Nq = \frac{B^2}{T(T - B)}$

c. The average waiting time before service is $W = \frac{Nq}{B}$

d. EXAMPLE: Cars arrive at a toll booth at the average rate of 3 cars per minute. The toll booth can serve (collect tolls from) 6 cars per minute on the average. In the formulas, B = 3, and T = 6

1) The average number of cars waiting in line or paying tolls at any time is

$$N = \frac{B}{T - B} = \frac{3}{6 - 3} = 1 \textit{ car}$$

2) The average number in the waiting line, not being serviced, is

$$Nq = \frac{B^2}{T(T - B)} = \frac{9}{6(6 - 3)} = .5 \textit{ car} \text{ (one car waiting half the time)}$$

3) The average waiting time is

$$W = \frac{Nq}{B} = \frac{(1/2)}{3} = .16\tfrac{2}{3} \textit{ minute, or } 10 \textit{ seconds}$$

3. Stop and review! You have completed the outline for this subunit. Study multiple-choice questions 20 through 28 beginning on page 264.

C. **Forecasting**. Forecasts are the basis for business plans, including budgets. In addition to intuition, many quantitative methods are useful in projecting the future from past experience.

1. Examples of forecasts include sales projections, inventory demand, cash flow, and future capital needs.

a. Most models are used in the forecasting process. They are used to make decisions that optimize future results.

b. The reliability of the forecast should be determined before using it. No objective method can determine the reliability of judgmental forecasts. When quantitative methods are used, however, measurement of reliability is usually possible, e.g., by calculating the standard error of the estimate.

2. **Correlation analysis** is used to measure the strength of the linear relationship between two or more variables. Correlation between two variables can be seen by plotting their values on a single graph to form a scatter diagram. If the points tend to form a straight line, correlation is high. If they form a random pattern, there is little correlation. Correlation measures only linear relationships.

   a. If the points form a curve, several possibilities exist.

      1) A linear relationship (a straight line) may be used to approximate a portion of the curve.

      2) A linear relationship exists between some other function of the independent variable x (e.g., log x) and the dependent variable y.

      3) No relationship exists.

   b. In standard notation, the coefficient of correlation is r; the coefficient of determination is r^2.

   c. The **coefficient of correlation** measures the relative strength of the linear relationship. It has the following properties:

      1) The magnitude of r is independent of the scales of measurement of x and y.

      2) $-1.0 \leq r \leq 1.0$

         a) A value of –1.0 indicates a perfectly inverse linear relationship between x and y.

         b) A value of zero indicates no linear relationship between x and y.

         c) A value of +1.0 indicates a direct relationship between x and y.

   d. **Scatter diagrams** may be used to demonstrate correlations. Each observation creates a dot that pairs the x and y values. The collinearity and slope of these observations are related to the coefficient of correlation by the above-stated rules.

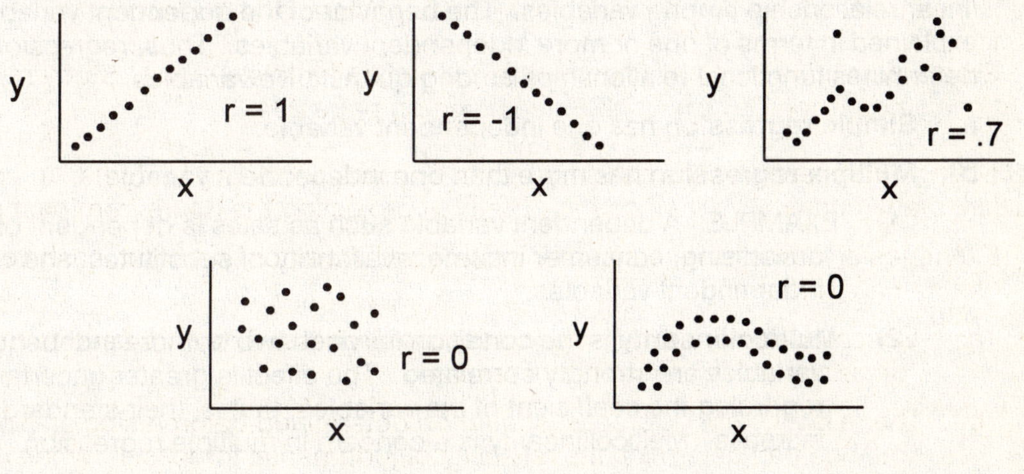

e. The **coefficient of determination (r^2)**, or the coefficient of correlation squared, may be interpreted as the proportion of the total variation in y that is explained or accounted for by the regression equation.

1) It is equal to 1 minus the quotient of the unexplained variation divided by the total variation. The following is the formula:

$$r^2 = 1 - \frac{\Sigma (y_i - \hat{y})^2}{\Sigma (y_i - \bar{y}_i)^2}$$

If: r^2 = the coefficient of determination
Σ = summation
y_i = an actual data point
$\hat{y}$ = a point on the regression line
$\bar{y}_i$ = the mean of the observed data points

2) EXAMPLE: The assertion that new car sales are a function of disposable income with a coefficient of correlation of .8 is equivalent to stating that 64% ($.8^2$) of the variation of new car sales (from average new car sales) can be explained by the variation in disposable income (from average disposable income).

3) Because r^2 increases as the number of independent variables increases, regardless of whether the additional variables are actually correlated with the dependent variable, r^2 may be adjusted (reduced) to allow for this effect. If k is the number of independent variables and n is the number of observations, the formula for adjusted r^2 is

$$r^2 - \frac{(k-1)}{(n-k)} \times (1 - r^2)$$

3. **Regression (least squares) analysis** extends correlation to find an equation for the linear relationship among variables. The behavior of the dependent variable is explained in terms of one or more independent variables. Thus, regression analysis determines functional relationships among quantitative variables.

a. **Simple regression** has one independent variable.

b. **Multiple regression** has more than one independent variable.

1) EXAMPLE: A dependent variable such as sales is dependent on advertising, consumer income, availability of substitutes, and other independent variables.

2) **Multicollinearity** is the condition in which two or more independent variables are strongly correlated. The effect is greater uncertainty regarding the coefficient of the variables; that is, their standard errors increase. Multicollinearity is a concern in multiple regression.

c. Regression analysis is used to find trend lines in business data such as sales or costs (time series analysis or trend analysis) and to develop models based on the association of variables (cross-sectional analysis, a method that is not time related as is trend analysis). Examples are

1) Trend in product sales
2) Trend in overhead as a percentage of sales
3) Relationship of direct labor hours to variable overhead
4) Relationship of direct material usage to accounts payable

d. Some reasonable basis should exist for expecting the variables to be related.

1) If they are obviously independent, any association found by regression is mere coincidence.

2) Regression does not determine causality, however. Although x and y move together, the apparent relationship may be caused by some other factor.

a) EXAMPLE: A strong correlation exists between car-wash sales volume and sunny weather, but sales volume does not cause sunny weather.

3) The statistical relationships revealed by regression and correlation analysis are valid **only** for the range of the data in the sample.

e. The simple regression equation is

$$y = a + bx + e$$

If: y = the dependent variable
a = the y-axis intercept (the fixed cost in cost functions)
b = the slope of the regression line (the variable portion of the total cost in cost functions)
x = the independent variable
e = the error term

1) Assumptions of the model are that

a) For each value of x, a distribution of values exists for y. The means of these distributions form a straight line. Hence, x and y are linearly related.

b) The error term (e) is normally distributed with a mean or expected value equal to zero.

i) The y-intercept (a) and the slope of the regression line (b) also have normal distributions.

c) Errors in successive observations are statistically independent.

i) Thus, the estimators are unbiased.

ii) **Autocorrelation (serial correlation)** occurs when the observations are not independent; in other words, later observations may be dependent on earlier ones.

d) The distribution of y around the regression line is constant for different values of x.

i) Thus, the observations are characterized by **homoscedasticity** or **constant variance**. The deviation of points from the regression line does not vary significantly with a change in the size of the independent variable.

- **Heteroscedasticity** is the condition in which the variance of the error term is not constant.

ii) Graphically, the model is represented by a series of normal distributions (subpopulations of y) around the regression line. As noted on the previous page, these subpopulations have the same variance.

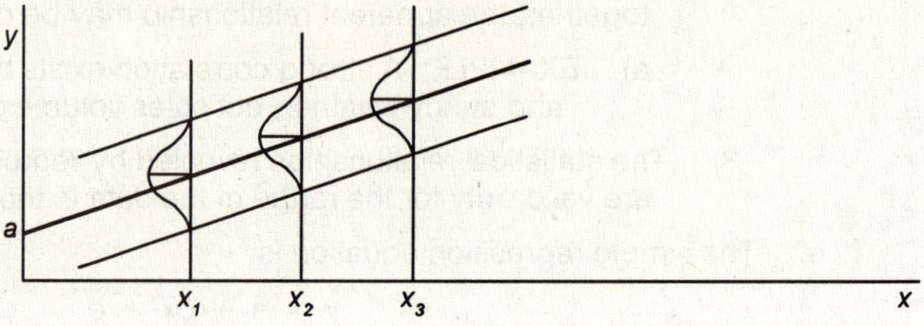

f. From linear algebra, the equation for a straight line may be stated as follows:

$$y = a + bx$$

If: a = the y-axis intercept
b = the slope of the line

1) Regression analysis uses the method of least squares, which minimizes the sum of the squares of the vertical distance between each observation point and the regression line.

2) EXAMPLE: Observations are collected on advertising expenditures and annual sales for a firm.

Sales ($000,000s)	Advertising ($000s)
28	71
14	31
19	50
21	60
16	35

a) According to the regression equation that results from using least squares computations, expected sales equal 4.2 plus .31 times the advertising expenditure.

$$y = 4.2 + .31(x)$$

b) The observations are graphed as follows:

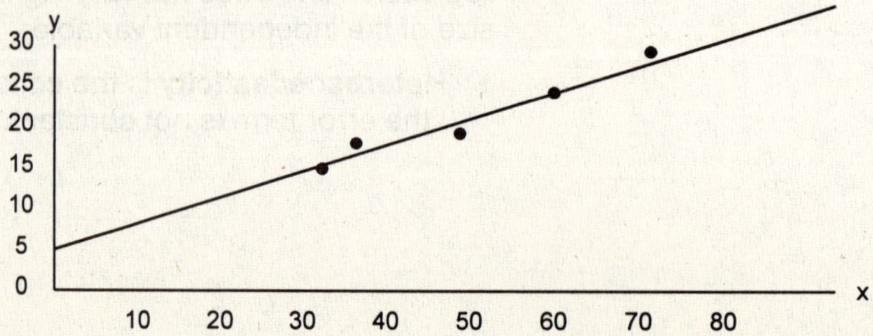

g. Regression analysis is particularly valuable for budgeting and cost accounting purposes. For instance, it is almost a necessity for computing the fixed and variable portions of mixed costs for flexible budgeting.

h. The following equations can be used to determine the equation for the least squares regression line (the equation for the line is in the form of y = a + bx):

$$\Sigma y = na + b(\Sigma x)$$
$$\Sigma xy = a(\Sigma x) + b(\Sigma x^2)$$

1) EXAMPLE: The use of the two equations can be illustrated with the following data based on a set of six paired observations (n = 6):

y	x
$ 6	2
7	3
5	2
4	1
8	3
6	2
Σy = $36	Σx = 13

Σxy	Σx²
6 × 2 = 12	4
7 × 3 = 21	9
5 × 2 = 10	4
4 × 1 = 4	1
8 × 3 = 24	9
6 × 2 = 12	4
83	31

a) Substituting into the two equations gives

$$36 = 6a + 13b$$
$$83 = 13a + 31b$$

b) Solving simultaneously for the two unknowns,

$$1116 = 186a + 403b$$
$$\underline{1079} = \underline{169a} + \underline{403b}$$
$$\underline{\underline{37}} = \underline{\underline{17a}}$$

c) Thus, a = 2.176. Solving for b in the second original equation gives

$$83 = 13(2.176) + 31b$$
$$83 = 28.288 + 31b$$
$$31b = 54.712$$
$$b = 1.765$$

d) Hence, future costs can be predicted using the following equation:

$$y = \$2.176 + \$1.765x$$

e) Alternative formulas that are ordinarily simpler to use are given below:

i) The slope may be expressed as

$$b = \frac{n\Sigma xy - \Sigma x\Sigma y}{n\Sigma x^2 - (\Sigma x)^2}$$

ii) The value of the y-intercept may be expressed as

$$a = \bar{y} - b(\bar{x})$$

i. The statistical significance of the slope of the regression line is important because, if its true value is zero, changes in the independent variable have no effect on the dependent variable.

1) Because the distribution of b is normal, the t-distribution (see Study Unit 4) may be used to determine whether b is significantly different from zero, that is, whether one can reject the null hypothesis that b equals zero.

a) One approach is to divide b by the standard error of the estimate of b. (The formula is not given here. The standard error is usually provided in the computer output.) If the result exceeds the critical value of t determinable from a standard table, the conclusion is that b is not zero. For example, this critical value is 2.0 for a sample of 60, 58 degrees of freedom (60 – 2 parameters of a and b estimated), and a 95% confidence level.

b) Another approach is to construct a precision interval (b ± t x the standard error of the estimate of b). If the interval does not contain zero, the null hypothesis may be rejected.

i) The value of t is the critical value for the given sample size, degrees of freedom, and confidence level used in the first approach.

j. The **high-low method** is used to generate a regression line by basing the equation on only the highest and lowest of a series of observations.

1) EXAMPLE: A regression equation covering electricity costs could be developed by using only the high-cost month and the low-cost month. If costs were $400 in April when production was 800 machine hours and $600 in September when production was 1,300 hours, the equation would be determined as follows:

High month	$600	for	1,300 hours
Low month	400	for	800 hours
Increase	$200		500 hours

Because costs increased $200 for 500 additional hours, the variable cost is $.40 per machine hour. For the low month, the total variable portion of that monthly cost is $320 ($.40 x 800 hours). Given that the total cost is $400 and $320 is variable, the remaining $80 must be a fixed cost. The regression equation is y = 80 + .4x.

2) The major criticism of the high-low method is that the high and low points may be abnormalities not representative of normal events.

4. **Time series or trend analysis** relies on past experience. Changes in the value of a variable (e.g., unit sales of a product) over time may have several possible components.

   a. In time series analysis, the dependent variable is regressed on time (the independent variable).

   b. The **secular trend** is the long-term change that occurs in a series. It is represented by a straight line or curve on a graph.

   c. **Seasonal variations** are common in many businesses. A variety of analysis methods includes seasonal variations in a forecasting model, but most methods make use of a seasonal index.

   d. **Cyclical fluctuations** are variations in the level of activity in business periods. Although some of these fluctuations are beyond the control of the firm, they need to be considered in forecasting. They are usually incorporated as index numbers.

   e. **Irregular or random variables** are any variations not included in the categories above. Business can be affected by random happenings (e.g., weather, strikes, fires, etc.).

5. **Exponential smoothing** is a technique used to level or smooth variations encountered in a forecast. This technique also adapts the forecast to changes as they occur.

   a. The simplest form of smoothing is the moving average, in which each forecast is based on a fixed number of prior observations. Exponential smoothing is similar to the moving average.

   b. Exponential means that greater weight is placed on the most recent data, with the weights of all data falling off exponentially as the data age. The selection of alpha (α), the smoothing factor, is important because a high alpha places more weight on recent data.

   c. The equation for the forecast (F) for period t+1 is

$$F_{t+1} = \alpha(x_t) + (1 - \alpha)F_t$$

   If: x_t = the observation for period t
   t = the most recent period
   α = the smoothing factor ($0 \leq \alpha \leq 1$)
   F_t = the forecast for period t

      1) This method weights the observation for period t by α and the forecast for period t by $(1 - \alpha)$.

6. **Learning curves** reflect the increased rate at which people perform tasks as they gain experience. The time required to perform a given task becomes progressively shorter, but this technique is only applicable to the early stages of production or of any new task.

   a. Ordinarily, the curve is expressed as a percentage of reduced time to complete a task for each doubling of cumulative production. Research has shown learning curve percentages to be between 60% and 80%. In other words, the time required is reduced by 20% to 40% each time cumulative production is doubled, with 20% being common.

1) One common assumption made in a learning curve model is that the **cumulative average time per unit** is reduced by a certain percentage each time production doubles.

a) The alternative assumption is that **incremental unit time** (time to produce the last unit) is reduced when production doubles.

2) EXAMPLE: Given an 80% learning curve model based on the first assumption stated above, the following performance is expected during the early stages of the manufacture of a new product:

Cumulative Number of Tasks	Cumulative Average Time per Unit
100	3.0
200	2.4 (3.0 × 80%)
400	1.92 (2.4 × 80%)
800	1.536 (1.92 × 80%)
1,600	1.228 (1.536 × 80%)

b. **Graphical presentation**

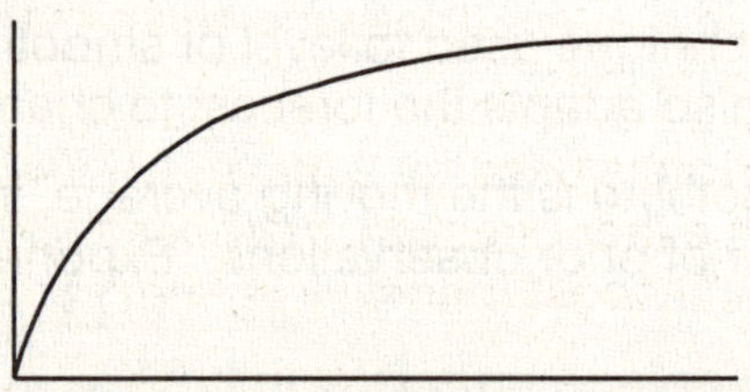

c. If the average time for 100 units in the example above were 3 minutes per unit, the total time would be 300 minutes. At an average time of 2.4 minutes for 200 units, the total time would be 480 minutes. In other words, the additional 100 units required only 180 minutes (480 – 300), or 1.8 minutes per unit.

1) This analysis reflects a typical CMA question requirement. The candidate must be able to calculate not only the average completion time but also the completion time for the last batch.

7. Stop and review! You have completed the outline for this subunit. Study multiple-choice questions 29 through 53 beginning on page 267.

D. Game Theory

1. Game (or decision) theory is a mathematical approach to decision making when confronted with an "enemy" or competitor.

2. Games are classified according to the number of players and the algebraic sum of the payoffs.

a. In a two-person game, if the payoff is given by the loser to the winner, the algebraic sum is zero and the game is called a **zero-sum game**.

b. If it is possible for both players to profit, however, it is called a **positive-sum game**.

3. Mathematical models have been developed to select optimal strategies for certain simple games.
   a. Few applications of the mathematics of game theory have been made in business, but the concepts can be useful.
   b. For example, labor negotiations can be viewed as a two-person, nonzero-sum game.
4. Game theorists have developed various decision rules.
   a. The **maximax criterion** is a decision rule adopted by risk-seeking, optimistic players who desire the largest possible payoff and are willing to accept high risk. The player determines the payoff for each state of nature expected to arise after each possible decision, ascertains the maximum payoff for each decision, and then chooses the decision with the maximum payoff.
   b. A player who uses the **minimax criterion** determines the maximum loss for each decision possibility and then chooses the decision with the minimum maximum loss. This rule produces the same result as the **maximin** technique, which determines the minimum payoff for each decision and then chooses the decision with the maximum minimum payoff. Minimax and maximin are conservative criteria used by risk-averse players (those for whom the utility of a gain is less than the disutility of an equal loss).
   c. The **minimax regret criterion** is used by a player who wishes to minimize the effect of a bad decision in either direction. It chooses the decision that has the lowest maximum opportunity cost (benefit forgone).
   d. The **insufficient reason (Laplace) criterion** may be used when the decision maker cannot assign probabilities to the states of nature arising after a decision. The reasoning is that, if no probability distribution can be assigned, the probabilities must be equal, and the expected value is calculated accordingly; i.e., for each decision, the payoffs for the various states of nature are simply added, and the decision with the highest total is chosen. This criterion is risk-neutral.
   e. An **expected value criterion** might be used by a risk-neutral player, that is, one for whom the utility of a gain is the same as the disutility of an equal loss.
5. Stop and review! You have completed the outline for this subunit. Study multiple-choice questions 54 through 63 beginning on page 277.

E. Markov Analysis

1. Markov processes are useful in decision problems in which the probability of the occurrence of a future state depends only on the current state.
   a. A characteristic of the Markov process is that the initial state matters less and less as time goes on because the process will eventually reach its steady state.
   b. EXAMPLE: A machine tool may be in one of two states, in adjustment or out of adjustment. The machine moves from one state to the other in one day with the following probabilities:

From \ To	In adjustment	Out of adjustment
In adjustment	.8	.2
Out of adjustment	.6	.4

   If the machine is in adjustment on day 1, the probabilities of its being in or out of adjustment are as follows:

			IN	OUT		Pr
On day 2	(1.0) × (.8)	=	.8	(1 –.8)	=	.2
On day 3	(.8) × (.8) + (.2) × (.6)	=	.76	(1 –.76)	=	.24
On day 4	(.76) × (.8) + (.24) × (.6)	=	.752	(1 –.752)	=	.248
On day 5	(.752) × (.8) + (.248) × (.6)	=	.7504	(1 –.7504)	=	.2496

   The process approaches a probability of .75 on day n of being in adjustment.
2. Stop and review! You have completed the outline for this subunit. Study multiple-choice questions 64 and 65 beginning on page 280.

F. Simulation and Sensitivity Analysis

1. **Simulation** is a technique for experimenting with logical and mathematical models using a computer. Despite the power of mathematics, many problems cannot be solved by known analytical methods because of the behavior of the variables and the complexity of their interactions, e.g.,
   a. Corporate planning models
   b. Financial planning models
   c. New product marketing models
   d. Queuing system simulations
   e. Inventory control simulations
2. Experimentation is neither new nor uncommon in business. Building a mockup of a new automobile, having one department try out new accounting procedures, and test-marketing a new product are all forms of experimentation. In effect, experimentation is organized trial and error using a model of the real world to obtain information prior to full implementation.
3. Models can be classified as either physical or abstract.
   a. Physical models include automobile mockups, airplane models used for wind-tunnel tests, and breadboard models of electronic circuits.

b. Abstract models may be pictorial (architectural plans), verbal (a proposed procedure), or logical-mathematical. Experimentation with logical-mathematical models can involve many time-consuming calculations. Computers have eliminated much of this costly drudgery and have led to the growing interest in simulation for management.

4. The simulation procedure has five steps.

a. **Define the objectives**. The objectives serve as guidelines for all that follows. The objectives may be to aid in the understanding of an existing system (e.g., an inventory system with rising costs) or to explore alternatives (e.g., the effect of investments on the firm's financial structure). A third type of objective is estimating the behavior of some new system such as a production line.

b. **Formulate the model**. The variables to be included, their individual behavior, and their interrelationships must be defined in precise logical-mathematical terms. The objectives of the simulation serve as guidelines in deciding which factors are relevant.

c. **Validate the model**. Some assurance is needed that the results of the experiment will be realistic. This assurance requires validation of the model -- often using historical data. If the model gives results equivalent to what actually happened, the model is historically valid. Some risk remains, however, that changes could make the model invalid for the future.

d. **Design the experiment**. Experimentation is sampling the operation of a system. For example, if a particular policy is simulated on an inventory model for two years, the results are a single sample. With replication, the sample size can be increased and the confidence level raised. The number of runs to be made, length of each run, measurements to be made, and methods for analyzing the results are all part of the design of the experiment.

e. **Conduct the simulation -- evaluate results**. The simulation should be conducted with care. The results are analyzed using appropriate statistical methods.

5. The **Monte Carlo technique** is often used in simulation to generate the individual values for a random variable. A random number generator is used to produce numbers with a uniform probability distribution (equal likelihoods of occurrence). The second step is to transform these numbers into values consistent with the desired distribution.

a. The performance of a quantitative model may be investigated by randomly selecting values for each of the variables in the model (based on the probability distribution of each variable) and then calculating the value of the solution. If this process is performed a large number of times, the distribution of results from the model will be obtained.

b. EXAMPLE: A new marketing model includes a factor for a competitor's introduction of a similar product within 1 year. Management estimates a 50% chance that this event will happen. For each simulation, this factor must be determined, perhaps by flipping a coin, or by putting two numbers in a hat and selecting one number. Random numbers between 0 and 1 could be generated. Numbers under .5 would signify introduction of a similar product; numbers over .5 would indicate the nonoccurrence of this event.

6. The advantages of simulation are as follows:
   a. Time can be compressed. A corporate planning model can show the results of a policy for 5 years into the future, using only minutes of computer time.
   b. Alternative policies can be explored. With simulations, managers can ask what-if questions to explore possible policies, providing management with a powerful new planning tool.
   c. Complex systems can be analyzed. In many cases, simulation is the only possible quantitative method for analyzing a complex system such as a production or inventory system, or the entire firm.
7. The limitations of simulation are as follows:
   a. Cost. Simulation models can be costly to develop. They can be justified only if the information to be obtained is worth more than the costs to develop the model and carry out the experiment.
   b. Risk of error. A simulation results in a prediction of how an actual system would behave. As in forecasting, the prediction may be in error.
8. **Sensitivity Analysis**. After a problem has been formulated into any mathematical model, it may be subjected to sensitivity analysis.
   a. A trial-and-error method may be adopted in which the sensitivity of the solution to changes in any given variable or parameter is calculated.
      1) The risk of the project being simulated may also be estimated.
      2) The best project may be one that is least sensitive to changes in probabilistic (uncertain) inputs.
   b. In linear programming problems, sensitivity is the range within which a constraint value, such as a cost coefficient or any other variable, may be changed without changing the optimal solution. Shadow price is the synonym for sensitivity in that context.
9. Stop and review! You have completed the outline for this subunit. Study multiple-choice questions 66 through 70 beginning on page 281.

MULTIPLE-CHOICE QUESTIONS

A. Project Scheduling

1. The primary difference between PERT and CPM is that

A. CPM uses probabilities on the activity times and PERT does not.

B. PERT considers activity costs and CPM does not.

C. PERT can assign probabilities to activity times and CPM does not.

D. CPM considers activity costs and PERT does not.

The correct answer is (D). *(CMA, adapted)*

REQUIRED: The primary difference between PERT and CPM.

DISCUSSION: Both PERT and CPM are network analysis techniques. But CPM was developed independently of PERT and is widely used in the construction industry. CPM may be thought of as a subset of PERT. Like PERT, it is a network technique, but, unlike PERT, it uses deterministic time and cost estimates. Its advantages include cost estimates plus the concept of "crash" efforts and costs. Activity times are estimated for normal effort and crash effort. Crash time is the time to complete an activity assuming that all available resources were devoted to the task (overtime, extra crew, etc.). Activity costs are also estimated for normal and crash efforts. These estimates allow the project manager to estimate the costs of completing the project if some of the activities are completed on a crash basis. The network diagram is constructed in the same manner as PERT diagrams. Once the diagram is constructed, the critical paths are found for normal and crash times. More than one critical path may exist for each diagram.

Answer (A) is incorrect because CPM specifies the activity times (uses deterministic estimates). Answer (B) is incorrect because CPM but not PERT uses activity costs and considers crash times. Answer (C) is incorrect because a less significant difference between PERT and CPM is that PERT uses probabilistic estimates of completion times. CPM times are deterministic. The 1:4:1 method is typically used in PERT. Under this method, the most optimistic and pessimistic estimates are weighted equally, but the most likely estimate is weighted four times more heavily than the others.

2. A Gantt chart

A. Shows the critical path for a project.

B. Is used for determining an optimal product mix.

C. Shows only the activities along the critical path of a network.

D. Does not necessarily show the critical path through a network.

The correct answer is (D). *(CMA, adapted)*

REQUIRED: The true statement about a Gantt chart.

DISCUSSION: A Gantt or bar chart is sometimes used in conjunction with PERT or CPM to show the progress of a special project. Time is shown on the horizontal axis, the length of a bar equals the length of an activity, and shading indicates the degree of completion. However, the Gantt chart is not as sophisticated as PERT or CPM in that it does not reflect the relationships among the activities or define a critical path.

Answer (A) is incorrect because the critical path is not shown. Answer (B) is incorrect because linear programming is used to determine an optimal product mix. Answer (C) is incorrect because a Gantt chart shows the activities to be completed but not their relationships (sequencing).

3. Manning Corporation is preparing a proposal for the government to produce a steam generator to be used in nuclear submarines. Manning has three factories that can initiate and complete the generator; however, completion times vary due to different technologies and older equipment within each of the three factories. The decision tool that would best assist Manning in its factory selection is

A. Queuing theory.

B. Time series analysis.

C. PERT-cost analysis.

D. Linear programming.

The correct answer is (C). *(CMA, adapted)*

REQUIRED: The decision tool used to select a production facility given different technologies.

DISCUSSION: PERT is a network analysis technique that uses probabilistic time estimates to aid managers in planning and controlling large-scale, complex projects. PERT diagrams are free-form networks showing each activity as a line between events. A sequence of lines shows interrelationships among activities. PERT-cost is a modification that permits the effect of any change in a given task to be determined in terms of cost as well as time. PERT-cost is simply an application of cost figures to traditional network analysis.

Answer (A) is incorrect because queuing theory is used to minimize the cost of waiting lines. Answer (B) is incorrect because time series analysis is based on past experience. In time series analysis, a dependent variable is regressed on time. Answer (D) is incorrect because linear programming is a tool for optimizing a cost or profit function given specified constraints.

4. When using PERT (Program Evaluation Review Technique), the expected time for an activity when given an optimistic time (A), a pessimistic time (B), and a most likely time (m) is calculated by which one of the following formulas?

A. $(b - a) \div 2$

B. $(a + b) \div 2$

C. $(a + 4m + b) \div 6$

D. $(4abm) \div 6$

The correct answer is (C). *(CMA, adapted)*

REQUIRED: The formula for calculating the expected time for an activity when using PERT.

DISCUSSION: PERT was developed to aid managers in controlling large, complex projects. PERT analysis includes probabilistic estimates of activity completion times. Three time estimates are made: optimistic, most likely, and pessimistic. The time estimates for an activity are assumed to approximate a beta probability distribution. PERT approximates the mean of the beta distribution by dividing the sum of the optimistic time, the pessimistic time, and four times the most likely time by six.

Answers (A) and (B) are incorrect because the most likely time estimate should be included in the formula. Answer (D) is incorrect because all time estimates are not weighted equally.

5. California Building Corporation uses the critical path method to monitor construction jobs. The company is currently 2 weeks behind schedule on Job #181, which is subject to a $10,500-per-week completion penalty. Path A-B-C-F-G-H-I has a normal completion time of 20 weeks, and critical path A-D-E-F-G-H-I has a normal completion time of 22 weeks. The following activities can be crashed:

Activities	Cost to Crash 1 Week	Cost to Crash 2 Weeks
BC	$ 8,000	$15,000
DE	10,000	19,600
EF	8,800	19,500

California Building desires to reduce the normal completion time of Job #181 and, at the same time, report the highest possible income for the year. California Building should crash

A. Activity BC 1 week and activity EF 1 week.

B. Activity BC 2 weeks.

C. Activity EF 2 weeks.

D. Activity DE 1 week and activity EF 1 week.

The correct answer is (D). *(CMA, adapted)*

REQUIRED: The activity that should be crashed (speeded up at additional cost) to maximize income.

DISCUSSION: Activities that are to be crashed in a CPM problem should be ones that are on the critical (longest) path. Thus, activity BC should not be selected because it is not on the critical path. To finish activity BC 2 weeks early would not reduce the total time to complete the project. Therefore, the only feasible choices are DE and EF on the critical path. The total cost to crash DE and EF for 1 week each is $18,800 ($10,000 + $8,800), which is less than the cost to crash either activity for 2 weeks. Thus, DE and EF should be crashed for 1 week each because the total cost is less than the $21,000 ($10,500 x 2) 2-week delay penalty.

Answers (A) and (B) are incorrect because BC is not on the critical path. Answer (C) is incorrect because the cost of crashing EF 2 weeks is $19,500, which is greater than the total cost to crash DE and EF for 1 week each.

6. PERT and the critical path method (CPM) are used for

A. Determining the optimal product mix.

B. Project planning and control.

C. Determining product costs.

D. Determining the number of servers needed in a fast-food restaurant.

The correct answer is (B). *(CMA, adapted)*

REQUIRED: The reason for using PERT and CPM.

DISCUSSION: The Program Evaluation and Review Technique and critical path methods are useful in the planning and control of a complex system or process. PERT and CPM both construct a network of time relationships between each event or subproject to identify the subprojects that have a direct effect on the completion date of the project as a whole. The critical path is the longest path through the network. If any activity on this path takes longer than expected, the entire project will be delayed (slack time is zero). Every project has at least one critical path. Some have more than one.

Answer (A) is incorrect because determining product mix can be accomplished by using linear programming. Answer (C) is incorrect because product costs are determined by cost accounting. Answer (D) is incorrect because queuing theory determines the number of servers needed in a restaurant.

7. In a PERT network, the optimistic time for a particular activity is 9 weeks, and the pessimistic time is 21 weeks. Which one of the following is the best estimate of the standard deviation for the activity?

A. 2

B. 6

C. 9

D. 12

The correct answer is (A). *(CMA, adapted)*

REQUIRED: The standard deviation for the completion time of a PERT project given an optimistic time and a pessimistic time.

DISCUSSION: PERT analysis includes probabilistic estimates of activity completion times. Three time estimates are made: optimistic, most likely, and pessimistic. The time estimates for an activity are assumed to approximate a beta probability distribution. In contrast to the normal distribution, this distribution has finite endpoints (the optimistic and pessimistic estimates) and is unimodal; that is, it has only one mode (the most likely time). PERT approximates the mean of the beta distribution by dividing the sum of the optimistic time, the pessimistic time, and four times the most likely time (the mode) by six. The standard deviation is approximated by dividing the difference between the pessimistic and optimistic times by six. The basis for the latter approximation is that various probability distributions have tails that lie about plus or minus three standard deviations from the mean. For example, 99.9% of observations in the normal distribution are expected to lie within this range. Accordingly, if the pessimistic and optimistic times are 21 and 9 weeks, respectively, the standard deviation is 2 weeks ($12 \div 6$).

Answer (B) is incorrect because 6 is the approximate number of standard deviations between the tails of the normal distribution. Answer (C) is incorrect because 9 weeks is the pessimistic estimate. Answer (D) is incorrect because 12 weeks is the difference between the optimistic and the pessimistic estimates.

Questions 8 through 10 are based on the following information. The following are the times required for various sequences of activities (paths through the network) shown in a PERT diagram:

Path	Time in Days
A-B-C	27
A-D-E-C	33
A-F-C	34
A-F-G-C	36

8. Which is the critical path?

A. A-B-C

B. A-D-E-C

C. A-F-C

D. A-F-G-C

The correct answer is (D). *(Publisher)*

REQUIRED: The critical path in a PERT diagram.

DISCUSSION: The critical path is the longest path in a network. The reason that path is critical is that any delays along the critical path will cause delays in the completion of the overall project. The longest path in the network is A-F-G-C (36 days).

Answers (A), (B), and (C) are incorrect because A-B-C, A-D-E-C, and A-F-C require 27, 33, and 34 days, respectively.

9. How much slack time is there in path A-D-E-C?

A. 1 day.

B. 3 days.

C. 6 days.

D. 33 days.

The correct answer is (B). *(Publisher)*

REQUIRED: The amount of slack in path A-D-E-C.

DISCUSSION: PERT diagrams are free-form networks showing each activity in a large project as a line between events. The critical path is the longest path in time through the network. That path is critical because, if any activity on the critical path takes longer than expected, the entire project will be delayed. Paths that are not critical have slack time. Slack is the number of days an activity can be delayed without forcing a delay for the entire project. Given that path A-D-E-C is 33 days long, and the critical path is 36 days long, the slack is the 3-day difference. In other words, work along path A-D-E-C can be delayed by 3 days without affecting the final completion time.

Answer (A) is incorrect because 1 day is the difference between A-D-E-C and the next longest path, not the longest path in the network. Answer (C) is incorrect because 6 days is the difference between A-D-E-C and the shortest path. Answer (D) is incorrect because 33 days is the time required to complete path A-D-E-C.

10. Path A-B-C is expected to require 27 days to complete. This estimate was calculated using the PERT formula based on a most likely completion time of 25 days and a most optimistic completion time of 24 days. What estimate was used for the most pessimistic completion time?

A. 25 days.

B. 27 days.

C. 32 days.

D. 38 days.

The correct answer is (D). *(Publisher)*

REQUIRED: The estimate of the most pessimistic completion time.

DISCUSSION: A PERT analysis incorporates uncertainty by using probabilistic estimates of activity completion times. The most optimistic, most likely, and most pessimistic times are weighted on a basis of 1 to 4 to 1, respectively. Given an expected completion time of 27 days, a most likely completion time of 25 days, and a most optimistic completion time of 24 days, the following formula may be solved for the most pessimistic completion time (X):

$$\begin{aligned}(1 \times 24) + (4 \times 25) + (1 \times X) &= (6 \times 27)\\ 24 + 100 + X &= 162\\ X &= 38\end{aligned}$$

Answer (A) is incorrect because 25 days is the most likely completion time. Answer (B) is incorrect because 27 days is the expected completion time. Answer (C) is incorrect because 32 days is based on a simple arithmetic average.

11. A PERT network has only two activities on its critical path. These activities have standard deviations of 6 and 8, respectively. The standard deviation of the project completion time is

A. 7

B. 10

C. 14

D. 48

The correct answer is (B). *(CMA, adapted)*

REQUIRED: The standard deviation of the project completion time given the standard deviations of the activity times on the critical path.

DISCUSSION: The mean time for the critical path is simply the sum of the means of the activity times. However, the standard deviation equals the square root of the sum of the variances (squares of the standard deviations) of the times for activities on the critical path. The standard deviation of the project completion time (time for the critical path) is therefore the square root of 100 ($6^2 + 8^2$), or 10.

Answer (A) is incorrect because 7 is the average standard deviation. Answer (C) is incorrect because 14 is the sum of the standard deviations. Answer (D) is incorrect because 48 is the product of the standard deviations.

12. The critical path through a network is the

A. Shortest path.

B. Path with the most nodes.

C. Longest path.

D. Path with the most slack.

The correct answer is (C). *(CMA, adapted)*

REQUIRED: The true statement about a critical path through a network.

DISCUSSION: The critical path is the longest path through the network. If any activity on this path takes longer than expected, the entire project will be delayed (slack time is zero). Every project has at least one critical path. Some have more than one.

Answer (A) is incorrect because the critical path is the longest path. Answer (B) is incorrect because whether a path is critical is a function of its completion time, not the number of nodes. Answer (D) is incorrect because a critical path has no slack.

13. Schippi Inc. is developing a new type of watch. The optimistic time to complete the project is 7 weeks, the most likely time is 13 weeks, and the pessimistic time is 16 weeks. Using PERT, how long is the expected duration for this activity?

A. 6 weeks.

B. 7 weeks.

C. 12.5 weeks.

D. 13 weeks.

The correct answer is (C). *(Publisher)*

REQUIRED: The expected duration of an activity in a PERT network.

DISCUSSION: The expected duration of an activity in a PERT network can be found by taking the sum of the optimistic time, the pessimistic time, and four times the most likely time and dividing it by 6. Hence, the expected duration is 12.5 weeks.

$$\{[7 + (4 \times 13) + 16] \div 6\} = 12.5 \text{ weeks}$$

Answer (A) is incorrect because 6 weeks is the result of failing to multiply the most likely time by 4. Answer (B) is incorrect because 7 weeks is the optimistic time. Answer (D) is incorrect because 13 weeks is the most likely time.

Questions 14 and 15 are based on the following information. The PERT network diagram and corresponding activity cost chart for a manufacturing project at Networks, Inc. is presented below. The numbers in the diagram are the expected times (in days) to perform each activity in the project.

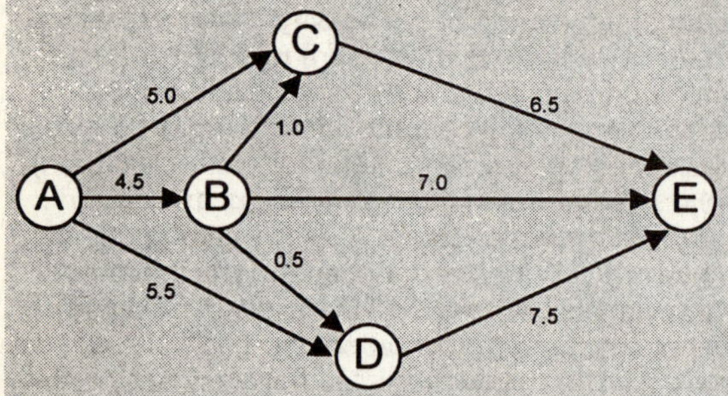

Activity	Normal Cost	Crash Time	Crash Cost
AB	$3,000	3.50 days	$4,000
AC	5,000	4.50	5,250
AD	4,000	4.00	4,750
BE	6,000	5.00	7,000
CE	8,000	5.00	9,200
DE	6,000	6.50	6,750
BC	2,500	.50	3,500
BD	2,000	.25	2,500

14. The expected time of the critical path is

A. 12.0 days.

B. 13.0 days.

C. 11.5 days.

D. 11.0 days.

The correct answer is (B). *(CMA, adapted)*

REQUIRED: The expected time of the critical path.

DISCUSSION: The critical path is the longest path. The longest path in the diagram is A-D-E, which requires 13 days (5.5 + 7.5) based on expected times.

Answers (A), (C) and (D) are incorrect because the expected time of the critical path is 13.0 days.

15. In order to keep costs at a minimum and decrease the completion time by 1½ days, Networks, Inc. should crash activity(ies)

A. AD and AB.

B. DE.

C. AD.

D. AB and CE.

The correct answer is (A). *(CMA, adapted)*

REQUIRED: The activity(ies) that can be crashed by a specified number of days at the least cost.

DISCUSSION: The critical (longest) path is A-D-E, which has an expected time of 13 days (see preceding question). However, to decrease the project's completion time by 1.5 days, paths A-B-C-E (4.5 + 1.0 + 6.5 = 12 days) and A-B-D-E (4.5 + .5 + 7.5 = 12.5 days) as well as A-D-E must also be shortened. Hence, A-D-E must be reduced by 1.5 days, A-B-C-E by .5 day, and A-B-D-E by 1.0 day. The only way to decrease A-D-E by 1.5 days is to crash activity AD (5.5 expective time – 4.0 crash time = 1.5 days). Crashing DE results in a 1.0-day saving (7.5 – 6.5) only. Crashing AB is the efficient way to reduce both A-B-C-E and A-B-D-E by the desired amount of time because it is part of both paths. The incremental cost of crashing AB is $1,000 ($4,000 crash cost – $3,000 normal cost) to shorten the completion time by 1.0 day (4.5 – 3.5). The alternatives for decreasing both A-B-C-E and A-B-D-E are more costly.

Answer (B) is incorrect because crashing activity DE saves only 1.0 day (7.5 – 6.5) on the critical path and does not reduce the time needed for A-B-C-E. Answer (C) is incorrect because crashing AD does not reduce the time necessary to complete A-B-C-E or A-B-D-E. Answer (D) is incorrect because AB and CE are not on the critical path.

16. When making a cost-time trade-off in PERT analysis, the first activity that should be crashed is the activity

A. With the largest amount of slack.

B. With the lowest unit crash cost.

C. On the critical path with the maximum possible time reduction.

D. On the critical path with the lowest unit crash cost.

The correct answer is (D). *(CMA, adapted)*

REQUIRED: The first activity that should be crashed when making a cost-time trade-off in PERT analysis.

DISCUSSION: When making a cost-time trade-off, the first activity to be crashed (have its completion time accelerated) is one on the critical path. To select an activity on another path would not reduce the total time of completion. The initial activity chosen should be the one with the completion time that can be accelerated at the lowest possible cost per unit of time saved.

Answer (A) is incorrect because eliminating an activity with slack will not reduce the total time of the project. Answer (B) is incorrect because the activity with the lowest unit crash cost may not be on the critical path. Answer (C) is incorrect because the time reduction should be related to its cost. The maximum time reduction may not be cost effective.

17. In PERT, slack is the

A. Uncertainty associated with time estimates.

B. Difference between the latest starting time and earliest finishing time.

C. Path that has the largest amount of time associated with it.

D. Number of days an activity can be delayed without forcing a delay for the entire project.

The correct answer is (D). *(CMA, adapted)*

REQUIRED: The meaning of slack in a PERT analysis.

DISCUSSION: PERT diagrams are free-form networks showing each activity in a large project as a line between events. The critical path is the longest path in time through the network. That path is critical in that, if any activity on the critical path takes longer than expected, the entire project will be delayed. Paths that are not critical have slack time. Slack is the number of days an activity can be delayed without forcing a delay for the entire project.

Answer (A) is incorrect because uncertainty is reflected in the use of probabilistic estimates of completion times. Answer (B) is incorrect because the difference between the latest starting time and earliest finishing time is irrelevant. Answer (C) is incorrect because the path with the largest amount of time associated with it is the critical path.

18. The network below describes the interrelationships of several activities necessary to complete a project. The arrows represent the activities. The numbers between the arrows indicate the number of months to complete each activity.

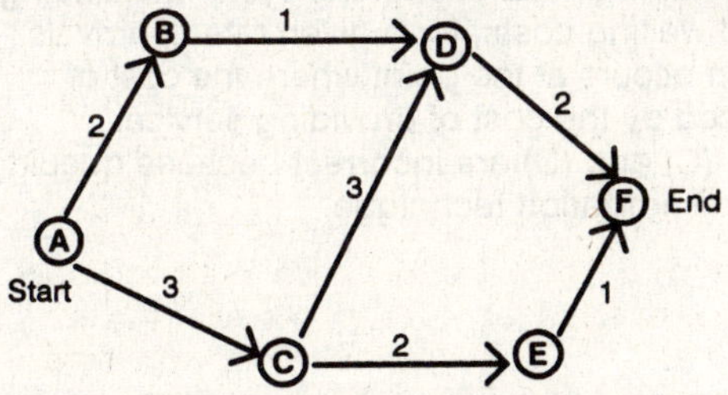

The shortest time to complete the project is

A. 5 months.

B. 6 months.

C. 8 months.

D. 14 months.

The correct answer is (C). *(CIA, adapted)*

REQUIRED: The shortest time to complete the project.

DISCUSSION: The longest, or critical, path in the network from node (A) to node (F) is path A-C-D-F. All other paths are shorter than path A-C-D-F, so the activities along those paths can be completed before the activities along path A-C-D-F. Thus, the shortest time to complete the project is 8 months (3 + 3 + 2).

Answers (A) and (B) are incorrect because the project cannot be completed in less than 8 months. Answer (D) is incorrect because no path through the network requires 14 months.

19. A shortest-route algorithm is used in network models to

A. Identify bottlenecks in a network and hence identify the longest path.

B. Minimize total travel time from one site to each of the other sites in a transportation system.

C. Maximize throughput in networks with distinct entry (source node) and exit (sink node) points.

D. Identify the set of connecting branches having the shortest combined length.

The correct answer is (B). *(CIA, adapted)*

REQUIRED: The purpose of a shortest-route algorithm used in network models.

DISCUSSION: Network models are used to solve managerial problems pertaining to project scheduling, information systems design, and transportation systems design. Networks consisting of nodes and arcs may be created to represent in graphic form problems related to transportation, assignment, and transshipment. The shortest-route, minimal spanning tree, and maximal flow problems are other applications of network models. A shortest-route algorithm minimizes total travel time from one site to each of the other sites in a transportation system.

Answer (A) is incorrect because the critical path method (CPM) is intended to identify bottlenecks in a network and hence identify the longest path. Answer (C) is incorrect because the maximal flow algorithm maximizes throughput in networks with distinct entry (source node) and exit (sink node) points. Examples of applications are highway transportation systems and oil pipelines. Flows are limited by capacities of the arcs (e.g., highways or pipes). Answer (D) is incorrect because the minimal spanning tree algorithm identifies the set of connecting branches having the shortest combined length. A spanning tree is a group of branches (arcs) that connects each node in the network to every other node. An example problem is the determination of the shortest telecommunications linkage among users at remote sites and a central computer.

B. Queuing Theory

20. Queuing models are concerned with balancing the cost of waiting in the queue with the

A. Cost of providing service.

B. Number of customers in the queue.

C. Average waiting time in the queue.

D. Utilization rate for the service being rendered.

The correct answer is (A). *(CMA, adapted)*

REQUIRED: The true statement about the objective of queuing models.

DISCUSSION: Two basic costs are involved in queuing (waiting-line) models: (1) the cost of providing service (including facility costs and operating costs) and (2) the cost of idle resources waiting in line. The latter may be a direct cost if paid employees are waiting or an opportunity cost in the case of waiting customers. The objective of the queuing theory is to minimize the total cost of the system, including both service and waiting costs, for a given rate of arrivals. This minimization occurs at the point where the cost of waiting is balanced by the cost of providing service.

Answers (B), (C) and (D) are incorrect because queuing theory is a cost minimization technique.

21. The operating condition that cannot be identified by using a queuing model is the

A. Average percentage of time that a service facility is idle.

B. Probability of a specified number of units in the queue.

C. Actual amount of time each unit spends in the queue.

D. Average number of units in the system.

The correct answer is (C). *(CMA, adapted)*

REQUIRED: The operating condition that cannot be identified by using a queuing model.

DISCUSSION: The objective of a queuing model is to minimize the total cost of a system, including both service and waiting costs, for a given rate of arrivals. The arrivals and services in a queuing model occur in accordance with a Poisson process; i.e., the probability of occurrence of an event is constant, and the occurrence is independent of what has happened immediately preceding the current observation. The Poisson probability distribution is used to predict the probability of the occurrence of a given number of events in a given time, given the expected number of events per unit of time and the length of time. The formula assumes a specified number of service facilities and random arrivals. Also, the formula calculates the average number in line or being serviced, the average waiting time, and the average percent of time that a service facility is idle. The actual waiting time cannot be determined from use of the model.

Answers (A) and (B) are incorrect because each can be calculated using the queuing model. Answer (D) is incorrect because the average number of units in the system (in line or being serviced) can be calculated.

22. A company is designing a new regional distribution warehouse. In order to minimize delays in loading and unloading trucks, an adequate number of loading docks must be built. The most relevant technique to assist in determining the proper number of docks is

A. Correlation and regression analysis.

B. Cost-volume-profit analysis.

C. PERT-cost analysis.

D. Queuing theory.

The correct answer is (D). *(CMA, adapted)*

REQUIRED: The technique for determining the optimal number of loading docks to minimize delays.

DISCUSSION: Queuing theory is a group of mathematical models for systems involving waiting lines. A queuing system consists of a waiting line and a service facility. The objective of queuing theory is to minimize total cost, including service and waiting costs, for a given rate of arrivals.

Answer (A) is incorrect because correlation and regression analysis is a means of determining the degree of the relationship among two or more variables. Answer (B) is incorrect because CVP analysis is used to analyze the relationship among fixed costs, variable costs, sales volume, and profit or loss. Answer (C) is incorrect because PERT-cost is a means of network analysis that assigns costs to all activities so that the effect of any change in a given task on cost as well as on time may be estimated.

23. A bank has changed from a system in which lines are formed in front of each teller to a one-line, multiple-server system. When a teller is free, the person at the head of the line goes to that teller. Implementing the new system will

A. Decrease the bank's wage expenses because the new system uses fewer tellers.

B. Decrease time customers spend in the line.

C. Increase accuracy in teller reconciliations at the end of the day because fewer customers are served by each teller.

D. Improve on-the-job training for tellers because each will perform different duties.

The correct answer is (B). *(CIA, adapted)*

REQUIRED: The effect of implementing the new queuing system.

DISCUSSION: When all customers must wait in a single queue, a decrease in waiting time is possible given multiple servers. An added effect is to increase customer satisfaction.

Answer (A) is incorrect because the number of employees is unlikely to change due to the new system. Answer (C) is incorrect because, assuming a Poisson process, the number of customers per teller will not change. Answer (D) is incorrect because tellers' duties will not change, so on-the-job training will not improve.

24. The drive-through service at a fast-food restaurant consists of driving up to place an order, advancing to a window to pay for the order, and then advancing to another window to receive the items ordered. This type of waiting-line system is

A. Single channel, single phase.
B. Single channel, multiple phase.
C. Multiple channel, single phase.
D. Multiple channel, multiple phase.

The correct answer is (B). *(CIA, adapted)*

REQUIRED: The type of waiting-line system described.

DISCUSSION: The drive-through represents a single queue (channel). Because this waiting line has three services in series, it may be said to be multiple phase. Another example is the typical factory assembly line. This terminology (channel, phase), however, is not used by all writers on queuing theory.

Answer (A) is incorrect because service by one ticket-seller at a movie theater is an example of a single-channel, single-phase system. Answer (C) is incorrect because supermarket checkout lines are a common example of multiple single-phase servers servicing multiple lines. Answer (D) is incorrect because an example of a multiple-channel, multiple-phase system is a set of supermarket checkout lines each of which is served in sequence by a cashier and a person who packs grocery bags.

25. A post office serves customers in a single line at one service window. During peak periods, the rate of arrivals has a Poisson distribution with an average of 100 customers per hour and service times that are exponentially distributed with an average of 60 seconds per customer. From this, one can conclude that the

A. Queue will expand to infinity.
B. Server will be idle one-sixth of the time.
C. Average rate is 100 customers per hour.
D. Average customer waiting time is 2.5 minutes.

The correct answer is (A). *(CIA, adapted)*

REQUIRED: The conclusion that can be reached about a queuing model.

DISCUSSION: One hundred customers arrive in line per hour and only 60 are serviced per hour. Accordingly, the queue will expand to infinity during peak periods.

Answers (B) and (D) are incorrect because insufficient information is given to determine overall idle time or average customer waiting time. The question gives only peak period data. Answer (C) is incorrect because peak customer service is only 60 per hour.

Questions 26 through 28 are based on the following information. A bank has two drive-in lanes to serve customers, one attached to the bank itself, the second on an island. One teller serves both stations. The bank is interested in determining the average waiting times of customers and has developed a model based on random numbers. The two key factors are the time between successive car arrivals and the time customers wait in line.

Assume that the analysis begins with cars just arriving at both service windows, both requiring 3 minutes of service time. Car 1 is at the attached window and car 2 at the island window. A car will always go to the window attached to the bank unless that window has more cars waiting than the island window. The lone teller will always serve the car that arrived first. If two cars arrive simultaneously, the one at the attached window will be served before the one at the island.

Based on a known probability distribution, the bank assigns random numbers to arrival and service times:

Random #	Time between Arrivals:	Random #	Service Time
1	1 minute	1,2	1 minute
2, 3	2 minutes	3	2 minutes
4, 5, 6	3 minutes	4, 5, 6	3 minutes
7, 8	4 minutes	7, 8, 9	4 minutes

The bank then selects random numbers for the next two cars as

Random Numbers Selected:

	Arrival	Service
Car 3	#3	#7
Car 4	#7	#8

26. The arrival time follows which probability distribution?

A. Binomial.

B. Chi-square.

C. Poisson.

D. Exponential.

The correct answer is (C). *(CIA, adapted)*

REQUIRED: The distribution of the arrival time.

DISCUSSION: Queuing models assume that arrivals follow a Poisson process: the events (arrivals) are independent, any number of events must be possible in the interval of time, the probability of an event is proportional to the length of the interval, and the probability of more than one event is negligible if the interval is sufficiently small. If λ is the average number of events in a given interval, k is the number of events, and e is the natural logarithm (2.71828...), the probability of k is

$$f(k) = \frac{\lambda^k e^{-\lambda}}{k!}$$

Answer (A) is incorrect because the binomial distribution is a discrete distribution in which each trial has just two outcomes. Answer (B) is incorrect because the chi-square distribution is a continuous distribution used to measure the fit between actual data and the theoretical distribution. Answer (D) is incorrect because service time has an exponential distribution. This distribution gives the probability of zero events in a given interval, i.e., the probability of a specified time between arrivals.

27. The time that car 3 will have to wait to be serviced (not including its own service time) is

A. 0-2 minutes.

B. 3 minutes.

C. 4 minutes.

D. 5+ minutes.

The correct answer is (C). *(CIA, adapted)*

REQUIRED: The time that car 3 will have to wait to be serviced (not including its own service time).

DISCUSSION: Car 1 is at the attached window and will require 3 minutes to service. Car 2 must wait for car 1 to be serviced (3 minutes in the queue + 3 minutes to be serviced = 6 minutes). Car 3 arrived at the attached window 2 minutes after cars 1 and 2. It must wait 1 minute for car 1 to be serviced and 3 minutes for car 2 to be serviced, a waiting time of 4 minutes.

Answers (A), (B), and (D) are incorrect because car 3 must wait for 4 minutes.

28. The time that car 4 will have to wait to be serviced (not including its own service time) is

A. 0-2 minutes.

B. 3 minutes.

C. 4 minutes.

D. 5+ minutes.

The correct answer is (C). *(CIA, adapted)*

REQUIRED: The time that car 4 will have to wait to be serviced (not including its own service time).

DISCUSSION: Car 4 arrives at the just-vacated island window 4 minutes after car 3. It must wait 4 minutes for car 3 to be serviced.

Answers (A), (B), and (D) are incorrect because car 4 must wait 4 minutes.

C. Forecasting

29. A regression equation

A. Estimates the dependent variables.

B. Encompasses factors outside the relevant range.

C. Is based on objective and constraint functions.

D. Estimates the independent variable.

The correct answer is (A). *(CMA, adapted)*

REQUIRED: The true statement about regression.

DISCUSSION: Regression analysis is used to find an equation for the linear relationship among variables. The behavior of the dependent variable is explained in terms of one or more independent variables. Regression analysis is often used to estimate a dependent variable (such as cost) given a known independent variable (such as production).

Answer (B) is incorrect because regression results are limited to observations within the relevant range. Answer (C) is incorrect because regression analysis does not use constraint functions. Answer (D) is incorrect because the dependent variable is estimated using regression analysis.

30. The correlation coefficient that indicates the weakest linear association between two variables is

A. −0.73

B. −0.11

C. 0.12

D. 0.35

The correct answer is (B). *(CMA, adapted)*

REQUIRED: The correlation coefficient that indicates the weakest linear association between two variables.

DISCUSSION: The correlation coefficient can vary from −1 to +1. A −1 relationship indicates a perfect negative correlation, and a +1 relationship indicates a perfect positive correlation. A zero correlation coefficient would indicate no association between the variables. Thus, the correlation coefficient that is nearest to zero would indicate the weakest linear association. Of the options given in the question, the correlation coefficient that is nearest to zero is −0.11.

Answer (A) is incorrect because −0.73 signifies a strong negative correlation. Answer (C) is incorrect because 0.12 is a slightly stronger correlation. Answer (D) is incorrect because 0.35 is a considerably stronger correlation.

31. Correlation is a term frequently used in conjunction with regression analysis and is measured by the value of the coefficient of correlation, **r**. The best explanation of the value **r** is that it

A. Is always positive.

B. Interprets variances in terms of the independent variable.

C. Ranges in size from negative infinity to positive infinity.

D. Is a measure of the relative relationship between two variables.

The correct answer is (D). *(CMA, adapted)*

REQUIRED: The best explanation of the coefficient of correlation (**r**).

DISCUSSION: The coefficient of correlation (**r**) measures the strength of the linear relationship between the dependent and independent variables. The magnitude of **r** is independent of the scales of measurement of x and y. The coefficient lies between −1.0 and +1.0. A value of zero indicates no linear relationship between the x and y variables. A value of +1.0 indicates a perfectly direct relationship, and a value of −1.0 indicates a perfectly inverse relationship.

Answer (A) is incorrect because the coefficient is negative if the relationship between the variables is inverse. Answer (B) is incorrect because the coefficient relates the two variables to each other. Answer (C) is incorrect because the size of the coefficient varies between −1.0 and +1.0.

32. Automite Company is an automobile replacement parts dealer in a large metropolitan community. Automite is preparing its sales forecast for the coming year. Data regarding both Automite's and industry sales of replacement parts as well as both the used and new automobile sales in the community for the last 10 years have been accumulated. If Automite wants to determine if its sales of replacement parts are patterned after the industry sales of replacement parts or to the sales of used and new automobiles, the company would employ

A. Simulation techniques.

B. Correlation and regression analysis.

C. Statistical sampling.

D. Time series analysis.

The correct answer is (B). *(CMA, adapted)*

REQUIRED: The technique to determine the variable to which sales are related.

DISCUSSION: Correlation and regression analysis can be used to determine if a relationship exists among two or more variables. The degree of that relationship is assessed by means of correlation analysis. Thus, regressing sales (the dependent variable) on both sales of replacement parts and sales of automobiles (independent variables) would determine the extent of the dependence.

Answer (A) is incorrect because simulation is a means of experimenting with logical or mathematical models using a computer. Answer (C) is incorrect because statistical sampling is a means of choosing and analyzing a sample to estimate population characteristics. Answer (D) is incorrect because time series or trend analysis regresses the dependent variable on time (the independent variable).

33. What coefficient of correlation results from the following data?

X	Y
1	10
2	8
3	6
4	4
5	2

A. 0

B. –1

C. +1

D. Cannot be determined from the data given.

The correct answer is (B). *(CIA, adapted)*

REQUIRED: The coefficient of correlation.

DISCUSSION: The coefficient of correlation (in standard notation, r) measures the strength of the linear relationship. The magnitude of r is independent of the scales of measurement of X and Y. Its range is –1.0 to 1.0. A value of –1.0 indicates a perfectly inverse linear relationship between X and Y. A value of zero indicates no linear relationship between X and Y. A value of +1.0 indicates a perfectly direct relationship between X and Y. As X increases by 1, Y consistently decreases by 2. Hence, a perfectly inverse relationship exists, and r must be equal to –1.0.

Answer (A) is incorrect because a perfect negative correlation exists. Answer (C) is incorrect because an inverse, not a direct, relationship exists. Answer (D) is incorrect because a linear relationship exists between X and Y.

34. In regression analysis, which of the following correlation coefficients represents the strongest relationship between the independent and dependent variables?

A. 1.03

B. –.02

C. –.89

D. .75

The correct answer is (C). *(CIA, adapted)*

REQUIRED: The correlation coefficient with the strongest relationship between independent and dependent variables.

DISCUSSION: Because the range of values is between –1.0 and 1.0, –.89 suggests a very strong inverse relationship between the independent and dependent variables. A value of –1.0 signifies a perfect inverse relationship, and a value of 1.0 signifies a perfect direct relationship.

Answer (A) is incorrect because 1.03 is an impossible value. Answer (B) is incorrect because –.02 is a very weak correlation coefficient. Answer (D) is incorrect because .75 is .25 from the maximum value, whereas –.89 is .11 from the minimum value.

35. The internal auditor of a bank has developed a multiple regression model which has been used for a number of years to estimate the amount of interest income from commercial loans. During the current year, the auditor applies the model and discovers that the r^2 value has decreased dramatically, but the model otherwise seems to be working okay. Which of the following conclusions are justified by the change?

A. Changing to a cross-sectional regression analysis should cause r^2 to increase.

B. Regression analysis is no longer an appropriate technique to estimate interest income.

C. Some new factors, not included in the model, are causing interest income to change.

D. A linear regression analysis would increase the model's reliability.

The correct answer is (C). *(CIA, adapted)*

REQUIRED: The implication of the decrease in r^2.

DISCUSSION: The coefficient of determination (r^2) is the amount of variation in the dependent variable (interest income) that is explained by the independent variables. In this case, less of the change in interest income is explained by the model. Thus, some other factor must be causing interest income to change. This change merits audit investigation.

Answer (A) is incorrect because cross-sectional regression analysis is inappropriate. The auditor is trying to estimate changes in a single account balance over time. Answer (B) is incorrect because regression analysis may still be the most appropriate methodology to estimate interest income, but the auditor should first understand the factors that may be causing r^2 to decrease. The reason may be a systematic error in the account balance. Answer (D) is incorrect because linear regression models are simpler models, but the auditor should be searching for a systematic error in the account balance or applying a more complex model.

Questions 36 through 39 are based on the following information. In preparing the annual profit plan for the coming year, Wilkens Company wants to determine the cost behavior pattern of the maintenance costs. Wilkens has decided to use linear regression by employing the equation $y = a + bx$ for maintenance costs. The prior year's data regarding maintenance hours and costs, and the results of the regression analysis are given below and in the opposite column.

Average cost per hour	$9.00
a	684.65
b	7.2884
Standard error of a	49.515
Standard error of b	.12126
Standard error of the estimate	34.469
r^2	.99724

	Hours of Activity	Maintenance Costs
January	480	$ 4,200
February	320	3,000
March	400	3,600
April	300	2,820
May	500	4,350
June	310	2,960
July	320	3,030
August	520	4,470
September	490	4,260
October	470	4,050
November	350	3,300
December	340	3,160
Sum	4,800	$43,200
Average	400	$3,600

36. In the standard regression equation $y = a + bx$, the letter b is best described as a(n)

A. Independent variable.

B. Dependent variable.

C. Constant coefficient.

D. Variable coefficient.

The correct answer is (D). *(CMA, adapted)*

REQUIRED: The meaning of the letter *b* in the standard regression equation.

DISCUSSION: In the standard regression equation, *b* represents the variable coefficient. For example, in a cost determination regression, *y* equals total costs, *b* is the variable cost per unit, *x* is the number of units produced, and *a* is fixed cost.

Answer (A) is incorrect because *x* is the independent variable. Answer (B) is incorrect because the dependent variable is *y*. Answer (C) is incorrect because the constant coefficient is *a*.

37. The letter *x* in the standard regression equation is best described as a(n)

A. Independent variable.

B. Dependent variable.

C. Constant coefficient.

D. Coefficient of determination.

The correct answer is (A). *(CMA, adapted)*

REQUIRED: The meaning of the letter *x* in the standard regression equation.

DISCUSSION: The letter *x* in the standard regression equation is the independent variable. For example, in a regression to determine the total cost of production, *x* equals units produced.

Answer (B) is incorrect because the dependent variable is *y*. Answer (C) is incorrect because the constant coefficient is *a*. Answer (D) is incorrect because r^2 is the coefficient of determination.

38. Based upon the data derived from the regression analysis, 420 maintenance hours in a month would mean the maintenance costs (rounded to the nearest dollar) would be budgeted at

A. $3,780

B. $3,600

C. $3,790

D. $3,746

The correct answer is (D). *(CMA, adapted)*

REQUIRED: The budgeted maintenance costs given the activity level.

DISCUSSION: Substituting the given data into the regression equation results in a budgeted cost of $3,746 (rounded to the nearest dollar).

$$y = a + bx$$
$$y = 684.65 + 7.2884(420)$$
$$y = \$3,746$$

Answers (A), (B), and (C) are incorrect because the budgeted maintenance costs are $3,746.

39. The percentage of the total variance that can be explained by the regression equation is

A. 99.724%

B. 69.613%

C. 80.982%

D. 99.862%

The correct answer is (A). *(CMA, adapted)*

REQUIRED: The percentage of the total variance that can be explained by the regression equation.

DISCUSSION: The coefficient of determination (r^2) measures the percentage of the total variance in cost that can be explained by the regression equation. If the coefficient of determination is .99724, 99.724% of the variance is explained by the regression equation. Thus, the values in the regression equation explain virtually the entire amount of total cost.

Answers (B), (C), and (D) are incorrect because the percentage of the total variance explained by the regression equation is 99.724%, which corresponds to the coefficient of determination (r^2), or .99724.

40. Jacob Corp. wishes to determine the fixed portion of its maintenance expense (a semivariable expense), as measured against direct labor hours, for the first 3 months of the year. The inspection costs are fixed; the adjustments necessitated by errors found during inspection account for the variable portion of the maintenance costs. Information for the first quarter is as follows:

	Direct Labor Hours	Maintenance Expense
January	34,000	$610
February	31,000	$585
March	34,000	$610

What is the fixed portion of Jacob's maintenance expense, rounded to the nearest dollar?

A. $283

B. $327

C. $258

D. $541

The correct answer is (B). *(Publisher)*

REQUIRED: The fixed portion of expense as measured against direct labor hours.

DISCUSSION: The high-low method can be used to determine the fixed and variable cost components of a mixed cost. The variable cost is found by dividing the change in total cost (TC) by the change in activity, e.g., DLH. The fixed cost is found by substituting the variable cost into either of the activity/cost functions. Alternatively, the fixed cost is the cost given a zero level of activity.

$$\frac{\text{Change in TC}}{\text{Change in DLH}} \quad \frac{\$25}{3{,}000} = \$.00833$$

$$FC = TC - VC$$
$$FC = \$585 - (31{,}000 \times \$.00833) = \$327$$
$$FC = \$610 - (34{,}000 \times \$.00833) = \$327$$

Answer (A) is incorrect because $283 equals variable cost for January or March. Answer (C) is incorrect because $258 equals variable cost for February. Answer (D) is incorrect because $541 equals variable cost for February and March.

41. An internal auditor for a large automotive parts retailer wishes to perform a risk analysis and wants to use an appropriate statistical tool to help identify stores that would be out of line compared to the majority of stores. The most appropriate statistical tool to use is

A. Linear time series analysis.

B. Cross-sectional regression analysis.

C. Cross tabulations with chi-square analysis of significance.

D. Time series multiple regression analysis to identify changes in individual stores over time.

The correct answer is (B). *(CIA, adapted)*

REQUIRED: The best statistical tool for identifying stores that deviate from the majority.

DISCUSSION: Time series data pertain to a given entity over a number of prior time periods. Cross-sectional data, however, pertain to different entities for a given time period or at a given time. Thus, cross-sectional regression analysis is the most appropriate statistical tool because it compares attributes of all stores' operating statistics at one moment in time.

Answer (A) is incorrect because linear time series analysis is inapplicable. It is a simple model that compares data for an individual store over time. Answer (C) is incorrect because cross tabulations have to be built on a model of expectations. Unless the model is built, the analysis is not useful. Answer (D) is incorrect because the objective is to compare stores at one moment in time. Multiple regression time series analysis compares the performance of an individual store over a period of time.

42. Automite Company is an automobile replacement parts dealer in a large metropolitan community. Automite is preparing its sales forecast for the coming year. Data regarding both Automite's and industry sales of replacement parts as well as both the used and new automobile sales in the community for the last 10 years have been accumulated. If Automite wants to determine if there is a historical trend in the growth of its sales as well as the growth of industry sales of replacement parts, the company would employ

A. Simulation techniques.

B. Queuing theory.

C. Statistical sampling.

D. Time series analysis.

The correct answer is (D). *(CMA, adapted)*

REQUIRED: The technique to determine if there is a historical trend in the growth of both company and industry sales.

DISCUSSION: Time series analysis would be appropriate because it relies on past experience. Changes in the value of a variable (such as sales) may have several possible components, e.g., a change in demand for the product or a change in market share. These changes are regressed on time (the independent variable).

Answer (A) is incorrect because simulation is a means of experimenting with logical or mathematical models using a computer. Answer (B) is incorrect because queuing theory is used to minimize the cost of waiting lines. Answer (C) is incorrect because statistical sampling is a means of estimating the characteristics of a population by examining a few of its units.

43. The four components of time series data are secular trend, cyclical variation, seasonality, and random variation. The seasonality in the data can be removed by

A. Multiplying the data by a seasonality factor.

B. Ignoring it.

C. Taking the weighted average over four time periods.

D. Subtracting a seasonality factor from the data.

The correct answer is (C). *(CMA, adapted)*

REQUIRED: The means by which seasonality can be removed from time series data.

DISCUSSION: Time series analysis relies on past experience. Changes in the value of a variable may have several possible components including secular trends, cyclical variation, seasonality, and random variation. Seasonal variations are common in many businesses. A variety of methods exist for including seasonal variations in a forecasting model, but most methods use a seasonal index. Alternatively, seasonal variations can be removed from data by using a weighted average of several time periods instead of data from individual periods.

Answer (A) is incorrect because adding a seasonality factor to, or subtracting it from, a forecast based on trend analysis is a means of adjusting for seasonality. Answer (B) is incorrect because seasonality factors cannot be ignored; they are reflected in the data and must be considered for a model to be accurate. Answer (D) is incorrect because the seasonality adjustment for a single season's data may be an increase or a decrease.

44. Violation of which assumption underlying regression analysis is prevalent in time series analysis?

A. Variance of error term is constant.

B. Error terms are independent.

C. Distribution of error terms is normal.

D. Expected value of error term equals zero.

The correct answer is (B). *(Publisher)*

REQUIRED: The assumption frequently violated in time series analysis.

DISCUSSION: Time series analysis is a regression model in which the independent variable is time. In time series analysis, the value of the next time period is frequently dependent on the value of the time period before that. Hence, the error terms are usually correlated or dependent on the prior period; i.e., they are characterized by autocorrelation (serial correlation).

Answers (A), (C), and (D) are incorrect because the other three major requirements of regression analysis -- constant variance of the error term, normal distribution of the error term, and an expected value of the error term equal to zero -- are usually met.

45. The moving-average method of forecasting

A. Is a cross-sectional forecasting method.

B. Regresses the variable of interest on a related variable to develop a forecast.

C. Derives final forecasts by adjusting the initial forecast based on the smoothing constant.

D. Includes each new observation in the average as it becomes available and discards the oldest observation.

The correct answer is (D). *(CIA, adapted)*

REQUIRED: The item that best describes the moving-average method of forecasting.

DISCUSSION: The simple moving-average method is a smoothing technique that uses the experience of the past N periods (through time period t) to forecast a value for the next period. Thus, the average includes each new observation and discards the oldest observation. The forecast formula for the next period (for time period t+1) is the sum of the last N observations divided by N.

Answer (A) is incorrect because cross-sectional regression analysis examines relationships among large amounts of data (e.g., many or different production methods or locations) at a particular moment in time. Answer (B) is incorrect because regression analysis relates the forecast to changes in particular variables. Answer (C) is incorrect because, under exponential smoothing, each forecast equals the sum of the last observation times the smoothing constant, plus the last forecast times one minus the constant.

46. A particular manufacturing job is subject to an estimated 90% learning curve. The first unit required 50 labor hours to complete. What is the cumulative average time per unit after four units are completed?

A. 50.0 hours.

B. 45.0 hours.

C. 40.5 hours.

D. 40.0 hours.

The correct answer is (C). *(Publisher)*

REQUIRED: The cumulative average completion time per unit after four units are produced.

DISCUSSION: The learning curve reflects the increased rate at which people perform tasks as they gain experience. The time required to perform a given task becomes progressively shorter. Ordinarily, the curve is expressed in a percentage of reduced time to complete a task for each doubling of cumulative production. One common assumption in a learning curve model is that the cumulative average time (and labor cost) per unit is reduced by a certain percentage each time production doubles. Thus, a 90% learning curve indicates that a doubling of production will reduce the cumulative average unit completion time by 10%. For example, if the first unit required 50 hours to complete, the average completion time after two units will be 45 hours (90% x 50 hours). If production is again doubled (to four units), the cumulative average completion time will be 40.5 hours (90% x 45 hours).

Answer (A) is incorrect because 50 hours is the completion time for the first unit. Answer (B) is incorrect because 45 is the cumulative average completion time after production of two units. Answer (D) is incorrect because 40.0 hours assumes two successive reductions of 5 hours [50 – (90% x 50)] each.

47. As part of a risk analysis, an auditor wishes to forecast the percentage growth in next month's sales for a particular plant using the past 30 months' sales results. Significant changes in the organization affecting sales volumes were made within the last 9 months. The most effective analysis technique to use would be

A. Unweighted moving average.

B. Exponential smoothing.

C. Queuing theory.

D. Linear regression analysis.

The correct answer is (B). *(CIA, adapted)*

REQUIRED: The most effective analysis technique to forecast the percentage growth in next month's sales.

DISCUSSION: Under exponential smoothing, each forecast equals the sum of the last observation times the smoothing constant, plus the last forecast times one minus the constant. Thus, exponential means that greater weight is placed on the most recent data, with the weights of all data falling off exponentially as the data age. This feature is important because of the organizational changes that affected sales volume.

Answer (A) is incorrect because an unweighted average will not give more importance to more recent data. Answer (C) is incorrect because queuing theory is used to minimize the cost of waiting lines. Answer (D) is incorrect because linear regression analysis determines the equation for the relationship among variables. It does not give more importance to more recent data.

Questions 48 and 49 are based on the following information. Moss Point Manufacturing recently completed and sold an order of 50 units that had costs as shown in the next column.

The company has now been requested to prepare a bid for 150 units of the same product.

Direct materials	$ 1,500
Direct labor (1,000 hours × $8.50)	8,500
Variable overhead (1,000 hours × $4.00)*	4,000
Fixed overhead**	1,400
	$15,400

*Applied on the basis of direct labor hours
**Applied at the rate of 10% of variable cost

48. If an 80% learning curve is applicable, Moss Point's total cost on this order would be estimated at

A. $26,400

B. $32,000

C. $38,000

D. $41,800

The correct answer is (A). *(CMA, adapted)*

REQUIRED: The total cost of a new order given a learning curve percentage.

DISCUSSION: Assuming that the cumulative average-time model applies, an 80% learning curve means that the cumulative average time per unit (and labor cost, given a constant labor rate) declines by 20% when unit output doubles in the early stages of production. The first lot size was 50 units, which was produced at a total cost of $15,400 ($1,500 for materials and $13,900 for labor and overhead). Materials costs are strictly variable and should remain proportional to production. The labor ($8,500) and variable overhead ($4,000) costs (labor-related), however, will be affected by the learning curve. The average cost per lot for labor and variable overhead after 100 units have been produced should be 80% of the costs of the first lot of 50 units. Thus, the average labor and variable overhead cost per 50-unit lot will be $10,000 (80% x $12,500). If production doubles again (to a total production of 200 units or four lots of 50 each), the cumulative average cost for labor and variable overhead will be $8,000 per lot (80% x $10,000). Given four lots of 50 each, at an average cost of $8,000 per lot, the total cost for labor and variable overhead must be $32,000. Adding $6,000 for raw materials ($1,500 per 50-unit lot) gives a total variable cost of $38,000 for 200 units. Fixed overhead is 10% of total variable cost, so total cost is $41,800. The total cost for the last 150 units is $26,400 ($41,800 – $15,400).

Answer (B) is incorrect because $32,000 is the total cost for labor and variable overhead for 200 units. Answer (C) is incorrect because $38,000 is the total variable cost for 200 units. Answer (D) is incorrect because $41,800 is the total cost for 200 units.

49. If Moss Point had experienced a 70% learning curve, the bid for the 150 units would

A. Show a 30% reduction in the total direct labor hours required with no learning curve.

B. Include increased fixed overhead costs.

C. Be 10% lower than the total bid at an 80% learning curve.

D. Include 6.40 direct labor hours per unit at $8.50 per hour.

The correct answer is (D). *(CMA, adapted)*

REQUIRED: The true statement about the bid for an incremental 150 units given a 70% learning curve effect.

DISCUSSION: The sum of the direct labor hours for the initial lot of 50 units was 1,000. A second lot of 50 would reduce the cumulative hours per lot to 700 (70% x 1,000). A doubling to four lots would reduce the cumulative hours per lot to 490 (70% x 700). Thus, for an output of 200 units, the total hours worked would be 1,960 (4 x 490). Subtracting the 1,000 hours required for the first 50 units from the 1,960-hour total gives 960 hours for the last 150 units. Dividing 960 hours by 150 units produces a per-unit time of 6.4 hours.

Answers (A) and (C) are incorrect because, with no learning curve effect, estimated total hours would be 4,000 instead of 1,960, a change of more than 50%. Answer (B) is incorrect because fixed costs applied per lot would decline because they are based on labor hours, which are declining.

50. The average labor cost per unit for the first batch produced by a new process is $120. The cumulative average labor cost after the second batch is $72 per product. Using a batch size of 100 and assuming the learning curve continues, the total labor cost of four batches will be

A. $4,320

B. $10,368

C. $2,592

D. $17,280

The correct answer is (D). *(CMA, adapted)*

REQUIRED: The total labor cost assuming the learning curve continues.

DISCUSSION: The learning curve reflects the increased rate at which people perform tasks as they gain experience. The time required to perform a given task becomes progressively shorter. Ordinarily, the curve is expressed in a percentage of reduced time to complete a task for each doubling of cumulative production. One common assumption in a learning curve model is that the cumulative average time (and labor cost) per unit is reduced by a certain percentage each time production doubles. Given a $120 cost per unit for the first 100 units and a $72 cost per unit when cumulative production doubled to 200 units, the learning curve percentage must be 60% ($72 ÷ $120). If production is again doubled to 400 units (four batches), the average unit labor cost should be $43.20 (60% x $72). Hence, total labor cost for 400 units is estimated to be $17,280 (400 x $43.20).

Answer (A) is incorrect because $4,320 equals the cost of the items in the fourth batch. Answer (B) is incorrect because $10,368 is based on the assumption that the cumulative average unit labor cost is reduced by the learning curve percentage with each batch, not each doubling of output. Answer (C) is incorrect because $2,592 represents the labor cost of 100 units at the unit rate expected after another doubling of production to eight batches.

51. A particular manufacturing job is subject to an estimated 80% learning curve. The first unit required 50 labor hours to complete. What is the cumulative average time per unit after eight units are completed?

A. 20.0 hours.

B. 25.6 hours.

C. 32.0 hours.

D. 40.0 hours.

The correct answer is (B). *(Publisher)*

REQUIRED: The cumulative average completion time per unit if 8 units are produced.

DISCUSSION: The learning curve reflects the increased rate at which people perform tasks as they gain experience. The time required to perform a given task becomes progressively shorter. Ordinarily, the curve is expressed in a percentage of reduced time to complete a task for each doubling of cumulative production. One common assumption in a learning curve model is that the cumulative average time (and labor cost) per unit is reduced by a certain percentage each time production doubles. Thus, an 80% learning curve indicates that a doubling of production will reduce the cumulative average unit completion time by 20%. For example, if the first unit required 50 hours to complete, the average completion time after two units will be 40 hours (80% x 50 hours). If production is again doubled (to four units), the average completion time will be 32 hours (80% x 40 hours). When production is doubled again to eight units, the cumulative average completion time will be 25.6 hours (80% x 32 hours).

Answer (A) is incorrect because 20.0 hours assumes three successive reductions of 10 hours [50 – (80% x 50)] each. Answer (C) is incorrect because 32 is the average completion time after four units have been produced. Answer (D) is incorrect because 40 is the average time after two units have been produced.

52. A particular manufacturing job is subject to an estimated 80% learning curve. The first unit required 50 labor hours to complete. If the learning curve is based on a cumulative average time per unit assumption, what is the time required to complete the second unit?

A. 30.0 hours.

B. 40.0 hours.

C. 45.0 hours.

D. 50.0 hours.

The correct answer is (A). *(Publisher)*

REQUIRED: The cumulative average completion time per unit after two units are produced.

DISCUSSION: The learning curve reflects the increased rate at which people perform tasks as they gain experience. The time required to perform a given task becomes progressively shorter. Ordinarily, the curve is expressed in a percentage of reduced time to complete a task for each doubling of cumulative production. One common assumption in a learning curve model is that the cumulative average time (and labor cost) per unit is reduced by a certain percentage each time production doubles. Thus, an 80% learning curve indicates that a doubling of production will reduce the cumulative average unit completion time by 20%. For example, if the first unit required 50 hours to complete, the average completion time after two units will be 40 hours (80% x 50 hours). If total production time is 80 hours (2 x 40 cumulative average time), and the first unit required 50 hours, the time to produce the second unit must be 30 hours.

Answer (B) is incorrect because 40.0 hours is based on the assumption that the time to produce the last unit (the incremental unit-time assumption) is reduced by 20%. Answer (C) is incorrect because 45.0 hours is based on a 90% learning curve and the incremental unit-time assumption. Answer (D) is incorrect because 50.0 hours is the completion time for the first unit.

53. Corrigon Industries is preparing a bid for a special project requiring the production of 35,000 units. The engineering personnel have advised that the units can be produced in groups with the first group consisting of 1,000 units. A review of prior experience indicates that the direct labor time needed per unit will be progressively smaller by a constant percentage rate as experience is gained in the production process. The quantitative method that would best estimate Corrigon's total cost for the project is

A. Linear programming.

B. Dynamic programming.

C. Learning curve analysis.

D. Time series analysis.

The correct answer is (C). *(CMA, adapted)*

REQUIRED: The quantitative method for estimating the decline in direct labor costs as production increases.

DISCUSSION: Learning curves reflect the increased rate at which people perform tasks as they gain experience. Thus, the time required to perform a given task becomes progressively shorter. Ordinarily, the learning curve is expressed as a percentage of reduced time to complete a task for each doubling of cumulative production.

Answer (A) is incorrect because linear programming is an optimizing model used to determine a minimum or maximum, e.g., of a cost or revenue function, given certain constraints on resources. Answer (B) is incorrect because dynamic programming is an approach to solving problems, not a particular algorithm. It divides a large mathematical model into smaller, more manageable pieces in such a way that, once the smaller problems have been solved, the result is the optimal solution to the overall model. Answer (D) is incorrect because time series analysis applies to data gathered at successive moments in time. It is a forecasting technique in which the dependent variable is regressed on time.

D. Game Theory

54. One of your firm's trucks was involved in an accident with a car, and both parties claimed the other was at fault. A court ruled that your firm was solely responsible for all damage. From a game theory perspective, this dispute can be described as a

A. Prisoner's dilemma.

B. Zero-sum game.

C. Saddle point.

D. Minimax regret.

The correct answer is (B). *(CIA, adapted)*

REQUIRED: The term for the dispute from a game theory perspective.

DISCUSSION: Game (or decision) theory is a mathematical approach to decision making when confronted with an "enemy" or competitor. Games are classified according to the number of players and the algebraic sum of the payoffs. In a two-person game, if the payoff is given by the loser to the winner, the algebraic sum is zero and the game is called a zero-sum game. If it is possible for both players to profit, however, it is called a positive-sum game.

Answer (A) is incorrect because the prisoner's dilemma is a special outcome of a partly competitive game. In these games, each player has a strategy that dominates all others, and the outcome from each player's choice of his/her dominant strategy is less favorable to both players than some other outcome. Answer (C) is incorrect because a saddle point is an optimal strategy point. Answer (D) is incorrect because the minimax regret decision criterion chooses the strategy that minimizes the maximum opportunity cost (benefit forgone).

55. Only two companies manufacture Product A. The finished product is identical regardless of which company manufactures it. The cost to manufacture Product A is $1, and the selling price is $2. One company considers reducing the price to achieve 100% market share but fears the other company will respond by further reducing the price. Such a scenario would involve a

A. No-win strategy.

B. Dual-win strategy.

C. One win-one lose strategy.

D. Neutral strategy.

The correct answer is (A). *(CIA, adapted)*

REQUIRED: The effect of a price war.

DISCUSSION: If both firms reduce the selling price of Product A, neither will gain sales, and the resultant price war will cause both firms to earn lower profits. This outcome is inevitable when reduced profit margins do not result in a significant increase in sales. The effect is a no-win strategy.

Answers (B), (C), and (D) are incorrect because both firms will experience lower profits.

56. The minimax regret decision rule is applicable to decision making under conditions of

A. Risk.

B. Uncertainty.

C. Subjective likelihoods.

D. Objective likelihoods.

The correct answer is (B). *(CMA, adapted)*

REQUIRED: The conditions under which the minimax regret decision rule is applicable.

DISCUSSION: The minimax regret decision rule is applicable under conditions of uncertainty, that is, when the probabilities of future consequences are not known. Using this criterion, management constructs a decision table based on opportunity costs for each possible action and state of nature that may exist and then selects the act that minimizes the maximum possible opportunity cost.

Answer (A) is incorrect because, in decision making under risk, the probabilities of the consequences of possible actions are known with a higher degree of confidence. Answers (C) and (D) are incorrect because probabilities can be determined objectively or subjectively.

57. The procedure for choosing the smallest maximum alternative loss is

A. Deterministic decision making.

B. Queuing theory.

C. Maximax.

D. Minimax.

The correct answer is (D). *(CMA, adapted)*

REQUIRED: The procedure for choosing the smallest maximum alternative loss.

DISCUSSION: In game theory, the minimax decision criterion selects the strategy that will minimize the maximum possible loss. It is a technique used by a risk-averse player. The maximin criterion, which chooses the strategy with the maximum minimum payoff, gives the same results as the minimax procedure.

Answer (A) is incorrect because deterministic decision making is based upon fixed (nonprobabilistic) inputs into the decision process. Answer (B) is incorrect because queuing theory is used to minimize the sum of waiting and service costs for a waiting line. Answer (C) is incorrect because maximax is a criterion chosen by a risk-seeking player who prefers the strategy that maximizes the maximum possible profit.

Questions 58 through 60 are based on the following information. A bank plans to open a branch in one of five locations (labeled L1, L2, L3, L4, L5). Demand for bank services may be high, medium, or low at each of these locations. Profits for each location-demand combination are presented in the payoff matrix in the right-hand column.

Payoff Matrix

Location	L1	L2	L3	L4	L5
Demand:					
High	15	21	17	26	29
Medium	12	8	14	10	4
Low	7	–2	4	–3	–6

58. If the bank uses the maximax criterion for selecting the location of the branch, it will select

A. L1.

B. L2.

C. L3.

D. L5.

The correct answer is (D). *(CIA, adapted)*

REQUIRED: The location selected assuming a maximax criterion.

DISCUSSION: Under the maximax criterion, the decision maker selects the choice that maximizes the maximum profit. The maximum profits for the five locations are

Location	L1	L2	L3	L4	L5
Maximum Profit	15	21	17	26	29

The location with the greatest potential profit is L5.

Answers (A), (B), and (C) are incorrect because the location with the greatest potential profit is L5.

59. If the bank uses the minimax regret criterion for selecting the location of the branch, it will select

A. L2.

B. L3.

C. L4.

D. L5.

The correct answer is (A). *(CIA, adapted)*

REQUIRED: The location selected using the minimax regret criterion.

DISCUSSION: Under the minimax regret criterion, the decision maker selects the choice that minimizes the maximum regret (opportunity loss). The maximum regret for each location is determined from the opportunity loss matrix (see Discussion of the next question). The maximum regret for each location is the highest number in each column as indicated below.

Location	L1	L2	L3	L4	L5
Maximum Regret	14	9	12	10	13

The location with the minimum regret is L2. If demand is low, L2 has a payoff of –2, whereas L1 has a payoff of 7.

Answers (B), (C), and (D) are incorrect because the location with the minimum regret is L2.

60. If, in addition to the estimated profits, management of the bank assesses the probabilities of high, medium, and low demands to be 0.3, 0.4, and 0.3, respectively, what is the expected opportunity loss from selecting location L4?

A. 5.50

B. 10.90

C. 5.67

D. 7.00

The correct answer is (A). *(CIA, adapted)*

REQUIRED: The expected opportunity loss from selecting location L4.

DISCUSSION: The opportunity loss matrix is as follows:

Location	L1	L2	L3	L4	L5
Demand:					
High	14 = 29–15	8 = 29–21	12 = 29–17	3 = 29–26	0 = 29–29
Medium	2 = 14–12	6 = 14–8	0 = 14–14	4 = 14–10	10 = 14–4
Low	0 = 7–7	9 = 7–(–2)	3 = 7–4	10 = 7–(–3)	13 = 7–(–6)

The expected opportunity loss from selecting location L4 is 5.5 [(3 x 0.3) + (4 x 0.4) + (10 x 0.3)].

Answers (B), (C), and (D) are incorrect because the expected opportunity loss from selecting location L4 is 5.5.

61. Following is a table of probabilities for two separate product lines, X and Y:

Probability	X profit	Y profit
.20	$500	$ 50
.70	300	400
.10	600	800

The product line to obtain maximum utility for a risk-averse decision maker is

A. X because it has the highest expected profit.

B. Y because it has the highest expected profit.

C. Y because it has the highest dispersion.

D. X because it has the lowest dispersion.

The correct answer is (D). *(CIA, adapted)*

REQUIRED: The product line to obtain maximum utility for a risk-averse decision maker.

DISCUSSION: For a risk-averse decision maker, the utility of a gain is less than the disutility of an equal loss. A risk-averse decision maker therefore tends to prefer a stable pattern of returns, in this case, a lower maximum return and a higher minimum return. Accordingly, X should be chosen because the dispersion of values is lower than that of Y.

Answers (A) and (B) are incorrect because the expected values of X and Y are equal. Answer (C) is incorrect because a risk-averse decision maker would want the least amount of risk for a given expected value of return. Given that the expected values are the same, the product with the least risk should be chosen.

62. The decision rule that selects the strategy with the highest utility payoff if the worst state of nature occurs is the

A. Minimize regret rule.

B. Maximize utility rule.

C. Maximin rule.

D. Maximax rule.

The correct answer is (C). *(CIA, adapted)*

REQUIRED: The decision rule that selects the strategy with the highest utility payoff if the worst state of nature occurs.

DISCUSSION: According to game theory, a player who uses the minimax decision criterion determines the maximum loss for each decision possibility and then chooses the decision with the minimum maximum loss. This rule produces the same result as the maximin technique, which determines the minimum payoff for each decision and then chooses the decision with the maximum minimum payoff. Minimax and maximin are conservative criteria adopted by risk-averse players, that is, those for whom the disutility of a loss exceeds the utility of an equal gain.

Answer (A) is incorrect because the minimize regret rule selects the action or strategy that minimizes the sum of the regrets for the strategy. Answer (B) is incorrect because the maximize utility rule is not a decision rule. Answer (D) is incorrect because the maximax rule selects the strategy or choice that provides the greatest payoff if the most favorable state of nature occurs.

63. Under conditions of uncertainty, the rational, economic decision maker will use which one of the following decision criteria?

A. Maximax.

B. Minimum regret.

C. Laplace.

D. Expected monetary value.

The correct answer is (D). *(CMA, adapted)*

REQUIRED: The decision criterion that will be used by a rational decision maker under conditions of uncertainty.

DISCUSSION: A rational economic decision maker (one completely guided by objective criteria) will use expected monetary value to maximize gains under conditions of uncertainty because (s)he is risk-neutral (the utility of a gain equals the disutility of a loss of the same absolute amount). For decisions involving variability in outcomes, the concept of expected value provides a rational means for selecting the best choice. Expected value represents the long-term average payoff for repeated trials. The best choice is the one having the highest expected value (sum of the products of the possible outcomes and their respective probabilities).

Answer (A) is incorrect because, in game theory, the maximax criterion is a decision rule adopted by risk-seeking, optimistic players who desire the largest possible payoff and are willing to accept high risk. The player determines the payoff for each state of nature expected to arise after each possible decision, ascertains the maximum payoff for each decision, and then chooses the decision with the maximum payoff. Answer (B) is incorrect because the minimum regret criterion is used by a player who wishes to minimize the effect of a bad decision in either direction. It chooses the decision that has the lowest maximum opportunity cost (benefit forgone). Answer (C) is incorrect because the insufficient reason (Laplace) criterion applies when the decision maker cannot assign probabilities to the states of nature arising after a decision. The reasoning is that, if no probability distribution can be assigned, the probabilities must be equal, and the expected value is calculated accordingly. This criterion is risk-neutral.

E. Markov Analysis

64. The manager of a car rental agency is planning for counter help during the next 2 weeks. He knows the approximate number of cars that will be rented, the probability of a car's being returned to his agency if rented from his agency, and the probability of its being returned to his agency if rented from another location. The quantitative method that the manager would find most helpful is

A. A transportation model.

B. Markov analysis.

C. Calculus-based optimization.

D. The critical path method.

The correct answer is (B). *(CMA, adapted)*

REQUIRED: The quantitative method helpful in determining the employees needed by a car rental agency.

DISCUSSION: Markov analysis is useful in decision problems in which the transition probability of the occurrence of a future state depends only on the current state. Thus, it is employed to analyze how certain systems evolve over repeated trials, with the initial state mattering less and less as time passes. The objective is to estimate the probability that the system will be in a given state at a moment in the future. This problem could probably also have been solved using queuing theory, but that was not one of the answer choices.

Answer (A) is incorrect because a transportation model is a form of linear programming used to allocate shipments most economically. Transition probabilities are not used in a transportation model. Answer (C) is incorrect because calculus is used to optimize a nonlinear model, for example, a cost function. Answer (D) is incorrect because the critical path method is used to plan and control large projects. The longest path through the network of activities is the critical path.

65. The Markov process

A. Assumes that the probabilities of an event depend only on the current state.

B. Assumes that the probabilities of an event are dependent on all previous events.

C. Is useful if sequencing of related decisions is necessary.

D. Is used to assess the implications of decisions in a competitive environment based on trend analysis.

The correct answer is (A). *(CMA, adapted)*

REQUIRED: The true statement about the Markov process.

DISCUSSION: Markov processes are useful in decision problems in which the probability of the occurrence of a future state depends only on the current state. A characteristic of the Markov process is that the initial state matters less and less as time goes on because the process will eventually reach its steady state.

Answer (B) is incorrect because Markov processes eventually reach a steady state; earlier probabilities are less important. Answer (C) is incorrect because sequencing is not important in a Markov process. Answer (D) is incorrect because earlier trends are weighted less than later trends.

F. Simulation and Sensitivity Analysis

66. A company is simulating the actions of a government agency in which 50% of the time a recall of a product is required, 40% of the time only notification of the buyer about a potential defect is required, and 10% of the time no action on its part is required. Random numbers of 1 to 100 are being used. An appropriate assignment of random numbers for the recall category would be

A. 1-40

B. 40-90

C. 61-100

D. 11-60

The correct answer is (D). *(CMA, adapted)*

REQUIRED: The appropriate random numbers to be assigned to an alternative possibility.

DISCUSSION: Given a 50% chance of a recall, 50 different numbers should be assigned to that alternative. Answer (D) is the only alternative with 50 numbers (11-60).

Answer (A) is incorrect because 1-40 is an appropriate assignment of random numbers for the notification category. Answer (B) is incorrect because 40-90 includes 51 numbers. Answer (C) is incorrect because 61-100 is an appropriate assignment of random numbers for the notification category.

67. Quick Response Plumbing (QRP), a wholesale distributor, supplies plumbing contractors and retailers throughout the Northeast on a next-day delivery basis. QRP has a centrally located warehouse to accept receipts of plumbing supplies. The warehouse has a single dock to accept and unload railroad freight cars during the night. It takes five hours to unload each freight car. QRP's prior records indicate that the number of freight cars that arrive in the course of a night range from zero to five or more, with no indicated pattern of arrivals. If more than two freight cars arrive on the same night, some freight must be held until the next day for unloading. QRP wants to estimate the wait time when more than two freight cars arrive in the same night. The appropriate technique to analyze the arrival of freight cars is

A. Integer programming.

B. Linear programming.

C. Monte Carlo simulation.

D. Regression analysis.

The correct answer is (C). *(CMA, adapted)*

REQUIRED: The appropriate technique to analyze the arrival of freight cars given no indicated pattern of arrivals.

DISCUSSION: The Monte Carlo simulation method is often used to generate the individual values for a random variable. The performance of a quantitative model under uncertainty may be investigated by randomly selecting values for each variable in the model (based on the probability distribution of each variable) and then calculating the value of the solution. If this process is performed many times, the distribution of results from the model will be obtained.

Answer (A) is incorrect because integer programming is a variation of linear programming that concerns problems in which some variables are not continuous. Integer programming problems are also known as discrete models because the variables take on discrete, noncontinuous values. Answer (B) is incorrect because linear programming is a technique used to maximize a revenue or profit function, or minimize a cost function, subject to constraints such as limited resources or minimum (or maximum) levels of production time. Answer (D) is incorrect because regression analysis is used to find an equation for the linear relationships among variables.

68. When simulating with the Monte Carlo technique, the average simulated demand over the long run should approximate the

A. Actual demand.

B. Real demand.

C. Sampled demand.

D. Expected demand.

The correct answer is (D). *(CMA, adapted)*

REQUIRED: The approximation of the average simulated long-term demand using the Monte Carlo technique.

DISCUSSION: The Monte Carlo simulation method is often used to generate the individual values for a random variable. The performance of a quantitative model under uncertainty may be investigated by randomly selecting values for each variable in the model (based on the probability distribution of each variable), and then calculating the value of the solution. If this process is performed many times, the distribution of results from the model will be obtained. Thus, random numbers are assigned so that the probability of an outcome in the model is the same as that in the actual circumstances. Average simulated long-term demand must therefore equal expected demand.

Answers (A) and (B) are incorrect because actual (real) demand is not known until it has occurred. Simulation is a forecasting tool. Answer (C) is incorrect because sampling actual demand for prior periods does not necessarily yield the expected demand given changes in future circumstances.

69. A widely used approach that managers use to recognize uncertainty about individual items and to obtain an immediate financial estimate of the consequences of possible prediction errors is

A. Expected value analysis.

B. Learning curve analysis.

C. Sensitivity analysis.

D. Regression analysis.

The correct answer is (C). *(CMA, adapted)*

REQUIRED: The approach that gives an immediate financial estimate of the consequences of possible prediction errors.

DISCUSSION: Sensitivity analysis determines how a result varies with changes in a given variable or parameter in a mathematical decision model. For example, in a present value analysis, a manager might first calculate the net present value or internal rate of return assuming that a new asset has a 10-year life. The NPV or IRR can then be recalculated using a 5-year life to determine how sensitive the result is to the change in the assumption.

Answer (A) is incorrect because expected value is the probabilistically weighted average of the outcomes of an action. Answer (B) is incorrect because learning curve analysis quantifies how labor costs decline as employees learn their jobs through repetition. Answer (D) is incorrect because regression, or least squares, analysis determines the average change in the dependent variable given a unit change in one or more independent variables.

70. Through the use of decision models, managers thoroughly analyze many alternatives and decide on the best alternative for the company. Often the actual results achieved from a particular decision are not what was expected when the decision was made. In addition, an alternative that was not selected would have actually been the best decision for the company. The appropriate technique to analyze the alternatives by using expected inputs and altering them before a decision is made is

A. Expected value analysis.

B. Linear programming.

C. Program Evaluation Review Technique (PERT).

D. Sensitivity analysis.

The correct answer is (D). *(CMA, adapted)*

REQUIRED: The technique that involves altering expected inputs during the decision process.

DISCUSSION: Sensitivity modeling can be used to determine the outcome of a variety of decisions. A trial-and-error method may be adopted, usually in a computer model, to calculate the sensitivity of the solution (variability of outcomes) to changes in a variable.

Answer (A) is incorrect because expected value analysis is used to determine an anticipated return or cost based upon probabilities of events and their related outcomes. Answer (B) is incorrect because linear programming optimizes a function given certain constraints. Answer (C) is incorrect because PERT is a network technique used to plan and control large projects.

Study Unit 7: INFORMATION SYSTEMS I

21 pages of outline
50 multiple-choice questions

Study Unit 7 is the first of two covering information systems, a major topic that has been assigned a relative weight range of 20% to 30% of Part 4. This study unit concerns systems concepts, hardware, software, and data storage.

A. Systems Overview

1. A system is a regularly interacting or interdependent group of items (subsystems) forming a unified whole.
   a. Each system, e.g., a business, in turn is part of a larger system, such as the national economy.
   b. The **boundary** of a system defines its physical limits and segregates it from the **environment.**
   c. A system performs its processing functions within the boundary.
      1) **Inputs** flow into and **outputs** flow out of the system through the boundary.
      2) A boundary shared with another system is an **interface**.
   d. To attain the system's objectives, it must have regulatory mechanisms (**controls**) to prevent, detect, and correct variances from standards.
      1) **Feedback loops** are essential to control.
      2) Another necessary aspect of control is the **law of requisite variety**: Countermeasures must be available to adjust out-of-control conditions; if the variety of countermeasures is not equal to the variety of deviations, increased **entropy** (disorder in the system) will result.
   e. The highest-order systems (a large business organization is an example) have substantial memory (data storage), the ability to learn (to modify its decision rules), and the capacity to choose and achieve new objectives.
   f. Feedback control systems are sometimes supplemented by **feedforward** control systems. The objective of a feedforward control system is to predict potential variations from plans so that adjustments can be made to prevent problems either before they occur or before they become too serious.
2. An **information system** transforms raw materials (**data**) into knowledge useful for decision making (**information**). The following activities are involved in the transformation:
   a. The system must **acquire** (capture) data from within or outside the entity.
   b. Quantitative data must be **measured**.
   c. Data are **recorded** on source documents.
   d. The data are **checked** for accuracy.
   e. **Classification** of data is necessary to determine the appropriate treatment.
   f. **Aggregation** or summarization may occur before data are processed.
   g. **Transmission** of data from the point of acquisition to a processing location.

h. **Transcription** of data from source documents to another medium is often a step preliminary to processing.

i. **Grouping** similar items may facilitate processing.

j. Within a group of items, **sorting** on the basis of primary and/or secondary keys is another processing step.

k. **Comparison and calculation** are logical and arithmetic steps, respectively, that transform quantitative data.

l. Before, during, and after processing, data must be temporarily or permanently **stored**, for example, in files or databases.

m. The system must be able to **retrieve** (obtain access to) stored data.

n. **Updating** keeps stored data current.

o. **Reports**, including replies to inquiries (whether or not interactive or unscheduled), contain the system's information output in a form understandable to users.

p. **Communication** of output to internal or external users is the ultimate purpose of the system.

q. At all stages, effective **control** should ensure the integrity of the data.

r. The following are the essential components of an information system model:

1) The **input** component of the system consists of inflows from the environment necessary for maintenance (such as heat and light), data to be processed, and control information (for example, feedback and computer program instructions).

2) **Output** includes outflows of information to users and waste; a reduction in the latter improves system efficiency.

3) The **transformation** of inputs into outputs is accomplished by a processing component with three subsystems.

a) **Storage** (memory) is necessary for retaining inputs (data and instructions), partial results, and finished output until processing is complete and transmittal is feasible.

b) An **arithmetic-logic unit (ALU)** manipulates the data in mathematical and logical ways (e.g., addition and comparison of numbers).

c) The **control subsystem** manages the transformation process by overseeing the storage of data and instructions, their transfer to and from the arithmetic-logic unit, and the ultimate transmittal of the results to the output component of the system.

3. A **management information system (MIS)** provides information for management decisions in an organization through the use of computer hardware and software, manual procedures, and models for analysis, planning, control, and decisionmaking.

a. An **integrated system** is planned as a group of subsystems within which applications are designed for compatibility and avoidance of redundancy.

1) An overall plan for the MIS should be developed to ensure that it avoids duplicate or inconsistent applications, provides for common standards and procedures, allows many users to share applications and data, and is subject to effective control.

b. An MIS segregates applications and their supporting data by creating a **database** to reduce data redundancy, exercise control, and improve data quality.

1) A database management system is the software that integrates the data to achieve these purposes (see Databases in Study Unit 8).

c. An MIS also includes embedded decision tables or **decision models** that permit processing of data to produce information directed toward particular decisions, such as those related to inventory control, budget analysis, or investment.

1) Models may be generalized and useful for many decisions or highly specific, but they are often most useful when they permit a manager the flexibility to use the interactive capacity of the system to explore the ramifications of a problem.

a) **Decision support systems (DSSs)** are extensions of the MIS concept that are primarily useful for semistructured problems, that is, those requiring the decision maker to exercise judgment in controlling the process but allowing for certain aspects of the problem to be preprogrammed. A DSS does not automate a decision but rather provides tools for the user to employ in applying his/her own insight and judgment.

i) The system should be interactive to permit the user to explore the problem by using the computational capacities, models, and data resources (internal and external) of the DSS.

ii) The system needs flexible access through reliable communications, ready availability of terminals, and possibly stand-alone microcomputers.

iii) Models tend to be relatively simple, and system development requires substantial involvement by the user because of the need for it to evolve as the user gains experience and new conditions emerge.

d. The trend in the development of MISs is from data processing to **information resources management**, which encompasses at least the overlapping activities of data processing, telecommunications, and office automation.

e. An **executive information system** (EIS) focuses on strategic (long-range) objectives and gives immediate information about an organization's critical success factors. Information is traditionally supplied from nontraditional computer sources. An EIS can be used on computers of all sizes.

f. MISs can be classified by function or by activity.

1) **Functions** include

a) **Marketing** -- sales analysis, forecasting, and planning

b) **Manufacturing** -- production planning, cost control, and quality control

c) **Logistics** -- planning and control of purchasing and inventory management

d) **Personnel** -- planning for the human resource function, evaluation of employees, and payroll

e) **Finance and accounting** -- capital budgeting, income measurement, and financial and cost analysis

f) **Information processing** -- planning systems development and cost-benefit analysis

g) **Top management** -- strategic planning and allocation of resources

2) **Activities** include

a) **Strategic planning** -- setting overall organizational objectives and drafting strategic plans

b) **Management control** -- budgeting and resource allocation

c) **Operational control** -- detailed scheduling of operations and performance supervision

d) **Transaction processing** -- receipts and disbursements, ordering and shipping

4. An **accounting information system (AIS)** is a subsystem of an MIS that processes financial and transactional data relevant to managerial as well as financial accounting.

a. An AIS is concerned not only with transactions with external parties (e.g., customers, suppliers, governments, owners, and creditors) reflected in financial statements prepared in conformity with GAAP, but also with the internal activities recorded in the cost accounting system and the preparation of related reports and analyses (e.g., production reports, pro forma financial statements, budgets, and cost-volume-profit analyses).

b. The major **components** of an AIS are the

1) **General ledger systems**, including general ledger accounting, budgeting, responsibility accounting, cost allocation, and profitability determination

2) **Cash receipts-disbursements systems**, including payroll, accounts receivable, and accounts payable

3) **Production systems**, including materials inventory, work-in-process, quality control, cost estimation, and scheduling

4) **Marketing**, including finished goods, order processing, and marketing (sales) analysis

5. The **open-systems** movement is an important consideration for the future development of information systems. Since the 1980s, government agencies have demanded that suppliers provide components with interfaces defined by public standards. These standards apply especially to operating systems and telecommunications protocols.

a. Thus, they permit systems from different suppliers to work together (**interoperability**). Moreover, open systems can be sized to an entity's needs (**scalability**), relocated to new platforms (**portability**), and used for years to come (**compatibility**).

b. Another result of product interchangeability is the formation of a commodity market and a decline in prices. Less product differentiation means that competition is increasingly based solely on price.

6. **Computer System Evolution**
   a. First-generation computers became commercially available around 1950. They used vacuum tubes and were large and slow. Entry of data and programs was by punched cards.
   b. Second-generation computers used transistors, buffers, magnetic core storage, and tape drives. The result was greater speed and reliability, lower cost, and smaller size.
   c. Third-generation computers incorporated integrated circuits. This generation of systems also made substantial use of magnetic disks, terminals, and data communications hardware.
   d. Fourth-generation computers employ very large-scale integrated circuits, multiprogramming, virtual storage, and microprocessors.
   e. Fifth-generation computers may use massively parallel processing, artificial intelligence concepts, and organic material in place of silicon chips.
7. Stop and review! You have completed the outline for this subunit. Study multiple-choice questions 1 through 13 beginning on page 304.

B. Systems Development and Design

1. The **systems development life-cycle approach** is the most common methodology applied to the development of large, highly structured application systems. The life-cycle approach is based on the idea that an information system has a finite life span that is limited by the changing needs of the organization. This cycle is analytically divisible into stages. A new system life cycle begins when the inadequacy of the current system leads to a decision to develop a new or improved system. This method is a structured process for controlling the creative activity required to devise, develop, and implement an information system. The process is described in varying terms by different writers, but the nature and sequence of the steps are essentially the same.
   a. The systems development cycle is divided into three steps: (1) definition, (2) development, and (3) installation and operation.
      1) **Definition**. The following phases are involved in this stage:
         a) A **proposal** for a new or modified application should be prepared indicating the need, the support for it within the organization, and scheduling considerations (timing of the project, employee availability, etc.).
         b) **Feasibility studies** should be conducted to determine whether the system is technically, economically, and operationally feasible.
            i) Availability of required technology, cost-benefit considerations, motivation of interested parties, timing of the project, existence of necessary resources, and prospects of successful implementation are aspects of feasibility.
         c) **Information requirements** must be ascertained, including the inquiries to be made by users, reports generated, database needs, and operating characteristics.
         d) The **general (conceptual) design** contains the users' description of the application, required inputs and outputs, the functions to be carried out, an overview of the processing flow (relationship of the most important programs, files, inputs, and outputs) and controls, and outlines of documentation.

2) **Development**

a) **Physical system design**

i) This step involves work by specialists to develop specifications for

- Work flow and programs (but not coding)
- Controls and points where they should be implemented
- Hardware
- Security measures, including backup
- Data communications
- Quality assurance testing for the balance of the development process

ii) One approach to design of the physical system is **top-down design**, which is the practice of defining a system by its general purpose and then progressively refining the level of detail in the form of a hierarchy.

- The top-down method begins with analysis of broad organizational goals, objectives, and policies as the bases for the design process. This step requires an understanding of the entity's environment and significant activities.
- The next step is to determine the decisions to be made by managers and the information required to make those decisions. The necessary reports, databases, inputs, processing methods, and equipment specifications can then be defined.
- The weakness of the top-down approach is that it tends to concentrate on managers' information needs at the expense of the design of efficient transaction processing at the operational level.

iii) **Structured design** of the physical system is a modular approach. Each module (subsystem) is functionally defined, and the degree of interdependence among the modules is minimized. This process simplifies development and enhances adaptability of the components of the system but requires careful definition of modules and the linkages (interfaces) between them.

iv) **HIPO** (Hierarchy-Input-Process-Output) is a documentation technique developed by IBM that relies on stylized charts depicting increasingly more detailed levels of the system.

b) **Physical database design** depends on the existing system.

i) New files or a new database may have to be designed.

ii) Modifying an existing database may be feasible.

iii) If the existing database provides for the new application, modification may not be necessary.

c) **Procedure development** includes writing technical manuals, forms, and other materials for all persons who will use, maintain, or otherwise work with the system.

   i) A **printer layout chart** is a gridded spacing chart that is an aid to designing documents and reports generated as hardcopy paper output by a printer.

d) **Flowcharting** is an essential aid in the development process. For information on flowcharting, see Study Unit 9, Internal Control.

   i) Other means of documenting the decision logic reflected in systems are matrices, decision tables, decision trees, and pseudocode.

e) **Program development** entails coding programs in accordance with the specifications in the physical design phase and then testing the results.

   i) **Structured programming** divides the system's set of programs into discrete modules by functional specifications. The objective is to create modules that are independent logical units, each of which has one entry and one exit point. Data sharing among modules should also be minimized. This method reduces the complexity created when instructions jump back and forth among different sections of the program.

   ii) Structured analysis relies on diagrams, not narrative. A typical tool is the data flow diagram, which shows the flows of data to, within, and from the system. A data flow diagram can be used to depict lower-level details as well as higher-level processes.

   iii) Each module can be coded by a separate team to

   - Facilitate security because no one group knows the complete set of programs
   - Expedite the development process because several programming teams can work simultaneously
   - Facilitate maintenance because a change or patch need only be module-specific, a less complicated procedure than fitting a patch to a complex, multifunction program

   iv) **Computer-aided systems engineering (CASE)** applies the computer to software design and development. It provides the capacity to maintain on the computer all of the system documentation, e.g., data flow diagrams, data dictionaries, and pseudocode (structured English); to develop executable input and output screens; and to generate program code in at least skeletal form. Thus, CASE facilitates the creation, organization, and maintenance of documentation and permits some automation of the coding process.

v) **Object-oriented programming** combines data and the related procedures into an object. Thus, an object's data can be operated on only within the object. If the procedures (called methods) for manipulating the data in an object are changed, no other parts of the program are affected.

- The basic concepts of object-oriented programming are class and inheritance. Programs are written not for objects but for categories of similar objects. One class of objects inherits the characteristics of a more general class. Thus, the basic advantage of object-oriented programming is that the code is reusable.
- However, the methods unique to a subclass may override an inherited method. When override occurs, the object responds to the same message but executes its method, not the inherited method. Accordingly, the advantage of having objects that inherit properties is that program code can be written once for the class instead of having to be rewritten for each instance of similar objects in the class.

vi) **Visual programming** creates programs by choosing objects from a library and arranging them in a desired sequence rather than by writing code.

3) **Installation and operation**

a) Training and educating system users is important not only for proper use of the system but also to offset the resistance of users whose jobs may have been substantially changed.

b) Acceptance testing by users of inputs, outputs, programs, and procedures is necessary to determine that the new system meets their needs.

c) Systems conversion is the final testing and switchover.

i) **Parallel operation** is the operation of the old and new systems simultaneously until satisfaction is obtained that the new system is operating as expected.

ii) **Pilot operation** (modular or phase-in conversion) is the conversion to the new or modified system by module or segment, e.g., one division, department, function, or branch of the company at a time. One disadvantage is the extension of the conversion time.

d) Systems follow-up or post-audit evaluation is a subsequent review of the efficiency and effectiveness of the system after it has operated for a substantial time (e.g., 1 year).

b. Another paradigm for the life-cycle approach is the following:

1) **Investigation**. The first step in systems development is identification and definition of a need relative to organizational objectives.

2) **Analysis**. The next step is to determine the scope of the required study and to proceed with a thorough analysis of the existing system.

3) **Systems design**. These steps lead to the general design of a new system. If the new system proves to be justified, the decision is then made to proceed with its implementation.

4) **Implementation**. Detailed systems design, including development and design of data files, is part of the implementation phase.

5) **Maintenance**. Following implementation, systems maintenance must be undertaken by systems analysts and applications programmers throughout the life of a system. Maintenance is the redesign of the system and programs to meet new needs or to correct design flaws. Ideally, these changes should be made as part of a regular program of preventive maintenance.

2. **Prototyping** (an experimental assurance process) is costly and time-consuming and thus is not currently the most common approach. It entails developing and putting into operation successively more refined versions of the system until sufficient information is obtained to produce a satisfactory design.

3. Systems development should be overseen by an **information systems steering committee** consisting of top-level managers representing the functional areas of the organization, such as information systems, accounting, and marketing.

   a. It provides overall guidance for information systems activities to assure that goals are consistent with those of the organization. Thus, the steering committee establishes priorities for implementing applications and either performs or approves high-level planning.

4. Stop and review! You have completed the outline for this subunit. Study multiple-choice questions 14 through 27 beginning on page 309.

C. Computer Hardware

1. **Hardware** is the configuration of electronic, magnetic, and mechanical devices that perform input, processing, output, storage, control, communications, and data preparation functions in a computer system.

2. **Central Processing Unit (CPU)**. All computers have at least one CPU that works in conjunction with peripheral devices.

   a. The CPU is the main element of a computer system. The major function of the CPU is to fetch stored instructions and data, decode the instructions, and carry out the instructions in the arithmetic-logic unit (ALU). The principal components of the CPU are the arithmetic-logic unit and the control unit.

   1) The control unit directs and coordinates the entire computer system.

   2) Primary storage is closely connected to the CPU in the central processor. It consists of electronic components that store letters, numbers, and special characters used in processing. The purpose of primary storage is to hold the operating system, part or all of a program being executed, and data used by the program.

   a) Internal primary storage includes **register memory** for very brief storage of small amounts of data and instructions immediately before use, **cache memory** for storage of frequently used data and programs, and **random-access memory (RAM)**, which is used to store large quantities of data. Data may be read from or written on RAM. A power interruption causes erasure of RAM.

   b) **Read-only memory (ROM)** is permanent storage used to hold the basic low-level programs and data. ROM can be read from but not written to; ROM chips are obtained from the manufacturer with programs already stored in them. A power interruption does not erase data written on ROM or on magnetic secondary storage devices. However, a power interruption may corrupt the data.

i) Programmable ROM (PROM) can be programmed once.

ii) Erasable programmable ROM (EPROM) can be erased using a special process and then reprogrammed.

b. Computer systems are typically classified by computing power dictated by CPU speed, memory capacity, and hardware architecture (and cost).

1) **Personal computers (microcomputers)** range in price and performance from low-end personal computers to powerful desktop models. **Workstations** are also desktop machines, but they are often classified separately from personal computers because of their enhanced mathematical and graphical abilities. In addition, workstations have the capacity to execute more complicated tasks simultaneously than a personal computer. Thus, they tend to be used by scientists and engineers, for example, for simulations and computer-aided design, and in the financial services industry.

a) Because of the large number of personal computers in use and aggressive pricing strategies, current personal computer prices have become very attractive. Moreover, personal computers have crossed into the minicomputer arena by providing comparable power and multi-user capabilities previously unavailable until recent technological improvements.

b) By adding a modem and communications software, the personal computer can also serve as an interface with other computers. Accordingly, many of the same control and security concerns that apply to larger computers also apply to a personal computer environment.

c) **Notebook, laptop, and palmtop computers** are the smallest forms of personal computers.

2) **Minicomputers** were historically used in multi-user or multi-terminal environments. The systems provide computing power and peripheral device access to a number of users and do not require a special environment for operation. Minicomputers have been used for industrial, general business, communications, and timesharing applications. Today, personal computers are far more powerful than the minicomputers of only a few years ago.

3) **Large, general-purpose (mainframe) computers** are characterized by large memory sizes, large secondary storage capacity, tape drives, high-speed printers, and other peripheral equipment. They often provide data communications, support many terminal users, are capable of handling large databases, and can execute many millions of instructions per second (MIPS). They usually need an air-conditioned, dust-free, humidity-controlled environment.

a) Super computers are the largest type of computer. Supercomputers can process billions of instructions per second.

3. **Peripheral devices** include equipment, other than the CPU, that provides for input, storage, output, outside communication, or additional facilities. They function much more slowly than the CPU, and a system is said to be input-output bound when the speed of peripheral devices determines the speed of task completion.

a. **Data entry devices**

1) A **data entry terminal** with a **keyboard** is online equipment that is a common input device. The terminal includes a monitor that allows display of user input and computer output.
   a) A **dumb terminal** has no other functions.
   b) A **smart terminal** has processing capacity.
   c) An **intelligent terminal** has programming capacity.
      i) A microcomputer can serve as a terminal if it is connected to another computer.
   d) A **terminal emulator** permits a microcomputer to interface with a mainframe.
2) A **magnetic ink character reader (MICR)** is used in the banking industry to read the magnetic ink on checks.
3) An **optical character reader (OCR)** does not need special ink and can read bar codes and alphanumeric characters.
   a) **Turnaround document** is a computer output prepared in such a way that it can eventually be used as a source document for an input transaction. For example, an optical character recognition document might be used as a sales invoice to be mailed to a customer and returned with the payment. Thus, no new document would have to be prepared to record the payment.
4) **Magnetic tape and disk drives** permit entry of data into primary storage from tapes and disks. These devices also serve an output function by transferring data from primary storage to tapes, disks, or output devices such as printers.
5) Point-of-sale terminals and automated teller machines are other input devices.
6) An alternative to keyboard and text-based input is the use of a **computer mouse**. Its point and click function controls a cursor on the video display screen. The user pushes a button on the mouse to choose the command indicated by the cursor.
7) **Touch screen** technology provides another limited alternative to keyboard input. Commands or data can be entered by touching the sensitized surface of a monitor.
8) **Digital scanners** convert documents and pictures into digital form.
9) **Voice recognition** input devices are still another alternative to keyboard input. These systems compare the speaker's voice patterns with prerecorded patterns. Advanced systems have large vocabularies that permit dictation and are therefore not limited to simple commands.
10) New forms of mobile data acquisition methods are available.
    a) **Palmtop computers** are miniature notebook computers with keyboard input.
    b) **Pen-based input** is possible with devices that have a flat-screen tablet on which the user can print directly with a special stylus.
    c) Other special devices may be used to gather inventory data, read meters, inspect goods, etc. Moreover, manufacturers may use infrared, ultraviolet, and other devices to automate the acquisition of production data.

b. **Output devices**

1) A **monitor** or **video display terminal** (also known as a cathode ray tube or CRT) can be used to display output.

2) **Printers** are the most common devices used solely for output. Printers operate at varying speeds and with varying quality.

3) A **plotter** is an output device that draws graphs on paper.

4) **COM (computer output to microfilm)** records output on microfilm.

c. **Storage devices**

1) **Floppy disk drives** serve as low-capacity backup devices.

2) **Hard disk drives** provide a permanent location for data and system files. A vast array of internal and external drives exist, providing gigabytes of flexible storage options.

3) **Magnetic tape drives** are primarily used for system backup and data transfer (streaming tape, DAT tape, etc.).

a) DAT (digital audiotape) is primarily used as a backup in imaging systems and as a master for CD-ROM.

4) **CD-ROM (compact disk-read only memory) drives**. CD-ROMs are laser-optical disks that provide capacities of over 660 MB. Write once, read many (WORM) versions are also available.

d. **Other peripheral equipment**

1) **Controllers** are hardware units designed to operate (control) specific input or output units, e.g., terminals and printers. These devices eliminate the need for the CPU to operate such devices.

2) **Channels** are hardware units located within or outside the CPU that are designed to handle the transfer of data into or out of primary storage (memory). Thus, the CPU need not handle the transfer of data.

3) A **console** consists of a VDT, a keyboard, and a printer. It is used for communication between an operator or engineer and the computer.

a) It permits the computer to exchange instructions with the operator, logs jobs, and provides a printout of activity that can be reviewed by auditors and the control group.

4) A **node** is a hardware device in a network that can act as a message buffer (hold part of the message until the rest is received), switch messages, and serve as an error control. A node can be a computer, controller, or multiplexor.

5) An **emulator** is a hardware device that permits one system to imitate another, that is, to use the same data and programs and obtain the same results as the other system.

6) A **performance monitor** is hardware or software that observes and records system activities, e.g., electrical activity or data about program execution.

4. Stop and review! You have completed the outline for this subunit. Study multiple-choice questions 28 through 35 beginning on page 313.

D. Computer Software

1. **Software** consists of computer programs, procedures, rules, and related documentation concerned with the operation of a computer system. It may be purchased from vendors or developed internally.
2. **Computer Languages**. Software is written in languages that are comprehensible by the computer either directly or after translation.
   a. First generation. **Machine language** is written in a **binary code** (a combination of ones and zeros) unique to the type of computer. It requires no translation.
      1) Each instruction specifies the operation to be performed and the data to be manipulated.
   b. Second generation. **Assembly (symbolic) languages** use mnemonic symbols to replace binary code.
      1) One assembly language instruction ordinarily corresponds to one machine language instruction. A program written in a language other than machine language is a source program.
      2) A **language translator** must be used to convert a source program into an **object program** in machine language.
      3) An **assembler** translates assembly language source programs into machine language.
   c. Third generation. **Procedural languages** consist of statements each of which corresponds to a group of machine language instructions. These languages are used to describe processing algorithms.
      1) Because each statement uses algebraic or English-like symbols to represent multiple machine language instructions, these languages are more readily learned, and coding is substantially simplified.
      2) Procedural languages also have the advantage of being adaptable to many different computers.
      3) The following are common procedural languages:
         a) **BASIC** is a widely used language for microcomputers but is not often used in large business application processing. BASIC stands for **B**eginner's **A**ll-Purpose **S**ymbolic **I**nstruction **C**ode.
         b) **COBOL** is a programming language consisting of a series of English-like statements. COBOL was designed to be easy to read and, when properly written, easy to maintain. COBOL stands for **CO**mmon **B**usiness **O**riented **L**anguage.
         c) **FORTRAN** is very effective for solving mathematical, engineering, and other types of problems, but not as useful for business applications. FORTRAN stands for **FOR**mula **TRAN**slation.
         d) **PL/1** is a cross between COBOL and FORTRAN, but it has not gained equal acceptance. PL/1 stands for **P**rogramming **L**anguage **1**.
         e) **RPG** stands for **R**eport **P**rogram **G**enerator. Its purpose is to produce business-related reports that do not require complex programming, such as for accounts that require substantial file updating (accounts receivable, accounts payable, and inventory are examples). RPG is not a standardized language.

f) **C** is a general-purpose programming language. UNIX is written mostly in C. C++ is a high-level programming language that adds object-oriented features to its predecessor, C.

g) **Java** is a high-level, object-oriented programming language developed by Sun Microsystems that is similar to C++ and, among other things, is used to write programs embedded in World Wide Web documents. It is designed to allow a user to download from a network only the data and the part of an application (an **applet**) needed to perform a given task. When processing is complete, the data and software are erased. Thus, software is stored on the network, and the user need not be concerned about compatibility of the software with the computer platform or with upgrades and installation. Moreover, separate versions of programs need not be written for different operating systems. Java is platform independent if each computer has a **Java Virtual Machine**, a program included in an operating system. It may also be incorporated into a browser.

h) **Pascal** is a language used on microcomputers because its compilation needs relatively little memory. It is often used to teach programming in computer courses.

i) **Ada** was developed by the Department of Defense and is used for all its system development work. It facilitates structured software design and is highly portable.

4) Procedural languages are often translated by a **compiler**. For example, COBOL and FORTRAN are always compiled. After the source program has been translated (compiled) into object code, a computer program known as a **linkage editor** (or linker) must be used to process the object code modules into a single module that can be loaded (entered) into the computer. The linkage editor creates an executable module by resolving the cross-references among dependent modules.

5) Some procedural languages are translated by an **interpreter**, which is a program that translates source code into object code as it is needed during processing; thus, no separate compiling run is necessary because each statement is executed immediately after translation.

a) Greater efficiency in translation and the elimination of the linkage edit step are offset, however, by slower processing speeds and an inability to save the full translation for audit purposes.

b) BASIC may be compiled or interpreted, and most fourth-generation languages (see below) are interpreted.

d. Fourth generation. **Problem-oriented or nonprocedural languages** (i.e., those that are not intended to express a procedure as a specific algorithm) provide still further simplification of the programming task.

1) These interactive, English-like languages permit the unsophisticated user to describe the problem to, and receive guidance from, the computer instead of specifying a procedure.

a) **Query languages** are most often used with databases. They permit reading and reorganization of data but not its alteration.

b) **Program generators** allow the user to create an application program based on requirements entered in a specific format. They differ from query languages because they permit data to be written or altered.

e. Specialized languages

1) Simulation and financial modeling languages have been created to support management decision making, e.g., for use in decision support systems.

2) Other languages have been developed for such purposes as database management and audits.

a) **Data-definition language** is used in database management systems (see Databases in subunit B in Study Unit 8). It defines the overall structure of a database (the **schema**) and the structure of the portion of the database that each application program is authorized to access (the **subschema** or external schema).

b) **Data-manipulation language** is also used in database management systems. It is for accessing and processing data from a database. The best known data-manipulation language is **Structured Query Language (SQL)**, which is used with relational database management systems.

c) **Generalized audit software** may allow not only parallel simulation but also a variety of other processing functions, such as extracting sample items, verifying totals, developing file statistics, and retrieving specified data fields.

3) **Job control language (JCL)** is used in job control statements that accompany jobs submitted in a batch processing (delayed update) mode. These statements describe the files to be entered, input and output devices to be used, and the programs to be executed.

f. **HTML (hypertext markup language)** is the authoring software language commonly used to create and link pages on the World Wide Web. Its emphasis is on document display.

g. **XML (extensible markup language)** was developed by an international consortium and released in 1998 as an open standard usable with many programs and platforms. XML codes all information in such a way that a user can determine not only how it should be presented but also what it is. Unlike HTML, XML uses codes that are extensible, not fixed. Thus, if an industry can agree on a set of codes, software for that industry can be written that incorporates those codes.

h. **Pseudocode** (structured English). In top-down coding, higher-level logic modules are coded in relatively broad form using an informal language (pseudocode). This procedure is necessary because the available programming languages are designed to represent more detailed forms of program design.

3. **Systems software** in machine or assembly language is necessary to facilitate the processing of application programs by the computer. It performs language translation, monitoring of data communications, job instruction, control of input and output, file management, data sorting, access control, etc.

a. The **operating system** of a personal computer mediates between the application programs and the computer hardware.

1) To communicate with the user, the operating system of a personal computer may include a **graphical user interface (GUI)**. It employs graphic icons to represent activities, programs, and files. The computer mouse is used to make selections. A GUI therefore simplifies computer usage compared with older systems that require the typing of text-based commands on a keyboard.

a) **Windows** is a GUI shell initially developed by Microsoft to run in conjunction with the DOS operating system. Newer operating systems also have this feature.

b) Windowing is the characteristic that allows a computer to display more than one program on the screen at the same time. Each program has its own section of the screen, but only one program is active.

2) An operating system also manages job scheduling and accounts for computer usage, controls input and output, assigns storage locations in main memory, and protects data in the event of a system malfunction.

3) Furthermore, an operating system may permit a single configuration of hardware to function in different modes.

a) **Batch mode (single thread)**. Programs are processed from beginning to end without interruption in processing.

b) **Multiprogramming**. The operating system processes a program until an input or output operation is required. Because input or output can be handled by peripheral hardware (channels and controllers), the CPU can begin executing another program's instructions while output is in progress and therefore increase its throughput. Several programs are being processed concurrently, but only one is actually being executed in the CPU.

i) **Throughput** is the quantity of work processed in a given period. It is used as a measuring tool to evaluate processors.

c) **Time sharing** differs from multiprogramming in that the CPU spends a fixed amount of time on each program.

d) **Multitasking** is multiprogramming on a single-user operating system, for example, in a microcomputer.

e) **Multiprocessing**. The operating system in conjunction with multiple CPUs actually executes instructions from more than one program simultaneously.

f) **Virtual storage**. The operating system automatically separates user programs into small fixed or variable length segments, allowing the user to have access to a memory virtually equivalent to the total of primary and secondary memories. To the user it appears as if unlimited memory is available for programs.

i) **Paging**. A page is a segment of a program. Paging is a memory-swapping technique. The major portion of a program or set of data may be kept in secondary storage while the remainder is held in main memory. This swapping means that the "virtual" capacity of primary storage is greatly expanded.

4) The following are popular operating systems:

a) **DOS** is used for older, 16-bit personal computers. It is based on the IBM standard. **MS-DOS** is a text-based desktop operating system made by Microsoft. **PC-DOS-2000** is a text-based desktop operating system made by IBM as an update of the older MS-DOS. Only low-end servers can run on these operating systems. Although text-based, DOS can run in connection with the GUI of Windows.

b) Microsoft's **Windows 95** is a 32-bit system with the ability to run programs using more than 640K of memory (the limit for DOS). It allows for multitasking and networking capabilities. **Windows 98** includes Microsoft's Web browser, Internet Explorer. It enhances the computer's multimedia capabilities. **Windows NT** is a 32-bit system that can be used by personal computers, workstations, and network servers. **Windows 2000** is the latest version of NT. It offers improved reliability and security. **Windows ME** offers improved network and multimedia support for Microsoft's line of home operating systems.

c) **UNIX** is a multi-user, multi-tasking operating system with different versions suitable for personal computers, workstations, minicomputers, and mainframes. Its capabilities extend to networking and mission-critical client/server applications, but it is complex and sometimes requires large amounts of memory.

d) **Linux** is a version of an UNIX-like operating system that runs on X86, Alpha, and Power PC processors. Unlike most UNIX distributions, Linux is free software, though full distributions with proprietary add-ons are available from companies like RedHat, Caldera, Debian, and S.u.S.E.

e) **OS/2** is a 32-bit system used with IBM and IBM-compatible personal computers and workstations and with Intel chips. **OS/2 Warp** is the latest version. It extends OS/2's capabilities to network computing, JAVA, and the Web.

f) **Macintosh OS X** (ten) is a desktop operating system based on Rhapsody. Macintosh OS X is made by Apple Computer and will run on Motorola/IBM PowerPC chips. Macintosh OS 8, OS 7, and OS 6 are desktop operating systems made by Apple Computer that run on Motorola/IBM PowerPC and Motorola 680x0 chips.

g) **MVS** is a mainframe operating system made by IBM.

h) **NetWare** "is a dedicated network operating system" that runs on Intel chips.

i) **BeOS** is a high-performance desktop operating system made by Be that runs on Motorola/IBM PowerPC and Intel Pentium. Only low-end servers can run on this operating system.

j) **AmigaOS** is an old but popular operating system that is being resurrected by Gateway Computers. Only low-end servers can run on this operating system.

b. **Access control software** protects files, programs, data dictionaries, processing, etc., from unauthorized access; restricts use of certain devices (e.g., terminals); and may provide an audit trail for access attempts.

c. **File access managers** provide for organizing and controlling data in a logical manner and for different modes of access (e.g., sequential and/or direct).

d. **Database management systems**. See Databases in subunit B in Study Unit 8.

e. **Library management software** stores, updates, and protects source programs, job command statements, and, in some cases, files and object programs.

1) This software also logs program changes.

f. **Teleprocessing monitors** control execution of application programs in an interactive (online) system.

g. **Online editors** are useful to programmers for creating and maintaining source programs, files, and job command statements.

h. **Communications controllers** allocate network resources and functions, control the connections between programs and terminals, transfer data between control nodes and terminals, and permit monitoring of network operations and changes in the network's configuration.

i. **Utility programs** are service programs that perform certain standard tasks, such as sorting, merging, copying, and printing file dumps.

j. **Operations management software** provides for tape and disk management and for job scheduling in a batch processing environment.

k. **Software monitor** is a program or package that performs functions similar to those of hardware monitors; thus, it records signals emitted by the system's components (e.g., the number and length of accesses to a disk). It may also perform other tasks such as taking snapshots of internal conditions and indicators at predetermined times. This function is useful for reconstruction of records after a system failure.

l. **Firmware** is software wired permanently into the hardware. An example is read-only memory (ROM).

m. **Kernel program** is a short program that is run on different systems by a potential buyer to provide information about their processing characteristics.

n. **Middleware** is software that oversees the interaction between disparate computers. Middleware allows each computer to continue operating independently without reprogramming and to cooperate with other systems.

o. **Business intelligence software** spans a category dedicated to extracting information from corporate databases for purposes of business analysis.

p. A **macro** is a program that records keystrokes and commands used in repetitious jobs, thereby reducing the necessary keystrokes.

q. **Antivirus software** is intended to prevent initial infections, stop replication attempts, detect infections after their occurrence, mark affected system components, and remove viruses from infected components. The major risk in relying on antivirus software is that it may not detect certain viruses. Antivirus software designed to identify and remove known viruses is sometimes known as a vaccine. A vaccine works only for known viruses and may not be completely effective for variants of those viruses.

r. **Groupware** provides support for the activities of work groups, for example, e-mail, electronic meetings, scheduling, and information sharing. Lotus Notes is the leading groupware product. Internet browsers can also be used for some groupware functions.

4. **Application software** includes the programs that perform the tasks required by ultimate users, e.g., standard accounting operations.

   a. Application programs may be developed internally or purchased from vendors.

      1) Vendor-produced software is in either source or object code, but vendors prefer to sell the latter.

      2) Application software production is obviously a vital aspect of system development, and control over its maintenance (changes to meet new user needs) after implementation is likewise crucial.

b. A **spreadsheet** is one type of application software. It displays a financial model in which the data are presented in a grid of columns and rows. An example is a financial statement spreadsheet.

1) The model is based on a set of mathematical relationships defined by the user, with specified inputs and outputs. The effects of changes in assumptions can be seen instantly because a change in one value results in an immediate recomputation of related values.

2) Thus, in designing a spreadsheet model, the first step is to define the problem. This is followed by an identification of relevant inputs and outputs, and the development of assumptions and decision criteria. Finally, formulas must be documented.

3) Quattro Pro, Lotus 1-2-3, and VP-Planner are common spreadsheet programs.

c. Software is copyrightable, but a substantial amount is in the public domain. Networks of computer users may share such software.

1) **Shareware** is software made available for a fee by the owners to users through a distributor (or electronic bulletin board services).

2) Software piracy is a problem for vendors. The best way to detect an illegal copy of application software is to compare the serial number on the screen with the vendor's serial number.

3) Use of unlicensed software increases the risk of introducing computer viruses into the organization. Such software is less likely to have been carefully tested.

4) Diskless workstations increase security by preventing the copying of software to a floppy disk from a workstation. This control not only protects the company's interests in its data and proprietary programs but also guards against theft of software licensed to the company by vendors.

5) To shorten the installation time for revised software in a network, a company can implement **electronic software distribution (ESD)**, which is the computer-to-computer installation of software on workstations. Instead of weeks, software distribution can be accomplished in hours or days and can be controlled centrally. Another advantage of ESD is that it permits tracking or metering of PC program licenses.

d. **Integrated packages** include two or more applications, e.g., spreadsheets, word processing, database management, and graphics. The advantage is that the package does not require redundant data entry and maintenance. However, the individual applications in an integrated package tend not to be as powerful as the stand-alone versions.

e. A **browser** is software that permits a computer to retrieve and recognize HTML files. Web browsers allow the user to display the graphical information that makes up much of the World Wide Web.

5. Stop and review! You have completed the outline for this subunit. Study multiple-choice questions 36 through 46 beginning on page 315.

E. Basic Data Structures

1. **Bit (binary digit)** is either 0 or 1 in the binary numbering system. The negative or positive polarity of an electrical charge on a magnetic medium is a means of representing a binary digit.
2. **Byte** is a group of bits (ordinarily six or eight) used to signify a character (a number, letter of the alphabet, or special character, such as a question mark or asterisk).
3. **Field** is an item in a record. It consists of a group of related characters (bytes) providing a unit of data about some entity (e.g., a customer or employee).
4. **Record** is a collection of related fields pertaining to some entity.
   a. **Keys** are attributes of logical records that permit them to be sorted. A primary key is the data item(s) that is the principal identifier, e.g., a vendor number in a file of vendors' invoices. A secondary key is an alternative identifier used either for sorting or special processing, e.g., the payment date on a vendor's invoice.
   b. **Logical records** are defined in terms of the information they contain. A logical record contains a number of fields relating to that record; e.g., a payroll record might include the Social Security number, the employee's name, the rate of pay, etc.
   c. The attributes of the **physical record** vary according to the physical storage device. Thus, a logical record may be divided among a number of physical records either on a magnetic tape or on a disk. A physical record may also contain one logical record.
   d. In addition, the various fields constituting a record may be distributed within a database and linked via pointers to maintain the relationship.
   e. **Fixed-length computer records** facilitate the handling of data on both disk and tape. They allow data to be blocked, moved, and stored in a standard-size format, thus permitting simpler treatment in programs. To handle variable-sized records, the program would have to modify itself while data are being read and processed.
   f. **Variable-length records** use space more efficiently because the storage size can vary with the record size. Data transmission is more efficient with variable-length records because they require less space.
   g. **Blocking** means putting a number of logical records into a physical record, i.e., to place a block of logical records between two interblock gaps, allowing the computer to read more data at one time.
   h. A **record layout** is used to describe the fields in each logical record of each file used in input, output, and storage. Layout sheets are preprinted forms used as documentation for record layouts.
   i. A **pointer** is an identifier that specifies the location of a data item.
   j. An **index** is a list of contents of a file or document together with references to their locations, for example, an index of keys and their related addresses.

5. A **file** is a logical collection of records, such as those for customer accounts.
   a. A **master file** is relatively permanent.
      1) A **transaction file** contains relatively temporary records processed together with a master file for updating purposes.
   b. File organization
      1) A **linked list** is a file organization in which each data record has a pointer field containing the address of the next record in the list.
      2) In a **direct file** organization, a randomizing formula or hashing scheme (a transform algorithm) converts a record key into a storage address. This method permits direct access without an index.
      3) The **indexed-sequential-access method (ISAM)** is a system in which records are stored sequentially in a direct-access file and organized by a primary key stored in an index record. It does not use pointers. The virtue of an ISAM system is that it permits sequential processing of large numbers of records while providing for occasional direct access.
   c. File management concerns creating, maintaining, retrieving information from, updating, and establishing control over data files. Updating totals, storing them in a record, comparing them periodically with record contents, and correcting discrepancies is an example of a file management control.
6. **Volatility** is a commonly used measure of the relative frequency of adds, deletes, and changes to a master file during a specified time period.
7. Stop and review! You have completed the outline for this subunit. Study multiple-choice questions 47 through 50 beginning on page 319.

MULTIPLE-CHOICE QUESTIONS

A. Systems Overview

1. Feedback, feedforward, and preventive controls are important types of control systems and procedures for an accounting information system. Which one of the following is in the correct order of feedback, feedforward, and preventive control systems?

A. Cash budgeting, capital budgeting, and hiring qualified employees.

B. Cash budgeting, cost accounting variances, and separation of duties.

C. Cost accounting variances, separation of duties, and cash planning.

D. Cost accounting variances, cash budgeting, and organizational independence.

The correct answer is (D). *(CMA, adapted)*

REQUIRED: The correct order of feedback, feedforward, and preventive control systems.

DISCUSSION: A feedback control system gathers information after the occurrence of an event, evaluates it, and initiates action to improve future activities. For example, calculation of variances in a standard cost accounting system is a feedback control. A feedforward control anticipates potential variations from plans so that adjustments can be made to prevent problems either before they occur or before they become too serious. A cash budget is an example of a feedforward control because it anticipates cash needs and allows for the provision of resources to meet those needs. A preventive control prevents an event from occurring. Separation of duties and organizational independence are examples of preventive controls because certain undesirable events are impossible if these controls are properly implemented.

Answer (A) is incorrect because cash budgeting is a feedforward control. Answer (B) is incorrect because cash budgeting is a feedforward control, and variance analysis is a feedback control. Answer (C) is incorrect because separation of duties is a preventive control, but cash planning is not.

2. Many organizations supplement feedback control systems with feedforward control systems. The major goal of most feedforward control systems is

A. To predict potential variations from plans so that adjustments can be made to prevent problems before they occur or become significant.

B. To predict potential variations from budgets so that variance analysis can be performed on problems that become significant.

C. To provide complete information by use of system controls on computer input.

D. To predict any future event in the organization.

The correct answer is (A). *(CMA, adapted)*

REQUIRED: The major goal of most feedforward control systems.

DISCUSSION: A feedforward system anticipates problems and strives for timely prevention rather than after-the-fact correction. Planning is a related concept. Preventive maintenance is an example of a feedforward control.

Answer (B) is incorrect because variance analysis is a feedback control. Answer (C) is incorrect because the goal of feedforward systems is to predict variations before they occur, not to provide information by use of system controls on computer input. Answer (D) is incorrect because feedforward controls are concerned with variations that will lead to problems, not with predicting all future events.

3. Which one of the following is not considered a typical risk associated with outsourcing (the practice of hiring an outside company to handle all or part of the data processing)?

A. Inflexibility.

B. Loss of control.

C. Loss of confidentiality.

D. Less availability of expertise.

The correct answer is (D). *(CMA, adapted)*

REQUIRED: The item that is not considered a typical risk associated with outsourcing.

DISCUSSION: Some companies have outsourced their data processing function because of the economies provided, superior service quality, avoidance of changes in the organization's information systems infrastructure, cost predictability, the freeing up of human and financial capital, avoidance of fixed costs, and the greater expertise offered by outside vendors. The risks of outsourcing include the inflexibility of the relationship, the loss of control, the vulnerability of important information, and often the dependency on a single vendor.

Answers (A), (B), and (C) are incorrect because they are risks associated with outsourcing the data processing function.

4. Which one of the following technological elements of computer-based information systems has the least effect in driving the changes that are currently occurring in the workplace?

A. Advances in microcomputer hardware and software.

B. Office automation and teleprocessing.

C. Decision support systems and artificial intelligence (AI).

D. Advances in disaster recovery systems.

The correct answer is (D). *(CMA, adapted)*

REQUIRED: The technological element of computer-based information systems with the least effect on current workplace changes.

DISCUSSION: Advances in disaster recovery systems are a response to, not a cause of, the technological changes in the workplace. Developments in computer hardware, software, and telecommunications devices have increased reliance on computer-based systems. Furthermore, the vulnerability of such systems has increased as technology has improved because more people are computing, more access points are available, and more opportunities for failure of system components have been created.

Answer (A) is incorrect because the number of workers who can be computing at the same time increased exponentially with the advent of personal computers and networks of desktop machines. Answer (B) is incorrect because office automation systems are knowledge-level systems that have improved the productivity of data workers by providing word processing, e-mail, voice mail, videoconferencing, electronic calendars, desktop publishing, and digital filing. Teleprocessing permits access to information and the performance of work at remote sites. Answer (C) is incorrect because a DSS is a management-level information system that facilitates semistructured and unstructured decision processes. AI is the capability of a computer to emulate human functions, for example, expert decision making.

5. Which one of the following terms best describes a decision support system (DSS)?

A. Management reporting system.

B. Formalized system.

C. Interactive system.

D. Accounting information system.

The correct answer is (C). *(CMA, adapted)*

REQUIRED: The best description of a DSS.

DISCUSSION: A decision support system is an interactive system that is useful in solving semistructured and unstructured problems, that is, those requiring a top-management or middle-management decision maker to exercise judgment and to control the process. A DSS does not automate a decision or provide structured information flows but rather provides tools for an end user to employ in applying insight and judgment. The system must be interactive to permit the user to explore the problem by using the computational capacities, models, assumptions, data resources, and display graphics of the DSS.

Answer (A) is incorrect because a management reporting system provides structured, routine information flows. Answer (B) is incorrect because the required flexibility of a DSS means that it cannot be highly formalized or structured. Answer (D) is incorrect because an accounting information system processes routine, highly structured financial and transactional data relevant for accounting purposes.

6. Which group of characteristics best describes decision support systems?

A. Analytical models, specialized databases, and interactive computer-based modeling processes.

B. Analytical models, specialized databases, interactive computer-based modeling processes, and the decision maker's own insights and judgments.

C. Analytical models, programming models, application models, and interactive computer-based modeling processes.

D. Expert systems, model-based information, electronic data interchange, and the decision maker's own insights and judgments.

The correct answer is (B). *(CMA, adapted)*

REQUIRED: The characteristics that best describe decision support systems.

DISCUSSION: Decision support systems (DSS) are extensions of the MIS concept that are useful for semistructured problems such as those requiring the decision maker to exercise judgment in controlling the process but allowing for certain aspects of the problem to be preprogrammed. A DSS does not automate a decision but rather provides tools for the user to employ in applying his or her own insight and judgment. The system should be interactive to permit the user to explore the problem by using the computational capacities, models, and data resources of the DSS. A DSS needs flexible access through reliable communications, ready availability of terminals, and possibly stand-alone microcomputers. Models tend to be relatively simple, and system development requires substantial involvement by the user because of the need for DSS to evolve with the user's experience and as new conditions emerge.

Answer (A) is incorrect because the decision maker's insight and judgment are important aspects of a DSS. Answer (C) is incorrect because a DSS does not include programming models. Answer (D) is incorrect because a DSS is not an expert system. An expert system actually makes the decision, whereas a DSS merely aids the decision maker.

7. Which one of the following statements about an executive information system (EIS) is incorrect? The EIS

A. Provides top executives with immediate and easy access to information in a highly interactive format.

B. Helps executives monitor business conditions in general and assists in strategic planning to control and operate the company.

C. Is designed to accept data from many different sources; to combine, integrate, and summarize the data; and to display this data in a format that is easy to understand and use.

D. Is likely to be one of the most widely used and the largest of the information subsystems in a business organization.

The correct answer is (D). *(CMA, adapted)*

REQUIRED: The incorrect statement regarding an executive information system (EIS).

DISCUSSION: An EIS focuses on strategic (long-range) objectives and gives immediate information about an organization's critical success factors. Information is ordinarily supplied from nontraditional computer sources. An EIS can be used on computers of all sizes. An EIS is typically used only by executives at the highest levels within the organization; as a result, it is not widely used within the organization. Information provided is often highly aggregated, but the details supporting the aggregated data are accessible.

Answers (A), (B), and (C) are incorrect because these statements about an EIS are true.

8. An executive information system (EIS) has all of the following characteristics except

A. Focusing on obtaining strategic objectives.

B. Giving immediate information about an organization's critical success factors.

C. Providing information from nontraditional computer sources.

D. Providing advice and answers to top management from a knowledge-based system.

The correct answer is (D). *(CMA, adapted)*

REQUIRED: The item that is not a characteristic of an executive information system (EIS).

DISCUSSION: An EIS serves the needs of top management for transaction processing (primarily inquiries and support for decisions), operational control, managerial control, and strategic planning. Top management needs access to the database and decision models, the use of large amounts of summarized internal and especially external data, and the capacity for ad hoc analysis. Thus, an EIS focuses on strategic (long-range) objectives and gives immediate information about a firm's critical success factors. Information is typically supplied from nontraditional computer sources. However, an EIS program can be used on computers of all sizes. An EIS is not a program for providing top management with advice and answers from a knowledge-based (expert) system.

Answer (A) is incorrect because an EIS does focus on obtaining strategic objectives. Answer (B) is incorrect because an EIS gives immediate information about an organization's critical (strategic) success factors. Answer (C) is incorrect because an EIS does not use traditional computer sources.

9. How is an accounting information system (AIS) distinguished from a management information system (MIS)?

A. An AIS deals with financial information; an MIS handles all other information.

B. An AIS may be either manual or computer based; an MIS is computer based.

C. An AIS is a subsystem within an MIS.

D. An AIS is control oriented; an MIS is used exclusively for planning.

The correct answer is (C). *(CMA, adapted)*

REQUIRED: The difference between an AIS and an MIS.

DISCUSSION: An information system transforms raw data into knowledge useful for decision making. An MIS provides information for management decisions. An AIS is a subsystem of an MIS and processes financial and transactional data relevant to managerial decisions as well as financial accounting.

Answer (A) is incorrect because an MIS provides financial information; an AIS is a subsystem within an MIS that is not limited to financial information. Answer (B) is incorrect because an AIS is a subsystem within an MIS, which is an integrated user-machine system that includes computer hardware and software. Answer (D) is incorrect because both types of systems are used to provide a wide range of information to management.

10. In an information system environment, many organizations combine key data processing cycles related to accounting and finance. Traditionally, these cycles are

A. Cash receipts, cash disbursements, and capital budgeting.

B. Capital budgeting and financial reporting.

C. Cash receipts and cash disbursements.

D. Cash receipts, cash disbursements, capital budgeting, and financial reporting.

The correct answer is (D). *(CMA, adapted)*

REQUIRED: The key data processing cycles related to accounting and finance.

DISCUSSION: An entity's financing and investing function, including capital budgeting, concerns the management of financial assets, acquisition and disposal of fixed assets, issuance of stock, payment of dividends, and borrowing. The accounting function manages the entity's financial records, accounts for the flow of funds (e.g., cash disbursements and receipts), and prepares financial statements for external reporting purposes. These functions are readily combined for data processing purposes. For example, the process of developing a budget may be divided into operating and financial (cash disbursements, cash receipts, capital budget, pro forma balance sheet, and cash flows statement) components linked by the pro forma income statement.

Answer (A) is incorrect because the financial reporting cycle is omitted. Answer (B) is incorrect because the cash receipts and cash disbursements cycles are omitted. Answer (C) is incorrect because the capital budgeting and financial reporting cycles are omitted.

11. An accounting information system (AIS) must include certain source documents in order to control purchasing and accounts payable. For a manufacturing organization, the best set of documents should include

A. Purchase requisitions, purchase orders, inventory reports of goods needed, and vendor invoices.

B. Purchase orders, receiving reports, and inventory reports of goods needed.

C. Purchase orders, receiving reports, and vendor invoices.

D. Purchase requisitions, purchase orders, receiving reports, and vendor invoices.

The correct answer is (D). *(CMA, adapted)*

REQUIRED: The best set of documents to be included in an AIS to control purchasing and accounts payable.

DISCUSSION: An AIS is a subsystem of a management information system that processes financial and transactional data relevant to managerial and financial accounting. The AIS supports operations by collecting and sorting data about an organization's transactions. An AIS is concerned not only with external parties but also with the internal activities needed for management decision making at all levels. An AIS is best suited to solve problems when reporting requirements are well defined. A manufacturer has well-defined reporting needs for routine information about purchasing and payables. Purchase requisitions document user department needs, and purchase orders provide evidence that purchase transactions were appropriately authorized. A formal receiving procedure segregates the purchasing and receiving functions and establishes the quantity, quality, and timeliness of goods received. Vendor invoices establish the liability for payment and should be compared with the foregoing documents.

Answer (A) is incorrect because receiving reports should be included. Answer (B) is incorrect because requisitions and vendor invoices should be included. Answer (C) is incorrect because purchase requisitions should be included.

12. Which one of the following statements about an accounting information system (AIS) is incorrect?

A. AIS supports day-to-day operations by collecting and sorting data about an organization's transactions.

B. The information produced by AIS is made available to all levels of management for use in planning and controlling an organization's activities.

C. AIS is best suited to solve problems where there is great uncertainty and ill-defined reporting requirements.

D. AIS is often referred to as a transaction processing system.

The correct answer is (C). *(CMA, adapted)*

REQUIRED: The incorrect statement about an accounting information system (AIS).

DISCUSSION: An AIS is a subsystem of a management information system that processes financial and transactional data relevant to managerial and financial accounting. The AIS supports operations by collecting and sorting data about an organization's transactions. An AIS is concerned not only with external parties but also with the internal activities needed for management decision making at all levels. An AIS is best suited to solve problems when reporting requirements are well defined. A decision support system is a better choice for problems in which decision making is less structured.

Answers (A), (B), and (D) are incorrect because they are attributes of an AIS.

13. Which one of the following features is least likely to apply to the transaction processing cycle of an accounting information system?

A. Data records are chiefly historical in nature.

B. Most of the sources of data are an organization's recurring transactions.

C. Data are usually financial in nature.

D. Data records are the basis of predictive systems.

The correct answer is (D). *(CMA, adapted)*

REQUIRED: The feature that is least likely to apply to the transaction processing cycle of an AIS.

DISCUSSION: An AIS is a subsystem of a management information system that processes financial and transactional data relevant to managerial and financial accounting. The AIS supports operations by collecting and sorting historical data about an organization's transactions. An AIS is concerned not only with external parties but also with the internal activities needed for management decision making at all levels. An AIS is best suited to solve problems when reporting requirements are well defined. An AIS does not typically use records based on predictive systems, which would be a feature of a decision support system.

Answers (A), (B), and (C) are incorrect because they are features commonly associated with an AIS.

B. Systems Development and Design

14. The three major activities of systems design are

A. User interface design, data manipulation, and output analysis.

B. Process design, output design, and output analysis.

C. User interface design, data design, and process design.

D. Data design, input validation, and processing.

The correct answer is (C). *(CMA, adapted)*

REQUIRED: The three major activities of systems design.

DISCUSSION: Systems design determines how information requirements will be met. It concerns how users will interact with the system to meet their needs, how data will be organized, and the formulation of processing steps.

Answer (A) is incorrect because data manipulation is not a part of systems design; it is an operational activity that occurs after a system has been installed. Answer (B) is incorrect because output analysis occurs after a system has been installed. Answer (D) is incorrect because input validation and processing are operational activities.

15. Ordinarily, the analysis tool for the systems analyst and steering committee to use in selecting the best system alternative is

A. Pilot testing.

B. User selection.

C. Decision tree analysis.

D. Cost-benefit analysis.

The correct answer is (D). *(CMA, adapted)*

REQUIRED: The analysis tool to use in selecting the best system alternative.

DISCUSSION: Feasibility studies should include an analysis of the cost-benefit ratio of any system alternatives. In many cases, the best possible system may not be cost effective. Thus, once the decision makers have determined that two or more systems alternatives are acceptable, the cost-benefit relationship should be used to select the best system for a particular application.

Answer (A) is incorrect because pilot testing determines only whether a system works, not how efficient it is in a particular application. Answer (B) is incorrect because users may not have the necessary systems knowledge to make a decision. Answer (C) is incorrect because decision tree analysis is probably more sophisticated than is necessary in choosing between a few systems alternatives.

16. In determining the need for system changes, several types of feasibility studies can be made. The most commonly recognized feasibility studies are

A. Legal, environmental, and economic.

B. Environmental, operational, and economic.

C. Technical, economic, legal, and practical.

D. Technical, operational, and economic.

The correct answer is (D). *(CMA, adapted)*

REQUIRED: The most commonly recognized types of system change feasibility studies.

DISCUSSION: A feasibility study consists of an investigation of the current system, determination of the information and processing requirements, evaluation of the possible applications of computer data processing, selection of the best option, and an evaluation of the proposed design choice's cost effectiveness and impact on the organization. A feasibility study must therefore consider technical, operational, and economic feasibility. The design choice must be within the range of available technology, meet the operational needs of users and otherwise be acceptable to them, and have a favorable cost-benefit ratio.

Answers (A) and (B) are incorrect because an information system is unlikely to have a significant effect on the environment. Answer (C) is incorrect because legal issues are least likely to be a concern in systems development.

17. Two phases of systems planning are project definition and project initiation. All of the following are steps in the project initiation phase except

A. Preparing the project proposal.

B. Informing managers and employees of the project.

C. Assembling the project team.

D. Training selected personnel.

The correct answer is (A). *(CMA, adapted)*

REQUIRED: The step not a part of the project initiation phase of systems planning.

DISCUSSION: The project initiation phase includes promptly informing managers and employees about the project, assembling the project team (possibly including systems analysts, programmers, accountants, and users), training selected personnel to improve necessary skills and enhance communication among team members, and establishing project controls (e.g., by implementing a project scheduling technique such as PERT). Preparing the project proposal is a part of the project definition phase, as are feasibility studies, determining project priority, and submitting the proposal for approval.

Answer (B) is incorrect because informing managers and employees of the project is a component of the project initiation phase. Answer (C) is incorrect because assembling the project team is a component of the project initiation phase. Answer (D) is incorrect because training selected personnel is a component of the project initiation phase.

18. In the systems development cycle, coding is

A. A form of testing and debugging.

B. Part of the detailed design phase.

C. Part of the data-flow diagram.

D. A form of program maintenance.

The correct answer is (B). *(CMA, adapted)*

REQUIRED: The true statement about coding in the system development cycle.

DISCUSSION: According to one paradigm for the life-cycle approach, the detailed design phase is part of the implementation of the system. Detailed design includes coding programs in accordance with the specifications established in the physical design phase. Testing the results is the next phase.

Answer (A) is incorrect because coding occurs prior to testing and debugging. Answer (C) is incorrect because data flows need to be known before coding can begin. Answer (D) is incorrect because program maintenance is the redesign of programs to meet new needs or to correct design flaws.

19. Systems analysts are the personnel within an organization who are responsible for the development of the company's information system. Which one of the following functions are least likely to be performed by a systems analyst?

A. Design of computer applications.

B. Preparation of specifications for computer programming.

C. Developing, coding, and testing computer programs.

D. Examining user information requirements.

The correct answer is (C). *(CMA, adapted)*

REQUIRED: The function least likely to be performed by a systems analyst.

DISCUSSION: A systems analyst plans and designs computer applications, often with the assistance of systems flowcharts. Systems analysis usually involves examining user information requirements, surveying the existing system, and preparing specifications for computer programming. The concept of systems analysis, however, does not include the development, coding, and testing of individual computer programs. The work of the systems analyst provides the design specifications that will guide the preparation of specific programs by programmers. In a good system of internal control, the systems analyst should not perform programming tasks or have access to computer equipment, production programs, data files, or input-output controls.

Answer (A) is incorrect because systems analysts design computer applications. Answer (B) is incorrect because analysts prepare specifications for computer programmers. Answer (D) is incorrect because the systems analyst must examine user information requirements to learn what type of application is needed.

20. The least risky strategy for converting from a manual to a computerized accounts receivable system would be a

A. Direct conversion.

B. Parallel conversion.

C. Pilot conversion.

D. Database conversion.

The correct answer is (B). *(CMA, adapted)*

REQUIRED: The least risky strategy for converting from a manual to a computerized accounts receivable system.

DISCUSSION: The least risky means of converting from a manual to a computerized system is a parallel conversion in which the old and new systems are operated simultaneously until satisfaction is obtained that the new system is operating as expected. Slightly more risky is a pilot conversion in which the new system is introduced by module or segment.

Answer (A) is incorrect because direct conversion is more risky than a parallel conversion. Answer (C) is incorrect because pilot conversion is more risky than a parallel conversion. Answer (D) is incorrect because database conversion is more risky than a parallel conversion.

21. Workwell Company operates in several regions, with each region performing its data processing in a regional data center. The corporate management information systems (MIS) staff has developed a database management system to handle customer service and billing. The director of MIS recommended that the new system be implemented in the Southwestern Region to ascertain if the system operates in a satisfactory manner. This type of conversion is called a

A. Parallel conversion.

B. Direct conversion.

C. Prototype conversion.

D. Pilot conversion.

The correct answer is (D). *(CMA, adapted)*

REQUIRED: The type of systems conversion process in which a new system is first implemented in one subunit of the organization.

DISCUSSION: A modular conversion approach entails switching to the new or improved system in organizational (division, region, product line, etc.) segments or system segments (accounts receivable, database, etc.). A pilot conversion is one in which the final testing and switchover are accomplished at one segment or division of the company.

Answer (A) is incorrect because parallel conversion operates the old and new systems simultaneously. Answer (B) is incorrect because direct conversion involves immediate conversion to the new system throughout the organization. Answer (C) is incorrect because a prototype conversion involves developing and putting into operation successively more refined versions of the system until sufficient information is obtained to produce a satisfactory design.

22. The graphic portrayal of the flow of data and the information processing of a system, including computer hardware, is best displayed in a

A. Data-flow diagram.

B. System flowchart.

C. Gantt chart.

D. Decision table.

The correct answer is (B). *(CMA, adapted)*

REQUIRED: The best method of displaying the flow of data and the information processing of a system.

DISCUSSION: A system flowchart is a pictorial representation of an information system at the macro level. It emphasizes inputs, processing steps, and outputs but not the details of execution.

Answer (A) is incorrect because a data-flow diagram would show only where data goes, not the total system. Answer (C) is incorrect because a Gantt chart is a bar chart used to monitor the progress of large projects. Answer (D) is incorrect because a decision table is used to show the various possibilities available in a given decision situation.

23. Errors are most costly to correct during

A. Programming.

B. Conceptual design.

C. Analysis.

D. Implementation.

The correct answer is (D). *(CMA, adapted)*

REQUIRED: The time when errors are most costly to correct.

DISCUSSION: Errors can be corrected most easily and clearly when they are found at an early stage of systems development. Their correction becomes more costly as the life cycle progresses. Because implementation is the last stage of the process listed, errors are most costly to correct when discovered at the implementation stage.

Answer (A) is incorrect because error correction at the programming level would be less costly than at the implementation stage. Answer (B) is incorrect because error correction at the conceptual design level would be less costly than at the implementation stage. Answer (C) is incorrect because error correction at the analysis level would be less costly than at the implementation stage.

24. The process of monitoring, evaluating, and modifying a system as needed is referred to as

A. Systems analysis.

B. Systems feasibility study.

C. Systems maintenance.

D. Systems implementation.

The correct answer is (C). *(CMA, adapted)*

REQUIRED: The term for the process of monitoring, evaluating, and modifying a system.

DISCUSSION: Systems maintenance must be undertaken by systems analysts and applications programmers continuously throughout the life of a system. Maintenance is the redesign of the system and programs to meet new needs or to correct design flaws. Ideally, these changes should be made as part of a regular program of preventive maintenance.

Answer (A) is incorrect because systems analysis is the process of determining user problems and needs, surveying the organization's present system, and analyzing the facts. Answer (B) is incorrect because a feasibility study determines whether a proposed system is technically, operationally, and economically feasible. Answer (D) is incorrect because systems implementation involves training and educating system users, testing, conversion, and follow-up.

25. The process of developing specifications for hardware, software, manpower, data resources, and information products required to develop a system is referred to as

A. Systems analysis.

B. Systems feasibility study.

C. Systems maintenance.

D. Systems design.

The correct answer is (D). *(CMA, adapted)*

REQUIRED: The process of developing specifications for the components of a system.

DISCUSSION: Detailed systems design involves developing specifications regarding input, processing, internal controls and security measures, programs, procedures, output, and databases.

Answer (A) is incorrect because systems analysis is the process of learning how the current system functions, determining the needs of users, and developing the logical requirements of a proposed system. Answer (B) is incorrect because a feasibility study determines the technical, operational, and economic feasibility of a system. Answer (C) is incorrect because systems maintenance is the process of monitoring, evaluating, and modifying a system.

26. The process of learning how the current system functions, determining the needs of users, and developing the logical requirements of a proposed system is referred to as

A. Systems maintenance.

B. Systems analysis.

C. Systems feasibility study.

D. Systems design.

The correct answer is (B). *(CMA, adapted)*

REQUIRED: The term referring to the process of learning how a system functions, determining the needs of users, and developing the logical requirements of a proposed system.

DISCUSSION: A systems analysis requires a survey of the existing system, the organization itself, and the organization's environment to determine (among other things) whether a new system is needed. The survey results determine not only what, where, how, and by whom activities are performed but also why, how well, and whether they should be done at all. Ascertaining the problems and informational needs of decision makers is the next step. The systems analyst must consider the entity's key success variables (factors that determine its success or failure), the decisions currently being made and those that should be made, the factors important in decision making (timing, relation to other decisions, etc.), the information needed for decisions, and how well the current system makes those decisions. Finally, the systems analysis should establish the requirements of a system that will meet user needs.

Answer (A) is incorrect because maintenance is the final stage of the life cycle in that it continues throughout the life of the system; maintenance includes the redesign of the system and programs to meet new needs or to correct design flaws. Answer (C) is incorrect because the systems feasibility study does not involve the process of learning how the current system works. Answer (D) is incorrect because systems design is the process of developing a system to meet specified requirements.

27. An information system (IS) project manager is currently in the process of adding a systems analyst to the IS staff. The new systems analyst will be involved with testing the new computerized system. At which stage of the systems development life cycle will the analyst be primarily used?

A. Cost-benefit analysis.

B. Requirements definition.

C. Flowcharting.

D. Implementation.

The correct answer is (D). *(CMA, adapted)*

REQUIRED: The stage of the systems development life cycle involving testing of a new system.

DISCUSSION: The systems development life-cycle approach is the oldest methodology applied to the development of medium or large information systems. The cycle is analytically divisible into stages: investigation, analysis, systems design, implementation, and maintenance. Testing, training, and conversion occur in the installation and operation, or implementation, stage of the life cycle.

Answer (A) is incorrect because cost-benefit analysis is a part of the feasibility study conducted early in the life cycle. Answer (B) is incorrect because requirements are defined during the analysis or systems study stage. Answer (C) is incorrect because flowcharting is a necessary activity in all early stages of the life cycle.

C. Computer Hardware

28. The main components of the central processing unit (CPU) of a computer are

A. Semiconductors, online devices, and memory.

B. Arithmetic-logic unit, control unit, and primary memory.

C. Random-access memory, read-only memory, and auxiliary storage.

D. Primary storage, input-output devices, and arithmetic-logic unit.

The correct answer is (B). *(CMA, adapted)*

REQUIRED: The main components of the CPU.

DISCUSSION: All computers have a CPU that works in conjunction with peripheral devices. The CPU is the main element of a computer system. Its principal components are primary storage, the arithmetic-logic unit, and the control unit.

Answer (A) is incorrect because online devices are peripheral units. Answer (C) is incorrect because auxiliary storage is not a part of the CPU. Answer (D) is incorrect because input-output devices are peripheral equipment.

29. The location in the central processing unit (CPU) where data and programs are temporarily stored during processing is the

A. Floppy disk drive.

B. Magnetic tape drive.

C. Random-access memory (RAM).

D. Magnetic disk drive.

The correct answer is (C). *(CMA, adapted)*

REQUIRED: The location in the CPU where data and programs are temporarily stored during processing.

DISCUSSION: The CPU is the part of the computer system that manipulates numbers, symbols, and letters. Primary storage is the area in the CPU where programs and data are temporarily stored during processing. Internal primary storage is also known as random-access memory because each memory location therein can be randomly accessed in the same amount of time.

Answer (A) is incorrect because a floppy disk drive is a random access storage medium for microcomputers that permits permanent storage of data. Answer (B) is incorrect because a magnetic tape drive is not for temporary storage. Answer (D) is incorrect because a magnetic disk drive is a secondary storage medium for permanent storage of data.

30. In a microcomputer system, the place where parts of the operating system program and language translator program are permanently stored is

A. Read-only memory (ROM).

B. Magnetic disk drive.

C. Random-access memory (RAM).

D. Magnetic tape drive.

The correct answer is (A). *(CMA, adapted)*

REQUIRED: The place where parts of the operating system and language translator are stored.

DISCUSSION: ROM consists of microinstructions hardwired into the computer. Data cannot be written on or erased from ROM. The operating system and the language translator (software that translates source programs into machine-readable instructions) are permanently stored in ROM to prohibit users from accidentally erasing or changing the system. Some microcomputers, however, have erasable, programmable ROM. This form of storage may be erased by an ultraviolet technique (but not by the microcomputer) after which new instructions may be entered.

Answers (B) and (D) are incorrect because drives are not storage devices. Answer (C) is incorrect because RAM is erasable.

31. The current trend in systems environments is to replace large mainframe computers with multiprocessor computer systems. All of the following are reasons this trend is occurring except that multiprocessor computer systems

A. Are generally more flexible than mainframe computer systems.

B. Are most applicable in installations having a large number of small application programs.

C. Are more effective at handling high-volume, online transaction processing.

D. Applications are generally easier to program than mainframe applications.

The correct answer is (C). *(CMA, adapted)*

REQUIRED: The item not a reason for replacing mainframe computers with multiprocessor computer systems.

DISCUSSION: Mainframes are the largest computers. They have very rapid processing ability and a massive amount of memory. Consequently, they are appropriate when great quantities of data must be processed.

Answer (A) is incorrect because a multiprocessor system is more flexible than a mainframe. It consists of machines that may be networked in a variety of ways or treated as stand-alone processors. Answer (B) is incorrect because small applications may be processed more cheaply on smaller, less expensive machines. Answer (D) is incorrect because programming tools available for smaller machines ordinarily simplify the programming process.

32. All of the following are computer input devices except a(n)

A. Plotter.

B. Mouse.

C. Magnetic ink character recognition device.

D. Light pen.

The correct answer is (A). *(CMA, adapted)*

REQUIRED: The item that is not a computer input device.

DISCUSSION: An input, or data entry, device communicates information to a computer. Examples include bar code readers, magnetic ink character recognition devices, light pens, keyboards, touch screens, a computer mouse, digital scanners, and voice recognition devices. A plotter is an output device that draws graphs directly on paper.

Answers (B), (C), and (D) are incorrect because a mouse, a magnetic ink character recognition device, and a light pen are input devices.

33. The input device used in a department store that allows a sales clerk to pass a light pen over the price tag to record the purchase is a(n)

A. Mark-sense reader.

B. Optical scanner.

C. Touch-tone device.

D. Laser bar code scanner.

The correct answer is (B). *(CMA, adapted)*

REQUIRED: The input device that allows a clerk to pass a light pen over the price tag to record a purchase.

DISCUSSION: An optical scanner, or optical character reader, passes a light pen over the information to be entered into a computer system. Special inks are not needed, and the scanner can read both bar codes and alphanumeric characters.

Answer (A) is incorrect because a mark-sense reader can only read marks that are in a proper place on a predesigned computer form. Answer (C) is incorrect because a touch-tone device is a telephone. Answer (D) is incorrect because a laser bar code scanner reads bar codes only.

34. Banks are required to process many transactions from paper documents (e.g., checks, deposit slips) during the course of an average business day. This requires a reliable, yet economical, form of input. The most common source automation device used by banks is

A. A disk pack.

B. Magnetic tape.

C. Bar coding.

D. Magnetic ink character recognition.

The correct answer is (D). *(CMA, adapted)*

REQUIRED: The most common source automation device used by banks.

DISCUSSION: Magnetic ink character recognition (MICR) is used by banks to read the magnetic ink on checks and deposit slips. MICR is a form of data entry device.

Answer (A) is incorrect because a disk pack is a storage device. Answer (B) is incorrect because magnetic tape is a storage device, not a data entry device. Answer (C) is incorrect because bar coding is a data entry technique often used by manufacturers, wholesalers, and retailers but is rarely used by banks.

35. Access time in relation to computer processing is the amount of time it takes to

A. Transmit data from a remote terminal to a central computer.

B. Complete a transaction from initial input to output.

C. Perform a computer instruction.

D. Retrieve data from memory.

The correct answer is (D). *(CMA, adapted)*

REQUIRED: The definition of access time.

DISCUSSION: Access time is the interval between the moment at which an instruction control unit initiates a call for data and the moment at which delivery of the data is completed. For example, direct access memory is faster than sequential access memory.

Answer (A) is incorrect because access time refers to the speed of data retrieval, not data transmittal. Answer (B) is incorrect because throughput time is the time to complete a transaction from initial input to output. Answer (C) is incorrect because access time is much slower than the time required to execute an instruction.

D. Computer Software

36. All of the following are examples of computer software except a(n)

A. Database management system.

B. Telephone modem.

C. Language translator.

D. Word processing package.

The correct answer is (B). *(CMA, adapted)*

REQUIRED: The item that is not an example of computer software.

DISCUSSION: Software consists of computer programs, procedures, rules, and related documentation concerned with the operation of the computer system. A modem, however, is hardware used to convert digital signals from terminals and the CPU into analog signals for transmission along telephone lines.

Answers (A), (C), and (D) are incorrect because a DBMS, a language translator, and a word processing package are examples of computer software.

37. Microcomputer systems have enhanced use with systems software and applications software. Which one of the following statements concerning microcomputer systems is false?

A. Database management systems are available for microcomputer systems.

B. An operating system program is a critical software package for microcomputers.

C. Electronic spreadsheet packages are types of applications software for microcomputers.

D. Integrated packages are examples of operating systems for microcomputers.

The correct answer is (D). *(CMA, adapted)*

REQUIRED: The false statement concerning microcomputer systems.

DISCUSSION: Integrated packages consist of two or more applications programs rather than operating system programs. These software packages are increasingly available from vendors at prices less than the cost of internal development. However, they may require expensive modifications.

Answer (A) is incorrect because many database management systems are available for microcomputers. Answer (B) is incorrect because an operating system is critical since it is the software that controls the overall functioning of the computer. Operating systems are also commercially available. Answer (C) is incorrect because electronic spreadsheet packages are types of business applications programs. An applications program is used to perform desired data processing tasks.

38. All of the following are characteristic of computer machine language except

A. Internal binary code.

B. Hexadecimal code.

C. Assembly language.

D. On/off electrical switches.

The correct answer is (C). *(CMA, adapted)*

REQUIRED: The item not characteristic of machine language.

DISCUSSION: An assembly language program may be written in one of several languages that use mnemonic codes, not the binary code in which a machine language program is written. In the hierarchy of programming languages, assembly language is one level higher than machine language. Machine language is a programming language made up of instructions that a computer can directly recognize and execute.

Answer (A) is incorrect because machine language is internally stored in binary form (the only characters used in binary notation are 0 and 1). Answer (B) is incorrect because machine language uses binary code for internal purposes, but printing out (dumping) the contents of storage can be done more efficiently if the printout is expressed in a number system with a base of 8 (actual) or 6 (hexadecimal). Answer (D) is incorrect because on/off electrical switches correspond to the binary digits 0 and 1 in which machine language is written.

39. Mainframe computer systems include several advanced processing procedures. Two of the most common processing procedures are multiprocessing and multiprogramming. Which one of the following statements about these processing procedures is false?

A. Multiprocessing usually involves two or more computers functioning simultaneously.

B. Multiprogramming allows multiple programs to be executed at exactly the same time.

C. Multiprogramming switches back and forth between programs during processing.

D. Multiprocessing allows the sharing of a central memory during processing.

The correct answer is (B). *(CMA, adapted)*

REQUIRED: The incorrect statement about multiprocessing and multiprogramming.

DISCUSSION: In multiprogramming, the operating system processes a program until an input/output operation is required. Because input or output can be performed by peripheral hardware, the CPU can thus begin executing another program's instructions while input or output is in progress. Several programs are being processed concurrently, but only one is actually being executed in the CPU. This is accomplished by the computer's switching back and forth between programs during processing. In multiprocessing, the operating system, in conjunction with multiple CPUs, actually executes instructions from more than one program simultaneously.

Answers (A), (C), and (D) are incorrect because they are true statements about multiprocessing and multiprogramming.

40. A software tool infrequently used to select or access items in the database is most likely a(n)

A. Report generator.

B. Program generator.

C. Application generator.

D. Query utility program.

The correct answer is (D). *(CMA, adapted)*

REQUIRED: The software tool infrequently used to select or access items in a database.

DISCUSSION: Utility programs are service programs that perform certain standard tasks, such as sorting, merging, copying, and printing file dumps. A query utility program could be used to access items in a database. However, such programs would not be used in normal operational circumstances. A query language permits interactive searching of a database. Data may be read and reorganized but not altered.

Answer (A) is incorrect because a report generator is frequently used. It generates reports based on special requirements by users, often in conjunction with query language. Answer (B) is incorrect because program generators allow the user to create application programs based on requirements entered in a specific format. They differ from query languages because they permit data to be written or altered. Answer (C) is incorrect because an application generator is frequently used.

41. Object oriented programming is characterized by an emphasis on objects and the procedures performed upon them. Which of the following programming languages is most closely associated with object oriented programming?

A. Basic.

B. Fortran.

C. C.

D. C++.

The correct answer is (D). *(Publisher)*

REQUIRED: The programming language most closely associated with object oriented programming.

DISCUSSION: C++ is sometimes referred to as "C with classes." As the name implies, C is its foundation. Classes provide a means of encapsulating the objects used by the program (which ordinarily represent the program's data) into a well organized, modular format that is easy to reuse and maintain. Through a process called inheritance, new objects can be derived from existing objects by adding new elements to the existing object design. C++ was specifically designed with these features in mind.

Answers (A), (B), and (C) are incorrect because Basic, Fortran, and C traditionally emphasize procedures rather than objects.

42. Computers containing more than one central processing unit (CPU) are increasingly common. This feature enables a computer to execute multiple instructions from multiple programs simultaneously. This process is

A. Time sharing.

B. Multitasking.

C. Multiprocessing.

D. Batch processing.

The correct answer is (C). *(Publisher)*

REQUIRED: The term for executing multiple programs with multiple CPUs.

DISCUSSION: Multiprocessing greatly increases system efficiency by executing multiple programs on the same computer at the same time. In systems with only one CPU, although multiple programs may be active simultaneously, program instructions can only be executed for one of these programs at a time.

Answer (A) is incorrect because, in time sharing, the CPU spends a fixed amount of time on each program. Answer (B) is incorrect because multitasking is multiprogramming on a single-user operating system. It is the process of having multiple programs active at a given time, although the CPU is executing instructions from only one program at a time. Answer (D) is incorrect because batch processing entails execution of a list of instructions from beginning to end without interruption.

43. Software offered to users on a "try before you buy" basis is called

A. Shareware.

B. Firmware.

C. Middleware.

D. Freeware.

The correct answer is (A). *(Publisher)*

REQUIRED: The software offered on a "try before you buy" basis.

DISCUSSION: Shareware is commercial software offered to users without initial charge. It is traditionally offered by less established software developers. Often it comes with some features disabled and can only be used without charge for a limited time. Users pay a fee to "register" it, in return for which they receive a license to use it with all features enabled.

Answer (B) is incorrect because firmware is permanently wired into the hardware. Answer (C) is incorrect because middleware oversees the interaction between disparate systems. Answer (D) is incorrect because freeware is software that, although usually copyrighted, may be used without any licensing fee on a much less limited basis than shareware.

44. Computer manufacturers are now installing software programs permanently inside the computer as part of its main memory to provide protection from erasure or loss if there is interrupted electrical power. This concept is known as

A. File integrity.

B. Software control.

C. Firmware.

D. Random-access memory (RAM).

The correct answer is (C). *(CMA, adapted)*

REQUIRED: The term for software installed permanently in the computer.

DISCUSSION: Firmware consists of software programs "hardwired" or permanently installed in the computer hardware. Firmware can be used to monitor internal conditions, e.g., by making signal counts (such as accesses to the computer) or taking "snapshots" of indicators. Thus, ROM (read-only memory) is firmware.

Answer (A) is incorrect because file integrity is achieved by implementing controls that protect the completeness, accuracy, and physical security of files. Answer (B) is incorrect because software control refers to library control of programs. Answer (D) is incorrect because RAM (as opposed to sequential-access memory) describes storage devices in which data locations can be directly accessed, thereby speeding data retrieval.

45. Which one of the following statements most accurately represents the basic steps in designing a spreadsheet model?

A. Define the problem, identify inputs and outputs, develop assumptions and decision criteria, and document formulas.

B. Identify inputs and outputs, develop assumptions and decision criteria, document formulas, and build macros.

C. Define the problem, develop assumptions and decision criteria, develop test cases, and build macros.

D. Develop assumptions and decision criteria, document formulas, develop test cases, and run test cases.

The correct answer is (A). *(CMA, adapted)*

REQUIRED: The statement that most accurately reflects the basic steps in the design of a spreadsheet model.

DISCUSSION: A spreadsheet is software that displays a financial model in which the data are presented in a grid of columns and rows. An example is a financial statement spreadsheet. The model is based on a set of mathematical relationships defined by the user, with specified inputs and outputs. The effects of changes in assumptions can be seen instantly because a change in one value results in an immediate recomputation of related values. Thus, in designing a spreadsheet model, the first step is to define the problem. Next, relevant inputs and outputs are identified and assumptions and decision criteria are developed. Finally, formulas are documented.

Answers (B) and (D) are incorrect because the definition of the problem is a necessary first step in designing a spreadsheet model. Answer (C) is incorrect because inputs and outputs must be identified at some point (early) in the spreadsheet design process.

46. A characteristic of a microcomputer system that displays more than one program on the screen at the same time, places each program in its own area of the screen, but permits only one program to be active is

A. Windowing.

B. Distributed processing.

C. Context switching.

D. File extension.

The correct answer is (A). *(CMA, adapted)*

REQUIRED: The characteristic of a microcomputer system that allows display of more than one program on the screen.

DISCUSSION: A graphical user interface is part of an operating system with which users may interact. It uses graphic icons to represent activities, programs, and files. The computer mouse is used to make selections. Windows is a graphical user interface shell initially developed by Microsoft to run in conjunction with DOS. Newer operating systems also have this feature. Thus, windowing is the characteristic that allows a computer to display more than one program on the screen at the same time. Each program has its own section of the screen, but only one program is active.

Answer (B) is incorrect because distributed processing is a means of assigning computer processing to various segments of a business, with some aspects centralized and some decentralized. Answer (C) is incorrect because context switching does not relate to the various segments on a computer screen. Answer (D) is incorrect because a file extension is a means of extending a logical collection of records.

E. Basic Data Structures

47. Block codes

A. Are generally used to identify missing items from a set of documents or records.

B. Allow a user to number items sequentially.

C. Allow a user to assign meaning to particular segments of a coding scheme.

D. Are randomly calculated groups of numbers used as a control check.

The correct answer is (C). *(CMA, adapted)*

REQUIRED: The true statement about block codes.

DISCUSSION: Coding of data is the assignment of alphanumeric symbols consistent with a classification scheme. Block coding assigns blocks of numbers in a sequence to classes of items. For example, in a chart of accounts, assets may be given numbers 100-199, liabilities the numbers 200-299, etc.

Answers (A) and (B) are incorrect because some items in a block code may be unassigned to allow for flexibility. Answer (D) is incorrect because block codes are assigned judgmentally, not at random.

48. Coding in data processing assigns a unique identification number or key to each data record. Which one of the following statements about coding is incorrect?

A. A primary key is the main code used to store and locate records within a file.

B. Records can be sorted, and temporary files created, using codes other than their primary keys.

C. Secondary keys are used when the primary keys cannot be found.

D. Secondary keys are used for alternative purposes including inverted files.

The correct answer is (C). *(CMA, adapted)*

REQUIRED: The statement about coding that is incorrect.

DISCUSSION: A primary key, such as a customer number in an accounts receivable file, is the main code used to store and locate records within a file. However, records can be sorted, and temporary files created, using codes other than the primary keys. Secondary keys are used to sort records based on an attribute other than the primary key, for example, the credit limit or current balance for an accounts receivable file. An inverted file is the result of such a process. Moreover, a given data record can have more than one secondary key. However, secondary keys are not substitutes for primary keys.

Answers (A), (B), and (D) are incorrect because they are true statements about coding.

49. The indexed-sequential-access method (ISAM) is an approach to file organization

A. In which each data record has a pointer field containing the address of the next record in the list.

B. In which an index of record pointers of some of the file attributes are maintained in a list.

C. That uses an algorithm to convert a record key into a storage address to assist with later retrieval.

D. In which records are stored sequentially in a direct-access file and organized by a primary key stored in an index record.

The correct answer is (D). *(CMA, adapted)*

REQUIRED: The true statement about indexed-sequential-access file organization.

DISCUSSION: ISAM is a system in which records are stored sequentially in a direct-access file and organized by a primary key stored in an index record. It does not use pointers. A pointer is a data element attached to a record that gives the address of the next logically related record. The virtue of an ISAM system is that it permits sequential processing of large numbers of records while providing for occasional direct access.

Answer (A) is incorrect because a linked list is a file organization in which each data record has a pointer field containing the address of the next record in the list. Answer (B) is incorrect because an ISAM system does not use pointers. Answer (C) is incorrect because, in a direct file organization, a randomizing formula or hashing scheme (a transform algorithm) converts a record key into a storage address. This method permits direct access without an index.

50. A commonly used measure of the relative frequency of adds, deletes, and changes to a master file during a specified time period is

A. Volatility.

B. The index ratio.

C. The frequency ratio.

D. The volume ratio.

The correct answer is (A). *(CMA, adapted)*

REQUIRED: The commonly used measure of the relative frequency of adds, deletes, and changes to a master file.

DISCUSSION: File volatility is the relative frequency with which records are added, deleted, or changed during a specified period.

Answers (B) and (C) are incorrect because changes to a master file are measured by volatility. Answer (D) is incorrect because the volume ratio is not a measure of changes to a master file.

Study Unit 8: Information Systems II

34 pages of outline
68 multiple-choice questions

Study Unit 8 is the second of two covering information systems, a major topic that has been assigned a relative weight range of 20% to 30% of Part 4. This study unit's heaviest emphasis is on internal control.

A. Processing Modes

1. **Batch Mode**. Batch processing is the accumulation and grouping of transactions for processing on a delayed basis. The batch approach is suitable for applications that can be processed at intervals and involve large volumes of similar items, e.g., payroll, sales, inventory, and billing.
2. **Online Mode**. An online processing system is in direct communication with the computer, giving it the capability to handle transactions as they are entered. An online system permits both immediate posting (updating) and inquiry of master files as transactions occur.
   a. **Real-time processing** involves processing an input record and receiving the output soon enough to affect a current decision-making process. In a real-time system, the user interacts with the system to control an ongoing activity.
   b. The term "online" often used with "real time" indicates that the decision maker is in direct communication with the computer. **Online, real-time systems** usually permit access to the main computer from multiple remote terminals.
3. Firms may find it beneficial to incorporate both batch and online processing modes into one system.
   a. A database may be established for information that must be obtained quickly, for instance, a sales processing system in which credit information must be available to sales personnel on an ongoing basis.
   b. However, other processing requirements may take advantage of the speed and control provided in a batch processing system. For example, payroll transactions can be processed quickly and efficiently in a batch mode.
4. A **timesharing** system allows many users to have access through remote terminals to a CPU. Timesharing is different from multiprogramming because a fixed time is devoted to each program in turn (see Study Unit 7).
5. **Service bureaus** perform batch processing for subscribers.
   a. This off-site mode of processing requires a user to prepare input and then transport it to the bureau, with attendant loss of time and increase in security problems.

6. **Totally Centralized Systems**. All data processing and systems development are done at one data processing center.
   a. All processing is done by one large computer.
   b. Remote users are serviced via data communications channels between themselves and the center.
   c. Terminals at the remote sites are usually dumb terminals (providing communications only, with no stand-alone processing capabilities).
   d. Requests for development of new systems are submitted for the consideration of the centralized systems development group.
   e. The centralized staff is large.
   f. Advantages of total centralization arise primarily from the economies of scale permitted and the strengthening of control.
7. **Totally Decentralized Systems.** Data processing functions are independently developed at each remote site. Each site has its own smaller computer and its own staff.
   a. In a completely decentralized system, each computer stands alone, independent of any centralized or other computer.
   b. The primary advantages of a decentralized system are that
      1) The individual units' personnel identify more closely with the system.
      2) Development projects are more easily accepted and meet local needs better.
8. Totally centralized and totally decentralized systems represent two ends of a continuum. The essential activities are the planning of data processing activities, systems development, programming, and operations. Any combination of these four is possible, ranging from complete centralization to total decentralization.
9. **Downsizing** consists of moving organization-wide applications to mid-range or networked computers. The purpose is to reduce costs by using less expensive systems that are more versatile than larger, more expensive systems.
10. **Distributed Data Processing**. The advent of cheaper and smaller computers has permitted the development of a somewhat different alternative to centralization or decentralization.
    a. In a distributed data processing system, the organization's processing needs are examined in their totality.
    b. The decision is not whether an application should be done centrally or locally but, rather, which parts of the application are better performed by small local computers and which parts are better performed at some other, possibly centralized, site.
    c. In essence, the best distribution of processing tasks within application areas is sought.
       1) An advantage of distributed processing is **fail-soft protection**, that is, the ability to continue processing at all sites except a nonfunctioning one.
    d. The key distinction between decentralized and distributed systems is the interconnection among the nodes (sites) in the network.

e. EXAMPLE: In processing a sales order, order entry may be handled by a terminal. Upon the completion of the sales order, the terminal will transmit it to a minicomputer in a local warehouse. This computer will determine if the item is available by interrogating an inventory file at the organization's manufacturing plant. If the item is available, the paperwork is produced locally at the warehouse, and the clerk at the terminal is notified. If it is not available, the plant's computer determines the probable delay before the item is available. This information is transmitted via the local warehouse computer to the terminal. If this delay is acceptable to the customer, the production paperwork is performed at the plant. Thus, the actual processing of the order entry is shared by the terminal, the warehouse minicomputer, and the manufacturing plant computer.

f. The increased interdependence among processing sites allows greater flexibility in systems design and the possibility of an optimal distribution of processing tasks. The process of deciding on the best distribution of processing capabilities (hardware), safeguards (controls), information (files and databases), and personnel is not easy.

g. Under this form of in-house system, the firm may hire the services of an outside company, which employs the people to run the firm's computer-based system.

h. **Cooperative processing** is a system in which computers in a distributed processing network can share the use of application programs belonging to another end user. The system assigns different machines the functions they perform best in executing a transaction-based application program.

   1) For example, a microcomputer might be used to enter and validate data for the application, and a mainframe might handle file input and output.

11. Stop and review! You have completed the outline for this subunit. Study multiple-choice questions 1 through 9 beginning on page 355.

B. Databases

1. A database is a series of related files combined to eliminate unnecessary redundancy of data items. A single integrated information system allows for increased data accessibility. When systems within the organization are not integrated, they not only may contain different data but also may define and update data in inconsistent ways. Thus, determining the location of data and ensuring their consistency are more difficult.

   a. EXAMPLE: The various files related to personnel in the conventional record systems of most organizations include payroll, work history, and permanent personnel data.

      1) An employee's name must appear in each of these files when they are stored and processed separately. The result is redundancy.

      2) When data are combined in a database, each item is usually stored only once.

   b. The data are stored physically on direct-access (e.g., magnetic disk) devices. They are also stored for efficient access.

      1) The most frequently accessed items are placed in the physical locations permitting the fastest access.

      2) When these items were stored in separate files in conventional systems, the physical locations were usually similar to the logical structure of the data. Items that logically belonged together were stored in physical proximity to one another.

3) A logical data model is a user view. It is the way a user describes the data and defines their interrelationships based on the user's needs, without regard to how the data are physically stored.

c. The logical structure of a database may be of various kinds.

1) A **tree** or **hierarchical structure** arranges data in a one-to-many relationship in which each record has one antecedent but may have an unlimited number of subsequent records.

a) Because all paths, directories, and indexes must be predefined, a hierarchical structure is inflexible and does not support ad hoc queries.

2) A **network structure** reduces redundancy by arranging data through development of many-to-many relationships; that is, each item may have multiple antecedents as well as successive relationships.

3) A **relational structure** organizes data in conceptual tables. One relation (table or file) can be joined together or related to another by the database management system (see 3. on the following page) without pointers or linked lists if each contains one or more of the same fields (also known as columns or attributes). The relational structure is the most popular because it is relatively easy to construct and is useful for unplanned, ad hoc queries. However, its processing efficiency is relatively low because many accesses may be necessary to execute the select, join, and project commands.

a) **Normalization** is the term for determining how groups of data items in a relational structure are arranged in records in a database. This process relies on **normal forms**, that is, conceptual definitions of data records and specified design rules. Normalization is intended to prevent inconsistent deletion, insertion, and updating of data items.

b) The three basic operations in the relational model are selecting, joining, and projecting.

i) **Selecting** creates a subset of records (rows) that meet certain criteria.

ii) **Joining** is the combining of relational tables based on a common data element.

iii) **Projecting** is the basic operation in a relational database that results in a subset consisting of columns (fields) in a table. This operation creates a new table containing only the required information.

c) **Cardinality** expresses the bounds (a minimum and a maximum) of the association between related entities.

i) EXAMPLE: Assuming the entities are the number of students and a college class, a cardinality limit of (3, 59) for the class entity in the student-class relationship means that a class may contain from 3 to 59 students.

d. The data in a database are subject to the constraint of **referential integrity**. Thus, if data are collected about something, e.g., a payment voucher, all reference conditions regarding it must be met; thus, for a voucher to exist, a vendor must also exist.

e. A distributed database is stored in two or more physical sites using either replication or partitioning.

1) The **replication** or **snapshot technique** makes duplicates to be stored at multiple locations. Changes are periodically copied and sent to each location. If a database is small, storing multiple copies may be cheaper than retrieving records from a central site.

2) **Fragmentation** or **partitioning** stores specific records where they are most needed. For example, a financial institution may store a particular customer's data at the branch where (s)he usually transacts his/her business. If the customer executes a transaction at another branch, the pertinent data are retrieved via communications lines.

f. A **deadly embrace** occurs when each of two transactions has a lock on a data resource that the other transaction needs to run to completion. When deadly embraces occur, the database system must have an algorithm for undoing the effects of one of the transactions and releasing the data resources it controls so that the other transaction can run to completion. Then, the other transaction is restarted and permitted to run to completion. If deadly embraces are not resolved, response time worsens or the system eventually fails.

2. In a tree or network database, the logical relationship among items is achieved through the use of **pointers** (the disk address of a related item).

a. EXAMPLE: Consider the item EMPLOYEE #. For employee #123462, pointers stored with the employee # indicate the location on the disk of other items logically related to this employee.

3. A **Database Management System (DBMS)** is an integrated set of computer programs that create the database, maintain the elements, safeguard the data from loss or destruction, and make the data available to applications programs and inquiries.

a. The DBMS allows programmers and designers to work independently of the physical and logical structure of the database.

b. If programmers and systems designers needed to consider the logical and physical structure of the database, creation of new applications and maintenance of existing applications would be extremely time consuming and therefore expensive.

c. The **schema** is a description of the overall logical structure of the database using **data-definition language**, which is the connection between the logical and physical structures of the database. A **subschema** describes a particular user's (application's) view of a part of the database using data definition language.

d. A fundamental characteristic of databases is that applications are independent of the database structure; when writing programs or designing applications to use the database, only the name of the desired item is necessary.

e. Reference is made to the item(s) using the **data manipulation language**, after which the DBMS locates and retrieves the desired item(s).

1) The data manipulation language is used to add, delete, retrieve, or modify data or relationships.

f. The physical structure of the database can be completely altered without having to change any of the programs using the data items. Thus, different users may define different views of the data (subschemas).

g. Database management software for personal computers includes dBASE IV(R), dBASE III PLUS(R), Paradox, Rbase, Foxbase, and Microsoft Access (PC). Mainframe DBMSs include DB2 and Oracle.

4. **Other Database Definitions**
   a. The **database administrator** (DBA) is the individual who has overall responsibility for developing and maintaining the database and for establishing controls to protect its integrity. Thus, only the DBA should be able to update data dictionaries. In small systems, the DBA may perform some functions of a DBMS. In larger applications, the DBA uses a DBMS as a primary tool.
   b. The **data dictionary** is a file, either computer or manual, that describes the use of data from the database in applications. It provides a mapping from the database to applications and vice versa.
   c. The **database mapping facility** is software that is used to evaluate and document the structure of the database
   d. The **data control language** specifies the privileges and security rules governing database users.
   e. **Data command interpreter languages** are symbolic character strings used to control the current state of DBMS operations.
5. Storing all related data on one storage device creates security problems.
   a. Should hardware or software malfunctions occur, or unauthorized access be achieved, the results could be disastrous.
   b. Greater emphasis on security is required to provide backup and restrict access to the database. For example, the system may employ dual logging, that is, use of two transaction logs written simultaneously on separate storage media. It may also use a snapshot technique to capture data values before and after transaction processing. The files that store these values can be used to reconstruct the database in the event of data loss or corruption.
   c. The responsibility for creating, maintaining, securing, and restricting access to the database belongs to the database administrator.
   d. A DBMS includes security features. Thus, a specified user's access may be limited to certain data fields or logical views depending on the individual's assigned duties.
6. Databases and the associated DBMS permit efficient storage and retrieval of data for formal system applications.
   a. They also permit increased ad hoc accessing of data (e.g., to answer inquiries for data not contained in formal system outputs) as well as updating of files by transaction processing.
   b. These increased abilities, however, require
      1) The use of sophisticated hardware (direct-access devices)
      2) Sophisticated software (the DBMS)
      3) Highly trained technical personnel (database administrator, staff)
      4) Increased security controls
   c. All of the abilities listed above result in increased cost.
7. Stop and review! You have completed the outline for this subunit. Study multiple-choice questions 10 through 20 beginning on page 357.

C. Data Communications and Networks

1. The movement of data among CPUs and remote devices requires special hardware and software and telecommunications technology.
   a. To connect computers and remote terminals, the following devices may be used:
      1) **Communications processors** are minicomputers that are also called front-end processors. They perform message switching, move data to primary storage, translate coded data, and otherwise relieve the CPU of certain communications control functions.
         a) They are located between the computer and the modem in the network.
         b) Nonprogrammable units are known as communications controllers.
      2) **Multiplexors** are switching devices that route or channel the flow of data. They intermix the two-way flow of data so that data may flow over one line. A multiplexor channel permits sending more than one message on a communication line (interleaving). Thus, several terminals may be able to share a communication line to a CPU.
         a) A **concentrator** is a programmable, more expensive alternative to a multiplexor.
      3) **Modem (modulator-demodulator)** is a hardware device to convert digital signals from terminals and the CPU into analog signals for transmission across data (usually telephone) lines. The receiving modem converts the analog signal back to digital form for use by the receiving terminal or CPU.
         a) If digital transmission facilities are available, however, a modem is not required. Instead, the user must employ a digital interface or data service unit as a connection with the digital transmission service.
            i) Digital transmission is less prone to error because it is less sensitive to electrical interference.
         b) One way in which modems may differ is in their **bit rates**, not to be confused with **baud rates**. The bit rate, usually measured in bits per second, is a measurement of the transmission speed. The baud rate is the number of signal changes or cycles per period of time and cannot exceed the bandwidth of the communication channel.
            i) At high speeds, more than one bit may be transmitted by a signal change. Hence, the bit rate may be greater than the baud rate.
            ii) A telecommunications medium's transmission capacity depends on its frequency, i.e., the number of signal changes or cycles per second that can be sent through the medium as measured in **hertz**. The **bandwidth** is the range of frequencies from highest to lowest that a given telecommunications channel can accommodate. As the bandwidth increases, the capacity of the medium also increases.

4) **Communications channels** differ from the data channels connecting the CPU and peripheral equipment. They are the communications media for transmitting data and are classified according to their capacity.

   a) Narrowband (baseband), e.g., telegraph lines

   b) Voiceband, e.g., telephone lines

   c) Broadband, e.g., microwave circuits and satellite channels

   d) Baseband network -- a type of LAN used solely for data communications

5) Other hardware items used to connect computers in a network include:

   a) A **hub** is a central connecting device in a network that joins communications lines together in a star configuration.

      i) Passive hubs are just connecting units that add nothing to the data passing through them.

      ii) Active hubs, also sometimes called multiport repeaters, regenerate the data bits in order to maintain a strong signal.

      iii) Intelligent hubs provide added functionality, such as network management, bridging, routing, and switching.

   b) A **switch** is another connecting device. Each port on a switch can give full bandwidth to a single server, client, or hub.

   c) **Bridges** connect two or more LAN segments together. The segments can be of similar or dissimilar types. Bridges improve network performance by keeping traffic contained within smaller segments.

   d) **Routers** are devices that route data packets from one local area network (LAN) or wide area network (WAN) to another. Routers read the network address in each transmitted frame and make a decision on how to send it based on the most expedient route.

b. Transmission media include the following:

1) **Twisted copper wire** is used for analog communication by telephone. It is a slow medium for data transmission, but new software and hardware have improved its capacity.

2) **Coaxial cable** is used for cable television. It consists of thickly insulated copper wire that is faster and more interference free than twisted wire.

3) **Fiber optic cable** is made of clear glass fibers bound into a cable. Data are transmitted by pulses of light. It is faster, lighter, and more durable than the wire media and is difficult to tap.

4) **Microwave systems** use high-frequency radio signals transmitted through the atmosphere. Satellites may serve as relay points.

c. Transmission modes may be asynchronous or synchronous.

1) **Asynchronous** or **start-stop transmission** is used for slow, irregular transmissions, such as from a keyboard terminal. Each character is marked by a start and stop bit.

2) **Synchronous transmission** is used when rapid, continuous transmission is desired. It transfers blocks of characters without start and stop bits but requires that the sending and receiving modems be synchronized.

d. The following are the types of transmission circuits:

1) **Simplex** transmission is in one direction only, such as for display purposes.

2) **Half-duplex** transmission is in both directions but not at the same time. It is appropriate when processing is online but a response is not required.

3) **Duplex** transmission is in both directions at once, which is a necessity for real-time processing.

e. **Teleprocessing** is computer processing via remote terminals. Communications software for teleprocessing is necessarily complex because of the multiple tasks to be performed when many terminals are in simultaneous use. The CPU, the front-end processor, and the concentrator may all have communications software.

1) This software performs, among other things, the following functions:

a) Receives input, locates the appropriate program, loads it into memory, transmits the input to the program, and passes the output to the user

b) Identifies and corrects errors and provides for security. Encryption is a typical security measure. A program codes data prior to transmission. Another program decodes it after transmission.

c) Maintains a log of activity and a database or file of updated records

d) Manages buffers (special storage areas) that hold input before processing

e) Manages the sequencing and proper routing of messages

2) A **protocol** is a set of rules for message transmission among the devices in the network. Each device should adhere to the same protocol.

2. **Types of Networks**

a. **Private networks** are dedicated facilities, e.g., satellites, microwave transmitters, or telephone lines, leased from a common carrier. Hence, no dial-up access is required, and security is enhanced. A **PBX (private branch exchange)** is a special computer at a company facility used to store, hold, transfer, and redial telephone calls. It can carry both voice and data and can switch digital data among computers and office equipment, e.g., printers, copiers, and fax machines. A PBX uses telephone lines, so its data transmission capacity is limited.

b. **Public-switched networks** use public telephone lines. This arrangement may be the most economical, but data transmission may be of lower quality, no connection may be available, and security measures may be ineffective.

c. **Value-added networks (VANs)** lease the communications medium of a common carrier (such as telephone lines) but add value by improving the efficiency of transmissions, for example, by packet switching. A VAN also provides error detection and correction services and electronic mailbox facilities for EDI purposes.

1) **Packet switching** divides a file into small packages that are sent independently by available communication channels through a network and then reassembled at the receiving end.

a) **Frame relay** is faster and less expensive but, unlike packet switching, does not involve error correction. Frames are similar to packets. They are best used for data rather than voice or video communication.

d. A **local area network (LAN)** is a local distributed computer system, e.g., within a single office. Computers, communication devices, and other equipment are linked by cable. Special software facilitates efficient data communication among the hardware devices. The channel technology may be baseband or broadband. Baseband allows one path for transmission of video, voice, text, or graphics, and only one data type may be transmitted. Broadband provides multiple paths.

1) The LAN's hardware consists of several microcomputers either attached to a host computer, linked as part of several LANs that may or may not communicate with a host computer, or connected together but not connected to a host computer.

a) A **peer-to-peer network** operates without a mainframe or server.

2) A LAN may also use wireless spread spectrum broadcasting rather than a direct cable link. However, a wireless LAN is likely to have slower response times.

3) A network interface card links microcomputers and printers in a local area network that is connected by coaxial cable, twisted pair, or optical fiber. The card creates an address for the microcomputer, transmits data, and monitors incoming messages.

4) A **server** is a computer in the LAN that functions as a librarian. It stores programs and files for users and determines who obtains access to what. It also contains the **network operating system**.

5) A **gateway** is a means of connecting otherwise incompatible networks, nodes, or devices. It converts one set of communication protocols to another.

6) **Ethernet** is a set of LAN standards that allows networking products from different vendors to communicate with each other. It is the most widely used LAN technology.

e. A **wide area network (WAN)** provides data communication and file sharing among remote offices. It is an enterprise-wide communications network that allows signals to be transmitted from a LAN via a public or private line to other LANs in distant locations.

f. The **Internet** is a series of networks throughout the world that facilitates information transfer among computers. Gateways allow mainframe computers to interface with personal computers. The graphics-rich environment of the **World Wide Web (WWW)** has been largely responsible for bringing the Internet out of the static realm of e-mail and text-only documents. The WWW continues to grow faster than any other segment of the Internet. Businesses, schools, government, and nonprofit organizations (in addition to millions of individuals) are flocking to the Internet to promote themselves and their products to an audience spanning the entire planet.

1) The three main parts of the World Wide Web are the server that holds the information, the client that is viewing the information, and protocols that connect the two.

2) The WWW is accessed via programs called **browsers**. They allow text, graphics, and sound to be integrated for the user.

3) An **intranet** is an internal network that applies Internet connectivity standards and WWW software. An intranet addresses the connectivity problems of an organization with many types of computers.

g. **Virtual private networks (VPN)**

1) Businesses have traditionally relied on private leased lines to link offices so that workers could share information over a WAN. However, while providing a high degree of privacy, leased lines are expensive to set up and maintain. For many companies, a leased line may be impractical, providing more bandwidth than is needed at too high a price.

2) VPN emerged as a relatively inexpensive way to solve this problem. Rather than maintain a point-to-point leased line, a company connects each office or LAN to a local Internet service provider and routes data through the Internet, thereby using shared, low-cost public bandwidth as the communications backbone.

3) However, the major concern in using a public network for electronic data exchange is security. Unprotected data sent across the Internet are susceptible to being viewed, copied, or modified by unauthorized parties. The success of VPNs will depend on the development of secure encryption products that protect data while in transit across the Internet.

h. Given the worldwide proliferation of computer networks, **connectivity** has become a major issue. The desire for greater connectivity favors open systems.

1) **Open systems** are those for which suppliers provide components whose interfaces are defined by public standards. For example, the U.S. government specifies that its suppliers adhere to the Unix operating system and the telecommunications protocols developed for the Internet. In contrast, a closed system's components are built to proprietary standards so that the equipment made by other suppliers could not interface with the existing system. Accordingly, converting to open systems increases the number of vendors from which substitutable components can be acquired, which increases price competition for equipment.

2) Although uniform standards for telecommunications, networking, operating systems, and user interfaces have not yet emerged, some standards have been created by governments, industry associations, and international organizations.

a) **Open Systems Interconnect (OSI)** has been developed by the International Organization for Standardization. It is a seven-layer reference model that allows different types of computers and networks to communicate.

b) **Integrated Services Digital Network (ISDN)** provides international standards for voice, video, and data communications over telephone lines.

3. **Applications**

a. **Electronic mail** permits transfer, receipt, and storage of messages within or between computer systems. The "mail" consists of electronically transmitted messages. A user's "mailbox" is the storage area allocated for messages. The advantages of electronic mail are high-speed transmission, reduction of message preparation costs, and the possibility of sending or reading messages at a convenient time.

1) A typical system permits a user to answer messages, compose or delete messages, edit, file, forward messages to other users, move items among files, read, retrieve from files, scan contents of files, send messages, and print.

b. **Voice mail** converts spoken messages from analog to digital form, transmits them over a network, and stores them on a disk. Messages are then converted back to analog form when the recipient desires to hear them. Afterward, they may be saved, forwarded, or deleted.

c. Conducting an electronic meeting among several parties at remote sites is **teleconferencing**. It can be accomplished by telephone or electronic mail group communication software.

1) **Videoconferencing** permits the conferees to see each other on video screens.

2) These practices have grown in recent years as companies have attempted to cut their travel costs.

d. A **fax machine** can scan a document, convert its text and graphics to digital form, and then transmit the result over telephone lines. The recipient's fax machine can then create a facsimile of the transmitted document.

e. An **electronic bulletin board** is a database into which computer users may dial to read or post messages.

f. **Electronic data interchange (EDI)**

1) See section E.

g. **Groupware**

1) See Study Unit 7.

4. **Network Configurations (Topologies)**

a. **Point-to-point** networks provide a separate, direct link between each remote terminal and the CPU.

b. **Multidrop** (or **BUS**) networks provide links for each terminal to a single communications line connected to the CPU. However, only one terminal may send or receive messages at one time. A superior but more costly alternative is to use a line-sharing device (a multiplexor or a concentrator) to connect the group of terminals to the CPU.

1) Ethernet is an example of a network technology that is based on a BUS topology.

c. **Ring networks** have no central computer. Each computer can communicate with every other computer, but their connection forms a closed loop in which data pass from one device to another and always in one direction. In a **token ring network**, a packet of data (the token) passes along the network. Each computer reads it and either accepts data from it, loads data onto the token, or allows it to pass without change.

d. **Completely connected networks** have direct links among all computer locations.

e. **Star networks** permit each remote computer a direct link to the central location but not to other remote computers.

5. **Basic Architectures for Desktop Computing**
   a. **Client-server model**. A client-server system divides processing of an application between a client machine on a network and a server. This division depends on which tasks each is best suited to perform.
      1) However, user interaction is ordinarily restricted to the client part of the application. This portion normally consists of the user interface, data entry, queries, and receipt of reports.
      2) The server customarily manages peripheral hardware and controls access to shared databases. Thus, a client-server application must be designed as separate software components that run on different machines but appear to be one application.
      3) Security for client-server systems may be more difficult than in a mainframe-based system because of the numerous access points. They also use distributed processing methods that result in heightened risk of unauthorized access to data and processing. New methods of accessing data and processing, such as remote procedure calls, are also available.
   b. **X-terminal model**. In this architecture, desktop machines that lack stand-alone processing power have access to remote computers in a network. To run an application, programs are downloaded to the terminal. These machines are relatively inexpensive because they have no disk drives.
6. Stop and review! You have completed the outline for this subunit. Study multiple-choice questions 21 through 32 beginning on page 361.

D. Internal Control

1. **Introduction**. The use of computers in information systems has fundamental effects on internal control but not on its objectives or basic philosophy. These effects flow from the characteristics that distinguish computer-based from manual processing.
   a. **Transaction trails**. A complete trail useful for audit purposes might exist for only a short time or in only computer-readable form.
   b. **Uniform processing of transactions.** Computer processing uniformly subjects like transactions to the same processing instructions and thus virtually eliminates clerical error. However, programming errors (or other similar systematic errors in either the hardware or software) will result in all like transactions being processed incorrectly when they are processed under the same conditions.
   c. **Segregation of functions.** Many controls once performed by separate individuals may be concentrated in computer systems. Hence, an individual who has access to the computer may perform incompatible functions. As a result, other control procedures may be necessary to achieve the control objectives ordinarily accomplished by segregation of functions. Other controls may include
      1) Adequate segregation of functions within the computer processing activities.
      2) Establishment of a control group to prevent or detect processing errors or fraud.
      3) Use of password controls to prevent incompatible functions from being performed by individuals with online access to assets and records.

d. **Potential for errors and fraud.** The potential for individuals, including those implementing controls, to gain unauthorized access to data, to alter data without visible evidence, or to gain access (direct or indirect) to assets may be greater in computer systems. Decreased human involvement in handling transactions can reduce the potential for observing errors and fraud. Errors or fraud in the design or modification of application programs can remain undetected for a long time.

e. **Potential for increased management supervision.** Computer systems offer management many analytical tools for review and supervision of operations. These additional controls may enhance internal control. For example, traditional comparisons of actual and budgeted operating ratios and reconciliations of accounts are often available for review on a more timely basis. Additionally, some programmed applications provide statistics regarding computer operations that may be used to monitor actual processing.

f. **Initiation or subsequent execution of transactions by computer.** Certain transactions may be automatically initiated or certain procedures required to execute a transaction may be automatically performed by a computer system. The authorization of these transactions or procedures may not be documented in the same way as those in a manual system, and management's authorization may be implicit in its acceptance of the design of the system.

g. **Dependence of other controls on controls over computer processing.** Computer processing may produce reports and other output that are used in implementing manual controls. The effectiveness of these controls can be dependent on the effectiveness of controls over the completeness and accuracy of computer processing. For example, the effectiveness of a manual review of a computer-produced exception listing is dependent on the controls over the production of the listing.

2. **Classification of Controls**. The broad categories of controls are defined below.

   a. **General controls** concern all computer activities. They relate to all or many computerized accounting activities and often include control over the development, modification, and maintenance of computer programs and control over the use of and changes to data maintained on computer files. General controls encompass

      1) The plan of organization and operation of the computer activity

         a) Organizational controls are concerned with the proper segregation of duties and responsibilities within the computer processing environment.

            i) The responsibilities of systems analysts, programmers, operators, file librarians, and the control group should be performed by different individuals, and proper supervision should be provided.

         b) Operating controls ensure efficient and effective operation within the computer department.

            i) These controls also assure proper procedures in case of data loss because of error or disaster.

            ii) Typical operating controls include the proper labeling of all files both internally (machine-readable file header and trailer labels) and externally, halt and error procedures, duplicate files, and reconstruction procedures for files.

2) The procedures for documenting, reviewing, testing, and approving systems or programs and changes thereto

a) Program development and documentation controls are concerned with the proper planning, development, writing, and testing of computer application programs.

i) These activities require proper documentation, including flowcharts, listings, and run manuals for programs already written.

ii) Controls over proper authorization of any changes in existing programs are also necessary.

3) Controls built into the equipment by the manufacturer (hardware controls)

a) Hardware controls assure the proper internal handling of data as they are moved and stored.

i) Hardware controls include parity checks, echo checks, read-after-write checks, and any other procedure built into the equipment to assure data integrity.

4) Controls over access to equipment and data files

a) Access controls provide assurance that only authorized individuals use the system and that usage is for authorized purposes.

i) Such controls include physical safeguards of equipment, proper library security, and passwords.

5) Other data and procedural controls affecting overall computer operations

b. **Application controls** relate to specific tasks performed by the system. They should provide reasonable assurance that the recording, processing, and reporting of data are properly performed. Application controls relate to individual computerized accounting applications, for example, programmed edit controls for verifying customers' account numbers and credit limits.

1) **Input controls** provide reasonable assurance that data received for processing have been properly authorized, converted into machine-sensible form, and identified, and that data (including data transmitted over communication lines) have not been lost, suppressed, added, duplicated, or otherwise improperly changed. Input controls also relate to rejection, correction, and resubmission of data that were initially incorrect.

2) **Processing controls** provide reasonable assurance that processing has been performed as intended for the particular application, i.e., that all transactions are processed as authorized, that no authorized transactions are omitted, and that no unauthorized transactions are added.

3) **Output controls** provide assurance that the processing result (such as account listings or displays, reports, magnetic files, invoices, or disbursement checks) is accurate and that only authorized personnel receive the output.

3. **Organizational and Operating Controls**

   a. **Segregation of duties**. This general control is vital because a segregation of functions (authorization, recording, and access to assets) may not be feasible in a computer environment. For example, a computer may print checks, record disbursements, and generate information for reconciling the account balance, which are activities customarily segregated in a manual system. If the same person provides the input and receives the output for this process, a significant control weakness exists. Accordingly, certain tasks should not be combined.

      1) **Systems analysts** are specifically qualified to analyze and design computer information systems. They survey the existing system, analyze the organization's information requirements, and design new systems to meet those needs. These design specifications will guide the preparation of specific programs by computer programmers.

         a) Systems analysts should not perform programming tasks or have access to computer equipment, production programs, data files, and input-output controls.

      2) **Programmers** design, write, test, and document the specific programs developed by the analysts.

         a) Programmers as well as analysts may be able to modify programs, data files, and controls and should therefore have no access to computer equipment and files or to copies of programs used in production.

      3) **Computer (console) operators** are responsible for the actual processing of data in accordance with the program run manual and messages received from the system (preferably in hardcopy form for review by the control group). They load data, mount storage devices, and operate the equipment.

         a) Console operators should not be assigned programming duties or responsibility for systems design and should have no opportunity to make changes in programs and systems as they operate the equipment. Ideally, computer operators should not have programming knowledge or access to documentation not strictly necessary for their work.

      4) **Data conversion operators** perform the tasks of data preparation and transmission, for example, conversion of source data to magnetic disk or tape and entry of transactions from remote terminals.

      5) **Librarians** should maintain control over and accountability for documentation, programs, and data files.

         a) Librarians should have no access to equipment or the skills to perpetrate fraud.

      6) The **data control group** must be independent of systems development, programming, and operations. It receives user input, logs it, transfers it to the computer center, monitors processing, reviews error messages, compares control totals, distributes output, and determines whether error corrections have been made by users.

b. **Backup and recovery policies and procedures**. A computer center should have a reconstruction and recovery plan that will allow it to regenerate important programs and data files. The center should create backup (duplicate) copies of data files, databases, programs, and documentation; store backup copies off-site; and plan for auxiliary processing at another site.

1) The organization must undertake contingency planning and risk analysis. During an early stage of contingency planning for information systems, management must determine how various processing disruptions may affect the entity. **Risk analysis** identifies and prioritizes critical applications, evaluates their organizational impact, determines recovery time frames and minimum hardware platform requirements, assesses insurance coverage, identifies exposures and their implications, and develops recovery plans.

2) In any information processing environment, losing or otherwise destroying data must be avoided. Not only is the loss of data a problem, but the organization may also require continuous processing without disruptions. For these reasons, all systems must have adequate backup and recovery procedures in the event of system failure, power loss, or other potential corruption of data. The procedures implemented will normally be a function of the specific computer environment, type of processing, or storage mode.

a) **Sequential processing** -- magnetic tape, magnetic disks, and batch processing

i) **Grandfather-father-son approach**. Over a 3-day period, three generations of master files are generated. These master files are retained during this period along with the transaction files. If the current (son) file is destroyed or damaged, the information can be reconstructed using the father and running the current transaction file against it. If both father and son are destroyed, the grandfather along with the previous and current transaction files can be used to reconstruct the information.

ii) **Checkpoint procedures** involve capturing all the values of data and program indicators at specified points and storing these values in another file. If processing is interrupted, it can be resumed at the last checkpoint rather than at the beginning of the run.

b) **Random processing** -- magnetic disks, online processing

i) **Rollback and recovery** involve the dumping of the master file's contents and associated data structures onto a backup file. In the event of a faulty run, the dump is used together with the transaction log or file to reconstruct the file.

c) **Database management systems** -- magnetic disks, online processing

i) **Database systems** require a more elaborate backup procedure. Normally, recovery and restart procedures must provide for continued operations during reconstruction of lost information.

ii) **Dual logging** involves the use of two transaction logs written simultaneously on two separate storage media.

iii) **Before-image/after-image** captures the data values before and after transaction processing and stores them in files. These files can be used to re-create the database in the event of data loss or corruption.

d) **Fully protected systems** have generator or battery backup to prevent data destruction and downtime from electrical power disturbances. Loss of electrical power or voltage fluctuations need not disturb the vulnerable contents of main memory if a noninterruptible system is in place.

3) **Hot-site and cold-site backup facilities**. A hot site is a service bureau. It is a fully operational processing facility that is immediately available, whereas a cold site is a shell facility where the user can quickly install computer equipment.

c. A **computer virus** is a program that infects another program by altering its logic. Infection often results in destruction of data.

1) A virus may be contracted from an infected disk or machine, downloaded software, an electronic bulletin board, etc.

2) Easy access makes a network susceptible to the rapid spread of computer viruses. Ways to minimize computer virus risk in a networked system include

a) Formal security policy

b) Restricted access

c) Passwords (updated regularly)

d) Formal backup-recovery plan

e) Periodic testing with virus detection software

f) Using antivirus software on all shareware prior to introducing it into the network

4. **Program Development and Documentation Controls**

a. **Documentation** is the collection of documents that support and explain computer applications, including systems development. It is helpful to operators and other users, control personnel, new employees, auditors, programmers, and analysts.

1) Documentation should be secured in a library with access controlled.

2) It should be subject to uniform standards regarding flowcharting techniques, coding, and modification procedures (including proper authorization).

3) **System documentation** includes narrative descriptions, flowcharts, the system definition used for development, input and output forms, file and record layouts, controls, change authorizations, and backup procedures.

4) **Program documentation** contains descriptions, program flowcharts and decision tables, program listings of source code, test data, input and output forms, detailed file and record layouts, change requests, operator instructions, and controls.

5) **Operating documentation** (computer run manual) provides information about setup, necessary files and devices, input procedures, console messages and responsive operator actions, run times, recovery procedures, disposal of output, and controls.

6) **Procedural documentation** includes the system's master plan and operations to be performed, documentation standards, procedures for labeling and handling files, and standards for systems analysis, programming, operations, security, and data definition.

7) **User documentation** describes the system and procedures for data entry, error checking and correction, and formats and uses of reports.

b. **Systems and program development controls**

1) Effective systems development requires participation by top management. This can be achieved through a steering committee composed of higher-level representatives of system users. The committee approves or recommends projects and reviews their progress.

2) Studies of the economic, operational, and technical feasibility of new applications will necessarily entail evaluations of existing as well as proposed systems.

3) Another necessary control is the establishment of standards for system design and programming. These standards represent user needs and system requirements determined during the systems analysis.

4) Changes in the computer system should be subject to strict controls. For example, a written request for an application program change should be made by a user department and authorized by a designated manager or committee.

a) The program should then be redesigned using a working copy, not the version currently in use. Also, the systems documentation must be revised.

b) Changes in the program will be tested by the user, the internal auditor, and a systems employee who was not involved in designing the change.

c) Approval of the documented change and the results of testing should be given by a systems manager. The change and test results may then be accepted by the user.

d) Unauthorized program changes can be detected by code comparison. The version in use should be periodically compared with a copy controlled by the auditors. Software can be used to perform this procedure.

i) Versions of critical application programs should be stored in object code. Most users lack the technical skill to modify object code.

5) Proposed programs should be tested with incorrect or incomplete data as well as typical data to determine if controls have been properly implemented in the program.

a) Test data should test all branches of the program, including the program's edit capabilities. The edit function includes sequence checks, valid field tests, reasonableness checks, and other tests of the input data.

b) Expected results should be calculated and compared with actual performance. These results should include both accurate output and error messages.

6) To avoid legal liability, controls should also be implemented to prevent use of unlicensed software not in the public domain. A **software licensing agreement** permits a user to employ either a specified or an unlimited number of copies of a software product at given locations, at particular machines, or throughout the company. The agreement may restrict reproduction or resale, and it may provide for subsequent customer support and product improvements.

5. **Hardware Controls**

a. **Boundary (storage) protection** protects programs or data from interference (unauthorized reading and/or writing) caused by activity related to other programs or data stored on the same medium. Primary storage locations in the CPU may be protected by features built into the hardware, but boundary protection for disk storage is effected through programming.

b. **Diagnostic routines** check for hardware problems. If built into the equipment, they permit the system itself to give notice of imminent failure.

c. **Dual read**. An input device, such as a tape drive, may read an input twice for comparison.

d. **Dual read-write heads**. A dual head first writes on the storage medium and then reads what was written. If the comparison shows that the data written differ from the data at the source of the transfer, the device will back up and rewrite the data. This process provides a check on recorded information.

e. **Duplicate circuitry**. Dual circuits in the arithmetic-logic unit of the CPU permit calculations to be performed twice and compared.

f. An **echo check** provides for a peripheral device to return (echo) a signal sent by the CPU. For example, the CPU sends a signal to the printer, and the printer, just prior to printing, sends a signal back to the CPU verifying that the proper print position has been activated.

g. **File protection**. All data storage media, except hard disks, have a ring, tab, or notch that can be used to prevent or allow writing.

h. A **parity check** adds the bits in a character or message and checks the sum to determine if it is odd or even, depending on whether the computer has odd or even parity. This check verifies that all data have been transferred without loss. For example, if the computer has even parity, a bit will be added to a binary coded character or message that contains an odd number of bits. No bit is added if a character or message in binary form has an even number of bits.

i. **Preventive maintenance**. Regular servicing avoids equipment failure.

j. **Read-write suppression**. A control on a disk drive may prevent reading from or writing on a disk, e.g., one containing production programs.

k. **Validity checks**. Hardware that transmits or receives data compares the bits in each byte to the permissible combinations in order to determine whether they constitute a valid structure.

6. **Access Controls**. Access controls, such as passwords, ID numbers, access logs, and device authorization tables, prevent improper use or manipulation of data files and programs. They ensure that only those persons with a bona fide purpose and authorization have access to data processing.

a. **Passwords and ID numbers**. The use of passwords and identification numbers is an effective control in an online system to prevent unauthorized access to computer files. Lists of authorized persons are maintained in the computer. The entry of passwords or identification numbers, a prearranged set of personal questions, and the use of badges, magnetic cards, or optically scanned cards may be combined to avoid unauthorized access.

1) A security card may be used with a microcomputer so that users must sign on with an ID and a password. The card controls the machine's operating system and records access data (date, time, duration, etc.).

2) Proper user authentication by means of a password requires password-generating procedures to assure that valid passwords are known only by the proper individuals. Thus, a password should not be displayed when entered at a keyboard.

3) Password security may also be compromised in other ways. For example, because logon procedures may be cumbersome and tedious, users often store logon sequences on their personal computers and invoke them when they want to use mainframe facilities. A risk of this practice is that anyone with access to the personal computers could logon to the mainframe.

b. A **device authorization table** may restrict file access to those physical devices that should logically need access. For example, because it is illogical for anyone to access the accounts receivable file from a manufacturing terminal, the device authorization table will deny access even when a valid password is used.

1) Such tests are often called **compatibility tests** because they ascertain whether a code number is compatible with the use to be made of the information. Thus, a user may be authorized to enter only certain kinds of data, have access only to certain information, have access but not updating authority, or use the system only at certain times. The lists or tables of authorized users or devices are sometimes called **access control matrices**.

c. A **system access log** records all attempts to use the system. The date and time, codes used, mode of access, and data involved are recorded.

d. **Encryption** involves using a fixed algorithm to manipulate plain text. The information is sent in its manipulated form and the receiver translates the information back into plain text. Although data may be accessed by tapping into the transmission line, the encryption key is necessary to understand the data being sent.

e. A **callback** feature requires the remote user to call the computer, give identification, hang up, and wait for the computer to call the user's authorized number. This control ensures acceptance of data transmissions only from authorized modems. However, call forwarding may thwart this control.

f. **Controlled disposal of documents**. One method of enforcing access restrictions is to destroy data when they are no longer in use. Thus, paper documents may be shredded and magnetic media may be erased.

g. **Biometric technologies**. These are automated methods of establishing an individual's identity using physiological or behavioral traits. These characteristics include fingerprints, retina patterns, hand geometry, signature dynamics, speech, and keystroke dynamics.

h. **Automatic log-off** (disconnection) of inactive data terminals may prevent the viewing of sensitive data on an unattended data terminal.

i. **Utility software restrictions**. Utility software may have privileged access and therefore be able to bypass normal security measures. Performance monitors, tape and disk management systems, job schedulers, online editors, and report management systems are examples of utility software. Management can limit the use of privileged software to security personnel and establish audit trails to document its use. The purpose is to gain assurance that its uses are necessary and authorized.

j. **Security personnel**. An organization may need to hire security specialists. For example, developing an information security policy for the organization, commenting on security controls in new applications, and monitoring and investigating unsuccessful access attempts are appropriate duties of the information security officer.

7. **Application Controls**

a. **Input controls**

1) **Edit checks** are programmed into the software. They include

a) **Error listing**. Editing (validation) of data should produce a cumulative automated error listing that includes not only errors found in the current processing run but also uncorrected errors from earlier runs. Each error should be identified and described, and the date and time of detection should be given. Sometimes, the erroneous transactions may need to be recorded in a suspense file. This process is the basis for developing appropriate reports.

b) **Field checks** are tests of the characters in a field to verify that they are of an appropriate type for that field. For example, the field for a Social Security number should not contain alphabetic characters.

c) **Financial totals** summarize dollar amounts in an information field in a group of records.

d) A **hash total** is a control total without a defined meaning, such as the total of employee numbers or invoice numbers, that is used to verify the completeness of data. Thus, the hash total for the employee listing by the personnel department could be compared with the total generated during the payroll run.

e) **Limit and range checks** are based on known limits for given information. For example, hours worked per week will not equal 200.

f) **Preformatting**. To avoid data entry errors in online systems, a screen prompting approach may be used that is the equivalent of the preprinted forms routinely employed as source documents. The dialogue approach, for example, presents a series of questions to the operator. The preformatted screen approach involves the display of a set of boxes for entry of specified data items. The format may even be in the form of a copy of a transaction document.

g) **Reasonableness (relationship) tests** check the logical correctness of relationships among the values of data items on an input and the corresponding master file record. For example, it may be known that employee John Smith works only in departments A, C, or D; thus, a reasonableness test could be performed to determine that the payroll record contains one of the likely department numbers. In some texts, the term reasonableness test is defined to encompass limit checks.

h) **Record count** is a control total of the number of records processed during the operation of a program.

i) **Self-checking digits** may be used to detect incorrect identification numbers. The digit is generated by applying an algorithm to the ID number. During the input process, the check digit is recomputed by applying the same algorithm to the code actually entered.

j) **Sequence checks** determine that records are in proper order. For example, a payroll input file is likely to be sorted into Social Security number order. A sequence check can then be performed to verify record order.

k) **Sign checks** assure that data in a field have the appropriate arithmetic sign. For example, hours worked in a payroll record should always be a positive number.

l) **Validity checks** are tests of identification numbers or transaction codes for validity by comparison with items already known to be correct or authorized. For example, Social Security numbers on payroll input records can be compared with Social Security numbers authorized by the personnel department.

m) An **overflow test** is a programmed control that checks computational results and issues a warning if the result exceeds the capacity of the storage location, which would result in the loss of data. For example, if 5428 were stored as 542, the 8 lost on overflow would be discovered.

2) **Key verification** entails rekeying input and comparing the results.

3) A **redundancy check** requires sending additional data items to serve as a check on the other transmitted data; for example, part of a customer name could be matched against the name associated with the transmitted customer number.

4) An **echo check** is an input control over transmission along communications lines. Data are sent back to the user's terminal for comparison with the transmitted data.

5) **Completeness checks** of transmission of data determine whether all necessary information has been sent. The software notifies the sender if something is omitted.

   a) A complementary transmission control numbers and dates each message sent from a given terminal. This procedure allows verification of the completeness of the sequence and establishment of control totals for each terminal.

b. **Processing controls**

1) Some input controls are also processing controls, e.g., limit, reasonableness, and sign tests.

2) Other tests of the logic of processing are posting, cross-footing, and zero-balance checks.

   a) Comparing the contents of a record before and after updating is a posting check.

   b) Cross-footing compares an amount to the sum of its components.

   c) A zero-balance check adds the positive and negative amounts posted. The result should be zero.

3) **Run-to-run control totals** (e.g., record counts or certain critical amounts) should be generated and checked at designated points during processing.

   a) **Proof account activity listing**. In an online system, the change in a file for the day can be compared with source information.

4) **Internal header and trailer labels** ensure that incorrect files are not processed.

   a) A matching test should make certain an updating transaction is matched with the appropriate master file.

5) Programs used in processing should be tested, for example, by reprocessing actual data with.a known result or by employing test data.

6) **End-of-file procedures** should be available to avoid errors such as prematurely closing the transaction file when the end of the current master file is reached. The transaction file may contain new records to be added to the master file.

7) **Concurrency controls** manage situations in which two or more programs attempt to use a file or database at the same time.

8) An **audit trail** should be created through the use of input-output control logs, error listings, transaction logs, and transaction listings.

9) **Key integrity checks** prevent the updating process from creating inaccuracies in keys. Keys are attributes of records that permit them to be sorted. A primary key is the data item(s) that is the principal identifier, e.g., the vendor number. A secondary key is an alternative used either for sorting or special processing, e.g., the payment date on the vendor's invoice.

c. **Output controls**

1) The **data control group** supervises output control.

a) The daily **proof account activity listings** (changes in master files) should be sent to users for review.

b) **Error listings** should be received directly from the system by the control group, which should make any necessary inquiries and send the errors to users for correction and resubmission.

c) The **console log** should be reviewed for unusual interruptions, interventions, or other activity.

d) Output should be distributed in accordance with distribution registers that list authorized users.

e) **End-of-job markers** on the last page of printed output permits verification that the entire report has been received.

f) **Spooler controls** prevent access to spooled output, i.e., to the results of processing that are temporarily stored in an intermediate file rather than immediately printed.

2) An important detective control is user review of output. Users should be able to determine when output is incomplete or not reasonable, particularly when the user prepared the input. Thus, users as well as computer personnel have a quality assurance function.

8. **Auditing** is a fundamental aspect of control. Moreover, the principles underlying other audits also apply to audits of computer information systems. However, certain audit techniques are specific to these systems. They are presented in the tabulation on the next three pages.

COMPUTER AUDIT TOOL AND TECHNIQUE TABULATION

Tool/Technique	Explanation	Advantages	Disadvantages	Type of System or Environment Used In
Control Flowcharting (Analytic Audit Flowchart)	*Consists of developing flowcharts of the overall system -- manual and computer processing	*Aids in review of internal controls *Familiarizes auditor with the system *Develops communication between auditor and auditee *Normally used by auditors; no special training *Can use flowcharting software	*Costly to develop and maintain	*Any type of system
Test Data (Test Deck)	*Involves the use of specifically prepared sets of input data that test application controls by running a variety of transactions to be compared with previously determined results *The ideal test deck includes every possible combination of transactions and master file situations.	*Can use test data generator software *Test data can be prepared by persons with little technical background.	*Difficult to anticipate all combinations of transactions *Must determine that routinely used edition of application program is used	*Batch processing
Integrated Test Facility (ITF or "Minicompany" Approach)	*Involves the use of a fictitious entity, such as a dummy customer in accounts receivable, against which data transactions are processed and results are compared with previously determined results. This procedure is used within the framework of regular production, frequently without computer operator knowledge.	*Enables testing of the system as it routinely operates *Low processing costs *No special processing	*Effects of transactions on operations (books) must be nullified. *Quantity of live data inputs may be limited when submitted with regular runs. *Possibility of contamination of database	*Online, real-time

Tool/Technique	Explanation	Advantages	Disadvantages	Type of System or Environment Used In
Parallel Simulation	*Involves the use of specially prepared application-type programs to process transactions which have also been run in routine processing. This in effect simulates routine processing in an effort to verify results. A relative of parallel simulation is controlled reprocessing, which entails use of a copy of the auditee's program to reprocess actual data.	*Testing can be done on a surprise basis. *Cost of preparing test data is eliminated. *Can process many of auditee's transactions, eliminating need for small samples *More thorough than sampling	*Usually simulation is for only selected portions of a total application. *Cost of developing program may be prohibitive. *Auditor may need special skills. *Does not have broad application	*Database Management Systems *Most effective when applied to calculations, decisions, and large quantity of transactions
Generalized Audit Software Package (GASP)	*Involves the use of computer software packages (programs) that may allow not only parallel simulation but also a variety of other processing functions, such as extracting sample items, verifying totals, developing file statistics, retrieving specified data fields	*Can process several files (and file types) *Enables use with limited training *Packages interface with many types of hardware and software. *Decreases auditor dependence on data processing personnel and time	*Limited application in online, real-time systems *Limited logical and mathematical capabilities	*Batch processing
Transaction Selection	*Involves reprocessing transaction files with specialized computer programs, which extract or select particular transactions. Transaction files are retained and given to the auditor.	*Programs completely independent of application programs *Enables efficient selection of transactions for further review *Enables selection of targeted types of transactions	*May be costly to develop and execute	*All systems
Embedded Data Collection	*A transaction selection approach incorporated within the regular production programs to routinely select/ extract transactions meeting specified criteria for further testing	*All system activity is subject to review. *Can be used with online systems *Not limited to input transactions	*Additional processing cost of extra audit module program steps that must be executed *Difficult to implement unless it can be developed along with the system	*Online systems
Extended Records	*Involves extending the record of an application to include the data fields necessary to complete an audit trail	*Reduces audit costs associated with tracing audit trails *Creates an audit trail, which was previously nonexistent in hard copy of system input	*May not be economical because of added storage costs *High cost of implementation	*Database Management Systems *Online systems *Systems which normally lack hard copy of system input or other audit trail

Tool/Technique	Explanation	Advantages	Disadvantages	Type of System or Environment Used In
Tracing	*Involves the generation of a complete audit trail to trace transactions through processing. It normally provides a trail of the program statements executed through processing.	*Aids in verifying application of internal controls *Allows tagging of certain live data through the system *Can be used with both test and live data	*Auditor must have significant knowledge of the program to follow logic. *Can increase program processing time	*Advanced
Snapshot	*Involves capturing the data used in processing at a specific point in the stream of processing	*No permanent effect on data *Can insert code quickly with minimum development time *Can help to generate hard copy transaction trail for a specified item *Can be used selectively to print out data that meet certain criteria, e.g., over limit *Can aid in debugging *Can aid in determining intermediate values of data in processing stream	*Code must be embedded in the application program. *Code adds to the processing time. *Requires skilled data processing personnel	*Effective in high-volume systems where complete transaction trail would generate too much information
Mapping	*Involves monitoring the execution of an application program to determine certain statistical information about the run, e.g., program lines not executed, CPU time for certain program lines, and the number of times certain lines were executed	*Can aid in evaluating how well test data tested a run *Can indicate lines of code which are extraneous or not often used	*High cost	*Advanced
System Control Audit Review File (SCARF) (Specific Implementation of Embedded Data Collection)	*Involves the incorporation of reasonableness tests into the normal processing of application programs, the results being reported to the auditor rather than to the user for the auditor's review and investigation	*All system activity is subject to review. *Can be used with online systems *Not limited to input transactions	*Additional processing cost of extra audit module program steps that must be executed *Difficult to implement unless it can be developed along with the system	*Advanced
Sample Audit Review File (SARF)	*Similar to SCARF but random selection of transactions rather than special edit or reasonableness tests	*All system activity is subject to review. *Can be used with online systems *Not limited to input transactions	*Additional processing cost of extra audit module program steps that must be executed *Difficult to implement unless it can be developed along with the system	*Advanced

9. **Internet Security**. The most important control is to install an organization-wide network security policy.

   a. It should seek to ensure the following:

      1) **Availability** is the ability of the intended and authorized users to access computer resources to meet organizational goals.

      2) **Confidentiality** is assurance of the secrecy of information that could adversely affect the organization if revealed to the public or competitors.

      3) **Integrity** is maintained by preventing the unauthorized or accidental modification of programs or data.

b. The network security plan should at the very least include a user account management system, installation of an Internet firewall, and methods such as encryption to ensure that only the intended user receives the information and that the information is complete and accurate.

c. **User account management** involves installing a system to ensure that

1) New accounts are added correctly and assigned only to authorized users.
2) Old and unused accounts are removed promptly.
3) Passwords are changed periodically, and employees are educated on how to choose a password that cannot be easily guessed (e.g., a password of at least six characters that do not form a word should be chosen).

d. A **firewall** separates an internal network from an external network (e.g., the Internet) and prevents passage of specific types of traffic. It identifies names, Internet Protocol (IP) addresses, applications, etc., and compares them with programmed access rules.

1) A firewall may have any of the following features:
   a) A **packet filtering system** examines each incoming IP packet.
   b) A **proxy server** maintains copies of web pages to be accessed by specified users. Outsiders are directed there, and more important information is not available from this access point.
   c) An **application gateway** limits traffic to specific applications.
   d) A **circuit-level gateway** is a filter that connects an internal device, e.g., a network printer, with an outside TCP/IP port. It can identify a valid TCP session.
   e) **Stateful inspection** stores information about the state of a transmission and uses it as background for evaluating messages from similar sources.
2) Firewall systems ordinarily produce reports on organization-wide Internet use, exception reports for unusual usage patterns, and system penetration-attempt reports. These reports are very helpful as a method of continuous monitoring, or logging, of the system.
   a) Firewalls do not provide adequate protection against computer viruses. Thus, an organization should include one or more virus controls in its network security policy.

e. **Encryption** technology converts data into a code. Unauthorized users may still be able to access the data but, without the encryption key, they will be unable to decode the information.

1) Encryption software uses a fixed algorithm to manipulate plain text and an encryption key to introduce variation. The machine instructions necessary to encrypt and decrypt data can constitute a 20%-to-30% increase in system overhead.

2) Encryption technology may be either hardware- or software-based. Two major types of encryption software exist.

a) **Public-key/private-key**, or asymmetric, encryption requires two keys: The public key for encrypting messages is widely known, but the private key for decrypting messages is kept secret by the recipient.

i) RSA, named for its developers (Rivest, Shamir, and Adelman), is the most commonly used public-key/private-key method.

b) **Secret-key**, or symmetric, encryption requires only a single key for each pair of parties that want to send each other encrypted messages.

i) Data Encryption Standard (DES), a shared private-key method developed by the U.S. government, is the most prevalent secret-key method.

f. **Other controls** include authentication and checksums.

1) **Authentication** measures verify the identity of the user to ensure that only the intended and authorized users gain access to the system.

a) Most firewall systems provide authentication procedures.
b) Access controls are the most common authentication procedures.

2) **Checksums** help ensure the integrity of data by checking whether the file has been changed. The system computes a value for a file and then proceeds to check whether this value equals the last known value for this file. If the numbers are the same, the file has likely remained unchanged.

10. Stop and review! You have completed the outline for this subunit. Study multiple-choice questions 33 through 51 beginning on page 365.

E. Electronic Data Interchange (EDI)

1. EDI is the communication of electronic documents directly from a computer in one entity to a computer in another entity, for example to order goods from a supplier or to transfer funds.

a. EDI was developed to enhance JIT (just-in-time) inventory management.

b. Advantages

1) Reduction of clerical errors
2) Speed
3) Elimination of repetitive clerical tasks
4) Elimination of document preparation, processing, and mailing costs

c. Risks

1) Security of information

a) End-to-end data encryption is a security procedure that encrypts data during EDI.

2) Loss of data

d. An extension of EDI is computer-stored records, which can be less expensive than traditional physical file storage.

2. **EDI Terms and Components**
   a. **Standards** concern procedures to convert written documents into a standard electronic document-messaging format to facilitate EDI.
   b. **Conventions** are the procedures for arranging data elements in specified formats for various accounting transactions, e.g., invoices, materials releases, and advance shipment notices.
   c. A **data dictionary** prescribes the meaning of data elements, including specification of each transaction structure.
   d. **Transmission protocols** are rules used to determine how each electronic envelope is structured and processed by the communications devices.
      1) Normally, a group of accounting transactions is combined in an electronic envelope and transmitted into a communications network.
      2) Rules are required for transmission and the separation of envelopes.
3. **Methods of Communication between Computers**
   a. A **point-to-point** system requires use of dedicated computers by all parties. Each must be designed to be compatible with the other(s). This system is very similar to a network within one company. Dedicated lines or modems are used.
   b. **Value-added networks** provide common interfaces across organizations, thereby eliminating the need for one organization to establish direct computer communication with a trading partner. They also provide mailbox services. Thus, they eliminate the need for dedicated computers waiting for incoming messages. By storing messages, companies can batch outgoing and incoming messages.
      1) The VAN also logs all transactions. However, companies must purchase their own software to translate their data to a national standard protocol for EDI purposes, either ANSI X.12 in the U.S. or EDIFACT in Europe and most of the rest of the world. Once the data are in the standard format, the VAN handles all aspects of the communication.
   c. **Third-party networks** provide mailbox services plus translation of the sender's protocol (data configuration) to the receiver's protocol. Thus, the sender and receiver do not have to conform to the same standards, formats, etc.
4. **Electronic funds transfer** (EFT) is a service provided by financial institutions worldwide that is based on electronic data interchange (EDI) technology. EFT transaction costs are lower than for manual systems because documents and human intervention are eliminated from the transactions process.
5. **Implications for Control**
   a. EDI eliminates the paper documents, both internal and external, that are the traditional basis for many controls, including internal and external auditing.

1) Moreover, an organization that has reengineered its procedures and processes to take full advantage of EDI may have eliminated even the electronic equivalents of paper documents. For example, the buyer's POS system may directly transmit information to the seller, which delivers on a JIT basis. Purchase orders, invoices, and receiving reports are eliminated and replaced with a long-term contract establishing quantities, prices, and delivery schedules; production schedules; advance ship notices; evaluated receipts settlements (periodic payment authorizations transmitted to the trading partner with no need for matching purchase orders, invoices, and receiving reports); and payments by EFT.

a) Accordingly, auditors must seek new forms of evidence to support assertions about EDI transactions, whether it exists at the auditee organization, the trading partner, or a third party, such as a value-added network. Examples of such evidence are the authorized paper purchase contract, an electronic completed production schedule image, and internal and external evidence of evaluated receipts settlements sent to the trading partner.

b) Auditors must evaluate electronic signatures and reviews when testing controls.

c) Auditors may need to consider other subsystems when testing a particular subsystem. Thus, production cycle evidence may be needed to test the expenditure cycle.

6. Stop and review! You have completed the outline for this subunit. Study multiple-choice questions 52 through 62 beginning on page 371.

F. Artificial Intelligence

1. Artificial intelligence (AI) is computer software designed to perceive, reason, and understand. In the business area, emphasis has been on the particular form of AI termed **expert systems**.
2. AI is largely based on powerful reasoning capabilities, but most human decision making hinges on knowledge (i.e., remembering relationships between variables based on experience).
   a. Human reasoning is based, for the most part, on IF-THEN processes (also known as production and situation-action rules), which are also being developed in AI software.
3. The advantage in a business environment is that, relative to human experts, AI systems
   a. Can work 24 hours a day
   b. Will not become ill, die, or be hired away
   c. Are extremely fast processors of data, especially if numerous rules (procedures) must be evaluated
4. The development of artificial intelligence and its subfield, expert systems, has been identified by the AICPA Future Issues Committee as one of the major concerns facing the accounting profession. Expert systems rely on a computer's ability to think and make decisions in a human way.
   a. An expert system is an interactive system that asks a series of questions and uses knowledge gained from a human expert to analyze answers and come to a decision.

b. Expert systems were originally developed to make decisions in areas that did not have enough human experts to make decisions. Some of the earliest expert systems were used by doctors to diagnose diseases.

c. Experimental work is being done with expert systems in taxation, financial accounting, managerial accounting, and auditing.

5. An expert system can be separated into six components -- knowledge database, domain database, database management system, inference engine, user interface, and knowledge acquisition facility. The knowledge database and domain database together are referred to as the knowledge domain.

a. **Knowledge database** contains the rules used when making decisions on certain topics. The facts about certain topics are contained in the **domain database**. These facts are used as a basis for comparison when the system matches a pattern of events or facts with a decision situation.

b. **Inference engine** is the heart of the expert system's processing. It performs the deductive thinking and logic portion of the processing.

c. Rules from the knowledge database, facts from the domain database, and input from the system user are all used by the inference engine.

d. Expert systems use symbolic processing based on **heuristics** rather than algorithms. A heuristic procedure is an exploratory problem-solving technique that uses self-education methods, e.g., the evaluation of feedback, to improve performance. These systems are often very interactive and provide explanations of their problem-solving behavior.

1) Algorithms are defined procedures used in typical computer programs.

6. An expert system is developed using a continuous process of revision. As new knowledge or decision-making strategies become available, prior systems must be revised.

7. The most time-consuming and costly part of developing an expert system is gathering the knowledge of experts and decision makers.

8. By using expert systems, accountants can perform their duties in less time and with more uniformity. An expert system makes it more likely that different decision makers will come to the same conclusions given the same set of facts.

a. An expert system is useful in applying financial accounting standards in a consistent manner when preparing financial statements or performing audits.

b. Expert systems make compliance with tax laws much easier because all rules can be programmed into the computer. Tax planning has also benefitted from the use of expert systems.

c. For management control, expert systems can be used to supplement management information systems by providing models of decisions used for planning and control.

d. For auditing, an expert system can be used to choose an audit program, select a test sample type and size, determine the level of error, perform analytical procedures, and then make a judgment decision based on the findings.

9. Expert systems are becoming so simple that virtually anyone can develop them, even without knowledge of computer programming. This result is accomplished through the use of programs known as **shells**. An expert system shell is a flexible system that provides the framework for developing a customized expert system. It is quite advantageous for a company that wishes to take its own expert knowledge and design an expert system to fit its needs.
10. The field of expert systems is an area of artificial intelligence that is open to development of new ideas and applications. These systems can provide accountants and auditors with a powerful tool for expanding and developing their activities. Even small companies will be able to perform activities that only large firms could handle in the past.
11. A **neural network** learns from its mistakes by changing its knowledge database when informed of the accuracy of its decisions.
12. **Fuzzy logic systems**, a form of artificial intelligence, deal with imprecise data and problems that have many solutions.
    a. Fuzzy logic, a departure from classical two-valued sets and logic, uses "soft" linguistic (e.g., large, hot, tall) system variables and a continuous range of truth values rather than strict binary (true or false) decisions and assignments.
    b. Formally, fuzzy logic is a structured, model-free estimator that approximates a function through linguistic input/output associations.
    c. Fuzzy rule-based systems apply these methods to solve many types of real-world problems, especially when a system is difficult to model, when it is controlled by a human operator or expert, or when ambiguity or vagueness is common.
    d. Fuzzy set theory allows objects to belong partly to multiple sets. Fuzzy logic is useful for describing the vagueness of things in the real world, where belonging to a set is really a matter of degree. An object's membership value, or degree to which it belongs to a set, can be any number between one and zero.
        1) Fuzzy logic is particularly useful in the design of industrial controls, in data retrieval, and in systems in which the user is not intimately familiar with all the data. It is also useful when absolute accuracy is costly or judgments of value must be synthesized from multiple inputs.
        2) Fuzzy logic has emerged as a profitable tool for the controlling of subway systems and complex industrial processes, as well as for household and entertainment electronics, diagnosis systems, and other expert systems.
    e. The key benefits of a fuzzy design are simplified and reduced development cycle, ease of implementation, and more user-friendly and efficient performance.
13. **Genetic algorithms** (adaptive computation) are problem-solving methods that use evolutionary processes, that is, the model of living things adapting through reproduction, mutation, and natural selection.
14. **Intelligent agents** are programs that apply a built-in or learned knowledge base to execute a specific, repetitive, and predictable task, for example, showing a computer user how to perform a task or searching web sites for particular financial information.
15. Stop and review! You have completed the outline for this subunit. Study multiple-choice questions 63 through 68 beginning on page 375.

MULTIPLE-CHOICE QUESTIONS

A. Processing Modes

1. Batch processing

A. Is not used by most businesses because it reduces the audit trail.

B. Processes individual transactions as they occur.

C. Allows users to inquire about groups of information contained in the system.

D. Accumulates transaction records into groups for processing against the master file.

The correct answer is (D). *(CMA, adapted)*

REQUIRED: The true statement about batch processing.

DISCUSSION: Batch processing is the accumulation and grouping of transactions for processing on a delayed basis. The batch approach is suitable for applications that can be processed against the master file at intervals and involve large volumes of similar items, such as payroll, sales, inventory, and billing.

Answer (A) is incorrect because batch processing provides as much of an audit trail as any computerized operation. Answer (B) is incorrect because individual transactions are grouped into batches, and the entire batch is processed together. Answer (C) is incorrect because batch processing refers to the input of data, not inquiry.

2. Transaction processing systems frequently support the inquiry of online database users. Inquiry processing includes all of the following characteristics except that

A. Either batch or real-time processing may be used.

B. It is dependent on the use of telecommunication networks and database management query languages.

C. Responses are in a prespecified format displayed on the end user's terminal.

D. End users are allowed to make changes to the records retrieved.

The correct answer is (D). *(CMA, adapted)*

REQUIRED: The item not a characteristic of inquiry processing.

DISCUSSION: Inquiry processing in a database system can be the result of either batch or real-time processing. An inquiry system requires the use of sophisticated hardware and software, including a database query language. Responses are in a prespecified format. End users receive responses concerning the results of transaction activities but are not allowed to make changes to the records retrieved.

Answer (A) is incorrect because either batch or real-time processing may be used to query a database system. Answer (B) is incorrect because a query system is dependent on the use of telecommunication networks and database management query languages. Answer (C) is incorrect because responses are in a prespecified format.

3. The concept of timeliness of data availability is most relevant to

A. Payroll systems.

B. General ledger systems.

C. Manual systems.

D. Online systems.

The correct answer is (D). *(CMA, adapted)*

REQUIRED: The systems in which the concept of timeliness of data availability is most relevant.

DISCUSSION: An online processing system is in direct communication with the computer, giving it the capability to handle transactions as they are entered. An online system permits both immediate posting (updating) and inquiry of master files as transactions occur. In an online system, data are immediately available to users upon entry.

Answers (A), (B), and (C) are incorrect because timeliness is not necessarily an element of any system unless it is online.

4. An interactive system environment is best characterized by

A. Data files with records arranged sequentially.

B. The processing of groups of data at regular intervals.

C. Sorting the transaction file before processing.

D. The processing of data immediately on input.

The correct answer is (D). *(CMA, adapted)*

REQUIRED: The characteristic of an interactive system environment.

DISCUSSION: In an interactive (inquiry) system, users employ interactive terminals to converse directly with the system. The system is characterized by online entry and processing, direct access, and timesharing.

Answer (A) is incorrect because an interactive system requires direct-access files. Answer (B) is incorrect because an interactive system permits immediate, online processing of single transactions. Answer (C) is incorrect because the transaction file does not have to be sorted before processing.

5. A system that has several minicomputers connected for communication and data transmission purposes but also enables each computer to process its own data is known as a

A. Distributed data processing network.

B. Centralized network.

C. Decentralized network.

D. Multidrop network.

The correct answer is (A). *(CMA, adapted)*

REQUIRED: The type of system having several minicomputers connected for communication and data transmission purposes but also enabling each computer to process its own data.

DISCUSSION: Distributed processing is characterized by a merger of computer and telecommunications technology. Distributed systems permit not only remote access to a computer but also the performance of local processing at local sites. The result is greater flexibility in systems design and the possibility of an optimal distribution of processing tasks.

Answer (B) is incorrect because, in a centralized network, processing occurs in one location. Answer (C) is incorrect because, in a pure decentralized system, the nodes are not interconnected. Answer (D) is incorrect because a multidrop network provides links for each terminal to a single communications line connected to a central processing unit; only one terminal can send or receive messages at a time.

6. In traditional information systems, computer operators are generally responsible for backing up software and data files on a regular basis. In distributed or cooperative systems, ensuring that adequate backups are taken is the responsibility of

A. User management.

B. Systems programmers.

C. Data entry clerks.

D. Tape librarians.

The correct answer is (A). *(CIA, adapted)*

REQUIRED: The persons responsible for ensuring that adequate backups are taken in distributed or cooperative systems.

DISCUSSION: In distributed or cooperative systems, the responsibility for ensuring that adequate backups are taken is the responsibility of user management. The systems are under the control of users, not a central information processing department.

Answer (B) is incorrect because distributed environments have no systems programmers comparable to those at central sites for traditional systems. Answer (C) is incorrect because distributed environments may not have data entry clerks. Users typically perform their own data entry. Answer (D) is incorrect because, in distributed environments, there are no tape librarians.

7. An insurance company that has adopted cooperative processing is planning to implement new standard software in all its local offices. The new software has a fast response time, is very user friendly, and was developed with extensive user involvement. The new software captures, consolidates, edits, validates, and finally transfers standardized transaction data to the headquarters mainframe. Local managers, who were satisfied with existing locally written microcomputer applications, opposed the new approach because they anticipated

A. Increased workloads.

B. Centralization of all processing tasks.

C. More accountability.

D. Less computer equipment.

The correct answer is (C). *(CIA, adapted)*

REQUIRED: The reason for opposing introduction of new software.

DISCUSSION: Cooperative processing implies a tighter coupling than previously existed between the microcomputers and the mainframe. The result may threaten the managers' perceived autonomy by increasing the control exercised by headquarters and therefore the accountability of local managers.

Answer (A) is incorrect because, given that only existing systems would be converted, the transaction volume would likely remain relatively constant. Answer (B) is incorrect because, in a cooperative processing environment, different computers execute different parts of an application. Answer (D) is incorrect because, compared with mainframe-only processing, cooperative processing typically requires more computer equipment at distributed locations.

8. The system that permits the computers in a distributed processing network to share the use of another end user's application program is

A. Electronic data interchange.

B. Interactive processing.

C. Executive support system.

D. Cooperative processing.

The correct answer is (D). *(CMA, adapted)*

REQUIRED: The system that permits the computers in a distributed processing network to share the use of another end user's application program.

DISCUSSION: Cooperative processing is a system in which computers in a distributed processing network can share the use of application programs belonging to another end user. The system assigns different machines the functions they perform best in executing a transaction-based application program. For example, a microcomputer might be used to enter and validate data for the application, and a mainframe might handle file input and output.

Answer (A) is incorrect because EDI is the communication of electronic documents directly from a computer in one entity to a computer in another entity. Answer (B) is incorrect because interactive processing does not allow for the use of another end user's application programs. Answer (C) is incorrect because an executive support system focuses on strategic objectives and gives immediate information about an organization's critical success factors.

9. Today organizations are using microcomputers for data presentation because microcomputer use, compared with mainframe use, is more

A. Controllable.

B. Conductive to data integrity.

C. Reliable.

D. Cost effective.

The correct answer is (D). *(CIA, adapted)*

REQUIRED: The reason for using microcomputers for data presentation.

DISCUSSION: In cooperative processing, microcomputers are more cost effective than mainframes for data entry and presentation. They are better suited to frequent screen updating and graphical user interfaces.

Answer (A) is incorrect because microcomputer use is less controllable than mainframe use. Answer (B) is incorrect because the difficulty of control in a microcomputer environment threatens data integrity. Answer (C) is incorrect because, given their decades of refinement, mainframes are usually more reliable than microcomputers.

B. Databases

10. All of the following are methods for distributing a relational database across multiple servers except

A. Snapshot (making a copy of the database for distribution).

B. Replication (creating and maintaining replica copies at multiple locations).

C. Normalization (separating the database into logical tables for easier user processing).

D. Fragmentation (separating the database into parts and distributing where they are needed).

The correct answer is (C). *(CIA, adapted)*

REQUIRED: The item not a method for distributing a relational database across multiple servers.

DISCUSSION: A distributed database is stored in two or more physical sites. The two basic methods of distributing a database are partitioning and replication. However, normalization is a process of database design, not distribution. Normalization is the term for determining how groups of data items in a relational structure are arranged in records in a database. This process relies on "normal forms," that is, conceptual definitions of data records and specified design rules. Normalization is intended to prevent inconsistent updating of data items. It is a process of breaking down a complex data structure by creating smaller, more efficient relations, thereby minimizing or eliminating the repeating groups in each relation.

Answers (A) and (B) are incorrect because the replication or snapshot technique makes duplicates to be stored at multiple locations. Changes are periodically copied and sent to each location. If a database is small, storing multiple copies may be cheaper than retrieving records from a central site. Answer (D) is incorrect because fragmentation or partitioning stores specific records where they are most needed. For example, a financial institution may store a particular customer's data at the branch where (s)he usually transacts his/her business. If the customer executes a transaction at another branch, the pertinent data are retrieved via communications lines.

11. In a database, there are often conditions that constrain database records. For example, a sales order cannot exist unless the corresponding customer exists. This kind of constraint is an example of

A. Normalization.

B. Entity integrity.

C. Internal schema.

D. Referential integrity.

The correct answer is (D). *(CIA, adapted)*

REQUIRED: The constraint exemplified by prohibiting preparation of a sales order unless it references an existing customer.

DISCUSSION: The data in a database are subject to the constraint of referential integrity. Thus, if data are collected about something, e.g., a payment voucher, all reference conditions regarding the data must be met. Thus, for a voucher to exist, a vendor must also exist.

Answer (A) is incorrect because normalization is the practice of decomposing database relations to remove data field redundancies and thus reduce the likelihood of update anomalies. Answer (B) is incorrect because, in a database, entity integrity means that each thing or relationship in the database is uniquely identified by a single key value. Answer (C) is incorrect because, in a database, the internal schema describes the ways the data are physically organized on the disk.

12. In a database system, locking of data helps preserve data integrity by permitting transactions to have control of all the data needed to complete the transactions. However, implementing a locking procedure could lead to

A. Inconsistent processing.

B. Rollback failures.

C. Unrecoverable transactions.

D. Deadly embraces (retrieval contention).

The correct answer is (D). *(CIA, adapted)*

REQUIRED: The potential disadvantage of a locking procedure.

DISCUSSION: In a distributed processing system, the data and resources a transaction may update or use should be held in their current status until the transaction is complete. A deadly embrace occurs when two transactions need the same resource at the same time. If the system does not have a method to cope with the problem efficiently, response time worsens or the system eventually fails. The system should have an algorithm for undoing the effects of one transaction and releasing the resources it controls so that the other transaction can run to completion.

Answer (A) is incorrect because inconsistent processing occurs when a transaction has different effects depending on when it is processed. Data locking ensures consistent processing. Answer (B) is incorrect because rollback failure is the inability of the software to undo the effects of a transaction that could not be run to completion. A rollback failure is not caused by data locking. However, data locking may lead to situations in which rollback is required. Answer (C) is incorrect because unrecoverable transactions are not a typical symptom of locking procedures.

13. In data modeling and database design, the nature and extent of a relationship between two entities is the

A. Domain.

B. Subschema.

C. Cardinality.

D. Referential path.

The correct answer is (C). *(CMA, adapted)*

REQUIRED: The nature and extent of a relationship between two entities.

DISCUSSION: Cardinality expresses the bounds (a minimum and a maximum) of the association between related entities. For example, assuming the entities are (1) the number of students and (2) a college class, a cardinality limit of (3, 59) for the class entity in the student-class relationship means that a class may contain from 3 to 59 students. The minimum cardinality means that an occurrence of the entity on one side of the relation (the class) must be linked to at least three instances of the entity on the other side (the number of students). The maximum cardinality means that an occurrence of the class entity must be linked to no more than 59 instances of the student entity.

Answer (A) is incorrect because the domain is the set of possible values of an attribute of one particular entity. Answer (B) is incorrect because subschema is a particular user's (application's) view of a part of the database using data definition language. Answer (D) is incorrect because a referential path is the connection between an unspecified number of relations connected by a chain of referential constraints.

14. One advantage of a database management system (DBMS) is

A. The responsibility and control assumed by each organizational unit for its own data.

B. The decrease in the cost of the data processing department as users become responsible for establishing their own data handling techniques.

C. A decreased vulnerability because the database management system has numerous security controls to prevent disasters.

D. The independence of the data from the application programs, which allows the programs to be developed for the user's specific needs without concern for data capture problems.

The correct answer is (D). *(CMA, adapted)*

REQUIRED: The advantage of a DBMS.

DISCUSSION: A fundamental characteristic of databases is that applications are independent of the database structure; when writing programs or designing applications to use the database, only the name of the desired item is necessary. Programs can be developed for the user's specific needs without concern for data capture problems. Reference can be made to the items using the data manipulation language, after which the DBMS takes care of locating and retrieving the desired items. The physical or logical structure of the database can be completely altered without having to change any of the programs using the data items; only the schema requires alteration.

Answer (A) is incorrect because each organizational unit develops programs to make use of elements of a broad database. Answer (B) is incorrect because data handling techniques are still the responsibility of the data processing department; it is the use of the data that is departmentalized. Answer (C) is incorrect because the DBMS is no safer than any other database system.

15. A flat file structure is used in database management systems (DBMS) when

A. A complex network structure is employed.

B. A network based structure is used and a complex database schema is developed.

C. A simple network structure is employed.

D. A relational database model is selected for use.

The correct answer is (D). *(CMA, adapted)*

REQUIRED: The situation in which a flat file structure is used with a DBMS.

DISCUSSION: A flat file structure is used with a relational database model. A relational structure organizes data in conceptual tables. One relation (table or file) can be joined together or related to another without pointers or linked lists if each contains one or more of the same fields (also known as columns or attributes). The relational structure is expected to become the most popular structure because it is relatively easy to construct.

Answer (A) is incorrect because a complex network structure requires something more intricate than a flat file structure. Answers (B) and (C) are incorrect because a network structure reduces redundancy by arranging data through development of many-to-many relationships; that is, each item may have multiple antecedent, as well as successive, relationships, which would preclude a flat file structure.

16. The database approach to systems and the resulting concept of database management systems have several unique characteristics not found in traditional systems, specifically file-oriented systems. Which one of the following statements does not apply to database-oriented systems?

A. Database systems have data independence; that is, the data and the programs are maintained separately except during processing.

B. Database systems contain a data definition language that helps describe each schema and subschema.

C. The database administrator is the part of the software package that instructs the operating aspects of the program when data are retrieved.

D. A primary goal of database systems is to minimize data redundancy.

The correct answer is (C). *(CMA, adapted)*

REQUIRED: The statement that does not apply to database-oriented systems.

DISCUSSION: A database management system (DBMS) involves an integrated set of computer programs that create the database, maintain the elements, safeguard the data from loss or destruction, and make the data available to application programs and inquiries. In a database system, the data and programs are maintained separately except during processing. The DBMS contains a description of the logical and physical structure of the database called the schema. The schema is the description of the structure or organization of the database using data description (definition) language. A primary goal of a DBMS is to minimize data redundance, and user interface is enhanced through increased accessibility and flexibility. The system is administered by a database administrator who is a person with overall responsibility for developing and maintaining the database.

Answers (A), (B), and (D) are incorrect because they are true statements about database management systems.

17. The financial accounting database has several critical relationships that must be properly maintained if the system is to function in an orderly manner. Which one of the following statements about the financial accounting database is incorrect?

A. The general ledger is a master file in which a record is maintained for each and every account in the organization's accounting system.

B. Subsidiary ledgers are master files containing accounting records by specific account categories.

C. Cash disbursements journals are complete records of each transaction that reduces cash.

D. Transaction records include cross-references between general ledger files, subsidiary account numbers, and source document numbers.

The correct answer is (C). *(CMA, adapted)*

REQUIRED: The false statement about the financial accounting database.

DISCUSSION: Although most cash disbursements are recorded in cash disbursements journals, such transactions can also be recorded in other journals, primarily the general journal.

Answer (A) is incorrect because the general ledger is a master file with a record for every account. Answer (B) is incorrect because subsidiary ledgers are master files containing accounting records by specific account categories. Answer (D) is incorrect because transaction records contain cross-references.

18. In an overall description of a database, the names of data elements, their characteristics, and their relationship to each other are defined by using a

A. Data definition language.

B. Data control language.

C. Data manipulation language.

D. Data command interpreter language.

The correct answer is (A). *(CIA, adapted)*

REQUIRED: The language used to define a database.

DISCUSSION: The data definition language defines the database structure and content, especially the schema (the description of the entire database) and subschema (logical views of the database). The schema specifies characteristics such as the names of the data elements contained in the database and their relationship to each other. The subschema defines the logical data views required for applications. Thus, it limits the data elements and functions available to each application.

Answer (B) is incorrect because the data control language specifies the privileges and security rules governing database users. Answer (C) is incorrect because data manipulation language provides application programs with a means of interacting with the database to add, retrieve, modify, or delete data or relationships. Answer (D) is incorrect because data command interpreter languages are symbolic character strings used to control the current state of database management system operations.

19. The increased use of database processing systems makes managing data and information a major information service function. Because the databases of an organization are used for many different applications, they are coordinated and controlled by a database administrator. The functions of a database administrator are

A. Data input preparation, database design, and database operations.

B. Database design, database operation, and database security.

C. Database design, database operation, and equipment operations.

D. Database design, software support, and database security.

The correct answer is (B). *(CMA, adapted)*

REQUIRED: The functions of a database administrator.

DISCUSSION: A database administrator (DBA) is the person who has overall responsibility for developing, designing, controlling, and maintaining the database. The DBA manages all database functions including design and maintenance of the schema that describes the structure of the database. The DBA also assigns user passwords and establishes other security measures. Control of changes in data items and in the programs that use the database is another responsibility of the DBA.

Answer (A) is incorrect because the user is responsible for input preparation. Answer (C) is incorrect because the manager of the EDP department is responsible for equipment (hardware) operations. Answer (D) is incorrect because the systems programming group is responsible for software support.

20. The function of a data dictionary is to

A. Mark the boundary between two consecutive transactions.

B. Organize and share information about objects and resources.

C. Specify systems users.

D. Specify privileges and security rules for objects and resources.

The correct answer is (B). *(CIA, adapted)*

REQUIRED: The function of a data dictionary.

DISCUSSION: According to the *Systems Auditability and Control Report* (The Institute of Internal Auditors, 1991, p. 5-35), a "data dictionary/directory is an organized and shared collection of information about the objects and resources used by the IS organization to deliver or exchange information internally and externally."

Answer (A) is incorrect because the database management system log contains checkpoint records that mark the boundary between two consecutive transactions. Answer (C) is incorrect because specification of system users is a function of the security features of a DBMS. Answer (D) is incorrect because the data control language specifies privileges and security rules for objects and resources.

C. Data Communications and Networks

21. The encryption technique that requires two keys, a public key that is available to anyone for encrypting messages and a private key that is known only to the recipient for decrypting messages, is

A. Rivest, Shamir, and Adelman (RSA).

B. Data encryption standard (DES).

C. Modulator-demodulator.

D. A cypher lock.

The correct answer is (A). *(CIA, adapted)*

REQUIRED: The encryption technique requiring two keys.

DISCUSSION: RSA is a potential encryption standard licensed to hardware and software vendors. Public-key encryption requires management of fewer keys for a given client-server environment than does private-key encryption. However, compared with DES, RSA entails more complex computations and therefore has a higher processing overhead. RSA requires two keys: the public key for encrypting messages is widely known, but the private key for decrypting messages is kept secret by the recipient.

Answer (B) is incorrect because DES is a shared private-key method developed by the U.S. government. It encrypts data into 64-bit blocks using a 56-bit key. DES requires only a single key for each pair of communicants that want to send each other encrypted messages. Answer (C) is incorrect because a modem is used for telecommunications. Answer (D) is incorrect because a cypher lock is a physical device.

22. The use of message encryption software

A. Guarantees the secrecy of data.

B. Requires manual distribution of keys.

C. Increases system overhead.

D. Reduces the need for periodic password changes.

The correct answer is (C). *(CIA, adapted)*

REQUIRED: The effect of message encryption software.

DISCUSSION: Encryption software uses a fixed algorithm to manipulate plain text and an encryption key (a set of random data bits used as a starting point for application of the algorithm) to introduce variation. The machine instructions necessary to encrypt and decrypt data constitute system overhead. As a result, processing speed may be slowed.

Answer (A) is incorrect because no encryption approach absolutely guarantees the secrecy of data. Answer (B) is incorrect because keys may also be distributed electronically via secure key transporters. Answer (D) is incorrect because periodic password changes are needed. Passwords are the typical means of validating users' access to unencrypted data.

23. An insurance firm uses a wide area network (WAN) to allow agents away from the home office to obtain current rates and client information and to submit approved claims using notebook computers and dial-in modems. In this situation, which of the following methods would provide the best data security?

A. Dedicated phone lines.

B. Call-back features.

C. Frequent changes of user IDs and passwords.

D. End-to-end data encryption.

The correct answer is (D). *(CIA, adapted)*

REQUIRED: The best data security method for a wide area network.

DISCUSSION: Encryption of data is a security procedure in which a program encodes data prior to transmission and another program decodes the data after transmission. Encoding is important when confidential data that can be electronically monitored are transmitted between geographically separated locations.

Answer (A) is incorrect because dedicated phone lines are not available to agents in the field. Answers (B) and (C) are incorrect because call-back features and frequent changes of user IDs and passwords are used to authenticate the user but do not otherwise protect the transmitted data.

24. An electronic meeting conducted between several parties at remote sites is referred to as

A. Teleprocessing.

B. Interactive processing.

C. Telecommuting.

D. Teleconferencing.

The correct answer is (D). *(CMA, adapted)*

REQUIRED: The process of holding an electronic meeting between several parties at remote sites.

DISCUSSION: Conducting an electronic meeting among several parties at remote sites is teleconferencing. It can be accomplished by telephone or electronic mail group communication software. Videoconferencing permits the conferees to see each other on video screens. These practices have grown in recent years as companies have attempted to cut their travel costs.

Answer (A) is incorrect because teleprocessing refers to connections in an online system. Answer (B) is incorrect because interactive processing allows users to converse directly with the system. It requires online processing and direct access to stored information. Answer (C) is incorrect because telecommuting refers to the practice of individuals working out of their homes by communicating with their office via the computer.

25. Large organizations often have their own telecommunications networks for transmitting and receiving voice, data, and images. Very small organizations, however, are unlikely to be able to make the investment required for their own networks and are more likely to use

A. Public switched lines.

B. Fast-packet switches.

C. Standard electronic mail systems.

D. The Internet.

The correct answer is (A). *(CIA, adapted)*

REQUIRED: The telecommunications networks likely to be used by small organizations.

DISCUSSION: Companies can use public switched lines (phone lines) on a per-transmission basis. This option is the most cost-effective way for low-volume users to conduct telecommunications.

Answer (B) is incorrect because fast-packet switches receive transmissions from various devices, break the data into packets, and route them over a network to their destination. They are typically installed by telecommunication utility companies and other large companies that have their own networks. Answer (C) is incorrect because electronic mail systems do not allow for voice and image transmissions. Answer (D) is incorrect because the Internet does not provide telephone connections.

26. A local area network (LAN) is best described as a(n)

A. Computer system that connects computers of all sizes, workstations, terminals, and other devices within a limited proximity.

B. System to allow computer users to meet and share ideas and information.

C. Electronic library containing millions of items of data that can be reviewed, retrieved, and analyzed.

D. Method to offer specialized software, hardware, and data handling techniques that improve effectiveness and reduce costs.

The correct answer is (A). *(CMA, adapted)*

REQUIRED: The best description of a local area network (LAN).

DISCUSSION: A LAN is a local distributed computer system, often housed within a single building. Computers, communication devices, and other equipment are linked by cable. Special software facilitates efficient data communication among the hardware devices.

Answer (B) is incorrect because a LAN is more than a system to allow computer users to share information; it is an interconnection of a computer system. Answer (C) is incorrect because a LAN is not a library. Answer (D) is incorrect because a LAN does not require specialized hardware.

27. Which of the following networks provides the least secure means of data transmission?

A. Value-added.

B. Public-switched.

C. Local area.

D. Private.

The correct answer is (B). *(CIA, adapted)*

REQUIRED: The network that provides the least secure means of data transmission.

DISCUSSION: Public-switched networks are wide area networks that use public telephone lines. This arrangement may be the most economical, but data transmission may be of lower quality, no connection may be available, and security measures may be ineffective.

Answer (A) is incorrect because value-added carriers provide data security and error detection and correction procedures. Answer (C) is incorrect because local area networks inherently limit data transmission exposures. Answer (D) is incorrect because private networks provide security through limited access and dedicated facilities.

28. If a system does not have a mainframe computer or a file server but does processing within a series of microcomputers, the network is a(n)

A. Offline processing system.

B. Expert system.

C. Direct access system.

D. Peer-to-peer system.

The correct answer is (D). *(CIA, adapted)*

REQUIRED: The system in which processing is done within a series of microcomputers.

DISCUSSION: A local area network is a user-controlled network that operates without the assistance of a common carrier. It can have several microcomputers attached to a host computer, can be linked as part of several LANs that may or may not communicate with a host computer, or can be connected together but not connected to a host computer (stand alone). A peer-to-peer network operates without a mainframe or server.

Answer (A) is incorrect because offline processing occurs when devices are not directly connected to the computer. Answer (B) is incorrect because expert systems are information systems that provide diagnostic and problem solving through the use of structured software and expert experience. Answer (C) is incorrect because direct access refers to the method for storing and retrieving data within a database.

29. Which of the following is considered to be a server in a local area network (LAN)?

A. The cabling that physically interconnects the nodes of the LAN.

B. A device that stores program and data files for users of the LAN.

C. A device that connects the LAN to other networks.

D. A workstation that is dedicated to a single user on the LAN.

The correct answer is (B). *(CIA, adapted)*

REQUIRED: The server in a local area network.

DISCUSSION: A server is a computer in a network that operates as a librarian. It stores programs and data files for users of the LAN and manages access to them.

Answer (A) is incorrect because the cabling that interconnects the nodes of the LAN is the telecommunications link. Answer (C) is incorrect because a device that connects the LAN to other networks is a network gateway. Answer (D) is incorrect because a workstation dedicated to a single user of the LAN is a client.

30. When connecting two or more electronic mail systems, which of the following is a major security issue?

A. Inability to encrypt messages going between network gateways.

B. Loss of critical text within messages.

C. Inability of receiving users to acknowledge receipt of messages automatically.

D. Inability to keep backup copies of messages.

The correct answer is (A). *(CIA, adapted)*

REQUIRED: The major security issue when connecting two or more electronic mail systems.

DISCUSSION: Loss of confidentiality is a major risk of an e-mail system. Thus, inability to encrypt messages going between network gateways is a major security issue. A gateway is a means of connecting otherwise incompatible networks, nodes, or devices. It performs this function by converting one set of communication protocols to another. Accordingly, even if all systems are secured, an unsecured gateway can be a security exposure.

Answers (B), (C), and (D) are incorrect because loss of critical text within messages, inability of receiving users to acknowledge receipt of messages automatically, and inability to keep backup copies of messages are less serious concerns than loss of confidentiality.

31. A company has a very large, widely dispersed internal audit department. Management wants to implement a computerized system to facilitate communications among auditors. The specifications require that auditors have the ability to place messages in a central electronic repository where all auditors can access them. The system should facilitate finding information on a particular topic. Which type of system would best meet these specifications?

A. Electronic data interchange (EDI).

B. Electronic bulletin board system (BBS).

C. Fax/modem software.

D. Private branch exchange (PBX).

The correct answer is (B). *(CIA, adapted)*

REQUIRED: The best system to facilitate communications among auditors.

DISCUSSION: Bulletin board systems function as a centralized information source and message switching system for a particular interest group. Users review and leave messages for other users, and communicate with other users on the system at the same time.

Answer (A) is incorrect because EDI is for the electronic transmission of business information and electronic mail, but it does not offer central repositories that store messages for many parties to read. Answer (C) is incorrect because, although fax/modem software can store images of faxes received, it does not meet the criterion of ease of access to information on a particular topic. Answer (D) is incorrect because a PBX is a telecommunications system that routes calls to particular extensions within an organization.

32. In distributed data processing, a ring network

A. Has all computers linked to a host computer, and each linked computer routes all data through the host computer.

B. Links all communication channels to form a loop, and each link passes communications through its neighbor to the appropriate location.

C. Attaches all channel messages along one common line with communication to the appropriate location via direct access.

D. Organizes itself along hierarchical lines of communication usually to a central host computer.

The correct answer is (B). *(CMA, adapted)*

REQUIRED: The true statement about a ring network in a distributed data processing system.

DISCUSSION: In a distributed system, an organization's processing needs are examined in their totality. The decision is not whether an application should be done centrally or locally but, rather, which parts are better performed by small local computers as intelligent terminals and which parts are better performed at some other, possibly centralized, site. The key distinction between decentralized and distributed systems is the interconnection among the nodes in the network. A ring network links all communication channels to form a loop and each link passes communications through its neighbor to the appropriate location.

Answer (A) is incorrect because a star network routes all data through the host computer. Answer (C) is incorrect because a bus network attaches all channel messages along one common line with communication to the appropriate location via direct access. Answer (D) is incorrect because a tree configuration is organized along hierarchical lines to a host computer.

D. Internal Control

33. Which of the following is not an advantage of using computers during the audit process?

A. Auditor can work independently of the auditee.

B. Working papers can be reviewed from an off-site location, thus saving on travel costs.

C. Auditor has access to records at remote sites.

D. Auditors can alter client data, if desired.

The correct answer is (D). *(Publisher)*

REQUIRED: The item that is not an advantage of using a computer in the audit process.

DISCUSSION: Auditors may use computers in many ways to simplify and enhance the audit process. The auditor can work independently of the auditee and can access records at remote sites without travel. Auditors should never alter client data.

Answers (A), (B), and (C) are incorrect because they are advantages of using computers in the internal audit process.

34. Preventive controls are

A. Usually more costly to use than detective controls.

B. Found only in general accounting controls.

C. Found only in accounting transaction controls.

D. Usually more cost beneficial than detective controls.

The correct answer is (D). *(CMA, adapted)*

REQUIRED: The true statement about preventive controls.

DISCUSSION: Preventive controls in a system anticipate and avoid deviations. Examples include training of personnel, segregation of duties, prenumbered forms, documentation, passwords, compatibility tests, and turnaround documents. Preventive controls are usually more cost beneficial as compared with other controls because they avoid the costs of deviations and the costs of correction.

Answer (A) is incorrect because preventive controls are usually cheaper to implement than other types of controls. Answer (B) is incorrect because preventive controls can be of many types other than general accounting controls. Answer (C) is incorrect because preventive controls can be of many types other than accounting transaction controls.

35. In the organization of the information systems function, the most important separation of duties is

A. Not allowing the data librarian to assist in data processing operations.

B. Assuring that those responsible for programming the system do not have access to data processing operations.

C. Having a separate information officer at the top level of the organization outside of the accounting function.

D. Using different programming personnel to maintain utility programs from those who maintain the application programs.

The correct answer is (B). *(CMA, adapted)*

REQUIRED: The most important separation of duties in the information systems function.

DISCUSSION: Separation of duties is a general control that is vital in a computerized environment. Some separation of duties common in noncomputerized environments may not be feasible in a computer environment. However, certain tasks should not be combined. Systems analysts, for example, should be separate from programmers and computer operators. Programmers design, write, test, and document specific programs required by the system developed by the analysts. Both programmers and analysts may be able to modify programs, data files, and controls and should therefore have no access to computer equipment and files or to copies of programs used in production. Operators should not be assigned programming duties or responsibility for systems design and should have no opportunity to make changes in programs and systems.

Answer (A) is incorrect because librarians maintain control over documentation, programs, and data files; they should have no access to equipment, but they can assist in data processing operations. Answer (C) is incorrect because a separate information officer outside of the accounting function would not be as critical a separation of duties as that between programmers and processors. Answer (D) is incorrect because programmers usually handle all types of programs.

36. A critical aspect of a disaster recovery plan is to be able to regain operational capability as soon as possible. In order to accomplish this, an organization can have an arrangement with its computer hardware vendor to have a fully operational facility available that is configured to the user's specific needs. This is best known as a(n)

A. Uninterruptible power system.

B. Parallel system.

C. Cold site.

D. Hot site.

The correct answer is (D). *(CMA, adapted)*

REQUIRED: The fully operational facility that is configured to the user's specific needs.

DISCUSSION: A disaster recovery plan may include a contract with an external contingency facility vendor. Depending on the organization's needs, the contingency facility may be a hot site or a cold site. A hot site is an arrangement with a vendor for a fully operational facility that is configured to the user's specific needs and that will be available within 24 hours. A hot site may also be fixed or portable and is recommended for an organization that cannot afford for its computer system to be down for even one day.

Answer (A) is incorrect because an uninterruptible power system is a system that is fully protected by a generator or battery backup to prevent data destruction and downtime from electrical power outages. Answer (B) is incorrect because a parallel system exists if a company maintains an identical system to the main system. Answer (C) is incorrect because a cold site is a cheaper alternative to a hot site. It is a shell facility suitable for the quick installation of computer equipment. It provides a prebuilt, environmentally controlled area with raised flooring, electrical power, and appropriate plumbing.

37. Most of today's computer systems have hardware controls that are built in by the computer manufacturer. Common hardware controls are

A. Duplicate circuitry, echo check, and internal header labels.

B. Tape file protection, cryptographic protection, and limit checks.

C. Duplicate circuitry, echo check, and dual reading.

D. Duplicate circuitry, echo check, tape file protection and internal header labels.

The correct answer is (C). *(CMA, adapted)*

REQUIRED: The common examples of hardware controls.

DISCUSSION: Hardware controls include parity checks, echo checks, dual read-write heads, dual circuitry, diagnostic routines, boundary protection, interlock, and a file protection ring.

Answers (A) and (D) are incorrect because an internal header label is not a hardware control. Answer (B) is incorrect because these are not hardware controls.

38. Online access controls are critical for the successful operation of today's computer systems. To assist in maintaining control over such access, many systems use tests that are maintained through an internal access control matrix consisting of

A. Authorized user code numbers, passwords, lists of all files and programs, and a record of the type of access each user is entitled to have to each file and program.

B. Authorized user code numbers and passwords.

C. A list of controls in the online system and a list of those individuals authorized to change and adjust these controls along with a complete list of files in the system.

D. A completeness test, closed loop verification, and a compatibility test.

The correct answer is (A). *(CMA, adapted)*

REQUIRED: The components of an internal access control matrix.

DISCUSSION: Access controls, such as passwords, ID numbers, access logs, and device authorization tables, prevent improper use or manipulation of data files and programs. They insure that only those persons with a bona fide purpose and authorization have access to data processing. Many systems use tests that are maintained through an internal access control matrix that consists of authorized user code numbers, passwords, lists of all files and programs, and a record of the type of access each user is entitled to have to each file and program.

Answer (B) is incorrect because user code numbers and passwords may limit access to the overall system but are not capable of allowing limited use in certain files. Answer (C) is incorrect because access codes are not concerned with adjusting controls in all files. Answer (D) is incorrect because a completeness test is not an access control but is used to determine whether all necessary information has been transmitted.

39. To properly control access to accounting database files, the database administrator should ensure that database system features are in place to permit

A. Read-only access to the database files.

B. Updating from privileged utilities.

C. Access only to authorized logical views.

D. User updates of their access profiles.

The correct answer is (C). *(CIA, adapted)*

REQUIRED: The purpose of access control for accounting database files.

DISCUSSION: A database management system's software includes security features. Thus, a specified user's access may be limited to certain data fields or logical views depending on the individual's assigned duties. A logical view consists of the fields available to a given user, function, or application. It may include all or part of a physical data file or a combination of fields from multiple physical data files.

Answer (A) is incorrect because, if the only access is read-only, no updating is possible. Answer (B) is incorrect because utility programs (e.g., performance monitors, job schedulers, and tape and disk management systems) may have privileged access capabilities. Thus, they may be able to bypass normal security measures. Permitting updating from privileged software is a breach of security. Answer (D) is incorrect because updating of users' access profiles should be a function of a security officer.

40. Data processed by a computer system are usually transferred to some form of output medium for storage. However, the presence of computerized output does not, in and of itself, assure the output's accuracy, completeness, or authenticity. For this assurance, various controls are needed. The major types of controls for this area include

A. Transaction controls, general controls, and printout controls.

B. Activity listings, echo checks, and pre-numbered forms.

C. Tape and disk output controls and printed output controls.

D. Input controls, tape and disk output controls, and printed output controls.

The correct answer is (D). *(CMA, adapted)*

REQUIRED: The major types of controls to assure that computerized output is accurate, complete, and authentic.

DISCUSSION: Input controls provide reasonable assurance that data received for processing have been properly authorized, converted into machine-sensible form, and identified, and that data have not been lost, suppressed, added, duplicated, or otherwise improperly changed. Input controls also relate to rejections, correction, and resubmission of data that were initially incorrect. Output controls provide assurance that the processing result is accurate and that only authorized personnel receive the output.

Answer (A) is incorrect because general, transaction, and print-out controls do not assure accuracy of inputs. Answer (B) is incorrect because an echo check, which is an input control over transmission along communications lines, does not assure proper authorization of data. Neither do the other techniques assure completeness of data. Answer (C) is incorrect because output controls are insufficient to assure completeness and accuracy of output. Input controls are also needed.

41. Data input validation routines include

A. Terminal logs.

B. Passwords.

C. Hash totals.

D. Backup controls.

The correct answer is (C). *(CMA, adapted)*

REQUIRED: The example of a data input validation routine.

DISCUSSION: Application controls, including input controls, are designed to assure the accuracy and completeness of data entered into the computer. Input controls provide assurance that data have not been lost, suppressed, added, duplicated, or otherwise improperly changed. A hash total is an example of a data input validation routine. A hash total is a control total without a defined meaning, such as the total of employee numbers or invoice numbers, that is used to verify the completeness of data. Thus, the hash total for the employee listing by the personnel department could be compared with the total generated during the processing run.

Answers (A) and (B) are incorrect because terminal logs and passwords are access controls. Answer (D) is incorrect because backup controls are general controls.

42. An accounting system identification code that uses a sum-of-digits check digit will detect all of the following errors except

A. Completeness errors.

B. Transcription errors.

C. Transposition errors.

D. Validity errors.

The correct answer is (C). *(CMA, adapted)*

REQUIRED: The error not detected by a sum-of-digits check digit.

DISCUSSION: Self-checking digits may be used to detect incorrect identification numbers. The digit is generated by applying an algorithm to the ID number. During the input process, the check digit is recomputed by applying the same algorithm to the code actually entered. If the check digit is merely a sum, transposition errors will not be detected because the sum will be unaffected.

Answer (A) is incorrect because completeness errors will be detected. The sum will be different if such errors occur. Answer (B) is incorrect because transcription errors will be detected. The sum will be different if such errors occur. Answer (D) is incorrect because validity errors will be detected. The sum will be different if such errors occur.

43. In order to prevent, detect, and correct errors and unauthorized tampering, a payroll system should have adequate controls. The best set of controls for a payroll system includes

A. Batch and hash totals, record counts of each run, proper separation of duties, passwords and user codes, and backup copies of activity and master files.

B. Employee supervision, batch totals, record counts of each run, and payments by check.

C. Passwords and user codes, batch totals, employee supervision, and record counts of each run.

D. Batch totals, record counts, user codes, proper separation of duties, and online edit checks.

The correct answer is (A). *(CMA, adapted)*

REQUIRED: The best set of controls over payroll.

DISCUSSION: Controls in a payroll system should include a proper separation of the functions of authorization, record keeping, and custody of assets; batch totals for such items as hours worked and payroll amounts; hash totals (e.g., of employee identification numbers) to test for completeness of processing; record counts for each run; special control over unclaimed checks (the person who distributes checks must not retain unclaimed checks); and backup copies of files to allow for reconstruction if information is lost.

Answers (B) and (C) are incorrect because separation of duties and backup procedures are not mentioned. Answer (D) is incorrect because special controls over unclaimed checks and backup procedures are omitted.

44. An employee in the receiving department keyed in a shipment from a remote terminal and inadvertently omitted the purchase order number. The best systems control to detect this error would be

A. Batch total.

B. Completeness test.

C. Sequence check.

D. Reasonableness test.

The correct answer is (B). *(CMA, adapted)*

REQUIRED: The best systems control to detect the omission of a purchase order number on a receiving report keyed in from a remote terminal.

DISCUSSION: A completeness test checks that all data elements are entered before processing. An interactive system can be programmed to notify the user to enter the number before accepting the receiving report.

Answer (A) is incorrect because a batch total is a total of one information field (such as sales on invoices) for all records in a batch. Answer (C) is incorrect because a sequence check tests for the ordering, not omission, of records. Answer (D) is incorrect because a limit or reasonableness test checks the values of data items against established limits.

45. Edit checks in a computerized accounting system

A. Are preventive controls.

B. Should be performed on transactions prior to updating a master file.

C. Must be installed for the system to be operational.

D. Should be performed immediately prior to output distribution.

The correct answer is (B). *(CMA, adapted)*

REQUIRED: The true statement about edit checks in a computerized accounting system.

DISCUSSION: Edit checks are those that are programmed into the software. They include error listings, field checks, financial totals, hash totals, limit and range checks, preformatting, reasonableness (relationship) tests, record counts, self-checking digits, sequence checks, sign checks, and validity checks. Such checks should be performed on transactions before the master file is updated.

Answer (A) is incorrect because edit checks also include detective and corrective controls. Answer (C) is incorrect because edit checks are not necessary for a system to run; they are purely for internal control purposes. Answer (D) is incorrect because edit checks are normally performed at the time of input or during manipulation of data, not at the time of output.

46. Which one of the following statements concerning concurrent auditing techniques is not correct?

A. They allow monitoring a system on a continuous basis for fraudulent transactions.

B. They are most useful in complex online systems in which audit trails have either become diminished or are very limited.

C. They allow faster detection of unauthorized transactions.

D. They are standard components of generic software packages.

The correct answer is (D). *(CMA, adapted)*

REQUIRED: The false statement concerning concurrent auditing techniques.

DISCUSSION: The primary use of generalized audit software (GAS) is to select and summarize a client's records for additional testing. These packages permit the auditor to audit through the computer; to extract, compare, analyze, and summarize data; and to generate output for use in the audit. They allow the auditor to exploit the computer to examine many more records than otherwise possible with far greater speed and accuracy. Hence, GAS facilitates analysis of all sources of potential error. However, concurrent auditing techniques are not included because they must be incorporated into the client's systems. For example, embedded audit data collection is a transaction selection approach incorporated within the regular production programs to routinely extract transactions meeting certain criteria for further testing. In effect, it provides a window through which the auditor can access the process.

Answers (A) and (C) are incorrect because dubious transactions can be immediately identified and reported to the auditor (rather than the user) for review and investigation, without waiting for the scheduled audit. Answer (B) is incorrect because embedded audit modules and the like compensate for the loss of the traditional paper audit trail.

47. In auditing computer-based systems, the integrated test facility (ITF)

A. Allows the auditor to assemble test transactions and run them through the computer system to test the integrity of controls on a sample data base.

B. Is a set of specialized software routines that are designed to perform specialized audit tests and store audit evidence.

C. Is a concurrent audit technique that establishes a special set of dummy master files and enters transactions to test the programs using the dummy files during regular processing runs.

D. Uses an audit log to record transactions and data having special audit significance during regular processing runs.

The correct answer is (C). *(CMA, adapted)*

REQUIRED: The true statement about an ITF.

DISCUSSION: An ITF involves the use of a fictitious entity, such as a dummy customer in accounts receivable, against which data transactions are processed. Results are compared with previously determined results. This procedure is used within the framework of regular production, frequently without computer operator knowledge. The use of an ITF enables testing of a system as it routinely operates. The cost of using an ITF is low. The disadvantages of the ITF include the need to later nullify the data put into the system and the possibility of contaminating a database.

Answer (A) is incorrect because an ITF includes a dummy entity as well as test data. Answer (B) is incorrect because generalized audit software is a set of specialized software routines that are designed to perform specialized audit tests and store audit evidence. Answer (D) is incorrect because the ITF does not use an audit log.

48. A company makes snapshot copies of some often-used data and makes them available in files on the mainframes. Authorized users can then download data subsets into spreadsheet programs. A risk associated with this means of providing data access is that

A. Data replicas may not be synchronized.

B. Data fragments may lack integrity.

C. Data transactions may be committed prematurely.

D. Data currency may not be maintained.

The correct answer is (D). *(CIA, adapted)*

REQUIRED: The risk associated with snapshot access to data.

DISCUSSION: Snapshot files are created at a fixed time. Thus, by the time an employee downloads the data subset, it could be obsolete. Snapshot data available to download into spreadsheets may contain old or erroneous information that was later corrected on the main file system.

Answer (A) is incorrect because data replication consists of copying all of the data on multiple devices for improved accessibility. Answer (B) is incorrect because data fragmentation involves distributing data across different locations so that the fragments from the different locations must be put together to create the complete file. Answer (C) is incorrect because the snapshot approach does not involve writing of transactions to files.

49. The firewall system that limits access to a computer network by routing users to replicated Web pages is

A. A packet filtering system.

B. Kerberos.

C. A proxy server.

D. An authentication system.

The correct answer is (C). *(Publisher)*

REQUIRED: The firewall system that routes users to replicated web pages.

DISCUSSION: A proxy server maintains copies of web pages to be accessed by specified users. Outsiders are directed there, and more important information is not available from this access point.

Answer (A) is incorrect because a packet filtering system examines each incoming IP packet. Answer (B) is incorrect because Kerberos is encryption and authentication software that uses DES encryption techniques. Answer (D) is incorrect because an authentication system verifies a user's identity and is often an application provided by a firewall system, but it is not a firewall system itself.

50. An Internet firewall is designed to provide adequate protection against which of the following?

A. A computer virus.

B. Unauthenticated logins from outside users.

C. Insider leaking of confidential information.

D. A Trojan horse application.

The correct answer is (B). *(Publisher)*

REQUIRED: The protection provided by an Internet firewall.

DISCUSSION: A firewall is a device that separates two networks and prevents passage of specific types of network traffic while maintaining a connection between the networks. Generally, an Internet firewall is designed to protect a system from unauthenticated logins from outside users, although it may provide several other features as well.

Answer (A) is incorrect because a firewall cannot adequately protect a system against computer viruses. Answer (C) is incorrect because industrial spies need not leak information through the firewall. A telephone or floppy disk are much more common means of sharing confidential information. Answer (D) is incorrect because a firewall cannot adequately protect against a Trojan horse (a program, such as a game, that appears friendly but that actually contains applications destructive to the computer system) or any other program that can be executed in the system by an internal user.

51. The encryption technique that requires two keys, a public key that is available to anyone for encrypting messages and a private key that is known only to the recipient for decrypting messages, is

A. Rivest, Shamir, and Adelman (RSA).

B. Data encryption standard (DES).

C. Modulator-demodulator.

D. A cypher lock.

The correct answer is (A). *(CIA, adapted)*

REQUIRED: The encryption technique requiring two keys.

DISCUSSION: RSA is a potential encryption standard licensed to hardware and software vendors. Public-key encryption requires management of fewer keys for a given client-server environment than does private-key encryption. However, compared with DES, RSA entails more complex computations and therefore has a higher processing overhead. RSA requires two keys: The public key for encrypting messages is widely known, but the private key for decrypting messages is kept secret by the recipient.

Answer (B) is incorrect because DES is a shared private-key method developed by the U.S. government. It encrypts data into 64-bit blocks using a 56-bit key. DES requires only a single key for each pair of parties that want to send each other encrypted messages. Answer (C) is incorrect because a modem is used for telecommunications. Answer (D) is incorrect because a cypher lock is a physical device.

E. Electronic Data Interchange (EDI)

52. Which of the following is likely to be a benefit of electronic data interchange (EDI)?

A. Increased transmission speed of actual documents.

B. Improved business relationships with trading partners.

C. Decreased liability related to protection of proprietary business data.

D. Decreased requirements for backup and contingency planning.

The correct answer is (B). *(CIA, adapted)*

REQUIRED: The benefit of EDI.

DISCUSSION: Electronic data interchange is the electronic transfer of documents between businesses. EDI was developed to enhance just-in-time (JIT) inventory management. Advantages include speed, reduction of clerical errors, and elimination of repetitive clerical tasks and their costs. Improved business relationships result because of the mutual benefits conferred by EDI. Accordingly, some organizations require EDI.

Answer (A) is incorrect because EDI transmits document data, not the actual document. Answer (C) is incorrect because liability for protection of a trading partner's proprietary business data is a major risk that must be addressed by the control structure. Answer (D) is incorrect because backup and contingency planning requirements are not diminished by use of EDI.

53. The emergence of electronic data interchange (EDI) as standard operating practice increases the risk of

A. Unauthorized third-party access to systems.

B. Systematic programming errors.

C. Inadequate knowledge bases.

D. Unsuccessful system use.

The correct answer is (A). *(CIA, adapted)*

REQUIRED: The risk increased by the emergence of EDI as standard operating practice.

DISCUSSION: EDI is the communication of electronic documents directly from a computer in one entity to a computer in another entity. EDI for business documents between unrelated parties has the potential to increase the risk of unauthorized third-party access to systems because more outsiders will have access to internal systems.

Answer (B) is incorrect because systematic programming errors are the result of misspecification of requirements or lack of correspondence between specifications and programs. Answer (C) is incorrect because inadequate knowledge bases are a function of lack of care in building them. Answer (D) is incorrect because a benefit of EDI is to improve the efficiency and effectiveness of system use.

54. A control a company could use to detect forged EDI messages is to

A. Acknowledge all messages initiated externally with confirming messages.

B. Permit only authorized employees to have access to transmission facilities.

C. Delay action on orders until a second order is received for the same goods.

D. Write all incoming messages to a write-once/read-many device for archiving.

The correct answer is (A). *(CIA, adapted)*

REQUIRED: The control to detect forged electronic data interchange messages.

DISCUSSION: If the company acknowledges messages initiated externally, the alleged sender would have the opportunity to recognize that it had not sent the message and could notify the company of the potential forgery. Then corrective action can be taken by the company.

Answer (B) is incorrect because permitting only authorized employees to have access to transmission facilities controls for unauthorized access to the facilities but would not detect forged EDI messages. Answer (C) is incorrect because delaying action on orders until a second order is received for the same goods defeats the purpose of using EDI, namely, rapid communication followed by rapid response. Answer (D) is incorrect because writing all incoming messages to a write-once/read-many device is a good practice, but it will not detect forgeries.

55. Before sending or receiving EDI messages, a company should

A. Execute a trading partner agreement with each of its customers and suppliers.

B. Reduce inventory levels in anticipation of receiving shipments.

C. Demand that all its suppliers implement EDI capabilities.

D. Evaluate the effectiveness of its use of EDI transmissions.

The correct answer is (A). *(CIA, adapted)*

REQUIRED: The process that should be performed before sending or receiving electronic data interchange messages.

DISCUSSION: Before sending or receiving EDI messages, a company should execute a trading partner agreement with its customers and suppliers. All parties should understand their responsibilities, the messages each will initiate, how they will interpret messages, the means of authenticating and verifying the completeness and accuracy of messages, the moment when the contract between the parties is effective, the required level of security, etc.

Answer (B) is incorrect because the company may intend to reduce inventory levels, but that intention is unrelated to the timing of its first EDI messages. Answer (C) is incorrect because the company may want to demand or encourage all its customers and suppliers to implement EDI capabilities, but that request is independent of sending and receiving messages. Answer (D) is incorrect because it is not possible to evaluate the effectiveness of EDI transmissions until after they occur.

56. The best approach for minimizing the likelihood of software incompatibilities leading to unintelligible messages is for a company and its customers to

A. Acquire their software from the same software vendor.

B. Agree to synchronize their updating of EDI-related software.

C. Agree to use the same software in the same ways indefinitely.

D. Each write their own version of the EDI-related software.

The correct answer is (B). *(CIA, adapted)*

REQUIRED: The best approach for minimizing the likelihood of software incompatibilities leading to unintelligible messages.

DISCUSSION: EDI entails the exchange of common business data converted into standard message formats. Thus, two crucial requirements are that the participants agree on transaction formats and that translation software be developed to convert messages into a form understandable by other companies. Thus, if one company changes its software, its trading partners must also do so.

Answer (A) is incorrect because the company and its customers may obtain their EDI-related software from the same vendor but still have software incompatibility problems if they do not synchronize their installation of updated versions. Answer (C) is incorrect because, as business requirements change, it may not be possible to use the same software in the same ways indefinitely. Answer (D) is incorrect because, even if the company and its customers each write their own versions, synchronization problems will arise from updates.

57. As organizations move to implement EDI, more of them are turning to the use of value-added networks (VANs). Which of the following would not normally be performed by a VAN?

A. Store electronic purchase orders of one organization to be accessed by another organization.

B. Provide common interfaces across organizations thereby eliminating the need for one organization to establish direct computer communication with a trading partner.

C. Maintain a log of all transactions of an organization with its trading partner.

D. Provide translations from clients' computer applications to a standard protocol used for EDI communication.

The correct answer is (D). *(CIA, adapted)*

REQUIRED: The function not performed by a VAN.

DISCUSSION: Companies must purchase their own software to translate their data to a national standard protocol for EDI purposes, either ANSI X.12 in the U.S. or EDIFACT in Europe and most of the rest of the world. Once the data are in the standard format, the VAN handles all aspects of the communication. VANs are privately owned telecommunications carriers that sell capacity to outside users. Among other things, a VAN provides a mailbox service permitting EDI messages to be sent, sorted, and held until needed in the recipient's computer system.

Answers (A), (B), and (C) are incorrect because VANs normally provide mailbox services, common communication interfaces, and logs of transactions.

58. In a review of an EDI application using a third-party service provider, the auditor should

I. Ensure encryption keys meet ISO standards.

II. Determine whether an independent review of the service provider's operation has been conducted.

III. Verify that only public-switched data networks are used by the service provider.

IV. Verify that the service provider's contracts include necessary clauses, such as the right to audit.

A. I and II.

B. I and IV.

C. II and III.

D. II and IV.

The correct answer is (D). *(CIA, adapted)*

REQUIRED: The auditor's procedures in a review of an EDI application.

DISCUSSION: An auditor should review trading partner agreements and contracts with third-party service providers. These documents should contain necessary clauses and appropriately limit liabilities. Moreover, legal counsel should have reviewed the agreements or contracts. An auditor should also determine whether the third-party service provider's operations and controls have been independently reviewed (for example, by public accountants).

Answers (A) and (B) are incorrect because using a third-party service provider does not require encryption. Answer (C) is incorrect because use of public-switched data networks is not a requirement of EDI.

59. After implementing EDI with suppliers, a company discovered a dramatic increase in the prices it paid the single supplier of some special materials for its primary product line. After consulting with the supplier, the company determined that the supplier had assumed the risk of not having inventory and raised its prices accordingly since the company was the only buyer for the special materials. The best approach for managing inventory in this situation is for the company to

A. Give the supplier more information about expected use of the materials.

B. Demand that the supplier reduce the prices of the materials.

C. Find another supplier to replace the one charging higher prices.

D. Change its product line so the special materials are no longer needed.

The correct answer is (A). *(CIA, adapted)*

REQUIRED: The best approach for managing inventory when the supplier increases its price to compensate for the higher risk of not having inventory.

DISCUSSION: If the company gives the supplier more information about use of the materials, the supplier may be able to plan its production more effectively. It could then reduce its inventory of the materials and its inventory costs, thus permitting it to charge a lower price.

Answer (B) is incorrect because the company could demand that the supplier reduce the price of the materials, but the supplier could then decline to supply them. Answer (C) is incorrect because other suppliers may also charge a high price. Answer (D) is incorrect because, if the special materials are needed in the primary product line, it is unlikely that the company would discontinue that line before investigating alternatives, e.g., working with the supplier to help the supplier manage its inventory.

60. If the cycle time for manual purchase orders is 25 days, composed of 4 days of preparation, 3 days in the mail, 14 days in process at the supplier, and 4 days for delivery of raw materials, the shortest possible cycle time if a company fully implemented EDI with suppliers would be

A. 21 days.

B. 18 days.

C. 4 days.

D. 1 day.

The correct answer is (C). *(CIA, adapted)*

REQUIRED: The shortest possible time for the purchase order cycle when an EDI is implemented.

DISCUSSION: The full implementation of an EDI system will eliminate the manufacturer's preparation time for purchase orders, the days in the mail, and processing by the supplier. The only time required will be the 4 days for physical delivery. An EDI system allows for the computer-to-computer exchange of transaction documents such as purchase orders, invoices, and shipping documents. It eliminates the printing and handling of paper by one party and the input of data by the other.

Answer (A) is incorrect because a cycle time of 21 days does not include reductions possible by using EDI to eliminate mail time and supplier processing time. Answer (B) is incorrect because a cycle time of 18 days does not include reductions possible by using EDI to eliminate supplier processing time. Answer (D) is incorrect because the cycle time cannot be reduced below the delivery time of 4 days with implementation of EDI alone. More efficient transportation would be required.

61. Which of the following risks is not greater in an electronic funds transfer (EFT) environment than in a manual system using paper transactions?

A. Unauthorized access and activity.

B. Duplicate transaction processing.

C. Higher cost per transaction.

D. Inadequate backup and recovery capabilities.

The correct answer is (C). *(CIA, adapted)*

REQUIRED: The risk not greater in an EFT environment than in a manual system using paper transactions.

DISCUSSION: EFT is a service provided by financial institutions worldwide that is based on EDI technology. EFT transaction costs are lower than for manual systems because documents and human intervention are eliminated from the transactions process.

Answers (A), (B), and (D) are incorrect because unauthorized access and activity, inaccurate transaction processing (including duplication), and inadequate backup and recovery capabilities are risks specific to EFT.

62. Which of the following is a risk that is higher when an electronic funds transfer (EFT) system is used?

A. Improper change control procedures.

B. Unauthorized access and activity.

C. Insufficient online edit checks.

D. Inadequate backups and disaster recovery procedures.

The correct answer is (B). *(CIA, adapted)*

REQUIRED: The higher risk in an EFT system.

DISCUSSION: Unauthorized access to money transfer activities or data is an inherent and unique risk of EFT systems. An unauthorized person may attempt to read, alter, or delete information in data files or to enter authorized fund transfers. Hence, in the financial services industry, protection of confidential customer transactions is especially important. Moreover, unauthorized transfers subject a financial institution to a direct risk of serious loss.

Answers (A), (C), and (D) are incorrect because improper change control procedures, insufficient online edit checks, and inadequate backups and disaster recovery procedures are risks common to all information technology environments. These risks are not higher than for other systems.

F. Artificial Intelligence

63. Expert systems consist of

A. Software packages with the ability to make judgment decisions.

B. A panel of outside consultants.

C. Hardware designed to make judgment decisions.

D. Hardware and software used to automate routine tasks.

The correct answer is (A). *(CIA, adapted)*

REQUIRED: The definition of expert systems.

DISCUSSION: Artificial intelligence and its subfield, expert systems, have been identified by the AICPA Future Issues Committee as one of the major issues the accounting profession will face in the future. Expert systems attempt to permit a computer to think and make decisions in a human way. An expert system is an interactive system that asks a series of questions and uses knowledge gained from a human expert to analyze answers and come to a decision, that is, to exercise judgment. They were originally developed to make decisions in areas that did not have enough human experts to make decisions. Some of the earliest expert systems were used by doctors to diagnose diseases.

Answer (B) is incorrect because expert systems do not require outside consultants. Answer (C) is incorrect because hardware does not make judgment decisions. Answer (D) is incorrect because automation of routine tasks is not the purpose of expert systems.

64. Which one of the following incorporates making the best decisions possible, using a logical approach to solving problems, using reasoning, having the capability to learn, and allowing subjective inputs and outputs?

A. Expert systems.

B. Decision support systems.

C. Multi-networking.

D. Neural networks.

The correct answer is (A). *(CMA, adapted)*

REQUIRED: The item based on a logical approach to making the best decisions possible by using reasoning, learning, and allowing subjective inputs and outputs.

DISCUSSION: An expert system is software designed to perceive, reason, and understand. An expert system is an interactive system that asks a series of questions and uses knowledge gained from a human expert to analyze answers and come to a decision. The system is developed by using a continuous process of revision. As new knowledge or decision-making strategies become available, prior systems must be revised.

Answer (B) is incorrect because decision support systems are extensions of the MIS concept that are primarily useful for semistructured problems, such as those requiring the decision maker to exercise judgment in controlling the process but allowing for certain aspects of the problem to be preprogrammed. A decision support system does not automate a decision but rather provides tools for the user to employ in applying his/her own insight and judgment. Answer (C) is incorrect because multi-networking involves using multiple networks with the same computer. Answer (D) is incorrect because neural networks, although similar to expert systems, can learn from making mistakes.

65. Prudent managers will recognize the limits within which expert systems can be effectively applied. An expert system would be most appropriate to

A. Compensate for the lack of certain technical knowledge within the organization.

B. Help make customer-service jobs easier to perform.

C. Automate daily managerial problem-solving.

D. Emulate human expertise for strategic planning.

The correct answer is (B). *(CIA, adapted)*

REQUIRED: The appropriate use for an expert system.

DISCUSSION: Expert systems allow even small companies to perform activities and provide services previously only available from larger firms. The use of expert systems has helped to improve the quality of customer service in applications such as maintenance and scheduling by automating them and making them easy to perform.

Answer (A) is incorrect because expert systems codify and apply existing knowledge, but they do not create knowledge that is lacking. Answer (C) is incorrect because expert systems do best in automating lower-level clerical functions. Answer (D) is incorrect because expert systems concern problems with relatively few possible outcomes that are all known in advance.

66. The processing in knowledge-based systems is characterized by

A. Algorithms.

B. Deterministic procedures.

C. Heuristics.

D. Simulations.

The correct answer is (C). *(CIA, adapted)*

REQUIRED: The characteristic of processing in knowledge-based systems.

DISCUSSION: Knowledge-based (expert) systems contain a knowledge base for a limited domain of human expertise and inference procedures for the solution of problems. They use symbolic processing based on heuristics rather than algorithms. A heuristic procedure is an exploratory problem-solving technique that uses self-education methods, e.g., the evaluation of feedback, to improve performance. These systems are often very interactive and provide explanations of their problem-solving behavior.

Answer (A) is incorrect because algorithms are defined procedures used in typical computer programs. Answer (B) is incorrect because deterministic procedures are procedures used in computer programs that permit no uncertainty in outcomes. Answer (D) is incorrect because simulations are computer programs that permit experimentation with logical and mathematical models.

67. A bank implemented an expert system to help account representatives consolidate the bank's relationships with each customer. The expert system has

A. A sequential control structure.

B. Distinct input/output variables.

C. A knowledge base.

D. Passive data elements.

The correct answer is (C). *(CIA, adapted)*

REQUIRED: The component of an expert system.

DISCUSSION: An expert system relies on a computer's ability to make decisions in a human way. There are six components to the expert system: knowledge base, domain database, database management system, inference engine, user interface, and knowledge acquisition facility. The knowledge base contains the rules used when making decisions.

Answer (A) is incorrect because traditional programs, e.g., in COBOL, have sequential control structures; expert systems do not. Answer (B) is incorrect because traditional programs, not expert systems, have distinct input/output variables. Answer (D) is incorrect because traditional programs, not expert systems, have passive data elements.

68. For which of the following applications would the use of a fuzzy logic system be the most appropriate artificial intelligence (AI) choice?

A. Assigning airport gates to arriving airline flights.

B. Forecasting demand for spare auto parts.

C. Ventilating expressway tunnels.

D. Diagnosing computer hardware problems.

The correct answer is (C). *(CIA, adapted)*

REQUIRED: The most appropriate use for fuzzy logic.

DISCUSSION: Fuzzy logic is a superset of conventional (Boolean) logic that has been extended to handle the concept of partial truth. Because they use nonspecific terms (membership functions) characterized by well-defined imprecision, fuzzy logic systems can create rules to address problems with many solutions. Fuzzy logic can be used when values are approximate or subject and data are incomplete or ambiguous. These systems have been applied successfully to applications such as ventilating expressway tunnels, backing a tractor-trailer into a parking space, reducing power usage in an air conditioner, selecting companies for business combinations, or detecting fraud in medical insurance claims.

Answer (A) is incorrect because assigning airport gates to arriving airline flights requires an expert system that uses precise data for quick and consistent decisions. Answer (B) is incorrect because neural networks provide the technology to undertake sophisticated forecasting and analysis. They emulate the processing patterns of the brain and therefore can learn from experience. Answer (D) is incorrect because diagnosing problems with computer hardware requires an expert system.

STUDY UNIT 9: INTERNAL CONTROL

20 pages of outline
64 multiple-choice questions

This study unit is the first of two pertaining to management controls, a major topic that has been assigned a relative weight range of 10% to 15% of Part 4. It begins with the AICPA's approach to internal control stated in SAS 78 (AU 319), *Consideration of Internal Control in a Financial Statement Audit*. This pronouncement incorporates the definition stated in *Internal Control-Integrated Framework* published by the Committee of Sponsoring Organizations (COSO). SAS 78 redefines internal control, broadens the auditor's responsibility to consider it, and integrates internal control principles with those of audit risk and audit evidence.

The balance of this study unit concerns matters that may have relevance beyond the scope of the financial statement audit. Also see Study Unit 10, Internal Auditing. Internal auditors and management accountants share the CPA's concern for reliable financial data and safeguarding of assets, but they are also interested in a much broader range of controls. Thus, SMA 2A, *Management Accounting Glossary*, defines internal control as follows:

> *The whole system of controls (financial and otherwise) established by management to carry on the business of the enterprise in an orderly and efficient manner, to ensure adherence to management policies, safeguard the assets, and ensure as far as possible the completeness and accuracy of the records.*

A. Internal Control According to the AICPA

1. **Introduction.** The crucial pronouncement is AU 319. Audits in accordance with generally accepted auditing standards (GAAS) require the auditor to obtain an understanding of the five components of internal control and to assess control risk.
   a. AU 319 revised the second standard of field work (GAAS) as follows:

      > *A sufficient understanding of internal control is to be obtained to plan the audit and to determine the nature, timing, and extent of tests to be performed.*

   b. As amended to reflect the COSO report, AU 319 defines internal control as follows:

      > *Internal control is a process -- effected by an entity's board of directors, management, and other personnel -- designed to provide reasonable assurance regarding the achievement of objectives in the following categories: (a) reliability of financial reporting, (b) effectiveness and efficiency of operations, and (c) compliance with applicable laws and regulations.*

1) Ordinarily, controls relevant to an audit pertain to the entity's objective of preparing financial statements for external purposes that are fairly presented in conformity with GAAP or another comprehensive basis of accounting.
2) Some are usually not relevant and need not be considered, for example, controls concerning the effectiveness and efficiency of certain management decision processes, such as those related to product pricing or some expenditures for R&D or advertising.

2. **Components of Internal Control**. For a financial statement audit, internal control consists of five components: the control environment, risk assessment, control activities, information and communication, and monitoring.
   a. "**The control environment** sets the tone of an organization, influencing the control consciousness of its people. It is the foundation of all other components of internal control, providing discipline and structure." It includes the following factors:
      1) Management's philosophy and operating style
      2) Organizational structure
      3) Board of directors or audit committee
      4) Assignment of authority and responsibility
      5) Integrity and ethical values
      6) Commitment to competence
      7) Human resource policies and practices
   b. "**Risk assessment** is the entity's identification and analysis of relevant risks to achievement of its objectives, forming a basis for determining how the risks should be managed."
      1) This process considers internal and external matters that may adversely affect the entity's ability to record, process, summarize, and report financial data consistent with financial statement assertions.
      2) The entity's risk assessment is distinct from the auditor's assessment of inherent and control risk.
   c. "**Control activities** are the policies and procedures that help ensure that management directives are carried out. They help ensure that necessary actions are taken to address risks to achievement of the entity's objectives." Control activities relevant to a financial statement audit consist of controls related to
      1) Performance reviews
      2) Information processing (general controls, application controls)
      3) Physical controls
      4) Segregation of duties
   d. "**Information and communication** are the identification, capture, and exchange of information in a form and time frame that enable people to carry out their responsibilities." An effective information system provides for
      1) Identifying and recording all valid transactions
      2) Timely and sufficiently detailed descriptions of transactions to permit proper classification
      3) Proper measurement of transactions
      4) Recording transactions in the appropriate periods
      5) Proper financial statement presentation

e. "**Monitoring** is a process that assesses the quality of internal control performance over time."

   1) Controls should be monitored by management to consider whether they are properly designed, operating as intended, and modified to reflect changing conditions.

   2) Internal auditors and external parties, such as customers, regulators, and external auditors, contribute to the monitoring process.

3. **Limitations of Internal Control**

   a. **Reasonable assurance**. Internal control provides only reasonable assurance that an entity's control objectives are achieved.

   b. **Inherent limitations** of internal control include the following:

      1) Human judgment may be faulty.
      2) Simple error or mistake may occur.
      3) Collusion or management override may circumvent internal control.
      4) The cost of internal control should not exceed its benefits.

4. **Understanding Internal Control in Planning an Audit**. The auditor should obtain a sufficient understanding of internal control components to plan the audit, including knowledge about the design of relevant controls and whether they have been placed in operation.

   a. Such knowledge should be used to

      1) Identify types of potential misstatements
      2) Consider factors that affect the risk of material misstatements
      3) Design substantive tests

   b. Placed in operation differs from operating effectiveness.

      1) **Placed in operation** means the entity is using the control.

      2) **Operating effectiveness** is concerned with how and by whom the control was applied and the consistency of application. For example, a budgetary reporting system may produce adequate reports that are not acted upon.

         a) Knowledge about operating effectiveness need not be obtained as part of the understanding of internal control.

   c. The understanding may raise doubts about auditability.

      1) Concerns about management integrity may lead to the conclusion that the risk of misrepresentations precludes an audit.

      2) The nature and extent of records may indicate that sufficient competent evidence is unlikely to be available.

   d. In obtaining the understanding of internal control, the auditor considers

      1) Knowledge obtained from other sources, such as previous audits and the understanding of the industry

      2) Assessments of inherent risk

      3) Judgments about materiality

      4) Complexity and sophistication of operations and systems, including whether data processing depends on the computer

e. **Understanding the control environment component** requires sufficient knowledge to understand the attitude, awareness, and actions of management and the board regarding the control environment.

f. **Understanding the risk assessment component** requires sufficient knowledge of how management considers risks relevant to financial reporting objectives, estimates their significance, assesses their likelihood, and decides about appropriate actions.

g. **Understanding the control activities component** requires gaining an understanding of the other components sufficient to provide some knowledge about control activities. The knowledge thus obtained should be considered in determining the additional attention required for understanding control activities.

h. **Understanding the information and communication component** requires knowledge of the information system relevant to financial reporting to understand

1) Major classes of transactions and how they are initiated
2) The accounting records, supporting documents, and specific accounts involved in processing and reporting
3) The accounting process from initiation to inclusion in the financial statements
4) The financial reporting process, including significant estimates and disclosures

i. **Understanding of the monitoring component** requires sufficient knowledge of the major activities used to monitor internal control, including how they are used to initiate corrective action.

j. **Documentation of the understanding** is required. Its form and extent are influenced by the size and complexity of the entity and the nature of internal control.

5. **Consideration of Internal Control in Assessing Control Risk**

a. **The risk of material misstatement** in a financial statement assertion consists of the following:

1) "**Inherent risk** is the susceptibility of an assertion to a material misstatement assuming there are no related controls."
2) "**Control risk** is the risk that a material misstatement that could occur in an assertion will not be prevented or detected on a timely basis by the entity's internal control."
3) "**Detection risk** is the risk that the auditor will not detect a material misstatement that exists in an assertion."

b. **The assessment of control risk** is an evaluation of the effectiveness of internal control in preventing or detecting material misstatements. It is

1) Made in terms of financial statement assertions
2) Stated quantitatively [such as from 0% to 100% (0 to 1.0)] or nonquantitatively (such as a range from maximum to minimum)

a) For example, the auditor may decide that the probability is 10% that the controls relevant to an assertion will not prevent or detect a possible material misstatement of that assertion. However, absent any relevant controls, control risk is 1.0 (100%); that is, failure to prevent or detect a material misstatement is certain.

c. After obtaining the understanding of internal control, the auditor may **assess control risk at the maximum** (1.0 or 100%) because

1) Controls are unlikely to pertain to an assertion,
2) Controls are unlikely to be effective, or
3) Evaluating effectiveness would be inefficient.

d. **Assessing control risk at below the maximum** requires

1) Identifying specific controls relevant to specific assertions that are likely to prevent or detect misstatement
2) **Tests of controls** directed toward effectiveness of their design and operation

e. The lower the assessed level of control risk (the more closely it approaches zero), the more assurance audit evidence must provide that controls relevant to an assertion are effective.

f. According to both AU 319 and AU 312, *Audit Risk and Materiality in Conducting an Audit*, the ultimate purpose of assessing control risk (and inherent risk) is to contribute to the evaluation of the overall risk that the financial statements contain material misstatements. The assessed levels of control risk and inherent risk determine the acceptable level of detection risk for financial statement assertions.

1) Based on the level to which the auditor seeks to restrict the risk of a material misstatement and the assessed levels of inherent risk and control risk, substantive tests restrict detection risk to an acceptable level.

a) The risk of material misstatement of an assertion (AR) can be expressed algebraically as the product of inherent risk (IR), control risk (CR), and the acceptable detection risk (DR). Accordingly, if the auditor decides to restrict the risk of misstatement (that is, the risk of incorrect acceptance) to 5% and assesses inherent risk and control risk at 50% and 30%, respectively, the following calculation determines the acceptable level of detection risk:

$$AR = IR \times CR \times DR$$

$$DR = \frac{AR}{IR \times CR}$$

$$DR = \frac{5\%}{50\% \times 30\%}$$

$$DR = 33\tfrac{1}{3}\%$$

b) AU 350, *Audit Sampling*, further divides detection risk for a given substantive test of details into

i) The risk that analytical procedures and other substantive tests will fail to detect misstatements equal to tolerable misstatement and
ii) The allowable risk of incorrect acceptance for the substantive test of details.

c) The acceptable level of detection risk determines the nature, timing, and extent of substantive tests (procedures to detect material misstatements in assertions).

i) Because its relation to control risk is inverse, a lower acceptable level of detection risk increases the assurance to be provided by substantive tests. Thus, the auditor may

- Change their nature to a more effective procedure, such as directing tests toward independent parties rather than parties within the entity
- Change timing, such as from interim dates to year-end
- Change their extent, such as using a larger sample

g. **Documentation of the assessed level of control risk** is required. When control risk for an assertion is assessed at the maximum, the auditor should document the conclusion but need not document its basis.

6. Stop and review! You have completed the outline for this subunit. Study multiple-choice questions 1 through 27 beginning on page 397.

B. Internal Control According to The IIA

1. Internal control is the basis for the field of internal auditing. The controls envisioned extend beyond those relevant to the traditional financial statement audit conducted by external auditors.
2. **General Standard 300** of the Standards for the Professional Practice of Internal Auditing defines the scope of work of internal auditors largely in terms of internal control.

The scope of internal auditing should encompass the examination and evaluation of the adequacy and effectiveness of the organization's system of internal control and the quality of performance in carrying out assigned responsibilities.

a. Adequacy concerns whether reasonable assurance is given that objectives and goals will be met efficiently and economically.

b. Effectiveness concerns whether the system is functioning as intended.

c. The five primary objectives of internal control listed in Standard 300 (CARES) must be considered by internal auditors.

1) ***C**ompliance with policies, plans, etc.*
2) ***A**ccomplishment of objectives and goals*
3) ***R**eliability and integrity of information*
4) ***E**conomical and efficient use of resources*
5) ***S**afeguarding of assets*

3. **SIAS 1**, *Control: Concepts and Responsibilities*, defines control with an emphasis on the roles of the participants, especially the management functions of planning, organizing, and directing.

A control is any action taken by management to enhance the likelihood that established objectives and goals will be achieved. Management plans, organizes, and directs the performance of sufficient actions to provide reasonable assurance that objectives and goals will be achieved. Thus, control is the result of proper planning, organizing, and directing by management.

a. In the aggregate, the internal auditors' evaluations of the processes of planning, organizing, and directing permit appraisal of the system of internal control.

4. Stop and review! You have completed the outline for this subunit. Study multiple-choice questions 28 through 32 beginning on page 407.

C. Flowcharting

1. Flowcharting is a method of analyzing and understanding the processes and procedures involved in operations, whether manual or computerized. Thus, a flowchart is a pictorial diagram of the definition, analysis, or solution of a problem in which symbols are used to represent operations, data flow, equipment, etc.
    a. Flowcharting is therefore useful in obtaining an understanding of internal control and in systems development.
    b. Flowcharting is based on standardized symbols. However, different systems of symbols have been devised, so care must be taken that usage is consistent.
2. Flowcharts may be categorized as horizontal or vertical. **Horizontal flowcharts** (sometimes called systems flowcharts) depict areas of responsibility (departments or functions) arranged horizontally across the page in vertical columns or areas. Accordingly, activities, controls, and document flows that are the responsibility of a given department or function are shown in the same column.
    a. **Vertical flowcharts** present successive steps in a top-to-bottom format.
3. Flowcharting is still employed as a tool for developing information systems (see Study Unit 7).
    a. A **systems flowchart** can document not only manual processes but also the physical design specifications for the entire system, including all input, output, processing steps, and important files.
        1) A **program flowchart** depicts the specific steps within a program and the order in which they will be executed. However, program flowcharting is no longer viewed as a good program design technique because other methods are more effective for modular development.
    b. The following are input-output symbols commonly found in systems flowcharts used to develop information systems:

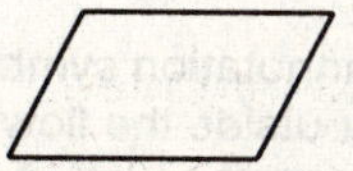
Generalized symbol for input or output used when the medium is not specified

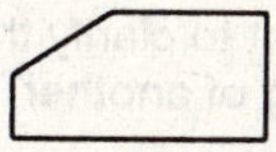
Punched card

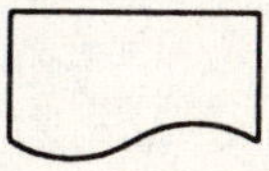
A document or report

Punched paper tape

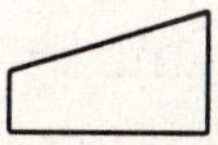
Manual input into a terminal or other online device

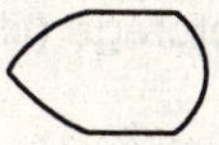
Display on a video terminal

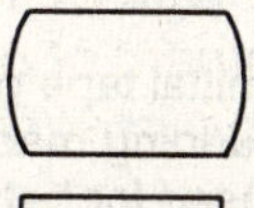
Keying operation

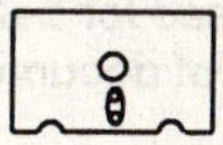
Floppy disk

Alternative possibilities

c. The following are common processing symbols used in systems flowcharts:

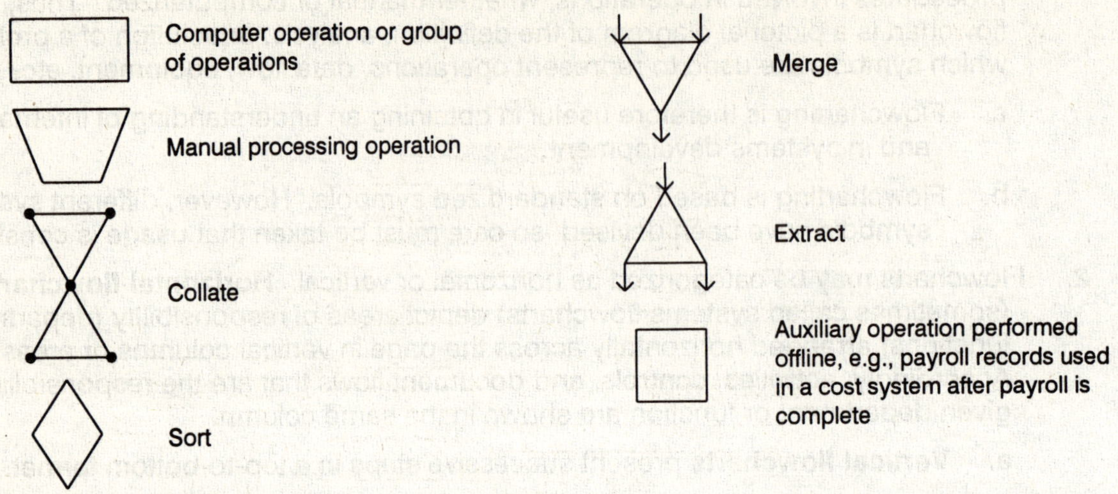

d. The following are other common symbols used in systems flowcharts:

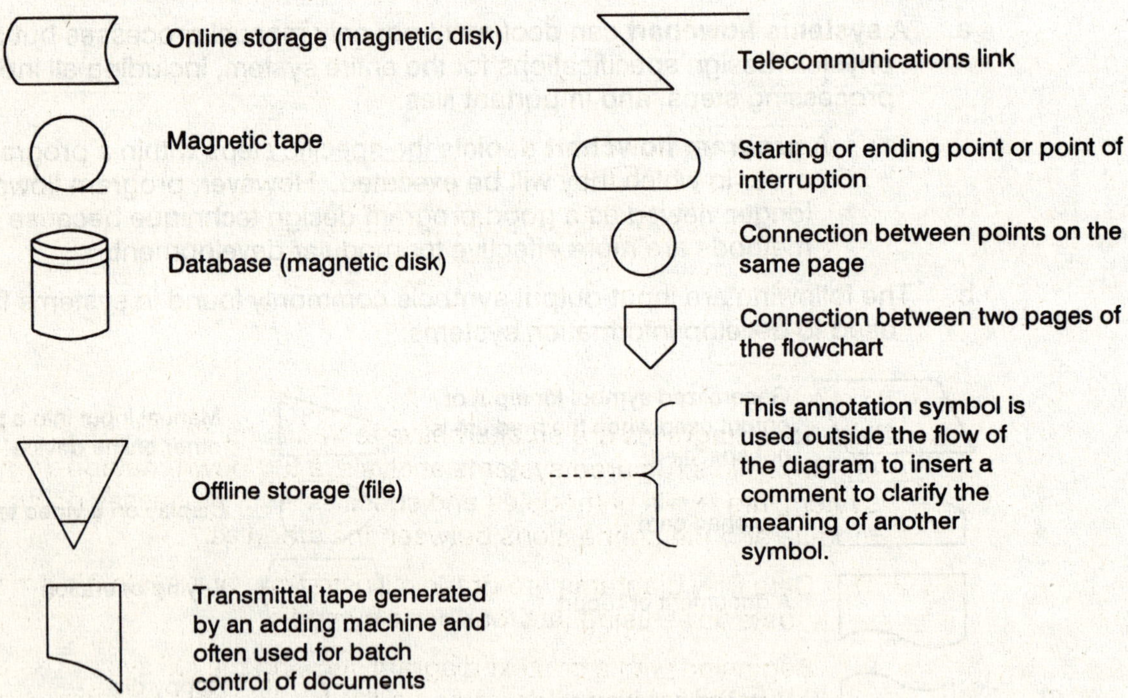

e. Program flowcharting symbols include some of those above (computer operation, general input-output, terminal point, connectors, and direction arrows). They also include

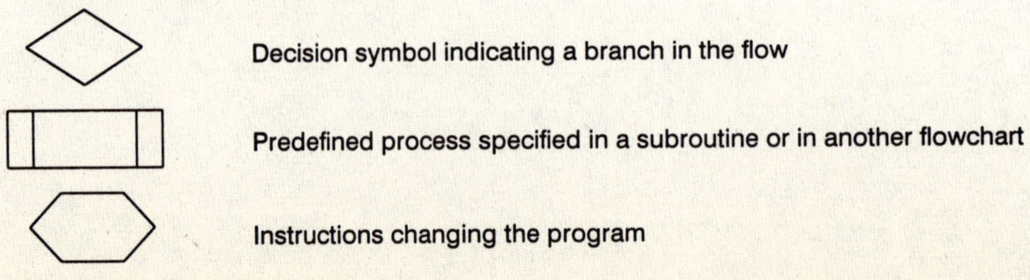

f. The following is an example of a systems flowchart for a computerized purchases-payables application:

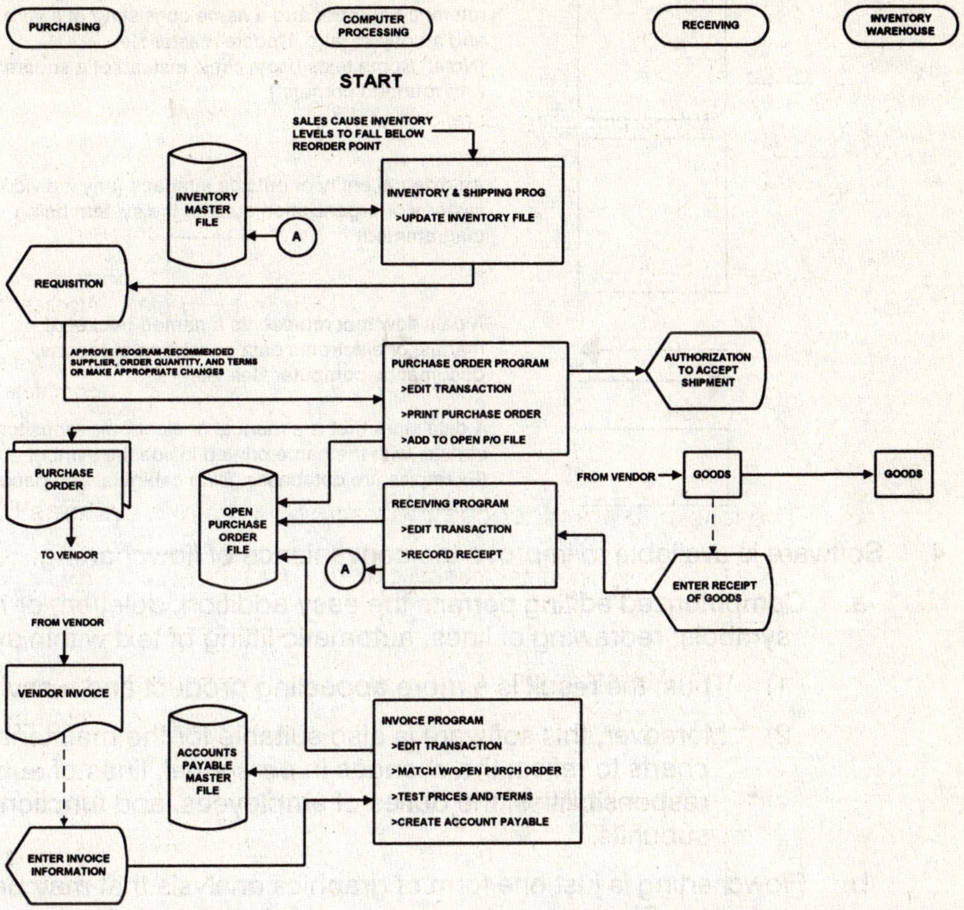

g. **Data flow diagrams** are an alternative to traditional flowcharting. They are employed in structured systems analysis, a top-down methodology that divides a system into levels of modules and specifies what processes occur in each module and the connections between the modules.

1) Data flow diagrams are graphic illustrations of a system's processes and data flows using just four basic symbols.

2) Beginning with a context diagram showing the full system as one process with its significant inputs and outputs, a complex system can be represented by a series of leveled diagrams that depict its components in successively greater detail.

3) The following are the symbols used in data flow diagrams:

A process or data transformation that has a reference number and a name consisting of a verb and an object (e.g., Update master file) [Note: Some texts use a circle instead of a square with rounded corners.]

An external entity or outside interface (any individual system, or organization outside the system being diagrammed)

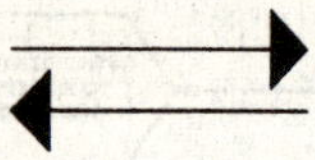

A data flow that represents a named packet of manual or electronic data consisting of reports, documents, computer files, etc.

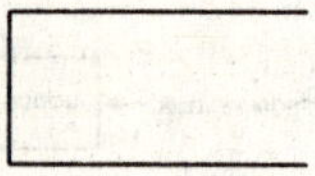

A data store that is a manual or electronic repository of data, with the name printed inside the symbol (Examples are databases, filing cabinets, and paper binders.)

4. Software is available to improve the convenience of flowcharting.
   a. Computerized editing permits the easy addition, deletion, or movement of symbols; redrawing of lines; automatic fitting of text within symbols; etc.
      1) Thus, the result is a more appealing product and a savings of audit time.
      2) Moreover, this software is also suitable for the maintenance of organizational charts to reflect the changes in personnel, lines of authority, reporting responsibilities, the duties of employees, and functions of organizational subunits.
   b. Flowcharting is just one form of graphics analysis that may be performed by software. Thus, a word-processing package should provide for the presentation of information in the form of tables, pictures, diagrams, bar charts, pie charts, etc. For example, floor plans might be generated for analysis of work flows, location of equipment, and potential fire hazards.
5. Stop and review! You have completed the outline for this subunit. Study multiple-choice questions 33 through 47 beginning on page 409.

D. Means of Achieving Control

1. Sawyer and Dittenhofer, *Sawyer's Internal Auditing* (Altamonte Springs, FL: The Institute of Internal Auditors, 4th ed., 1996, pages 102-108), defines control and provides a description the means of achieving control that is useful to management accountants as well as to internal auditors. Sawyer defines control as

   The employment of all the means devised in an enterprise to promote, direct, restrain, govern, and check upon its various activities for the purpose of seeing that enterprise objectives are met. These means of control include, but are not limited to, form of organization, policies, systems, procedures, instructions, standards, committees, charts of accounts, forecasts, budgets, schedules, reports, records, checklists, methods, devices, and internal auditing.

2. The means of control are

   a. **Organization**. Organization, as a means of control, is an approved intentional structuring of roles assigned to people within the enterprise so that the enterprise can achieve its objectives efficiently and economically.

      1) Responsibilities should be divided so that no one person will control all phases of any transaction.

      2) Managers should have the authority to take the action necessary to discharge their responsibility.

      3) Individual responsibility should always be clearly defined so that it can be neither sidestepped nor exceeded.

      4) An official who assigns responsibility and delegates authority to subordinates should have an effective system of follow-up for making sure that tasks assigned are properly carried out.

      5) The individuals to whom authority is delegated should be required to exercise that authority without close supervision. But they should check with their superiors in case of exception.

      6) People should be required to account to their superiors for the manner in which they have discharged their responsibilities.

      7) The organization should be flexible enough to permit changes in its structure when operating plans, policies, and objectives change.

      8) Organizational structures should be as simple as possible.

      9) Organization charts and manuals should be prepared that will help plan and control changes in, as well as provide better understanding of, the organization, chain of authority, and assignment of responsibilities.

   b. **Policies**. A policy is any stated principle that requires, guides, or restricts action. Policies should follow certain principles.

      1) Policies should be clearly stated in writing in systematically organized handbooks, manuals, or other publications, and properly approved.

      2) Policies should be systematically communicated to all officials and appropriate employees of the organization.

      3) Policies must conform with applicable laws and regulations, and they should be consistent with objectives and general policies prescribed at higher levels.

      4) Policies should be designed to promote the conduct of authorized activities in an effective, efficient, and economical manner and to provide a satisfactory degree of assurance that the resources of the enterprise are suitably safeguarded.

      5) Policies should be periodically reviewed, and they should be revised when circumstances change.

c. **Procedures**. Procedures are methods employed to carry out activities in conformity with prescribed policies. The same principles applicable to policies are also applicable to procedures. In addition,

1) To reduce the possibility of fraud and error, procedures should be so coordinated that one employee's work is automatically checked by another who is independently performing separate prescribed duties. In determining the extent to which automatic internal checks should be built into the system of control, such factors as degree of risk, cost of preventive procedures, availability of personnel, operational impact, and feasibility should be considered.
2) For nonmechanical operations, prescribed procedures should not be so detailed as to stifle the use of judgment.
3) To promote maximum efficiency and economy, prescribed procedures should be as simple and as inexpensive as possible.
4) Procedures should not be overlapping, conflicting, or duplicative.
5) Procedures should be periodically reviewed and improved as necessary.

d. **Personnel**. People hired or assigned should have the qualifications to do the jobs assigned to them. The best form of control over the performance of individuals is supervision. Hence, high standards of supervision should be established. The following practices help improve control.

1) New employees should be investigated as to honesty and reliability.
2) Employees should be given training and refresher courses that provide the opportunity for improvement and keep them informed of new policies and procedures.
3) Employees should be given information on the duties and responsibilities of other segments of the organization so they may better understand how and where their jobs fit into the organization as a whole.
4) The performance of all employees should be periodically reviewed to see whether all essential requirements of their jobs are being met. Superior performance should be given appropriate recognition. Shortcomings should be discussed with employees so that they are given an opportunity to improve their performance or upgrade their skills.

e. **Accounting**. Accounting is the indispensable means of financial control over activities and resources. It furnishes a framework that can be fitted to assignments of responsibility. Moreover, it is the financial scorekeeper of the enterprise. The problem lies in what scores to keep. Here are some basic principles for accounting systems.

1) Accounting should fit the needs of managers for rational decision making rather than the dictates of some textbook or canned check list.
2) Accounting should be based on lines of responsibility.
3) Financial reports of operating results should parallel the organizational units responsible for carrying out operations.
4) Accounting should be such that controllable costs can be identified.

f. **Budgeting**. A budget is a statement of expected results expressed in numerical terms. As a control, it sets a standard for input of resources and what should be achieved as output and income.

1) Those who are responsible for meeting a budget should participate in its preparation.

2) Those responsible for meeting a budget should be provided with adequate information that compares budgets with actual events and shows reasons for any significant variances.

3) All subsidiary budgets should tie into the overall budget for the enterprise.

4) Budgets should set measurable objectives; they are meaningless unless managers know what they are budgeting for.

5) Budgets should help sharpen the organizational structure because objective budgeting standards are difficult to set in a confused combination of subsystems. Budgeting is therefore a form of discipline and coordination.

g. **Reporting**. In most organizations, management functions and makes decisions on the basis of reports it receives. Therefore, reports should be timely, accurate, meaningful, and economical. Here are some principles for establishing a satisfactory internal reporting system.

1) Reports should be made in accordance with assigned responsibilities.

2) Individuals or units should be required to report only on those matters for which they are responsible.

3) The cost of accumulating data and preparing reports should be weighed against the benefits to be obtained from them.

4) Reports should be as simple as possible, and consistent with the nature of the subject matter. They should include only information that serves the needs of the readers. Common classifications and terminology should be used as much as possible to avoid confusion.

5) When appropriate, performance reports should show comparisons with predetermined standards of cost, quality, and quantity. Controllable costs should be segregated.

6) When performance cannot be reported in quantitative terms, the reports should be designed to emphasize exceptions or other matters requiring management attention.

7) For maximum value, reports should be timely. Timely reports based partly on estimates may be more useful than delayed reports that are more precise.

8) Report recipients should be polled periodically to see if they still need the reports they are receiving, or if the report could be improved.

3. Stop and review! You have completed the outline for this subunit. Study multiple-choice questions 48 through 58 beginning on page 414.

E. Legal and Professional Influences

1. The **Foreign Corrupt Practices Act (FCPA)** was enacted in 1977 as an amendment to the Securities Exchange Act of 1934. The FCPA is designed to prevent secret payments of corporate funds for purposes that Congress has determined to be contrary to public policy.

   a. The act prohibits a domestic concern, including any person acting on its behalf, whether or not doing business overseas and whether or not registered with the SEC, from offering or authorizing corrupt payments to any

      1) Foreign official
      2) Foreign political party or official thereof
      3) Candidate for political office in a foreign country

   b. Only political payments to foreign officials are prohibited. Payment to foreign business owners or corporate officers are not addressed by the FCPA.

   c. **Corrupt payments** are payments the purpose of which is to induce the recipient to act or refrain from acting so that the domestic concern might obtain or retain business.

      1) The FCPA prohibits a mere offer or promise of a bribe, even if it is not consummated.

      2) The FCPA prohibits payment of anything of value. *De minimis* gifts and tokens of hospitality are excepted.

      3) Payments are prohibited if the person making them knew or should have known that some or all of them would be used to influence a governmental official.

   d. Foreign officials do not include clerical or ministerial employees.

      1) EXAMPLE: Payments made to a clerk to expedite the processing of goods through customs may not be prohibited by the act.

      2) Such payments are not prohibited as long as the recipient has no discretion in carrying out a governmental function.

   e. Regardless of whether they have foreign operations, all public companies must make and keep books, records, and accounts in reasonable detail that accurately and fairly reflect transactions and dispositions of assets.

   f. All public companies must also devise and maintain a system of internal accounting control sufficient to provide **reasonable assurance** that

      1) Transactions are executed in accordance with management's general or specific authorization.

      2) Transactions are recorded as necessary

         a) To permit preparation of financial statements in conformity with GAAP or any other criteria applicable to such statements

         b) To maintain accountability for assets

      3) Access to assets is permitted only in accordance with management's general or specific authorization.

      4) The recorded accountability for assets is compared with the existing assets at reasonable intervals, and appropriate action is taken with respect to any differences.

2. **Treadway Report**. The Treadway Commission (named after its chair) was sponsored by the American Accounting Association, the AICPA, the Financial Executives Institute, The Institute of Internal Auditors, and the National Association of Accountants (now the IMA). The commission was established in 1985 and submitted its final report in 1987 regarding allegations of widespread financial reporting fraud by public companies.

   a. The objectives were to

      1) Consider the causes and effects of fraudulent financial reporting and how to prevent and detect it

      2) Examine the roles of both the internal auditor and the independent auditor in preventing and detecting fraudulent financial reporting

      3) Identify those attributes of corporate structure that may contribute to fraudulent reporting or to failure in detection

   b. As a result of research, public hearings, etc., the commission drafted a report containing recommendations regarding

      1) Public companies
      2) Independent public accountants
      3) The SEC and other regulatory bodies

         a) Improvement in the regulatory and legal environment

      4) Education

   c. **Recommendations for the public company**

      1) The tone at the top

         a) Management should perform an ongoing fraud-risk assessment.

         b) All public companies should maintain internal controls that provide reasonable assurance of the prevention or early detection of fraudulent financial reporting.

         c) All public companies should establish effective written codes of conduct including guidelines on

            i) Conflicts of interest
            ii) Compliance with domestic and foreign laws
            iii) Confidentiality of proprietary information

         d) All public companies should maintain accounting functions that are designed to meet their financial reporting obligations.

         e) The SEC should require that management acknowledge in the annual report to shareholders its responsibility for the financial statements and internal controls.

      2) The internal audit function

         a) The director of internal auditing should have unrestricted and direct access to both the audit committee and the CEO.

         b) Internal auditors should coordinate their work with the independent public accountants.

3) The audit committee

a) The board of directors of every public company should develop and approve a written charter describing the audit committee's duties and responsibilities.

b) Audit committees should review the independence of the independent public accountant.

c) The SEC should require that the annual report to shareholders include a letter from the audit committee chair describing the committee's responsibilities and activities during the year.

d) Audit committees should monitor compliance with codes of conduct.

e) Audit committees should have the resources necessary for meeting their responsibilities.

f) The audit committee should oversee the quarterly reporting process.

g) Audit committees should monitor instances in which management seeks second opinions on significant accounting issues.

4) Standards for internal control

a) The Treadway Commission recommends that its sponsoring organizations set up an interdisciplinary body to integrate internal control concepts and develop a common reference point.

b) Currently, varying interpretations and philosophies on internal control exist, which cause managements, independent public accountants, and internal auditors to disagree sometimes about the adequacy of internal controls.

d. **Recommendations for the independent public accountant**

1) The Auditing Standards Board of the AICPA should clarify the independent public accountant's responsibility for fraud detection.

2) To enhance their ability to detect fraud, CPAs should be required to use analytical procedures in all audit engagements.

3) Existing quality assurance procedures should be improved to recognize the risk inherent in audits of new clients.

4) Guidance as to qualifications of concurring, or second, partners and timing of their reviews should be enhanced.

5) Independent public accounting firms should recognize and control the pressures that might reduce audit quality.

6) The auditor's standard report should be revised to clarify the auditor's responsibility for fraud detection. (NOTE: The auditor's standard report does not contain such language.)

7) The auditor's standard report should be revised to explain the extent of the auditor's review and evaluation of the system of internal accounting control. (NOTE: The auditor's standard report does not contain such language.)

8) The AICPA should reorganize the Auditing Standards Board to provide full-time members.

e. **Recommendations to the SEC and others**

1) SEC resources should be increased to enhance its enforcement activities and its oversight of the financial reporting system.

2) The SEC should seek to expand its enforcement remedies (including increased criminal prosecution) to provide more severe penalties and more effective tailoring of penalties to individual cases.

3) When a public company changes independent public accountants, the SEC should require public disclosure of the nature of any material accounting or auditing issues discussed with its old or new auditors during the 3-year period preceding the change.

4) The SEC should require independent public accounting firms that audit public companies to be members of professional organizations that have peer review and independent oversight functions.

5) Federal agencies responsible for public companies not reporting to the SEC should be encouraged to implement the commission's SEC recommendations.

6) The exchange of information between financial institution regulatory agencies and independent public accountants should be improved.

7) State boards of accountancy should be encouraged to establish programs for monitoring the quality of services provided by accountants they license.

8) The SEC should reconsider, for independent directors, its position that corporate indemnification of directors for liabilities arising under the Securities Act of 1933 is against public policy and thus unenforceable.

9) The implications of the perceived liability crisis on long-term audit quality should be assessed by parties charged with responding to various tort reform initiatives.

f. **Education**

1) Educators should foster knowledge and understanding of the factors that may cause fraudulent financial reporting and the strategies that can lead to reduction in its incidence.

2) The business and accounting curricula should promote a better understanding of the function and the importance of internal controls, including the control environment.

3) Business and accounting students should be well-informed about the regulation and enforcement activities by which government and private bodies safeguard the financial reporting system and thereby protect the public interest.

4) The business and accounting curricula should help students develop stronger analytical, problem-solving, and judgment skills.

5) Business schools should encourage business and accounting faculty to develop their own personal competence as well as classroom materials for conveying information, skills, and ethical values that can help prevent, detect, and deter fraudulent financial reporting.

6) Professional certification examinations should be revised to test for an understanding of the complexities involved in fraudulent financial reporting.

7) Continuing professional education programs should address fraudulent financial reporting.

g. Among the many responses to the Treadway Commission's report are the following:

   1) The Committee of Sponsoring Organizations (COSO) was formed and subsequently issued *Internal Control-Integrated Framework*. This publication's definition and descriptions of internal control (including the components thereof) have been adapted for inclusion in the AICPA's basic pronouncement on internal control (see the introduction for this study unit).

   2) The AICPA has issued new pronouncements requiring a CPA firm to have a system of quality control for its accounting and auditing practice.

   3) A CPA firm must be enrolled in an approved practice-monitoring program, and the AICPA has therefore issued standards governing peer reviews. These reviews are administered by state CPA societies in conjunction with the AICPA.

   4) Continuing professional education requirements for CPAs have been strengthened.

   5) The Auditing Standards Board has issued the following Statements on Auditing Standards (among others):

      a) SAS 54 (AU 317), *Illegal Acts by Clients*

      b) SAS 56 (AU 329), *Analytical Procedures*

      c) SAS 59 (AU 341), *The Auditor's Consideration of an Entity's Ability to Continue as a Going Concern*

      d) SAS 60 (AU 325), *Communication of Internal Control-Related Matters Noted in an Audit*

      e) SAS 61 (AU 380), *Communication with Audit Committees*

      f) SAS 78 (AU 319), *Consideration of Internal Control in a Financial Statement Audit*

      g) SAS 82 (AU 316), *Consideration of Fraud in a Financial Statement Audit*

   6) In response to strong interest by Congress, regulatory bodies, and other parties, the AICPA has issued a Statement on Standards for Attestation Engagements that provides guidance for engagements to report on management's written assertion about the effectiveness of an entity's internal control over financial reporting.

3. **The Private Securities Litigation Reform Act of 1995** included provisions regarding auditor requirements for fraud detection and disclosure.

   a. Audits should provide reasonable assurance of detecting illegal acts having a direct and material effect on financial statement amounts, be designed to identify material related party transactions, and include an evaluation as to whether there is a substantial doubt about the issuer's ability to continue as a going concern.

1) Accountants must report illegal acts to the appropriate level of management and the audit committee unless they are clearly inconsequential.

2) If senior management and the board fail to take action on reported material illegal acts, and this failure will result in a departure from a standard report or resignation from the audit, the accountants should report their conclusions to the board immediately. The board must then, within 1 business day, notify the SEC.

3) If the accountants do not receive a copy of the notice within the 1-day period, they must furnish the SEC with a copy of their report within 1 business day.

4. **SAS 82** (AU 316), *Consideration of Fraud in a Financial Statement Audit*, addresses an auditor's responsibilities, the nature and characteristics of fraud, and risk factors.

a. The auditor should plan and perform the audit to obtain reasonable assurance about whether the financial statements are free of material misstatement, whether caused by error or fraud. However, an auditor cannot obtain absolute assurance that material misstatements will be detected. Because of the concealment aspects of fraudulent activity (e.g., collusion or falsification of documents) and the need to apply professional judgment in identification and evaluation of fraud risk factors, even a properly planned audit may not detect a material misstatement resulting from fraud.

b. **Description and characteristics of fraud.** Fraud differs from error because it is intentional. It typically involves pressures or incentives to commit fraud and a perceived opportunity to do so. The following are the two types of fraud relevant to the auditor:

1) **Misstatements arising from fraudulent financial reporting** are intentional misstatements or omissions to deceive financial statement users such as

a) Manipulation, falsification, or alteration of accounting records or documents

b) Misrepresentation in, or intentional omission from, the financial statements of events, transactions, or other significant information

c) Intentional misapplication of accounting principles

2) **Misstatements arising from misappropriation of assets** (sometimes termed defalcation) include theft, embezzlement, or an action that causes an entity to pay for items not received that causes the financial statements to be materially misstated.

c. **Assessment of risk**. The scope of the audit is determined in part by the consideration of audit risk, and audit risk includes the risk of material misstatement due to fraud. Thus, the risk of material misstatement due to fraud must be specifically assessed.

1) As part of the assessment, the auditor considers fraud risk factors relating to misstatements arising from

a) **Fraudulent reporting**. The auditor should consider the following categories of fraud risk factors related to fraudulent reporting:

i) Management characteristics and influence over the control environment, including management's abilities, pressures, style, and attitude

ii) Industry conditions, including the economic and regulatory environment

iii) Operating characteristics and financial stability of the organization, including its nature and complexity, transactions, financial condition, and profitability

b) **Misappropriation of assets**. The auditor should consider the following categories of fraud risk factors related to misappropriation of assets:

i) Susceptibility of assets to misappropriation. These fraud risk factors concern the nature of the organization's assets and their exposure to theft.

ii) Controls. These fraud risk factors relate to the extent of controls intended to prevent or detect misappropriation of assets.

2) The assessment also includes inquiries of management

a) To obtain management's understanding about the risk of fraud

b) To determine whether management knows of fraud perpetrated on or within the organization

3) SAS 82 (AU 316), *Consideration of Fraud in a Financial Statement Audit*, provides many examples of fraud risk factors for each of the five categories, outlined in c.1) a) and b), that must be considered in an audit. However, these examples are not exhaustive, and their applicability and significance will vary with the circumstances. Thus, professional judgment is necessary for determining the relevance and significance of fraud risk factors and their implications for the conduct of the audit.

d. Disclosure of possible illegal acts to outside parties ordinarily is not the auditor's responsibility and would violate the duty of confidentiality. However, the auditor may need

1) To comply with certain legal and regulatory requirements. For example, "reportable events" must be disclosed to the SEC, and reports to the SEC on illegal acts may be required by the *Private Securities Litigation Reform Act of 1995*.

2) To communicate with a successor auditor in accordance with AU 315, *Communications between Predecessor and Successor Auditors*

3) To respond to a subpoena

4) To report to a funding or other specified agency in accordance with governmental audit requirements, such as those established by the *Single Audit Act*

5. Stop and review! You have completed the outline for this subunit. Study multiple-choice questions 59 through 64 beginning on page 418.

MULTIPLE-CHOICE QUESTIONS

A. Internal Control According to the AICPA

1. When an organization has strong internal control, management can expect various benefits. The benefit least likely to occur is

A. Reduced cost of an external audit.

B. Elimination of employee fraud.

C. Availability of reliable data for decision-making purposes.

D. Some assurance of compliance with the Foreign Corrupt Practices Act of 1977.

The correct answer is (B). *(CMA, adapted)*

REQUIRED: The least likely benefit from a strong internal control.

DISCUSSION: Even the best internal control cannot guarantee the complete elimination of employee fraud. Effective internal control will reduce the amount of employee fraud and probably detect losses on a timely basis.

Answers (A), (C), and (D) are incorrect because each is a benefit of strong internal control. The cost of the external audit will be lower because of the reduction of the audit effort related to substantive testing. Also, management will have better data for decision-making purposes and some assurance of compliance with the FCPA.

2. One of the auditor's major concerns is to ascertain whether internal control is designed to provide reasonable assurance that

A. Profit margins are maximized, and operational efficiency is optimized.

B. The chief accounting officer reviews all accounting transactions.

C. Corporate morale problems are addressed immediately and effectively.

D. Financial statements are fairly presented.

The correct answer is (D). *(CMA, adapted)*

REQUIRED: The objective of internal control.

DISCUSSION: Internal control is designed to provide reasonable assurance of the achievement of objectives in the categories of (1) reliability of financial reporting, (2) effectiveness and efficiency of operations, and (3) compliance with laws and regulations. Controls relevant to an audit ordinarily pertain to the objective of preparing external financial statements that are fairly presented in conformity with GAAP or another comprehensive basis of accounting (AU 319).

Answer (A) is incorrect because many factors beyond the purview of the auditor affect profits, and the controls related to operational efficiency are usually not directly relevant to an audit. Answer (B) is incorrect because the chief accounting officer need not review all accounting transactions. Answer (C) is incorrect because controls relevant to a financial statement audit do not concern the treatment of corporate morale problems.

3. Which of the following best describe the interrelated components of internal control?

A. Organizational structure, management philosophy, and planning.

B. Control environment, risk assessment, control activities, information and communication systems, and monitoring.

C. Risk assessment, backup facilities, responsibility accounting, and natural laws.

D. Legal environment of the firm, management philosophy, and organizational structure.

The correct answer is (B). *(CMA, adapted)*

REQUIRED: The components of internal control.

DISCUSSION: Internal control includes five components: the control environment, risk assessment, control activities, information and communication, and monitoring. The control environment sets the tone of an organization, influences control consciousness, and provides a foundation for the other components. Risk assessment is the identification and analysis of relevant risks to achievement of objectives. Control activities help ensure that management directives are executed. Information and communication are the identification, capture, and exchange of information in a form and time frame that allow people to meet their responsibilities. Monitoring assesses the performance of internal control over time (AU 319).

Answer (A) is incorrect because planning is not a component of internal control. Organizational structure and management philosophy are factors in the control environment component. Answer (C) is incorrect because risk assessment is the only component listed. Answer (D) is incorrect because the legal environment of the firm, management philosophy, and organizational structure are factors in the control environment component.

4. Control activities constitute one of the five components of internal control. Control activities do not encompass

A. Performance reviews.

B. Information processing.

C. Physical controls.

D. An internal audit function.

The correct answer is (D). *(Publisher)*

REQUIRED: The item not belonging to the control activities component.

DISCUSSION: Control activities are policies and procedures that help ensure that management directives are carried out. They are intended to ensure that necessary actions are taken to address risks to achieve the entity's objectives. Control activities have various objectives and are applied at various organizational and functional levels. However, an internal audit function is part of the monitoring component.

Answers (A), (B), and (C) are incorrect because performance reviews, information processing, and physical controls are categories of control activities.

5. Monitoring is an important component of internal control. Which of the following items would not be an example of monitoring?

A. Management regularly compares divisional performance with budgets for the division.

B. Data processing management regularly generates exception reports for unusual transactions or volumes of transactions and follows up with investigation as to causes.

C. Data processing management regularly reconciles batch control totals for items processed with batch controls for items submitted.

D. Management has asked internal auditing to perform regular audits of the controls over cash processing.

The correct answer is (C). *(CIA, adapted)*

REQUIRED: The item not an example of monitoring.

DISCUSSION: Monitoring assesses the quality of internal control over time. Management considers whether internal control is properly designed and operating as intended and modifies it to reflect changing conditions. Monitoring may be in the form of separate, periodic evaluations or of ongoing monitoring. Ongoing monitoring occurs as part of routine operations. It includes management and supervisory review, comparisons, reconciliations, and other actions by personnel as part of their regular activities. However, reconciling batch control totals is a processing control.

Answer (A) is incorrect because budgetary comparison is a typical example of a monitoring control. Answer (B) is incorrect because investigation of exceptions is a monitoring control used by lower-level management to determine when their operations may be out of control. Answer (D) is incorrect because internal auditing is a form of monitoring. It serves to evaluate management's other controls.

6. A small entity may use less formal means to ensure that internal control objectives are achieved. For example, extensive accounting procedures, sophisticated accounting records, or formal controls are least likely to be needed if

A. Management is closely involved in operations.

B. The entity is involved in complex transactions.

C. The entity is subject to legal or regulatory requirements also found in larger entities.

D. Financial reporting objectives have been established.

The correct answer is (A). *(Publisher)*

REQUIRED: The situation in which less formal means of achieving control objectives are appropriate.

DISCUSSION: Effective management involvement may obviate the need for more formal means of ensuring that internal control objectives are met. Thus, a smaller entity may not have formal policies regarding credit approval, information security, or competitive bidding. It also may not have a written code of conduct. Instead, a smaller entity may develop a culture emphasizing integrity and ethical behavior through management example. Moreover, an effective control environment may not require outside members on the board. In a small company, less detailed control activities are possible when management retains authority for specific authorization of transactions and oversees employees performing incompatible tasks. Communication in a small company is also easier because of management's greater visibility and availability.

Answers (B) and (C) are incorrect because complex transactions and legal or regulatory requirements may necessitate the more formal arrangements found in larger entities. Answer (D) is incorrect because all entities should establish financial reporting objectives. However, they may be recognized implicitly rather than explicitly in smaller entities. Management can assess the risks related to these objectives through direct personal involvement rather than a formal assessment process.

7. Internal control can provide only reasonable assurance of achieving entity control objectives. One factor limiting the likelihood of achieving those objectives is that

A. The auditor's primary responsibility is the detection of fraud.

B. The board of directors is active and independent.

C. The cost of internal control should not exceed its benefits.

D. Management monitors internal control.

The correct answer is (C). *(Publisher)*

REQUIRED: The true statement about the limitation of internal control.

DISCUSSION: AU 319 states, "Another limiting factor is that the cost of an entity's internal control should not exceed the benefits that are expected to be derived. Although the cost-benefit relationship is a primary criterion that should be considered in designing internal control, the precise measurement of costs and benefits usually is not possible."

Answer (A) is incorrect because the auditor's responsibility is "to plan and perform the audit to obtain reasonable assurance about whether the financial statements are free of material misstatement, whether caused by error or fraud" (AU 110). Answer (B) is incorrect because an active and independent board strengthens the control environment. Answer (D) is incorrect because the absence of monitoring weakens internal control.

8. The primary responsibility for establishing and maintaining internal control rests with

A. The external auditor.

B. Management.

C. The controller.

D. The treasurer.

The correct answer is (B). *(CMA, adapted)*

REQUIRED: The person primarily responsible for establishing and maintaining internal control.

DISCUSSION: Establishing and maintaining internal control is the responsibility of management. Internal control is intended to provide reasonable assurance that the entity's objectives are achieved. Achievement of these objectives is the basic function of management.

Answer (A) is incorrect because auditors must consider internal control, but they do not establish and maintain it. Answers (C) and (D) are incorrect because the controller and the treasurer are responsible only to the extent that they are a part of the management team.

9. The situation most likely to be regarded as a strength in internal control by the external auditor is

A. The performance of financial audits by internal auditors.

B. The performance of operational audits by internal auditors.

C. The routine supervisory review of production planning.

D. The existence of a preventive maintenance program.

The correct answer is (A). *(CMA, adapted)*

REQUIRED: The activity most likely regarded as a strong internal control by the external auditor.

DISCUSSION: Internal auditing examines and evaluates the adequacy and effectiveness of an organization's controls. Its scope of work includes reviewing the reliability and integrity of financial data. The internal audit function is part of the monitoring component of internal control and therefore may have an important effect on the entity's ability to record, process, summarize, and report financial data.

Answer (B) is incorrect because operational audits are concerned with operational efficiency and effectiveness. Answer (C) is incorrect because routine supervisory review of production planning is a concern of management but does not directly affect the fairness of the financial statements. Answer (D) is incorrect because the existence of a preventive maintenance program is not directly relevant to a financial statement audit.

10. Control activities include procedures that pertain to physical controls over access to and use of assets and records. A departure from the purpose of such procedures is that

A. Access to the safe-deposit box requires two officers.

B. Only storeroom personnel and line supervisors have access to the raw materials storeroom.

C. The mail clerk compiles a list of the checks received in the incoming mail.

D. Only salespersons and sales supervisors use sales department vehicles.

The correct answer is (B). *(Publisher)*

REQUIRED: The departure from the purpose of control activities that limit access to assets.

DISCUSSION: Storeroom personnel have custody of assets, while supervisors are in charge of execution functions. To give supervisors access to the raw materials storeroom is a violation of the essential internal control principle of segregation of functions.

Answer (A) is incorrect because it is appropriate for two officers to be required to open the safe-deposit box. One supervises the other. Answer (C) is incorrect because mail room clerks typically compile a prelisting of cash. The list is sent to the accountant as a control for actual cash sent to the cashier. Answer (D) is incorrect because use of sales department vehicles should be limited to sales personnel unless proper authorization is obtained.

11. Which of the following features of a large manufacturing company's organization structure is a control weakness?

A. The information systems department is headed by a vice president who reports directly to the president.

B. The chief financial officer is a vice president who reports to the chief executive officer.

C. The audit committee of the board consists of the chief executive officer, the chief financial officer, and a major shareholder.

D. The controller and treasurer report to the chief financial officer.

The correct answer is (C). *(CIA, adapted)*

REQUIRED: The control weakness in a large manufacturing company's organization structure.

DISCUSSION: The audit committee has a control function because of its oversight of internal as well as external auditing. It should consist of directors who are independent of management. The authority and independence of the audit committee strengthen the position of internal auditing. The board should concur in the appointment or removal of the director of internal auditing, who should have direct, regular communication with the board (Standard 110).

Answer (A) is incorrect because this reporting relationship is a strength. It prevents the information systems operation from being dominated by a user. Answers (B) and (D) are incorrect because each is a normal and appropriate reporting relationship.

12. Which of the following controls could be used to detect bank deposits that are recorded but never made?

A. Establishing accountability for receipts at the earliest possible time.

B. Linking receipts to other internal accountabilities (i.e., collections to either accounts receivable or sales).

C. Consolidating cash receiving points.

D. Having bank reconciliations performed by a third party.

The correct answer is (D). *(CIA, adapted)*

REQUIRED: The control to detect failure to make recorded bank deposits.

DISCUSSION: Having an independent third party prepare the bank reconciliations would reveal any discrepancies between recorded deposits and the bank statements. A bank reconciliation compares the bank statement with company records and resolves differences caused by deposits in transit, outstanding checks, NSF checks, bank charges, errors, etc.

Answers (A), (B), and (C) are incorrect because these controls are implemented before deposits are prepared and recorded in the company's books. The problem here is the detection of the diversion of funds that have been properly recorded upon receipt.

13. An adequate system of internal controls is most likely to detect a fraud perpetrated by a

A. Group of employees in collusion.

B. Single employee.

C. Group of managers in collusion.

D. Single manager.

The correct answer is (B). *(CIA, adapted)*

REQUIRED: The fraud most likely to be detected by an adequate system of internal controls.

DISCUSSION: Segregation of duties and other control procedures serve to prevent or detect a fraud committed by an employee acting alone. One employee may not have the ability to engage in wrongdoing or may be subject to detection by other employees in the course of performing their assigned duties. However, collusion may circumvent controls. For example, comparison of recorded accountability with assets may fail to detect fraud if persons having custody of assets collude with record keepers.

Answer (A) is incorrect because a group has a better chance of successfully perpetrating a fraud than does an individual employee. Answer (C) is incorrect because management can override controls. Answer (D) is incorrect because even a single manager may be able to override controls.

14. Factors that should be considered when evaluating audit risk in a functional area include

1. Volume of transactions
2. Degree of system integration
3. Years since last audit
4. Significant management turnover
5. (Dollar) value of assets at risk
6. Average value per transaction
7. Results of last audit

Factors that best define the materiality of audit risk are

A. 1 through 7.

B. 2, 4, and 7.

C. 1, 5, and 6.

D. 3, 4, and 6.

The correct answer is (C). *(CIA, adapted)*

REQUIRED: The factors that best define the materiality of audit risk.

DISCUSSION: Audit risk is the risk that the audit will not detect material misstatements. Materiality is a function of quantitative and qualitative factors, of which the former are obviously more readily defined. Factors 1, 5, and 6 can all be quantified.

Answer (A) is incorrect because factors 2, 3, 4, and 7 are not quantifiable in dollars. Answer (B) is incorrect because factors 2, 4, and 7 are not quantifiable in dollars. Answer (D) is incorrect because factors 3 and 4 are not quantifiable in dollars.

15. According to AU 312, *Audit Risk and Materiality in Conducting an Audit*, the concepts of audit risk and materiality are interrelated and must be considered together by the auditor. Which of the following is true?

A. Audit risk is the risk that the auditor may unknowingly express a modified opinion when in fact the financial statements are fairly stated.

B. The phrase in the auditor's standard report "present fairly, in all material respects, in conformity with generally accepted accounting principles" indicates the auditor's belief that the financial statements taken as a whole are not materially misstated.

C. If misstatements are not important individually but are important in the aggregate, the concept of materiality does not apply.

D. Material fraud but not material errors cause financial statements to be materially misstated.

The correct answer is (B). *(Publisher)*

REQUIRED: The true statement about the significance of audit risk and materiality.

DISCUSSION: The opinion paragraph of the standard report explicitly refers to materiality. Hence, financial statements that are presented fairly, in all material respects, in conformity with GAAP are not materially misstated. Material misstatement can result from errors or fraud.

Answer (A) is incorrect because audit risk is the risk the auditor may unknowingly fail to appropriately modify the opinion on financial statements that are materially misstated. Answer (C) is incorrect because the concept of materiality recognizes that some misstatements, either individually or in the aggregate, are important for the fair presentation of financial statements. Qualitative as well as quantitative factors affect materiality judgments. Answer (D) is incorrect because both material errors and material fraud cause financial statements to be materially misstated.

16. According to AU 312 and AU 319, the auditor uses the assessed level of control risk (together with the assessed level of inherent risk) to determine the acceptable level of detection risk for financial statement assertions. As the acceptable level of detection risk decreases, the auditor may do one or more of the following except change the

A. Nature of substantive tests to more effective procedures.

B. Timing of substantive tests, such as performing them at year-end rather than at an interim date.

C. Extent of substantive tests, such as using larger sample sizes.

D. Assurances provided by substantive tests to a lower level.

The correct answer is (D). *(Publisher)*

REQUIRED: The improper action when the acceptable level of detection risk decreases.

DISCUSSION: The overall allowable audit risk of material misstatement in a financial statement assertion equals the product of inherent risk, control risk, and detection risk (expressed as probabilities). The audit risk formula in AU 350 further divides detection risk for a substantive test of details into (1) the risk that analytical procedures and other substantive tests will fail to detect misstatements equal to tolerable misstatement and (2) the allowable risk of incorrect acceptance for the substantive test of details. After determining the level to which (s)he wishes to restrict the risk of material misstatement and the assessed levels of control risk and inherent risk, the auditor performs substantive tests to restrict detection risk to the acceptable level. Accordingly, the level of detection risk that an auditor may accept is inversely related to control risk and inherent risk. If either increases, the acceptable level of detection risk decreases, and the auditor should change the nature, timing, or extent of substantive tests to increase the assurance they provide.

Answers (A), (B), and (C) are incorrect because use of more effective substantive tests, changing the timing of substantive tests, and changing the extent of testing are possible responses to a decrease in the acceptable level of detection risk.

17. Some account balances, such as those for pensions or leases, are the results of complex calculations. The susceptibility to material misstatements in these types of accounts is defined as

A. Audit risk.

B. Detection risk.

C. Sampling risk.

D. Inherent risk.

The correct answer is (D). *(CMA, adapted)*

REQUIRED: The susceptibility to material misstatements in account balances resulting from complex calculations.

DISCUSSION: Inherent risk is the susceptibility of an assertion to a material misstatement in the absence of related controls. This risk is greater for some assertions and related balances or classes than others. For example, complex calculations are more likely to be misstated than simple ones, and cash is more likely to be stolen than an inventory of coal. Inherent risk exists independently of the audit (AU 312).

Answer (A) is incorrect because audit risk is the risk that the auditor may unknowingly fail to appropriately modify an opinion on financial statements that are materially misstated. Answer (B) is incorrect because detection risk is the risk that the auditor will not detect a material misstatement that exists in an assertion. Answer (C) is incorrect because sampling risk is the risk that a particular sample may contain proportionately more or fewer monetary misstatements or deviations from controls than exist in the population as a whole (AU 350).

18. Audit risk consists of inherent risk, control risk, and detection risk. Which of the following statements is true?

A. Cash is more susceptible to theft than an inventory of coal because it has a greater inherent risk.

B. The risk that material misstatement will not be prevented or detected on a timely basis by internal control can be reduced to zero by effective controls.

C. Detection risk is a function of the efficiency of an auditing procedure.

D. The existing levels of inherent risk, control risk, and detection risk can be changed at the discretion of the auditor.

The correct answer is (A). *(Publisher)*

REQUIRED: The true statement about audit risk.

DISCUSSION: Inherent risk is the susceptibility of an assertion to material misstatement in the absence of related controls. Some assertions and related balances or classes of transactions have greater inherent risk. Thus, cash has a greater inherent risk than less liquid assets.

Answer (B) is incorrect because some control risk will always exist. Internal control has inherent limitations. Answer (C) is incorrect because detection risk is a function of auditing effectiveness (achieving results), not efficiency. Answer (D) is incorrect because the actual levels of inherent risk and control risk are independent of the audit process. Acceptable detection risk is a function of the desired level of overall audit risk and the assessed levels of inherent risk and control risk. Hence, detection risk can be changed at the discretion of the auditor, but inherent risk and control risk cannot. However, the auditor's preliminary judgments about inherent risk and control risk may change as the audit progresses.

19. The audit risk against which the auditor and those who rely on his/her opinion require reasonable protection is a combination of three separate risks at the account-balance or class-of-transactions level. The first risk is inherent risk. The second risk is that material misstatements will not be prevented or detected by internal control. The third risk is that

A. The auditor will reject a correct account balance as incorrect.

B. Material misstatements that occur will not be detected by the audit.

C. The auditor will apply an inappropriate audit procedure.

D. The auditor will apply an inappropriate measure of audit materiality.

The correct answer is (B). *(Publisher)*

REQUIRED: The component of audit risk in addition to inherent risk and control risk.

DISCUSSION: According to AU 312, one component of audit risk is detection risk, which is the risk that the auditor will not detect a material misstatement that exists in an assertion. Detection risk for a substantive test of details has two elements: (1) the risk that analytical procedures and other relevant substantive tests will fail to detect misstatements at least equal to tolerable misstatement and (2) the allowable risk of incorrect acceptance for the substantive test of details. The auditor assesses control risk (the second component) when considering the client's internal control. This assessment, the assessment of inherent risk, and the level to which the auditor wishes to restrict overall audit risk are the factors that the auditor uses to determine the acceptable level of detection risk.

Answers (A), (C), and (D) are incorrect because the components of audit risk are inherent risk, control risk, and detection risk.

20. The equation in AU 350 for the overall allowable audit risk (AR = IR x CR x AP x TD) is sometimes solved for TD (the allowable risk of incorrect acceptance associated with a test of details) because

A. The most important element is TD.

B. This version of the formula assists in planning a specific substantive test of details.

C. The overall allowable audit risk cannot be determined.

D. Auditors always consider tests of details first.

The correct answer is (B). *(Publisher)*

REQUIRED: The reason the allowable audit risk formula is solved for TD.

DISCUSSION: The auditor first establishes the overall allowable audit risk (AR) with respect to a particular balance or class of transactions. After considering internal control, (s)he can assess control risk (CR) as well as inherent risk (IR). After applying analytical procedures and considering the results of other substantive tests, (s)he can then assess the risk (AP) that those procedures and tests did not detect misstatements in an assertion equal to tolerable misstatement. The auditor can then calculate the allowable risk of incorrect acceptance (TD) for a particular substantive test. Determination of this level of risk is necessary for planning the nature, timing, and extent of the substantive test.

Answer (A) is incorrect because the overall allowable audit risk is the most important element in planning appropriate audit tests. Answer (C) is incorrect because the auditor's professional judgment and experience is used to determine overall allowable audit risk. Answer (D) is incorrect because overall audit risk should be established and some analytical procedures should be performed at an early stage. Also, the assessment of control risk should ordinarily be made before the planning of most tests of details.

21. AU 350 gives a formula for risk relationships. Overall allowable audit risk (AR) is the risk that monetary misstatements equal to tolerable misstatement may remain undetected. Control risk (CR) is the auditor's assessment of the risk that internal control may not prevent or detect material misstatements. Inherent risk (IR) is the susceptibility of an assertion to material misstatement given no related controls. In the audit risk formula, AP is the auditor's assessment of the risk that analytical procedures and other relevant substantive tests will fail to detect material misstatements not detected by the relevant controls. TD is the allowable risk of incorrect acceptance for a substantive test of details given that material misstatements occur in an assertion and are not detected by internal control or by analytical procedures and other substantive tests. Which model represents the overall allowable audit risk?

A. $AR = IR \times CR \times AP \times TD$.

B. $AR = IR + CR + AP + TD$.

C. $AR = IR + CR - (AP + TD)$.

D. $AR = IR + CR - (AP \times TD)$.

The correct answer is (A). *(Publisher)*

REQUIRED: The model associated with the overall allowable audit risk.

DISCUSSION: AU 350 states that the model for the overall allowable audit risk is not intended to be a mathematical formula including all factors that may influence the determination of individual risk components. However, the model is sometimes useful in considering and planning appropriate risk levels. AR is equal to the joint probability that material misstatements will occur in an assertion, that internal control will not prevent or detect material misstatements, and that subsequent procedures will also not detect them. Hence, AR is expressed as the product of IR, CR, AP, and TD.

Answers (B), (C), and (D) are incorrect because each is a nonsensical relationship.

22. A proper consideration of the client's internal control is an integral part of the independent external audit. The results

A. Must be reported to the shareholders.

B. Bear no relationship to the extent of substantive testing to be performed.

C. Are not reported to client management.

D. May be used as the basis for determining the acceptable level of detection risk.

The correct answer is (D). *(CMA, adapted)*

REQUIRED: The true statement about the results of an external auditor's consideration of a client's internal control.

DISCUSSION: The second standard of field work requires that the independent auditor obtain a sufficient understanding of internal control to plan the audit and determine the nature, timing, and extent of tests. After obtaining this understanding and assessing both control risk and inherent risk for specific financial statement assertions, the auditor determines the acceptable level of detection risk in the light of the level to which (s)he wishes to restrict the risk of a material misstatement in the financial statements (AU 319).

Answer (A) is incorrect because the results are not required to be reported to anyone but management. Answer (B) is incorrect because the consideration determines the extent of future audit testing. Answer (C) is incorrect because management is responsible for internal control and should thus receive the results.

23. The primary reason an auditor assesses control risk in conjunction with financial statement audits is to

A. Identify the causes of errors or fraud.

B. Identify the results of errors or fraud.

C. Fulfill the role of adviser to management by designing effective and efficient controls for management.

D. Determine the nature, timing, and extent of substantive tests.

The correct answer is (D). *(CMA, adapted)*

REQUIRED: The primary reason an auditor assesses control risk when auditing financial statements.

DISCUSSION: The assessed levels of control risk and inherent risk are used to determine the acceptable level of detection risk for financial statement assertions. This level of detection risk is then used to determine the nature, timing, and extent of the auditing procedures to detect material misstatements in financial statement assertions. Procedures designed to detect these misstatements are substantive tests. As the acceptable level of detection risk decreases, the assurance to be provided by substantive tests increases.

Answers (A) and (B) are incorrect because the ultimate purpose of the assessment of control risk in a financial statement audit is to determine the degree of audit effort to be devoted to substantive tests. Answer (C) is incorrect because advice to management is only a by-product of a financial statement audit.

24. According to SAS 78 (AU 319), *Consideration of Internal Control in a Financial Statement Audit*, only some of an entity's controls are relevant to a financial statement audit. Which one of the following would most likely be considered in such an audit?

A. Timely reporting and review of quality control results.

B. Maintenance of control over unused checks.

C. Marketing analysis of sales generated by advertising projects.

D. Maintenance of statistical production analyses.

The correct answer is (B). *(CMA, adapted)*

REQUIRED: The item most likely relevant to a financial statement audit.

DISCUSSION: The controls most likely to be relevant to a financial statement audit pertain to the entity's objective of preparing external financial statements that are fairly presented in conformity with GAAP or another comprehensive basis of accounting. Maintenance of control over unused checks is relevant because the objective is to safeguard cash. The auditor must understand the controls relevant to the assertions about cash in the financial statements. (S)he must then assess control risk for those assertions; that is, (s)he must evaluate the effectiveness of the controls in preventing or detecting material misstatements.

Answers (A), (C), and (D) are incorrect because quality control, marketing, and production analyses concern the effectiveness and efficiency of certain management decision processes that ordinarily are not relevant to a financial statement audit.

25. After gaining an understanding of internal control, the auditor may attempt to assess control risk at less than the maximum. For this purpose, the auditor should (1) identify specific controls that are likely to prevent or detect material misstatements in the relevant financial statement assertions and (2) perform tests of controls. The purpose of these tests is to

A. Assure that the auditor has an adequate understanding of internal control.

B. Evaluate the effectiveness of such controls.

C. Provide recommendations to management to improve internal control.

D. Evaluate inherent risk.

The correct answer is (B). *(Publisher)*

REQUIRED: The purpose of tests of controls.

DISCUSSION: The purpose of tests of controls is to evaluate the effectiveness of the design or operation of controls in preventing or detecting material misstatements. The auditor tests whether controls are suitably designed to prevent or detect material misstatements in specific assertions. The auditor also tests how a control was applied, by whom it was applied, and whether it was applied consistently during the audit period (AU 319).

Answer (A) is incorrect because the auditor gains an understanding of internal control primarily through previous experience with the entity, inquiries, inspection of documents and records, and observation of activities. Answer (C) is incorrect because the auditor is not obligated to search for reportable conditions but should communicate those of which (s)he becomes aware. Answer (D) is incorrect because inherent risk is the susceptibility of an assertion to a material misstatement in the absence of related controls.

26. Tests of controls are least likely to be omitted with regard to

A. Accounts believed to be subject to ineffective controls.

B. Accounts representing few transactions.

C. Accounts representing many transactions.

D. Subsequent events.

The correct answer is (C). *(Publisher)*

REQUIRED: The circumstances in which tests of controls are least likely to be omitted.

DISCUSSION: For high-volume accounts, the auditor usually must test controls because cost-benefit considerations preclude the review of all transactions. If the control risk for such accounts can be assessed at less than the maximum as a result of testing controls, the acceptable level of detection risk will be increased. The effect will be to reduce the assurance required by substantive tests.

Answer (A) is incorrect because, if, as a result of obtaining the understanding of internal control, the auditor believes that controls are unlikely to be effective, (s)he may assess control risk at the maximum and omit tests of controls. Answer (B) is incorrect because, given few transactions, examining all transactions is more efficient than testing controls. Answer (D) is incorrect because each subsequent event that requires consideration by management and evaluation by the independent auditor should be examined. Hence, tests of relevant controls are likely to be omitted.

27. Auditors document their understanding of internal control with questionnaires, flowcharts, and narrative descriptions. A questionnaire consists of a series of questions concerning controls that auditors consider necessary to prevent or detect errors and fraud. The most appropriate question designed to contribute to the auditors' understanding of the completeness of the expenditure cycle would concern the

A. Internal verification of quantities, prices, and mathematical accuracy of sales invoices.

B. Use and accountability of prenumbered checks.

C. Disposition of cash receipts.

D. Qualifications of accounting personnel.

The correct answer is (B). *(CMA, adapted)*

REQUIRED: The most appropriate question designed to contribute to the auditors' understanding of the completeness of the expenditure cycle.

DISCUSSION: A completeness assertion concerns whether all transactions and accounts that should be presented in the financial statements are so presented. The exclusive use of sequentially numbered documents facilitates control over expenditures. An unexplained gap in the sequence alerts the auditor to the possibility that not all transactions have been recorded. A failure to use prenumbered checks would therefore suggest a higher assessment of control risk. If a company uses prenumbered checks, it should be easy to determine exactly which checks were used during a period.

Answer (A) is incorrect because determination of proper amounts of sales invoices concerns the valuation assertion. Also, sales invoices are part of the sales-receivables (revenue) cycle. Answer (C) is incorrect because cash receipts are part of the revenue cycle. Answer (D) is incorrect because consideration of the qualifications of accounting personnel is not a test of controls over the completeness of any cycle. This procedure is appropriate during the consideration of the control environment.

B. Internal Control According to The IIA

28. SIAS 1, *Control: Concepts and Responsibilities*, adds a guideline to the Standards for the Professional Practice of Internal Auditing. Which of the following is a summary of that guideline?

A. Control is the result of proper planning, organizing, and directing by management.

B. Controls are the broadest statements of what the organization chooses to accomplish.

C. Control is provided when cost-effective actions are taken to restrict deviations to a tolerable level.

D. Control accomplishes objectives and goals in an accurate and timely fashion with minimal use of resources.

The correct answer is (A). *(CIA, adapted)*

REQUIRED: The summary of the guideline added by SIAS 1.

DISCUSSION: According to SIAS 1, "A control is any action taken by management to enhance the likelihood that established objectives and goals will be achieved. Management plans, organizes, and directs the performance of sufficient actions to provide reasonable assurance that objectives and goals will be achieved. Thus, control is the result of proper planning, organizing, and directing by management."

Answer (B) is incorrect because objectives are the broadest statements of what the organization chooses to accomplish. Answer (C) is incorrect because reasonable assurance is provided when cost-effective actions are taken to restrict deviations to a tolerable level. Answer (D) is incorrect because efficient performance accomplishes objectives and goals in an accurate and timely fashion with minimal use of resources.

29. Controls that are designed to provide management with assurance of the realization of specified minimum gross margins on sales are

A. Directive controls.

B. Preventive controls.

C. Detective controls.

D. Output controls.

The correct answer is (A). *(CIA, adapted)*

REQUIRED: The control that provides management with assurance of the realization of specified minimum gross margins on sales.

DISCUSSION: According to SIAS 1, "A control is any action taken by management to enhance the likelihood that established objectives and goals will be achieved." The objective of directive controls is to cause or encourage desirable events to occur, e.g., providing management with assurance of the realization of specified minimum gross margins on sales.

Answer (B) is incorrect because preventive controls deter undesirable events from occurring. Answer (C) is incorrect because detective controls detect and correct undesirable events. Answer (D) is incorrect because output controls relate to the accuracy and reasonableness of information processed by a system, not to operating controls.

30. According to The IIA, internal controls are designed to provide reasonable assurance that

A. Material errors or fraud will be prevented or detected and corrected within a timely period by employees in the course of performing their assigned duties.

B. Management's plans have not been circumvented by worker collusion.

C. The internal auditing department's guidance and oversight of management's performance is accomplished economically and efficiently.

D. Management's planning, organizing, and directing processes are properly evaluated.

The correct answer is (A). *(CIA, adapted)*

REQUIRED: The purpose of internal controls.

DISCUSSION: According to The IIA's SIAS 1, "Reasonable assurance is provided when cost-effective actions are taken to restrict deviations to a tolerable level. This implies, for example, that material errors and improper or illegal acts will be prevented or detected and corrected within a timely period by employees in the normal course of performing their assigned duties. The cost-benefit relationship is considered by management during the design of systems. The potential loss associated with any exposure or risk is weighed against the cost to control it."

Answer (B) is incorrect because collusion is an inherent limitation of internal control. Answer (C) is incorrect because the board of directors or a similar body is responsible for the guidance and oversight of management. Answer (D) is incorrect because the examination and evaluation of management processes is a function of the internal auditing department.

31. Which of the following features of a large manufacturing company's organization structure would be a control weakness?

A. The information systems department is headed by a vice president who reports directly to the president.

B. The chief financial officer is a vice president who reports to the chief executive officer.

C. The audit committee of the board consists of the chief executive officer, the chief financial officer, and a major shareholder.

D. The controller and treasurer report to the chief financial officer.

The correct answer is (C). *(CIA, adapted)*

REQUIRED: The control weakness in a large manufacturing company's organization structure.

DISCUSSION: The audit committee has a control function because of its oversight of internal as well as external auditing. It should be made up of directors who are independent of management. The authority and independence of the audit committee strengthen the position of internal auditing. The board should concur in the appointment or removal of the director of internal auditing, who should have direct, regular communication with the board (Standard 110).

Answer (A) is incorrect because this reporting relationship is a strength. It prevents the information systems operation from being dominated by a user. Answers (B) and (D) are incorrect because each is a normal and appropriate reporting relationship.

32. Corporate management has a role in the maintenance of internal control. In fact, management sometimes is a control. Which of the following involves managerial functions as a control device?

A. Supervision of employees.

B. Use of a corporate policies manual.

C. Maintenance of a quality control department.

D. Internal auditing.

The correct answer is (A). *(CIA, adapted)*

REQUIRED: The item that involves managerial functions as a control device.

DISCUSSION: SIAS 1 states, "Effective control is present when management directs systems in such a manner as to provide reasonable assurance that the organization's objectives and goals will be achieved." Directing includes "authorizing and monitoring performance, periodically comparing actual with planned performance, and documenting these activities to provide additional assurance that systems operate as planned." Monitoring "encompasses supervising, observing, and testing activities and appropriately reporting to responsible individuals. Monitoring provides an ongoing verification of progress toward achievement of objectives and goals."

Answer (B) is incorrect because the manual advises but does not control. Answer (C) is incorrect because a quality control department is a form of internal review. The manager of quality control should be independent of the operations reviewed. Answer (D) is incorrect because internal reviews (such as internal auditing) should be independent of the operations reviewed and are not a managerial function.

C. Flowcharting

33. Which one of the following best reflects the basic elements of a data flow diagram?

A. Data sources, data flows, computer configurations, flowchart, and data storage.

B. Data source, data destination, data flows, transformation processes, and data storage.

C. Data flows, data storage, and program flowchart.

D. Data flows, program flowchart, and data destination.

The correct answer is (B). *(CMA, adapted)*

REQUIRED: The best description of the basic elements of a data flow diagram.

DISCUSSION: Structured analysis is a graphical method of defining the inputs, processes, and outputs of a system and dividing it into subsystems. It is a top down approach that specifies the interfaces between modules and the transformations occurring within each. Data flow diagrams are used in structured analysis. The basic elements of a data flow diagram include data source, data destination, data flows, transformation processes, and data storage.

Answer (A) is incorrect because computer configuration is not an element of a data flow diagram. Answers (C) and (D) are incorrect because a program flowchart is not an element of a data flow diagram.

34. An auditor reviews and adapts a systems flowchart to understand the flow of information in the processing of cash receipts. Which of the following statements is true regarding the use of such flowcharts? The flowcharts

A. Show specific control procedures used, such as edit tests that are implemented and batch control reconciliations.

B. Are a good guide to potential segregation of duties.

C. Are generally kept up to date for systems changes.

D. Show only computer processing, not manual processing.

The correct answer is (B). *(CIA, adapted)*

REQUIRED: The use of a systems flowchart.

DISCUSSION: Systems flowcharts are overall graphic analyses of the flow of data and the processing steps in an information system. Accordingly, they can be used to show segregation of duties and the transfer of data between different segments in the organization.

Answer (A) is incorrect because a program flowchart will identify the specific edit tests implemented. Answer (C) is incorrect because the flowcharts are usually not kept up to date for changes. Thus, the auditor will have to interview key personnel to determine changes in processing since the flowchart was developed. Answer (D) is incorrect because a systems flowchart should show both manual and computer processing.

35. The symbol employed to determine if an employee's wages are above or below the maximum limit for FICA taxes is

A. 

B. 

C. 

D. 

The correct answer is (B). *(CMA, adapted)*

REQUIRED: The symbol to determine if an employee's wages are within a limit.

DISCUSSION: The question implies a decision, for which a diamond is the flowcharting symbol.

Answer (A) is incorrect because a rectangle is the general symbol for a process or operation. Answer (C) is incorrect because a trapezoid symbolizes a manual operation. Answer (D) is incorrect because a square represents an auxiliary operation performed by a machine other than a computer.

36. The symbol employed to represent the printing of the employees' paychecks by the computer is

A. 

B. 

C. 

D. 

The correct answer is (C). *(CMA, adapted)*

REQUIRED: The flowchart symbol for the printing of paychecks by a computer.

DISCUSSION: The printing of paychecks by the computer is an operation depicted by the general processing symbol, which is a rectangle.

Answer (A) is incorrect because a trapezoid depicts a manual operation. Answer (B) is incorrect because a square is an auxiliary operation performed by a machine other than a computer. Answer (D) is incorrect because this symbol indicates manual input, e.g., entry of a proper code through a computer console.

37. The symbol employed to represent the employees' checks printed by the computer is

A. 

B. 

C. 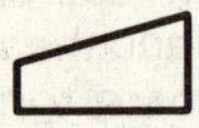

D. 

The correct answer is (D). *(CMA, adapted)*

REQUIRED: The symbol used to represent employee checks printed by a computer.

DISCUSSION: Employee checks printed by the computer are depicted by the document symbol, which resembles the top of a grand piano.

Answer (A) is incorrect because a parallelogram is the general symbol for input or output. Answer (B) is incorrect because a trapezoid indicates a manual operation. Answer (C) is incorrect because this symbol indicates manual input.

38. The symbol used to represent the physical act of collecting employees' time cards for processing is

A. 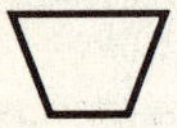

B. 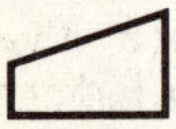

C. 

D. 

The correct answer is (A). *(CMA, adapted)*

REQUIRED: The symbol used to represent the physical act of collecting employees' time cards for processing.

DISCUSSION: Collecting employees' time cards is a manual operation represented by a trapezoid with equal nonparallel sides.

Answer (B) is incorrect because this symbol represents manual input. Answer (C) is incorrect because a rectangle is the general symbol for processing. Answer (D) is incorrect because a parallelogram is the general symbol for input or output.

39. The symbol used to represent the employees' payroll records stored on magnetic tape is

A. 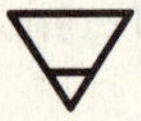

B. 

C. 

D. 

The correct answer is (D). *(CMA, adapted)*

REQUIRED: The symbol representing employees' payroll records stored on magnetic tape.

DISCUSSION: The magnetic tape symbol (a circle with a tangent at its base) indicates storage on magnetic tape.

Answer (A) is incorrect because a triangle with a mid-line parallel to its base depicts offline storage. Answer (B) is incorrect because this symbol represents online storage. Answer (C) is incorrect because this symbol represents punched paper tape.

40. The symbol used to represent the weekly payroll register generated by the computer is

A. 

B. 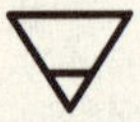

C. 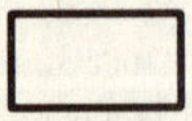

D. 

The correct answer is (D). *(CMA, adapted)*

REQUIRED: The symbol used to represent a weekly payroll register printed by a computer.

DISCUSSION: The weekly payroll register on a computer printout is represented by a document symbol, which resembles the top of a grand piano.

Answer (A) is incorrect because a circle with a tangent at its base represents magnetic tape input-output or storage. Answer (B) is incorrect because a triangle with a mid-line parallel to its base depicts offline storage. Answer (C) is incorrect because a rectangle is the general symbol for a process.

41. The symbol used to represent the file of hard-copy, computer-generated payroll reports kept for future reference is

A. 

B. 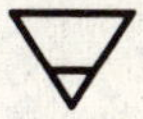

C. 

D. 

The correct answer is (B). *(CMA, adapted)*

REQUIRED: The symbol used to represent the kept file of hard-copy, computer-generated payroll reports.

DISCUSSION: Hard-copy, computer-generated payroll reports are kept in offline storage, which is symbolized by a triangle with a mid-line parallel to its base.

Answer (A) is incorrect because a circle with a tangent at its base represents a magnetic tape. Answer (C) is incorrect because this symbol represents online storage. Answer (D) is incorrect because a parallelogram is the general symbol for input or output.

42. The correct labeling, in order, for the following flowchart symbols is

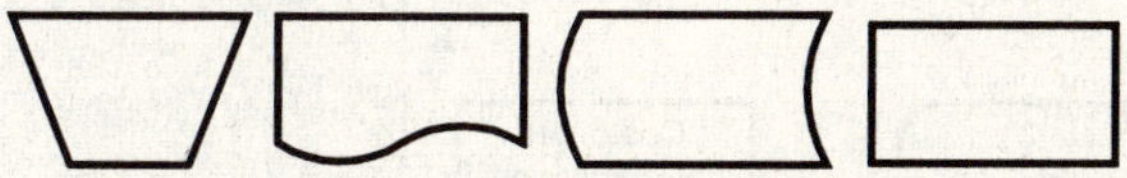

A. Document, display, online storage, and entry operation.

B. Manual operation, processing, offline storage, and input-output activity.

C. Display, document, online storage, and entry operation.

D. Manual operation, document, online storage, and entry operation.

The correct answer is (D). *(CMA, adapted)*

REQUIRED: The correct sequence of labels of four flowchart symbols.

DISCUSSION: The first symbol indicates a manual operation, which is an offline process. The second symbol represents a document, while the third symbol indicates online storage (e.g., a disk drive). The final symbol represents an operation. An operation is defined as a process resulting in a change in the information or the flow direction. In other words, it can be an entry operation.

Answer (A) is incorrect because the first symbol, a trapezoid, is for a manual operation. Answer (B) is incorrect because the third symbol is for online storage. Answer (C) is incorrect because the first symbol does not represent display.

Questions 43 through 47 are based on the following information. This flowchart depicts the processing of daily cash receipts for Rockmart Manufacturing.

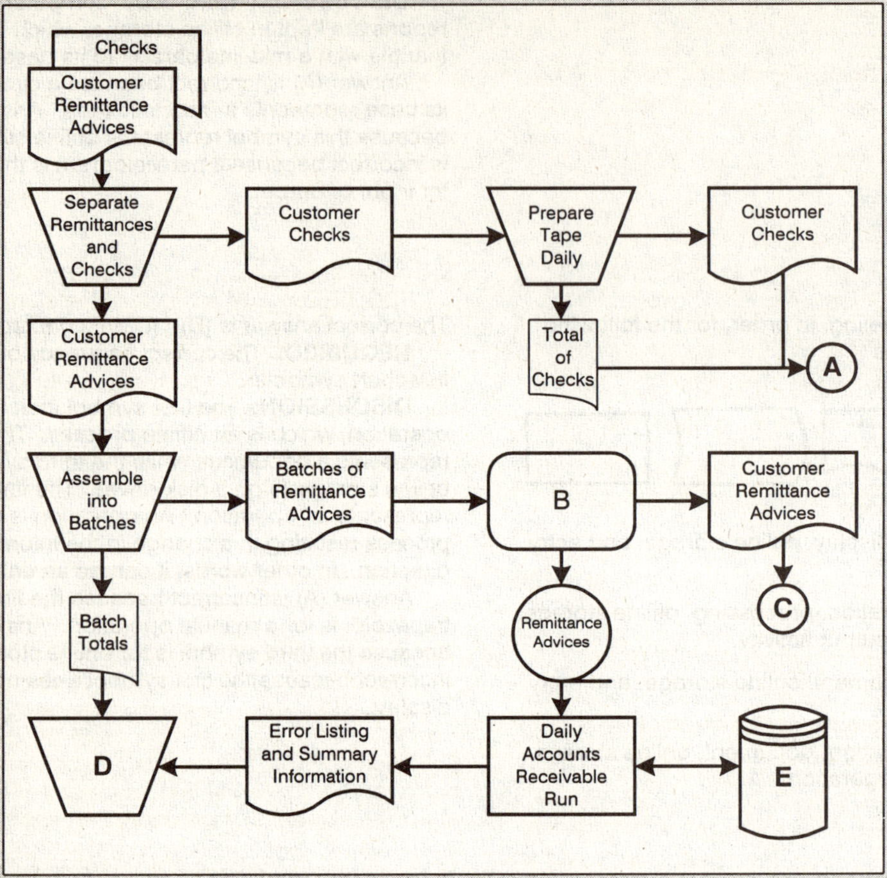

43. The customer checks accompanied by the control tape (refer to symbol A) would be

A. Forwarded daily to the billing department for deposit.

B. Taken by the mail clerk to the bank for deposit daily.

C. Forwarded to the treasurer for deposit daily.

D. Accumulated for a week and then forwarded to the treasurer for deposit weekly.

The correct answer is (C). *(CMA, adapted)*

REQUIRED: The proper procedure for handling customer checks and the related control tape.

DISCUSSION: Symbol A is a connector between a point on this flowchart and another part of the flowchart not shown. The checks and the adding machine control tape should flow through symbol A to the treasurer's office. The treasurer is the custodian of funds and is responsible for deposit of daily receipts.

Answer (A) is incorrect because record keepers perform functions that should be separate from custody of assets. Answer (B) is incorrect because the mail clerk should prepare a list of checks received before they are forwarded to the treasurer for deposit. Answer (D) is incorrect because daily receipts should be deposited intact daily and then reconciled with the bank deposit records. Prompt deposit also safeguards assets and avoids loss of interest income.

44. The appropriate description that should be placed in symbol B would be

A. Keying and verifying.

B. Error correction.

C. Collation of remittance advices.

D. Batch processing.

The correct answer is (A). *(CMA, adapted)*

REQUIRED: The appropriate description for symbol B.

DISCUSSION: Since the figure below symbol B signifies magnetic tape, the operation represented by symbol B must be keying the information onto the tape. Verifying the keyed data would also occur at this step.

Answer (B) is incorrect because error correction would occur subsequently except for keying errors. Answer (C) is incorrect because collation has already occurred. Answer (D) is incorrect because batch processing describes the entire system.

45. The next action to take with the customer remittance advices (refer to symbol C) would be to

A. Discard them immediately.

B. File them daily by batch number.

C. Forward them to the internal audit department for internal review.

D. Forward them to the treasurer to compare with the monthly bank statement.

The correct answer is (B). *(CMA, adapted)*

REQUIRED: The action taken with the customer remittance advices at symbol C.

DISCUSSION: All activity with respect to the paper documents most likely ceases at symbol C. Therefore, the batched documents must be filed.

Answer (A) is incorrect because the documents should be kept for reference and audit. Answer (C) is incorrect because internal auditors cannot feasibly review all documents regarding transactions even in an audit. Answer (D) is incorrect because comparison by the treasurer would be inappropriate. (S)he has custody of cash.

46. The appropriate description that should be placed in symbol D would be

A. Attach batch total to report and file.

B. Reconcile cash balances.

C. Compare batch total and correct as necessary.

D. Proof report.

The correct answer is (C). *(CMA, adapted)*

REQUIRED: The appropriate description for symbol D.

DISCUSSION: This flowcharting symbol indicates a manual operation or offline process. Since the input to this operation consists of an adding machine tape containing batch totals and a document containing summary information about the accounts receivable update and an error listing, the operation apparently involves comparing these items.

Answer (A) is incorrect because no filing symbol is given. Answer (B) is incorrect because the flowchart concerns daily receipts, not the reconciliation of cash balances. Answer (D) is incorrect because symbol D indicates a comparison, not output in the form of a report.

47. The appropriate description that should be placed in symbol E would be

A. Accounts receivable master file.

B. Bad debts master file.

C. Remittance advice master file.

D. Cash projection file.

The correct answer is (A). *(CMA, adapted)*

REQUIRED: The appropriate description of symbol E.

DISCUSSION: The flowcharting figure at symbol E indicates magnetic disk storage. Since it is an input and output for the daily computer processing of accounts receivable, it must be the accounts receivable master file.

Answer (B) is incorrect because bad debts are not a part of processing daily receipts. Answer (C) is incorrect because the remittance advice master file was not used for the daily accounts receivable run. Answer (D) is incorrect because the cash projection file was not used for the daily accounts receivable run.

D. Means of Achieving Control

48. Internal control should follow certain basic principles to achieve its objectives. One of these principles is the segregation of functions. Which one of the following examples does not violate the principle of segregation of functions?

A. The treasurer has the authority to sign checks but gives the signature block to the assistant treasurer to run the check-signing machine.

B. The warehouse clerk, who has the custodial responsibility over inventory in the warehouse, may authorize disposal of damaged goods.

C. The sales manager has the responsibility to approve credit and the authority to write off accounts.

D. The department time clerk is given the undistributed payroll checks to mail to absent employees.

The correct answer is (A). *(CMA, adapted)*

REQUIRED: The situation that does not violate the principle of segregation of functions.

DISCUSSION: Internal control has five components: the control environment, risk assessment, control activities, information and communication, and monitoring. Control activities include segregation of duties to reduce the risk that any person may be able to perpetrate and conceal errors or fraud in the normal course of his/her duties. Different persons should authorize transactions, record transactions, and maintain custody of assets. The treasurer's department should have custody of assets but should not authorize or record transactions. Because the assistant treasurer reports to the treasurer, the treasurer is merely delegating an assigned duty related to asset custody. The use of the check-signing machine does not conflict with any other duty of the assistant treasurer and does not involve authorization or recording of transactions.

Answer (B) is incorrect because authorization to dispose of damaged goods could be used to cover thefts of inventory for which the warehouse clerk has custodial responsibility. Transaction authorization is inconsistent with asset custody. Answer (C) is incorrect because the sales manager could approve credit to a controlled company and then write off the account as a bad debt. The sales manager's authorization of credit is inconsistent with his/her indirect access to assets. Answer (D) is incorrect because the time clerk could conceal the termination of an employee and retain that employee's paycheck. Record keeping is inconsistent with asset custody.

49. If internal control is well designed, two tasks that should be performed by different persons are

A. Approval of bad debt write-offs, and reconciliation of the accounts payable subsidiary ledger and controlling account.

B. Distribution of payroll checks and approval of sales returns for credit.

C. Posting of amounts from both the cash receipts journal and cash payments journal to the general ledger.

D. Recording of cash receipts and preparation of bank reconciliations.

The correct answer is (D). *(CMA, adapted)*

REQUIRED: The tasks that should be performed by different persons if internal control is well designed.

DISCUSSION: Recording of cash establishes accountability for assets. The bank reconciliation compares that recorded accountability with actual assets. The recording of cash receipts and preparation of bank reconciliations should therefore be performed by different individuals since the preparer of a reconciliation could conceal a cash shortage. For example, if a cashier both prepares the bank deposit and performs the reconciliation, (s)he could embezzle cash and conceal the theft by falsifying the reconciliation.

Answer (A) is incorrect because there is no conflict between writing off bad debts (accounts receivable) and reconciling accounts payable, which are liabilities. Answer (B) is incorrect because distribution of payroll checks and approval of sales returns are independent functions. People who perform such disparate tasks are unlikely to be able to perpetrate and conceal a fraud. In fact, some companies use personnel from an independent function to distribute payroll checks. Answer (C) is incorrect because posting both ledgers would cause no conflict as long as the individual involved did not have access to the actual cash. If a person has access to records but not the assets, there is no danger of embezzlement without collusion.

50. Which one of the following situations represents an internal control weakness in the payroll department?

A. Payroll department personnel are rotated in their duties.

B. Paychecks are distributed by the employees' immediate supervisor.

C. Payroll records are reconciled with quarterly tax reports.

D. The timekeeping function is independent of the payroll department.

The correct answer is (B). *(CMA, adapted)*

REQUIRED: The internal control weakness in the payroll department.

DISCUSSION: Paychecks should not be distributed by supervisors because an unscrupulous person could terminate an employee and fail to report the termination. The supervisor could then clock in and out for the employee and keep the paycheck. A person unrelated to either payroll record keeping or the operating department should distribute checks.

Answer (A) is incorrect because periodic rotation of payroll personnel inhibits the perpetration and concealment of fraud. Answer (C) is incorrect because this analytical procedure may detect a discrepancy. Answer (D) is incorrect because timekeeping should be independent of asset custody and employee records.

51. Which of the following activities represents both an appropriate personnel department function and a deterrent to payroll fraud?

A. Distribution of paychecks.

B. Authorization of overtime.

C. Authorization of additions and deletions from the payroll.

D. Collection and retention of unclaimed paychecks.

The correct answer is (C). *(CIA, adapted)*

REQUIRED: The activity that is both a personnel department function and a fraud deterrent.

DISCUSSION: The payroll department is responsible for assembling payroll information (record keeping). The personnel department is responsible for authorizing employee transactions such as hiring, firing, and changes in pay rates and deductions. Segregating the recording and authorization functions helps prevent fraud.

Answer (A) is incorrect because the treasurer should perform the asset custody function regarding payroll. Answer (B) is incorrect because authorizing overtime is a responsibility of operating management. Answer (D) is incorrect because unclaimed checks should be in the custody of the treasurer until they can be deposited in a special bank account.

52. Which one of the following situations represents a strength of internal control for purchasing and accounts payable?

A. Prenumbered receiving reports are issued randomly.

B. Invoices are approved for payment by the purchasing department.

C. Unmatched receiving reports are reviewed on an annual basis.

D. Vendors' invoices are matched against purchase orders and receiving reports before a liability is recorded.

The correct answer is (D). *(CMA, adapted)*

REQUIRED: The strength in internal control relevant to purchasing and accounts payable.

DISCUSSION: A voucher should not be prepared for payment until the vendor's invoice has been matched against the corresponding purchase order and receiving report. This procedure provides assurance that a valid transaction has occurred and that the parties have agreed on the terms, such as price and quantity.

Answer (A) is incorrect because prenumbered receiving reports should be issued sequentially. A gap in the sequence may indicate an erroneous or fraudulent transaction. Answer (B) is incorrect because invoices should not be approved by purchasing. That is the job of the accounts payable department. Answer (C) is incorrect because annual review of unmatched receiving reports is too infrequent. More frequent attention is necessary to remedy deficiencies in internal control.

53. The most appropriate method to control the frequent movement of trailers loaded with valuable metal scrap from the manufacturing plant to the company scrap yard about 10 miles away would be to

A. Perform complete physical inventory of the scrap trailers before leaving the plant and upon arrival at the scrap yard.

B. Require existing security guards to log the time of plant departure and scrap yard arrival. The elapsed time should be reviewed by a supervisor for fraud.

C. Use armed guards to escort the movement of the trailers from the plant to the scrap yard.

D. Contract with an independent hauler for the removal of scrap.

The correct answer is (B). *(CIA, adapted)*

REQUIRED: The most appropriate method to control the frequent movement of trailers loaded with valuable metal scrap.

DISCUSSION: Having the security guards record the times of departure and arrival is a cost-effective control because it entails no additional expenditures. Comparing the time elapsed with the standard time allowed and investigating material variances may detect a diversion of part of the scrap.

Answer (A) is incorrect because performing a complete physical inventory of the scrap at both locations would not be economically feasible. Answer (C) is incorrect because hiring armed guards to escort the scrap trailers is unlikely to be necessary unless the scrap is extremely valuable. Logging departures and arrivals will be sufficient in most cases. Answer (D) is incorrect because using an independent hauler would provide no additional assurance of prevention or detection of wrongdoing.

54. A utility company with a large investment in repair vehicles would most likely implement which internal control to reduce the risk of vehicle theft or loss?

A. Review insurance coverage for adequacy.

B. Systematically account for all repair work orders.

C. Physically inventory vehicles and reconcile the results with the accounting records.

D. Maintain vehicles in a secured location with release and return subject to approval by a custodian.

The correct answer is (D). *(CIA, adapted)*

REQUIRED: The internal control to reduce the risk of vehicle theft or loss.

DISCUSSION: Physical control of assets is a preventive control that reduces the likelihood of theft or other loss. Keeping the vehicles at a secure location and restricting access establishes accountability by the custodian and allows for proper authorization of their use.

Answer (A) is incorrect because insurance provides for indemnification if loss or theft occurs. It reduces financial exposure but does not prevent the actual loss or theft. Answer (B) is incorrect because an internal control designed to ensure control over repair work performed has no bearing on the risk of loss. Answer (C) is incorrect because taking an inventory is a detective, not a preventive, control.

55. To minimize the risk that agents in the purchasing department will use their positions for personal gain, the organization should

A. Rotate purchasing agent assignments periodically.

B. Request internal auditors to confirm selected purchases and accounts payable.

C. Specify that all items purchased must pass value-per-unit-of-cost reviews.

D. Direct the purchasing department to maintain records on purchase prices paid, with review of such being required each 6 months.

The correct answer is (A). *(CIA, adapted)*

REQUIRED: The control to minimize the risk that agents in the purchasing department will use their positions for personal gain.

DISCUSSION: The risk of favoritism is increased when buyers have long-term relationships with specific vendors. Periodic rotation of buyer assignments will limit the opportunity to show favoritism. This risk is also reduced if buyers are required to take vacations.

Answer (B) is incorrect because confirmation does not enable internal auditors to detect inappropriate benefits received by purchasing agents or deter long-term relationships. Answer (C) is incorrect because value-per-unit-of-cost reviews could be helpful in assuring value received for price paid but do not directly focus on receipt of inappropriate benefits by purchasing agents. Answer (D) is incorrect because review of records every 6 months does not enable the organization to detect receipt of inappropriate amounts by an agent or deter relationships that could lead to such activity.

56. Upon receipt of purchased goods, receiving department personnel match the quantity received with the packing slip quantity and mark the retail price on the goods based on a master price list. The annotated packing slip is then forwarded to inventory control and goods are automatically moved to the retail sales area. The most significant control strength of this activity is

A. Immediately pricing goods for retail sale.

B. Matching quantity received with the packing slip.

C. Using a master price list for marking the sale price.

D. Automatically moving goods to the retail sales area.

The correct answer is (C). *(CIA, adapted)*

REQUIRED: The most significant control strength of the procedure described.

DISCUSSION: Use of the master price list assures that the correct retail price is marked.

Answer (A) is incorrect because timing is not as important as the accuracy of prices. Answer (B) is incorrect because matching quantity received with the packing slip does not ensure receipt of the quantity ordered. Answer (D) is incorrect because goods may or may not be needed in retail sales.

57. Management can best strengthen internal control over the custody of inventory stored in an off-site warehouse by implementing

A. Reconciliations of transfer slips to/from the warehouse with inventory records.

B. Increases in insurance coverage.

C. Regular reconciliation of physical inventories to accounting records.

D. Regular confirmation of the amount on hand with the custodian of the warehouse.

The correct answer is (C). *(CIA, adapted)*

REQUIRED: The best method to strengthen control over off-site inventory.

DISCUSSION: The most effective control over off-site inventory is the periodic comparison of the recorded accountability with the actual physical inventory.

Answer (A) is incorrect because examination of documents is a less effective procedure than actual observation of the inventory. Answer (B) is incorrect because increasing insurance coverage helps protect the business against losses but does not strengthen internal control over the custody of inventory. Answer (D) is incorrect because confirming with the custodian the amount of inventory on hand does not verify that the inventory is actually at the warehouse.

58. A restaurant food chain has over 680 restaurants. All food orders for each restaurant are required to be input into an electronic device which records all food orders by food servers and transmits the order to the kitchen for preparation. All food servers are responsible for collecting cash for all their orders and must turn in cash at the end of their shift equal to the sales value of food ordered for their I.D. number. The manager then reconciles the cash received for the day with the computerized record of food orders generated. All differences are investigated immediately by the restaurant. Corporate headquarters has established monitoring controls to determine when an individual restaurant might not be recording all its revenue and transmitting the applicable cash to the corporate headquarters. Which one of the following would be the best example of a monitoring control?

A. The restaurant manager reconciles the cash received with the food orders recorded on the computer.

B. All food orders must be entered on the computer, and segregation of duties is maintained between the food servers and the cooks.

C. Management prepares a detailed analysis of gross margin per store and investigates any store that shows a significantly lower gross margin.

D. Cash is transmitted to corporate headquarters on a daily basis.

The correct answer is (C). *(CIA, adapted)*

REQUIRED: The best example of a monitoring control.

DISCUSSION: Monitoring is a process that assesses the quality of internal control over time. It involves assessment by appropriate personnel of the design and operation of controls and the taking of corrective action. Monitoring can be done through ongoing activities or separate evaluations. Ongoing monitoring procedures are built into the normal recurring activities of an entity and include regular management and supervisory activities. Thus, analysis of gross margin data and investigation of significant deviations is a monitoring process.

Answer (A) is incorrect because the manager's activity is an example of a reconciliation control applied at the store level. Monitoring is an overall control that determines whether other controls are operating effectively. Answer (B) is incorrect because the division of duties is an operational control. Answer (D) is incorrect because daily transmission of cash is an operational control.

E. Legal and Professional Influences

59. Firms subject to the reporting requirements of the Securities Exchange Act of 1934 are required by the Foreign Corrupt Practices Act of 1977 to maintain satisfactory internal control. The role of the independent auditor relative to this act is to

A. Report clients with unsatisfactory internal control to the SEC.

B. Provide assurances to users as part of the traditional audit attest function that the client is in compliance with the present legislation.

C. Express an opinion on the sufficiency of the client's internal control to meet the requirements of the act.

D. Attest to the financial statements.

The correct answer is (D). *(Publisher)*

REQUIRED: The role of the independent auditor relative to the Foreign Corrupt Practices Act.

DISCUSSION: Whether a client is in conformity with the Foreign Corrupt Practices Act is a legal question. Auditors cannot be expected to provide clients or users of the financial statements with legal advice. The role of the auditor is to assess control risk in the course of an engagement to attest to the fair presentation of the financial statements.

Answer (A) is incorrect because the auditor is not required to report violations of the act to the SEC, although a duty to disclose outside the client may exist in some circumstances; e.g., the client's failure to take remedial action regarding an illegal act may constitute a disagreement that it must report on Form 8-K (AU 317). Answer (B) is incorrect because the traditional attest function does not involve compliance auditing. Answer (C) is incorrect because the FCPA contains no requirement that an auditor express an opinion on internal control.

60. The requirement of the Foreign Corrupt Practices Act of 1977 to devise and maintain adequate internal control is assigned in the act to the

A. Chief financial officer.

B. Board of directors.

C. Director of internal auditing.

D. Company as a whole with no designation of specific persons or positions.

The correct answer is (D). *(CMA, adapted)*

REQUIRED: The person in a company responsible for compliance with the FCPA.

DISCUSSION: The accounting requirements apply to all public companies that must register under the Securities Exchange Act of 1934. The responsibility is thus placed on companies, not individuals.

Answers (A), (B), and (C) are incorrect because compliance with the FCPA is not the specific responsibility of the chief financial officer, the board of directors, or the director of internal auditing.

61. The National Committee on Fraudulent Financial Reporting (Treadway Commission) recommended that

A. All public companies have an audit committee made up of members of top management to assist the internal auditor in identifying potential areas of external auditor concern.

B. Internal auditors perform many of the functions of the external auditor in order to minimize audit fees while increasing the effectiveness of audits.

C. Internal audit departments engage in activities that enhance the objectivity of their function with the assistance of management and the audit committee.

D. Privately held companies have an internal audit staff with an adequate number of qualified personnel appropriate for the size of the company.

The correct answer is (C). *(CMA, adapted)*

REQUIRED: The recommendation of the Treadway Commission.

DISCUSSION: The Treadway Commission issued its report in 1987 in response to allegations of widespread financial reporting fraud by public companies. It recommended that (1) management perform an ongoing fraud-risk assessment, maintain effective internal control, establish written codes of conduct, and design appropriate accounting functions that meet reporting obligations; (2) an effective internal audit function exist in which auditors have unrestricted and direct access to the audit committee and the CEO and coordinate their work with that of the public accountants; (3) every public company have an audit committee composed of outside directors; and (4) the sponsoring organizations set up an interdisciplinary body to develop an integrated internal control framework.

Answer (A) is incorrect because the audit committee should consist only of outside directors. Answer (B) is incorrect because the extent to which the external auditor makes use of the work of the internal auditor is entirely at the discretion of the external auditor; however, internal and external audit efforts should be coordinated. Answer (D) is incorrect because the Treadway Report concerned public companies.

62. In relation to nonfinancial internal audits, the Treadway Commission recommended that

A. Internal auditors not be involved in any nonfinancial audits because their findings in financial audits might be biased.

B. Internal auditors be fully involved to gain greater knowledge of the company and a more informed perspective.

C. The public accountant review completely the work performed by internal auditors.

D. The public accountants review the nonfinancial audits prepared by internal auditors and include the internal auditors' findings in their reports.

The correct answer is (B). *(CMA, adapted)*

REQUIRED: The recommendation of the Treadway Commission with respect to nonfinancial internal audits.

DISCUSSION: The 1987 Treadway Commission Report examined the roles of the internal as well as external auditors in preventing and detecting fraudulent financial reporting. Thus, it emphasized that the internal audit function should have unrestricted and direct access to the CEO and the audit committee and should coordinate its work with that of the external auditors. The report also indicated that nonfinancial internal audits perform an educational role. Internal auditors are better able to detect fraudulent financial reporting if they have a better knowledge of company operations.

Answer (A) is incorrect because the scope of work of internal auditors extends to nonfinancial as well as financial audits. Answer (C) is incorrect because external auditors should obtain an understanding of the internal audit function, determine whether the internal auditors work is relevant to the audit and whether considering that work further is efficient, and, if the work is relevant and considering it further is efficient, assess the competence and objectivity of the internal auditors in the light of the effect of their work on the audit. Thus, external auditors do not consider the work of the internal auditors that is irrelevant to the audit. Answer (D) is incorrect because the external auditor is engaged to report on a financial statement audit.

63. In a financial statement audit, the auditor should consider categories of fraud risk factors relating to misstatements arising from (1) fraudulent financial reporting and (2) misappropriation of assets. Which of the following is a category of risk factors that should be considered in relation to misstatements arising from misappropriation of assets?

A. Industry conditions.

B. Operating characteristics.

C. Management's characteristics.

D. Controls.

The correct answer is (D). *(Publisher)*

REQUIRED: The category of risk factors that should be considered in relation to misstatements arising from misappropriation of assets.

DISCUSSION: The auditor must specifically assess the risk of material misstatement due to fraud, a risk that is part of audit risk. The assessment is considered in designing audit procedures. Accordingly, AU 316 states that the auditor should consider three categories of risk factors related to fraudulent reporting: management's characteristics and influence over the control environment, industry conditions, and operating characteristics and financial stability. The two categories of risk factors related to misappropriation of assets are controls and susceptibility of assets to misappropriation.

Answers (A), (B), and (C) are incorrect because industry conditions, operating characteristics, and management's characteristics relate to fraudulent reporting.

64. Auditing standards require that auditors be aware of relevant factors relating to fraudulent reporting. Which of the following statements is false concerning fraudulent reporting?

A. Fraud frequently involves a pressure or an incentive to commit fraud and a perceived opportunity to do so.

B. Two types of fraud relevant to the auditor include material misstatements arising from fraudulent financial reporting and material misstatements arising from misappropriation of assets.

C. Fraud involves actions of management but excludes the actions of employees or third parties.

D. An audit rarely involves the authentication of documentation; thus, fraud may go undetected by the auditor.

The correct answer is (C). *(Publisher)*

REQUIRED: The false statement concerning fraudulent reporting.

DISCUSSION: Misappropriation of assets may be accompanied by false or misleading records and may involve one or more individuals among management, employees, or third parties.

Answer (A) is incorrect because the two conditions are ordinarily present in fraud. Answer (B) is incorrect because misstatements arising from fraudulent reporting are intentional misstatements or omissions to deceive financial statement users, and misstatements arising from misappropriation of assets involve theft, the effect of which is nonconformity of the financial statements with GAAP. Answer (D) is incorrect because auditors are not trained or expected to be experts in authentication, and there is some risk that fraud may go undetected.

STUDY UNIT 10: INTERNAL AUDITING

17 pages of outline
45 multiple-choice questions

The relative weight range assigned to this study unit is 10% to 15% of Part 4. Management accountants require an understanding of internal auditing because it performs an essential management control function. For a full treatment of this topic, see Gleim's *CIA Review*. Also, internal control concepts relevant to internal auditing are presented in Study Unit 9, Internal Control.

A. Nature of Internal Auditing

1. Internal auditing is a management-oriented discipline that has evolved rapidly since World War II from a function primarily concerned with financial and accounting matters to one that addresses the entire range of operating activities. The increased size and decentralization of organizations, the greater complexity of their operations, and the resulting need for means of monitoring their numerous activities fostered the growth of internal auditing. The Institute of Internal Auditors (The IIA) has defined internal auditing as follows:

   Internal auditing is an independent, objective assurance and consulting activity designed to add value and improve an organization's operations. It helps an organization accomplish its objectives by bringing a systematic, disciplined approach to evaluate and improve the effectiveness of risk management, control, and governance processes.

2. Public policy considerations have also contributed to the improved status and broadened scope of internal auditing. Organizations are expected to maintain reasonably detailed and accurate accounting records and a reasonably effective system of internal auditing control.
3. Various governmental initiatives have provided reasons for upgrading internal auditing departments. Ordinarily, public companies are expected to have an internal auditing function and an audit committee composed of nonmanagement directors.
4. Internal auditing should encompass every phase and sector of the organization.
   a. It should not be denied access to any area of the organization. It extends beyond accounting and financial control to all operations of the organization.
   b. It should have unlimited access to any documents, records, or properties of the organization relevant to the subject under review.

5. **Reason for Internal Auditing**. Middle and executive-level managers may have little personal knowledge of current operations and the effectiveness with which objectives are met, policies are followed, and control systems are functioning.
   a. Thus, modern internal auditing evolved to assist management in discharging its responsibilities efficiently and effectively. Internal auditors assist management in regard to the following, among other things:
      1) Adequacy and effectiveness of accounting, financial, and operating controls
      2) Evaluation of the quality of performance
      3) Compliance with established policies, plans, procedures, laws, regulations, and contracts
      4) Risk assessment
      5) Safeguarding of assets
      6) The reliability and integrity of information
      7) Accomplishment of established objectives and goals for operations or programs
      8) Economical and efficient use of resources
      9) Prevention and detection of fraud
      10) Coordination with the external auditor
6. The **organizational status** of the internal auditing department must be sufficient for internal auditing to accomplish its objectives. Proper organizational status enhances the independence and objectivity of internal auditing. Without the support of management and the board of directors, auditees may be uncooperative.
   a. The director of internal auditing should be responsible to an individual with sufficient authority to promote and ensure
      1) Internal auditing independence
      2) Broad audit coverage
      3) Adequate consideration of audit reports
      4) Appropriate action on audit recommendations
   b. The board of directors should concur in appointment and removal of the director of internal auditing. Ideally, (s)he should report directly to the audit committee of the board of directors.
      1) The director of internal auditing should communicate regularly with the board, including annual activity reports.
   c. A formal, written charter should define internal auditing's purpose, authority, and responsibilities.
      1) The charter should
         a) Establish the department's position within the organization
         b) Define the scope of internal auditing activities
         c) Authorize access to records, personnel, and physical properties relevant to the performance of audits

2) Reasons for a charter include

   a) Documenting a commitment by executive management to internal auditing

   b) Defining the organizational and reporting relationships between internal auditing and others within the organization

   c) Defining the authority or responsibilities of internal auditing

   d) Providing a basis for organizing and staffing the internal auditing function

   e) Providing an authoritative guide to audit planning, including the setting of goals and objectives

   f) Providing a basis for evaluating overall audit effectiveness

3) Topics to be included in the charter are the

   a) Primary goal, or objective, of internal auditing
   b) Approval of senior management or the audit committee
   c) Authority granted to internal auditing
   d) Basic responsibilities of internal auditing (scope of work)
   e) Organizational status
   f) Relationships, both internal and external, of internal auditing with others
   g) Reporting relationships
   h) Responsibility to follow up on audit findings and recommendations

d. The director of internal auditing is responsible for the entire internal auditing function, including

   1) Accomplishment of audit objectives set by management and the board

   2) Development of a charter approved by management and the board

   3) Development of plans to meet internal auditing objectives

   4) Development of a program to select and develop audit personnel

   5) Development of written policies and procedures to guide audit staff

   6) Coordination with external auditors

   7) Review and approval of all audit reports

   8) Establishment and maintenance of a quality assurance program

   9) Conformity of audit work with Standards for the Professional Practice of Internal Auditing

7. **Authority and Responsibility of Internal Auditors**

   a. The internal auditor's responsibility is to review and appraise policies, procedures, plans, and records for the purpose of informing and advising management.

   b. Internal auditors have neither authority for, nor responsibility over, any operating activities.

      1) If they had either authority or responsibility, their objectivity would be impaired.

      2) Thus, only through their organizational status (defined in the charter) do internal auditors have the authority to conduct audits.

         a) They must have full access to all relevant records, properties, and personnel.

      3) Persons transferred or temporarily assigned to internal auditing should not, for a reasonable period, audit functions they previously performed.

   c. The auditor's findings, recommendations, etc., are advisory only.

      1) Auditors can only bring disputed or ignored findings to the attention of top management and the board of directors.

         a) The board has the authority to implement the recommendations.

      2) The correct procedure is to bring all recommendations to the appropriate level of management for review and implementation.

   d. The auditor may, however, recommend control standards for systems.

      1) The auditor may review procedures prior to their implementation.
      2) The auditor may not draft procedures.
      3) The auditor may not design, install, or operate systems.

8. Stop and review! You have completed the outline for this subunit. Study multiple-choice questions 1 through 6 beginning on page 438.

B. Types of Internal Auditing Services. Internal auditors assist management in discharging its responsibilities by examining and evaluating the adequacy and effectiveness of controls and the quality of performance in carrying out assigned responsibilities.

1. **Adequacy and Effectiveness of the Organization's System of Internal Control**

   a. For more about the internal auditors' services with respect to internal control, see Statement of Internal Auditing Standards (SIAS) 1, *Control: Concepts and Responsibilities*, issued by The IIA.

   b. The IIA's internal control concepts and terminology currently differ from those in AICPA pronouncements.

2. **Reliability and Integrity of Information**
   a. Financial and operating reports provided to management should be accurate, reliable, timely, complete, and useful.
   b. Internal auditors evaluate the adequacy and effectiveness of controls over financial and operating records and reports.
   c. Internal auditors' responsibility with respect to the reliability and integrity of information is more extensive than that of external auditors.
      1) External auditors are primarily concerned with the entity's ability to record, process, summarize, and report financial data consistent with the assertions embodied in the financial statements.
      2) Internal auditors are also concerned with matters that ordinarily are not relevant to a financial statement audit, such as the policies and procedures concerning the effectiveness, economy, and efficiency of management decision making.
3. **Safeguarding of Assets**
   a. Auditors emphasize avoidance of losses due to theft, fire, improper or illegal activities, and exposure to the elements.
   b. The primary control is accountability. Assets are recorded when acquired, and their existence is verified periodically.
4. **Compliance with Policies, Plans, Procedures, Laws, Regulations, and Contracts**
   a. Internal auditors determine if employees are doing what they are supposed to do, that is, following specified policies, plans, etc.
   b. In the absence of compliance, the internal auditor determines
      1) The reasons for noncompliance
         a) Procedures may be faulty.
         b) Better procedures may be available.
         c) Conditions may have changed.
      2) The cost of noncompliance
      3) How to obtain compliance if desirable
5. **Accomplishment of Established Objectives and Goals for Operations or Programs**
   a. Goals, objectives, and control procedures are developed by management. Internal auditors determine whether they conform with the overall goals and objectives of the organization and whether they are being met.
      1) Organizational goals and objectives are established by the board of directors and senior management.
   b. Internal auditors ascertain whether operations or programs are being carried out as planned.

6. **Economical and Efficient Use of Resources**
   a. Efficiency is evaluated in terms of operating standards.
   b. Auditors should determine that standards have been set, are understood, and are being met.
   c. Auditors should identify and report inefficiency, e.g., underused facilities, nonproductive work, over- or understaffing, and uneconomical procedures.
7. **Risk Assessment**
   a. Internal auditors identify areas with high expected or suspected risk and notify management to determine if additional auditing procedures should be applied.
   b. Risk areas are identified by
      1) Prior audits
      2) Indications from audits elsewhere in the organization
      3) Auditor's experience (educational and on-the-job)
      4) Problems in other companies and related industries
      5) Interaction with top management and the board of directors
      6) Other procedures as appropriate
   c. The internal auditing department's work schedule is established, in part, with regard to possible loss and risk.
8. **Detection and Prevention of Fraud**. See the next sideheading.
9. **Coordination with External Auditors**
   a. Internal auditors may provide the external auditor with direct assistance in performing both substantive tests and tests of controls provided that the external auditor considers the internal auditor's competence and objectivity, supervises and tests their work, and makes all judgments that affect the financial statements.
   b. The objective of coordination is to minimize duplicate efforts and encourage broad audit coverage.
   c. To the extent that internal auditing of accounting control provides a basis for reliance thereon by external auditors, the cost of internal auditing is offset by the reduced cost of external auditing.
   d. Because internal auditing has a broader scope, duplication of efforts is of concern only for areas addressed in a financial statement audit.
   e. Coordination of audit effort includes
      1) Periodic meetings to discuss matters of mutual interest
      2) Access to each other's audit programs and working papers
      3) Exchange of audit reports and management letters
      4) Common understanding of audit techniques, methods, and terminology
10. Many of the above services are performed in operational audits rather than financial audits.
11. Stop and review! You have completed the outline for this subunit. Study multiple-choice questions 7 through 9 beginning on page 440.

C. **Detection and Prevention of Fraud**

1. **General Responsibilities**. The internal auditor
   a. Should not be regarded as an insurer (guarantor) against the existence of fraud in an organization
   b. Examines and evaluates the adequacy and effectiveness of control. Controls should be designed to prevent or deter the forms of fraud widely known to be possible.
   c. Is responsible for seeking to identify areas of risk where fraud may be likely to occur
   d. Is responsible for exercising the care and skill of a reasonably prudent and competent professional
2. **Auditor Objectives Concerning Fraud**
   a. Prevention through effective control is preferable to detection.
      1) Use of control systems is more cost beneficial than 100% auditor verification.
3. When auditors suspect wrongdoing, the proper authorities within the organization should be informed.
   a. The auditor may recommend the type of investigation.
      1) Ordinarily, specialists, e.g., security personnel, should conduct investigations (especially questioning of those suspected).
      2) Auditors should avoid personal contact with suspects to avoid hindering further investigation or giving cause for legal action against the auditor for libel, slander, false imprisonment, etc.
   b. The auditor's interest in fraud is to determine
      1) How fraud occurred
      2) How to correct the faults in internal control that led to fraud
      3) How to eliminate future occurrences
4. **Factors Contributing to or Permitting Fraud**
   a. Insufficient internal controls, such as
      1) Insufficient separation of functional responsibilities of authorization, custodianship, and record keeping
      2) Not limiting access to assets
      3) Insufficient accountability arising from not recording transactions
      4) Failing to compare existing assets with recorded accountability
      5) Executing transactions without proper authorization
      6) Nonimplementation of prescribed controls because of lack of personnel or unqualified personnel
   b. Collusion among employees
   c. Existence of liquid assets, such as cash, securities payable to bearer, or highly marketable merchandise

5. The IIA has issued a pronouncement on the internal auditor's responsibilities with regard to fraud. The summary of **SIAS 3**, *Deterrence, Detection, Investigation, and Reporting of Fraud*, is quoted below.

> *This statement interprets the Standards and establishes guidelines for internal auditors regarding their responsibility for deterring, detecting, investigating, and reporting of fraud. It does not provide guidance on specific audit procedures used in performing audits; rather, it establishes guidelines by which internal auditors conform their activities with the stated concepts of due professional care. As used in this statement, the term management includes anyone in an organization with responsibilities for setting and/or achieving objectives.*
>
> *Major conclusions of this statement are:*
>
> ***Deterrence of fraud**. This is the responsibility of management. Internal auditors are responsible for examining and evaluating the adequacy and the effectiveness of actions taken by management to fulfill this obligation.*
>
> ***Detection of fraud**. Internal auditors should have sufficient knowledge of fraud to be able to identify indicators that fraud might have been committed.*
>
> *If significant control weaknesses are detected, additional tests conducted by internal auditors should include tests directed toward identification of other indicators of fraud.*
>
> *Internal auditors are not expected to have knowledge equivalent to that of a person whose primary responsibility is to detect and investigate fraud. Also, audit procedures alone, even when carried out with due professional care, do not guarantee that fraud will be detected.*
>
> ***Investigation of fraud**. Fraud investigations may be conducted by or involve participation of internal auditors, lawyers, investigators, security personnel, and other specialists from inside or outside the organization.*
>
> *Internal auditing should assess the facts known relative to all fraud investigations in order to:*
>
> - *Determine if controls need to be implemented or strengthened.*
> - *Design audit tests to help disclose the existence of similar frauds in the future.*
> - *Help meet the internal auditor's responsibility to maintain sufficient knowledge of fraud.*
>
> ***Reporting of fraud**. A written report should be issued at the conclusion of the investigation phase. It should include all findings, conclusions, recommendations, and corrective action taken.*

6. Stop and review! You have completed the outline for this subunit. Study multiple-choice questions 10 through 14 beginning on page 441.

D. Audit Committees

1. The audit committee of a company's board of directors is a subcommittee made up of outside directors, i.e., directors who are independent of corporate management. They help to
   a. Keep external and internal auditors independent of management
   b. Assure that the directors are exercising due care
2. **Audit Committee Functions**
   a. Select an external auditor and review the audit fee and the engagement letter
   b. Review the external auditor's overall audit plan
   c. Review preliminary annual and interim financial statements
   d. Review results of external and internal audits, e.g., restrictions, audit findings and recommendations
   e. Review the internal audit work schedule, budget, reports, etc.
   f. Meet regularly with the internal auditing director
   g. Review evaluations of internal control
   h. Review the company's accounting, financial, and operating controls
   i. Review policies on unethical and illegal procedures
   j. Review financial statements for regulatory agencies
   k. Review observations of company personnel
   l. Participate in the selection of accounting policies
   m. Review the impact of new or proposed legislation or governmental regulations
   n. Review the company's insurance program
   o. Review the external auditor's management letter
3. The AICPA has recognized the importance of internal reporting to audit committees in **SAS 61** (AU 380), *Communication with Audit Committees*. The IIA's *Standards for the Professional Practice of Internal Auditing* require the director of internal auditing to have direct communication with the board.
4. Stop and review! You have completed the outline for this subunit. Study multiple-choice questions 15 and 16 on page 443.

E. Purpose and Functions of the Internal Audit Program

1. Audit programs are required as part of the planning for each audit.
   a. Outline what is to be accomplished
   b. Specify how it is to be done
   c. Provide for a record of what has been done
   d. Facilitate supervision and control of the audit
2. The emphasis and extent of the audit program depends on the nature (scope) of the review to be undertaken.
   a. The review can be an all-inclusive, in-depth audit of controls throughout an entire department or activity.
   b. However, the review may have a narrower scope if its focus is on one aspect of the internal audit objectives. Thus, the internal auditor's program may be restricted to the procedures necessary to perform one type of audit service (see subunit B.).

3. The scope of the audit program is determined during the first step in the planning process: establishing audit objectives and scope of work.
   a. Objectives are specific audit goals, and procedures are the detailed audit steps to achieve them.
   b. The director of internal auditing, with the guidance of the board and top management, must determine the relative emphasis of each type of service to be performed during each audit project.
      1) The review emphasis depends on the type of operation being audited. The following should be considered:
         a) The operation's objectives
         b) Cost-benefit analysis
         c) Constraints, such as the availability of audit personnel
4. The audit program is prepared after the preliminary survey and before the start of field work.
   a. It is usually a rewrite of programs used for previous or related audits, or a pro forma program.
      1) Pro forma audit programs are generalized audit programs that have been developed for both financial and operational audits.
      2) The audit program must then be tailored to each specific client, engagement, situation, etc.
         a) A shift in audit emphasis from audit to audit may be the result of past audits, changed operations, etc.
            i) The board and top management may request a special emphasis to help in long- or short-range planning.
   b. Substantial attention should always be directed to known or suspected areas of risk.
   c. Audit programs are tentative because deviations from expected results discovered during performance of the audit should be pursued.
      1) Exceptions should be analyzed to determine the control weakness that
         a) Permitted the exception
         b) Allowed the exception to go undetected
      2) Cost-benefit analysis must be applied to
         a) Determine the materiality of the exception
         b) Evaluate the advisability of strengthened or new controls
5. The audit program should indicate the objectives of the operation being audited, as disclosed during the preliminary survey.
   a. The audit program should also indicate the controls in effect or needed to achieve the objectives of the operation.
   b. Matching the operation's objectives with its existing controls may reveal that
      1) One or more controls necessary to attain objectives are missing.
      2) One or more controls are unnecessary, redundant, inefficient (too expensive for the benefit gained).
   c. The program should provide for tests of transactions to obtain evidence of whether the controls are functioning as intended and desired results (objectives) are being achieved.

6. The execution of the audit program should yield evidence to support audit results. The IIA Standard 420 requires audit evidence to be sufficient, competent, relevant, and useful.
   a. **Sufficient** information is factual, adequate, and convincing so that a prudent, informed person would reach the same conclusions as the auditor.
   b. **Competent** information is reliable and the best obtainable through the use of appropriate audit techniques.
   c. **Relevant** information supports audit findings and recommendations and is consistent with the objectives for the audit.
   d. **Useful** information helps the organization meet its goals.
7. Stop and review! You have completed the outline for this subunit. Study multiple-choice questions 17 through 28 beginning on page 444.

F. Operational Auditing

1. Sawyer and Dittenhofer, in *Sawyer's Internal Auditing* (Altamonte Springs, FL: The Institute of Internal Auditors, 4th ed., 1996, page 4), define operational auditing as "the comprehensive review of the varied functions within an enterprise to appraise the efficiency and economy of operations and the effectiveness with which those functions achieve their objectives." An operational audit is thus a thorough examination of a department or division. Its purpose is to appraise managerial organization, performance, and techniques.
   a. It attempts to determine the extent to which organizational objectives have been achieved.
   b. It is a control technique that provides management with a method for evaluating the effectiveness of operating procedures and internal controls.
   c. The focus is on efficiency, effectiveness, and economy (these terms are sometimes called the "three E's of operational auditing").
   d. The audit report resulting from an operational audit consists primarily of specifying where problems exist or emphasizing the absence of problems.
   e. The auditor compares a department's operations with company policies and procedures, industry averages, and departmental trends.
      1) Common business sense is an important aspect of the auditor's work.
   f. The basic tools of the auditor include
      1) Financial analysis
      2) Observation of departmental activities
      3) Questionnaire interviews of departmental employees
   g. The operational audit evolved as an extension of the typical financial audit in that it goes beyond what is ordinarily considered to be the accounting function, e.g.,
      1) Reviewing purchasing policies
      2) Appraising compliance with company policies and procedures
      3) Appraising safety standards and maintenance of equipment
      4) Reviewing production controls and scrap reporting
      5) Reviewing adequacy of facilities

2. Some writers have contrasted financial (external) auditing and operational auditing. Based on the success of The Institute of Internal Auditors and its leadership in this area (including issuance of the Statement of Responsibilities and the Standards for the Professional Practice of Internal Auditing), your authors conclude that internal auditing includes, but is not limited to, operational auditing; i.e., internal auditing is the broader and preferable term.

3. Reviews for operating controls include audits of the control loops throughout the organization.

   a. The controls relevant to a financial statement audit concern the reliability and integrity of information and the safeguarding of assets.

   b. The controls relevant to an operational audit concern the adequacy and effectiveness of internal control; compliance with policies, plans, procedures, laws, and regulations; and accomplishment of established objectives.

4. The control loop consists of

   a. Establishing standards for the operation to be controlled
   b. Measuring performance against the standards
   c. Examining and analyzing deviations
   d. Taking corrective action
   e. Reappraising the standards, based on experience

5. The control loop

   a. Begins with planning
   b. Requires organizing
   c. Monitors directing and motivating
   d. Is the essence of controlling

6. Stop and review! You have completed the outline for this subunit. Study multiple-choice questions 29 through 32 beginning on page 448.

G. Internal Audit Reports

1. **Types of Reports**

   a. Formal -- with carefully structured formats

   b. Informal -- in letters or memoranda to operating management

   c. Interim (progress) -- with brief statements of conditions requiring immediate attention

   d. Oral -- with formal audiovisual presentations or informal comments

   e. Summary -- intended for levels of management above the audited unit, with highlights that omit some details

   f. Deficiency -- including comments on only those matters calling for corrective action

   g. Financial -- with statements of account balances and opinions on whether they fairly reflect results of operations

   h. Operational -- with discussions of the adequacy and effectiveness of the operating controls in any sector of the enterprise

2. The report format depends on the

   a. Type of audit

   b. Results of the audit

   c. Needs of management

d. Nature of the company and its degree of centralization

e. Acceptance by managers, at various levels, of internal auditing's position in the company

3. According to Standard 430 and SIAS 2, audit reports must include the purpose, scope, and results of the audit and, if appropriate, an opinion.

   a. Audit reports may include

      1) Background information
      2) Summaries
      3) Status of findings, conclusions, and recommendations from prior reports
      4) An indication of whether the report covers a scheduled audit or the response to a request
      5) Recommendations for potential improvements
      6) Acknowledgments of satisfactory performance and corrective action
      7) Auditee comments

   b. Results may include

      1) Findings
      2) Conclusions (opinions)
      3) Recommendations

   c. The attributes of findings include

      1) Criteria -- the standards used in the evaluation
      2) Condition -- the facts found
      3) Cause -- the reason for the variance, if any, between the criteria and the condition
      4) Effect -- the risk or exposure resulting from the difference between the criteria and the condition

4. **Preparation of Internal Audit Reports**

   a. Except for particularly simple reports, the auditor should prepare an outline of the proposed audit report.

      1) The outline should be brief.
      2) The auditor should first divide the outline into the main headings, such as

         a) Capsule summary

            i) Provides an overview of the report

         b) Foreword

            i) Regular or special exam
            ii) Organizations or functions reviewed
            iii) Relevant prior examinations
            iv) Size of auditee, e.g., volume of transactions
            v) Any other necessary explanation

         c) Purpose

            i) Audit objectives

d) Scope

i) Any limitation on the examination
ii) Any special audit procedures

e) Opinion

i) Constitutes an overall summary of the auditor's assessment of auditee operations

f) Findings

i) Favorable or unfavorable for each condition

ii) Unfavorable findings may require the following points of explanation:

- Capsule summary
- Criteria
- Condition (facts found)
- Cause
- Effect
- Recommendation
- Corrective action taken

3) Thereafter, the details under each major heading may be developed in capsule form.

4) Internal audit reports should be

a) Objective (factual, unbiased)
b) Clear (easily understood and logical)
c) Concise (avoiding unneeded detail)
d) Timely (enabling prompt, effective action)
e) Constructive (helpful and leading to needed improvements)

5. **Oral Reports and Interim Reports**

a. Oral reports should be used to complement and support written audit reports.

1) Oral reports cannot substitute for written reports, but they do have the following purposes (advantages):

a) They offer timely feedback.

i) Some deficiencies need immediate auditee action.

ii) The auditor can provide instant response to auditee questions, positions, suggestions, etc.

b) Oral reports support participative auditing and can improve auditor-auditee relationships.

i) The auditor can become better known to the auditee and reduce auditee tensions.

ii) The auditee can express his/her views.

iii) The oral reports can reduce auditee resistance to auditor recommendations.

c) Oral reports provide the auditor with additional information.

i) The auditor can assess the strength of auditee attitudes and convictions.

ii) The auditee can point out auditor misunderstandings and errors.

2) Although oral reports allow for flexibility, careful planning is necessary, which may include the following steps:

a) Gaining familiarity with the participants

b) Determining the objectives

c) Listing the materials, topics, etc., relevant to the objectives

d) Ordering the material

e) Preparing visual aids, handouts, etc.

f) Practicing the presentation

g) Anticipating and preparing for questions and arguments from the auditee

b. Interim reports are used to communicate information needing immediate action, to communicate a change in audit scope, or to advise management about progress when the audit is lengthy.

1) Interim reports are preliminary reports and should indicate that

a) Only current information, i.e., an incomplete study, is the basis for the report.

b) The final report will follow up on the topics covered in the interim report.

c) The final report will also cover the other topics on the audit program.

2) Interim reports prepared by audit staff should be reviewed by the director of internal auditing or other supervisory personnel.

3) Interim reports may also be used to report the status of long, sensitive, or otherwise special audits to the auditees and top management.

6. **Report Reviews and Distribution**. Reviews of report drafts with the auditee are a courtesy to the auditee and a form of insurance for the audit. According to Standard 430, "The internal auditor should discuss conclusions and recommendations at appropriate levels of management before issuing final written reports." SIAS 2 observes that this discussion usually occurs during the audit or at a postaudit meeting (an exit interview or closing conference).

a. The auditee may have discussed all matters in the report during the review.

1) As a matter of courtesy, however, the auditee should be given the opportunity to read what will be sent to his/her superior.

2) The formal report may convey a different impression.

b. In complex operations, it is easy for the auditor to miss a point or report it incorrectly. Reviewing the report with the auditee can bring out any such inaccuracies before the final report is issued.

c. The auditor should carefully consider the following matters before scheduling the report review:

1) The person(s) with whom the draft report should be reviewed

2) Whether each review should be done separately or at least partially on a group basis

3) The timing of the reviews

4) The order of the reviews

5) Whether to try expediting a review by sending copies of the report in advance of the meeting

6) Whether in-person discussions are necessary. Sending copies of the report and receiving written feedback may be sufficient.

d. The auditor should be in charge of the report review conference.

e. The auditor should be prepared for conflicts and questions.

1) When the auditor has previously experienced difficulty with an individual, that individual's superior may be invited to attend.

2) To be able to answer questions promptly, the auditor may wish to prepare notes, e.g., marginal references on a draft of the report and on pertinent working papers.

3) The auditor should be flexible regarding matters that do not affect the substance of the report.

a) But the auditor should never negotiate the audit opinion.

f. If the reviews result in significant changes in the report, the other people with whom the draft was reviewed should have an opportunity to see, or be told of, the revisions.

g. The auditor should maintain careful records of the review conferences, of any objections, and of the manner in which conflicts were resolved.

h. When drafts are to be mailed to the people who should review them, the auditor should

1) Ask for the timely return of the drafts with any comments considered appropriate.

2) Set a specific due date for the return of the draft.

3) Offer to meet with those who wish to discuss the draft further.

7. Reports should be distributed to all those having a direct interest in the audit. Each report should contain a distribution sheet listing the distributees and indicating with whom the report has been reviewed in draft. Distributees include

a. The executive to whom the internal audit function reports
b. The person or persons to whom replies to the report will be addressed
c. Persons responsible for the activity or activities reviewed
d. Persons required to take corrective action

8. Information that is privileged, proprietary, or related to improper or illegal acts should be disclosed in a separate report and distributed to the audit committee.

9. Stop and review! You have completed the outline for this subunit. Study multiple-choice questions 33 through 40 beginning on page 449.

H. Auditor Follow-Up

1. The IIA Standard 440 requires internal auditors to follow up to ascertain that appropriate action is taken on deficiency findings. The internal auditor should determine that
   a. Corrective action being taken has the desired results, or
   b. Management or the board has assumed the risk of not taking corrective action.
2. In following up, the auditor should
   a. Receive all auditee replies to the audit report
   b. Evaluate the adequacy of those replies
   c. Be convinced that the action taken will cure the defects
3. The auditor is in the best position to carry out this responsibility.
   a. The auditor is better acquainted with the facts than higher management or other control centers in the organization.
   b. (S)he is more objective than the operating manager who must take the corrective action.
4. The responsibility for determining whether corrective action is adequate should be coupled with the authority to evaluate the adequacy of replies to audit reports. The internal auditor should
   a. Report to management when corrective actions are not timely or effective.
   b. Submit periodic reports to management on open audit findings.
5. The adequacy of a response depends on the circumstances in each case. In general, a satisfactory response
   a. Addresses itself to the complete problem -- not only to those specific items included in the auditor's sample
   b. Shows that action has also been taken to prevent a recurrence of the deficient condition
6. In evaluating the reply, the auditor should be satisfied that the action promised is actually taken. The auditor should
   a. Obtain copies of revised procedures issued to correct conditions.
   b. Make any field tests needed to provide assurance that the condition has been corrected.
7. A formal system should be designed to keep audit projects open until adequate corrective action is assured. For example,
   a. Provisions should be made for formal opening and closing of audit projects.
   b. The auditors should issue a formal statement of closure, supported by copies of replies to audit reports and explanations of the action taken to ensure the adequacy and effectiveness of corrective measures.
      1) Closure reports are directed to the director of internal auditing.
   c. Audit projects should not be removed from the audit department's open project listing until all required corrective actions have been taken and evaluated.
8. Stop and review! You have completed the outline for this subunit. Study multiple-choice questions 41 through 45 beginning on page 452.

MULTIPLE-CHOICE QUESTIONS

A. Nature of Internal Auditing

1. The primary objective of internal auditing is to

A. Locate errors and fraud.

B. Attest to the fairness of financial statements.

C. Assist members of the organization in the effective discharge of their responsibilities.

D. Provide audit assistance and guidance to the external accountant.

The correct answer is (C). *(CMA, adapted)*

REQUIRED: The primary objective of internal auditing.

DISCUSSION: Internal auditing is an independent appraisal function that examines and evaluates organizational activities. The Statement of Responsibilities of Internal Auditing states that the objective of internal auditing is to assist members of the organization in the effective discharge of their responsibilities by furnishing an analysis of internal control activities.

Answers (A) and (D) are incorrect because each is a lesser responsibility of the auditor. Answer (B) is incorrect because the internal auditor does not attest to the fairness of financial statements.

2. It has been established that an internal auditing charter is one of the more important factors positively affecting the internal auditing department's independence. The Standards help clarify the nature of the charter by providing guidelines as to the contents of the charter. Which of the following is not suggested in the Standards as part of the charter?

A. The department's access to records within the organization.

B. The scope of internal auditing activities.

C. The length of tenure for the internal auditing director.

D. The department's access to personnel within the organization.

The correct answer is (C). *(CIA, adapted)*

REQUIRED: The item not included in the internal auditing charter.

DISCUSSION: The Standards state that the independence of internal auditing is enhanced when the board concurs in the appointment or removal of the director but otherwise do not discuss the length of the director's employment.

Answers (A), (B), and (D) are incorrect because the charter establishes the department's position within the organization; authorizes access to records, personnel, and physical properties; and defines the scope of internal audit activities.

3. In considering the internal auditing department's independence, which of the following facts, by themselves, could contribute to a lack of internal auditing independence?

I. The CEO accused the new director of not operating "in the best interests of the organization."

II. The majority of audit committee members come from within the organization.

III. The internal auditing charter has not been approved by the board or the audit committee.

A. I only.

B. II only.

C. II and III only.

D. I, II, III.

The correct answer is (D). *(CIA, adapted)*

REQUIRED: The factor(s) contributing to a lack of audit independence.

DISCUSSION: The CEO's statement suggests that the internal auditors lack the support of management and the board. Furthermore, the lack of outside directors may contribute to a loss of auditor independence. The failure to approve the charter may have the same effect. The charter enhances the auditor's independence because it clearly specifies, in advance, the authority, scope, and responsibility of the internal auditing function.

Answer (A) is incorrect because lack of support by the CEO and lack of outside directors weaken the internal auditors' position. Answer (B) is incorrect because lack of support by the CEO and lack of a charter weaken the internal auditors' position. Answer (C) is incorrect because lack of support by the CEO weakens the internal auditors' position.

4. The auditor has planned an audit of the effectiveness of the quality assurance function as it affects the receiving of goods, the transfer of the goods into production, and the scrap costs related to defective items. The auditee argues that such an audit is not within the scope of the internal auditing function and should come under the purview of the quality assurance department only. What would be the most appropriate audit response?

A. Refer to the audit department charter and the approved audit plan that includes the area designated for audit in the current time period.

B. Because quality assurance is a new function, seek the approval of management as a mediator to set the scope of the audit.

C. Indicate that the audit will examine the function only in accordance with the standards set by, and approved by, the quality assurance function before beginning the audit.

D. Terminate the audit because an operational audit will not be productive without the auditee's cooperation.

The correct answer is (A). *(CIA, adapted)*

REQUIRED: The most appropriate response to an assertion that an audit of the effectiveness of the quality assurance function is beyond the scope of internal auditing.

DISCUSSION: The charter should define the purpose, authority, and responsibility of the internal auditing department. Among other matters, it should define the scope of internal auditing activities. Furthermore, the director should submit annually to management for approval and to the board for its information a summary of the department's audit work schedule, staffing plan, and financial budget (Standard 110).

Answer (B) is incorrect because the auditee does not determine the scope of the audit. Answer (C) is incorrect because other objectives may be established by management and the auditor. The audit should not be limited to the specific standards set by the quality assurance department, but it should consider such standards in the development of the audit program. Answer (D) is incorrect because the auditor should conduct the audit and communicate any scope limitations to management and the board.

5. From a modern internal auditing perspective, which one of the following statements represents the most important benefit of an internal audit department to management?

A. Assurance that published financial statements are correct.

B. Assurance that fraudulent activities will be detected.

C. Assurance that the organization is complying with legal requirements.

D. Assurance that there is reasonable control over day-to-day operations.

The correct answer is (D). *(CMA, adapted)*

REQUIRED: The most important benefit from an internal audit department.

DISCUSSION: Internal auditing is an independent appraisal activity within an organization for the review of operations as a service to members of the organization. It is a management control that functions by examining and evaluating the efficiency and effectiveness of other controls, i.e., to see that day-to-day operations are under reasonable control.

Answer (A) is incorrect because published financial statements are only required to be fairly presented. An audit cannot assure correctness. Answer (B) is incorrect because the internal auditor's responsibility is limited to determining that the system has adequate controls to prevent or deter forms of fraud generally known to be possible. Answer (C) is incorrect because the internal auditor is not an attorney and accordingly cannot assure compliance with legal requirements.

6. Management has requested the internal auditing department to perform an operational audit of the telephone marketing operations of a major division and to recommend procedures and policies for improving management control over the operation. The auditor should

A. Not accept the engagement because recommending controls would impair future objectivity of the department regarding this auditee.

B. Not accept the engagement because audit departments are presumed to have expertise on accounting controls, not marketing controls.

C. Accept the engagement but indicate to management that, because recommending controls impairs audit independence, future audits of the area will be impaired.

D. Accept the audit engagement because independence will not be impaired.

The correct answer is (D). *(CIA, adapted)*

REQUIRED: The acceptability of an operational audit engagement to recommend standards of control.

DISCUSSION: The director of internal auditing should accept the engagement; assign staff with the knowledge, skills, and disciplines essential to the performance of this audit; and make appropriate recommendations. Recommending standards of control does not impair objectivity (Standard 120).

Answer (A) is incorrect because the auditor should accept the engagement. Recommending controls is not considered a violation of the auditor's independence or objectivity. Answer (B) is incorrect because the auditor should accept the engagement. Auditors should have control knowledge that is not limited to accounting controls. Answer (C) is incorrect because audit independence is not impaired by making control recommendations.

B. Types of Internal Auditing Services

7. In assessing relative risks, internal auditors should be least concerned with

A. Reliability and integrity of information.

B. Compliance with internal and external rules and regulations.

C. Statistical sampling techniques.

D. Safeguarding of assets.

The correct answer is (C). *(CMA, adapted)*

REQUIRED: The area of risk with which an internal auditor should be least concerned.

DISCUSSION: According to authoritative pronouncements of The IIA, the scope of work of internal auditors extends to "the examination and evaluation of the adequacy and effectiveness of the organization's system of internal control and the quality of performance in carrying out assigned responsibilities." The primary objectives of internal control are to ensure compliance with policies, plans, procedures, laws, and regulations; accomplishment of established objectives and goals; reliability and integrity of information; economical and efficient use of resources; and safeguarding of assets. However, risk associated with statistical sampling (sampling risk) is a lesser concern of an internal auditor because it can be measured and controlled.

Answers (A), (B), and (D) are incorrect because reliability and integrity of information, compliance with internal and external rules and regulations, and safeguarding of assets are primary objectives of internal control.

8. While assisting the external auditor in the performance of substantive tests or tests of controls, the internal auditor should

A. Establish limits of materiality that are below the usual limits set by the external auditor.

B. Establish limits of materiality that are above the usual limits set by the external auditor.

C. Be supervised by the external auditor.

D. Be independent of the external auditor.

The correct answer is (C). *(CMA, adapted)*

REQUIRED: The relationship between an internal auditor and an external auditor in the performance of tests.

DISCUSSION: The independent external auditor may make use of internal auditors to provide direct assistance in performing both substantive tests and tests of controls provided that (s)he considers their competence and objectivity, supervises and tests their work, and makes all judgments regarding matters that affect the report on the financial statements.

Answers (A) and (B) are incorrect because the external auditor must establish limits of materiality, not the internal auditor. Answer (D) is incorrect because the internal auditor should not be independent of the external auditor when working under his/her supervision.

9. During an audit of a company's financial statements by an external auditor, the audit procedure that is most likely to be performed by a member of the company's internal audit staff under the supervision of the external auditor is the determination of the

A. Legitimacy of confirmation exceptions received during accounts receivable confirmation.

B. Sample size for the confirmation of accounts receivable.

C. Effect of weaknesses in the credit sales system.

D. Extent of procedures used to test the validity of accounts receivable.

The correct answer is (B). *(CMA, adapted)*

REQUIRED: The audit procedure most likely to be performed by an internal auditor under the supervision of an external auditor.

DISCUSSION: Because the ultimate responsibility for expression of an opinion rests with the external auditor, (s)he must make all decisions that require judgment. Thus, the internal auditor might select the sample size once the external auditor has chosen the confidence level. The selection of sample size is essentially a clerical task once risk levels have been assessed by the external auditor.

Answers (A), (C), and (D) are incorrect because judgments as to control risk, sufficiency of tests performed, materiality of transactions, and other matters affecting the report on the financial statements must be those of the independent external auditor (AU 322).

C. Detection and Prevention of Fraud

10. If an internal auditor suspects that a bookkeeper for a small plant was engaging in lapping, the internal auditor should

A. Prepare a schedule of interbank transfers.

B. Compare remittance advices and duplicate deposit slips to postings in the cash receipts journal and the accounts receivable subsidiary ledger cards.

C. Prepare a proof of cash.

D. Control all cash receipts for a business day on an unannounced basis.

The correct answer is (B). *(CMA, adapted)*

REQUIRED: The procedure to detect lapping of receivables.

DISCUSSION: Lapping is the delayed recording of cash receipts to cover a cash shortage, such as when receipts from accounts which were actually paid yesterday are reported as today's receipts. The best protection is for the customers to send payments directly to the company's depository bank. This procedure precludes client personnel from having the opportunity to "borrow" the money. Lapping may be detected by comparing details of bank deposits with the client's record of cash receipts. Since the theft of a payment from one customer may be covered (lapped) with a payment from another customer, a comparison of remittance advices with the subsidiary accounts receivable ledger may be helpful. Also, if the auditor suspects the duplicate deposit slips have been tampered with, (s)he should compare them with the originals held by the bank.

Answer (A) is incorrect because a schedule of interbank transfers is used to uncover kiting, not lapping. Kiting is the recording of a deposit from an interbank transfer in the current period while failing to record the related disbursement until the next period. Answer (C) is incorrect because a proof of cash would not uncover lapping since it does not entail an examination of receivables. Answer (D) is incorrect because controlling cash receipts will mean only that the one day's receipts will be properly recorded; a lapper may not work every day.

11. When comparing perpetrators who have embezzled company funds with perpetrators of financial statement fraud (falsified financial statements), those who have falsified financial statements would be less likely to

A. Have experienced an autocratic management style.

B. Be living beyond their obvious means of support.

C. Rationalize the fraudulent behavior.

D. Use company expectations as justification for the act.

The correct answer is (B). *(CIA, adapted)*

REQUIRED: The least likely characteristic of those who have falsified financial statements.

DISCUSSION: Living beyond one's means has been linked to employee fraud (embezzlement), not to financial statement fraud. Fraud perpetrated for the benefit of the organization ordinarily benefits the wrongdoer indirectly, whereas fraud that is detrimental to the organization provides immediate, direct benefits to the employee (SIAS 3).

Answer (A) is incorrect because autocratic management styles have been linked to management (financial statement) fraud. Answer (C) is incorrect because rationalization is common to all fraud. Answer (D) is incorrect because high expectations are often given as a motivating factor by those who have committed financial statement fraud.

12. Which of the following policies is most likely to result in an environment conducive to the occurrence of fraud?

A. Budget preparation input by the employees who are responsible for meeting the budget.

B. Unreasonable sales and production goals.

C. Frequent rejection of adequately trained applicants by the division's hiring process.

D. The application of some accounting controls on a sample basis.

The correct answer is (B). *(CIA, adapted)*

REQUIRED: The policy most likely to result in an environment conducive to the occurrence of fraud.

DISCUSSION: Unrealistically high sales or production quotas can be an incentive to falsify the records or otherwise take inappropriate action to improve performance measures so that the quotas appear to have been met.

Answer (A) is incorrect because participatory budgeting can reduce antagonism to budgets and reduce the likelihood of inappropriate means of meeting the budget. Answer (C) is incorrect because hiring policies should be based on factors other than adequate training, such as the applicants' personal integrity. Furthermore, hiring all adequately trained applicants is unlikely to be necessary. Answer (D) is incorrect because, under the reasonable assurance concept, the cost of controls should not exceed their benefits. The cost of applying controls to all relevant transactions rather than a sample may be greater than the resultant savings.

13. Interviewing techniques are used frequently by internal auditors. When considering the potential use of interviewing techniques to gather audit evidence, auditors should be aware that interviews

A. Are more objective than questionnaires in gathering data.

B. Provide a systematic format to ensure audit coverage.

C. Should be corroborated by gathering objective data.

D. Are best suited to reaching audit conclusions.

The correct answer is (C). *(CIA, adapted)*

REQUIRED: The true statement about interviews.

DISCUSSION: Oral evidence is presumptively less reliable than other forms of evidence, such as that obtained from independent sources outside the entity or from the auditor's direct experience. Consequently, it should be corroborated.

Answer (A) is incorrect because interviews are not more objective than questionnaires. Answer (B) is incorrect because interviews are often unstructured. Answer (D) is incorrect because the need for corroboration prevents treating the evidence from interviews as conclusive.

14. Which of the following is true about interviewing an individual during the investigation of suspected fraud?

A. The internal auditor's role involves collecting facts.

B. Internal auditors should be empowered to confine fraud suspects to the office but only for the purpose of interviewing them.

C. The internal auditor's role involves attempting to obtain confessions of guilt.

D. Internal auditors are authorized to waive punishment of the employee if the employee restores the item(s) stolen.

The correct answer is (A). *(CIA, adapted)*

REQUIRED: The internal auditor's role during the investigation of suspected fraud.

DISCUSSION: According to SIAS 3, "Investigation consists of performing extended procedures necessary to determine whether fraud, as suggested by the indicators, has occurred. It includes gathering sufficient evidential matter about the specific details of a discovered fraud. Internal auditors, lawyers, investigators, security personnel, and other specialists from inside or outside the organization are the parties that usually conduct or participate in fraud investigations." Hence, internal auditors are fact gatherers. However, internal auditors are not normally trained as interrogators of suspected perpetrators of fraud.

Answer (B) is incorrect because confining a suspect is considered false imprisonment. Answer (C) is incorrect because obtaining confessions is the role of an investigator. Answer (D) is incorrect because waiving punishment is considered to be compounding a felony. The right to punish or forgive a criminal act is reserved to the state.

D. Audit Committees

15. Audit committees have been identified as a major factor in promoting both the internal and external auditor's independence. Which of the following is the most important limitation on the effectiveness of audit committees?

A. Audit committees may be composed of independent directors. However, those directors may have close personal and professional friendships with management.

B. Audit committee members are compensated by the organization and thus favor a shareholder's view.

C. Audit committees devote most of their efforts to external audit concerns and do not pay much attention to internal auditing and the overall control environment.

D. Audit committee members do not normally have degrees in the accounting or auditing fields.

The correct answer is (A). *(CIA, adapted)*

REQUIRED: The most important limitation on the effectiveness of audit committees.

DISCUSSION: The audit committee is a subcommittee made up of outside directors who are independent of corporate management. Its purpose is to help keep external and internal auditors independent of management and to assure that the directors are exercising due care. However, if independence is impaired by personal and professional friendships, the effectiveness of the audit committee may be limited.

Answer (B) is incorrect because the compensation audit committee members receive is usually minimal. They should be independent and therefore not limited to a shareholder's perspective. Answer (C) is incorrect because, although audit committees are concerned with external audits, they also devote attention to the internal audit function. Answer (D) is incorrect because audit committee members do not need degrees in accounting or auditing to understand audit reports.

16. A medium-sized publicly owned corporation operating in Country X has grown to a size that the directors of the corporation believe warrants the establishment of an internal auditing department. Country X has legislated internal auditing requirements for government-owned companies. The company changed the corporate bylaws to reflect the establishment of the internal auditing department. The directors decided that the director of internal auditing must be a certified internal auditor and will report directly to the newly established audit committee of the board of directors. Which of the items discussed above will contribute the most to the new audit director's independence?

A. The establishment of the internal auditing department is documented in corporate bylaws.

B. Country X has legislated internal auditing requirements.

C. The director will report to the audit committee of the board of directors.

D. The director is to be a certified internal auditor.

The correct answer is (C). *(CIA, adapted)*

REQUIRED: The item that contributes most to an audit director's independence.

DISCUSSION: According to Standard 100, independence is achieved through organizational status and objectivity. The director should be responsible to an individual with sufficient authority to promote independence. The board of directors is the highest authority in the organization.

Answer (A) is incorrect because documentation in the bylaws does little to promote independence. Answer (B) is incorrect because legislated internal auditing requirements in Country X do not promote independence. Answer (D) is incorrect because independence is achieved through organizational status and objectivity.

E. Purpose and Functions of the Internal Audit Program

17. The scope of an internal audit is initially defined by the

A. Audit objectives.

B. Scheduling and time estimates.

C. Preliminary survey.

D. Audit program.

The correct answer is (A). *(CIA, adapted)*

REQUIRED: The factor initially defining the scope.

DISCUSSION: According to Standard 410, internal auditors should plan each audit. Planning should be documented and should include, as a first step, establishing audit objectives and scope of work.

Answer (B) is incorrect because the scheduling and time estimates are based on the audit objectives and the scope of the audit. Answer (C) is incorrect because the preliminary survey is performed after the audit objectives are determined. Answer (D) is incorrect because the audit program is developed after the preliminary survey and is based on the audit objectives and the scope of the audit.

18. An internal auditor has just completed an on-site survey to become familiar with the company's payroll operations. Which of the following should be performed next?

A. Assign audit personnel.

B. Establish initial audit objectives.

C. Write the audit program.

D. Conduct field work.

The correct answer is (C). *(CIA, adapted)*

REQUIRED: The step following the on-site survey.

DISCUSSION: The audit program is normally prepared after the on-site survey. The on-site survey allows the auditor to become familiar with the auditee and therefore provides input to the audit program.

Answer (A) is incorrect because audit personnel are usually assigned before the on-site survey. Answer (B) is incorrect because initial audit objectives are established at the beginning of the planning process. They should be specified before the on-site survey. Answer (D) is incorrect because field work can be performed only after the audit program has been written. Thus, field work cannot immediately follow the on-site survey.

19. Which of the following statements is an audit objective?

A. Observe the deposit of the day's cash receipts.

B. Analyze the pattern of any cash shortages.

C. Evaluate whether cash receipts are adequately safeguarded.

D. Recompute each month's bank reconciliation.

The correct answer is (C). *(CIA, adapted)*

REQUIRED: The audit objective.

DISCUSSION: Objectives are specific audit goals, and procedures are the detailed audit steps to achieve them. Evaluating whether cash receipts are adequately safeguarded is an audit objective because it states what the audit is to accomplish.

Answer (A) is incorrect because observation is an audit procedure. Answer (B) is incorrect because analysis is an audit procedure. Answer (D) is incorrect because recomputation is an audit procedure.

20. A standardized internal audit program is not appropriate for which situation?

A. A stable operating environment undergoing only minimal changes.

B. A complex or changing operating environment.

C. Multiple locations with similar operations.

D. Subsequent inventory audits performed at the same location.

The correct answer is (B). *(CIA, adapted)*

REQUIRED: The situation in which a standardized internal audit program is not appropriate.

DISCUSSION: A standard program is not appropriate for a complex or changing operating environment. The audit objectives and related work steps may no longer be relevant.

Answer (A) is incorrect because a standard audit program is appropriate for use in a minimally changing operating environment. It may save effort and provide continuity. Answer (C) is incorrect because a standard audit program can be used to audit multiple locations with similar operations if the same objectives and controls are present. Answer (D) is incorrect because a standard audit program is acceptable for conducting subsequent inventory audits at the same location if the inventory functions performed have not varied substantially.

21. Audit programs testing internal controls should

A. Be tailored for the audit of each operation.

B. Be generalized to fit all situations without regard to departmental lines.

C. Be generalized so as to be usable at all locations of a particular department.

D. Reduce costly duplication of effort by ensuring that every aspect of an operation is examined.

The correct answer is (A). *(CIA, adapted)*

REQUIRED: The true statement about audit programs.

DISCUSSION: A written audit program prescribes the nature, timing, and extent of work to be done. It sets forth in reasonable detail the specific audit procedures the auditor believes are necessary to accomplish the audit objectives. It is thus a useful tool in scheduling and controlling the audit. However, an audit program must be adapted to the specific needs of the audit after the auditor establishes the audit objectives and scope, determines the resources required, and conducts a preliminary survey.

Answer (B) is incorrect because a generalized program cannot take into account variations resulting from changing circumstances and varied conditions. Answer (C) is incorrect because a generalized program cannot take into account variations in circumstances and conditions. Answer (D) is incorrect because every aspect of an operation need not be examined, only those aspects likely to conceal problems and difficulties.

22. In conducting internal audits, secondary evidence is used to support primary evidence. Secondary evidence may include a copy of written evidence or oral evidence. Which one of the following is the weakest form of supportive evidence?

A. Direct evidence.

B. Circumstantial evidence.

C. Corroborative evidence.

D. Conclusive evidence.

The correct answer is (B). *(CMA, adapted)*

REQUIRED: The weakest form of supportive evidence.

DISCUSSION: Circumstantial evidence is usually considered to be the weakest form of evidence. It tends to prove a primary fact by proving other intermediate events or circumstances that provide a basis for a reasonable inference that the primary fact occurred. Hence, the proof is indirect.

Answer (A) is incorrect because direct evidence is proof without presumption or inference. Answer (C) is incorrect because corroborative evidence is additional evidence of a different nature from the evidence it supplements. Answer (D) is incorrect because conclusive evidence is, by definition, incontrovertible.

23. To determine the reasonableness of financial data, auditors use analytical reviews. Which one of the following is least likely to be considered an analytical reasonableness review?

A. Trend analysis.

B. Physical inventories.

C. Comparison with budgets and forecasts.

D. Ratio analyses.

The correct answer is (B). *(CMA, adapted)*

REQUIRED: The item least likely to be considered an analytical reasonableness review.

DISCUSSION: Analytical auditing procedures are performed by study and comparison of plausible relationships among both financial and nonfinancial data. The premise is that, absent known contrary circumstances, certain relationships among information may reasonably be expected to continue. The result of analytical procedures is an assessment of information collected in an audit in relation to expectations developed by the auditor. Thus, a physical inventory is not a form of analytical procedure because it does not involve predictable relationships among information. Instead, it is a form of direct, observational evidence.

Answer (A) is incorrect because trend or time series analysis uses past experience as a predictor. Answer (C) is incorrect because comparing actual with budgeted amounts may indicate the need for further investigation. Answer (D) is incorrect because ratio analyses are an analytical means of observing relationships among amounts.

24. Audit information is usually considered relevant when it is

A. Derived through valid statistical sampling.

B. Objective and unbiased.

C. Factual, adequate, and convincing.

D. Consistent with the audit objectives.

The correct answer is (D). *(CIA, adapted)*

REQUIRED: The circumstance in which audit information is usually considered relevant.

DISCUSSION: "Information should be sufficient, competent, relevant, and useful to provide a sound basis for audit findings and recommendations. Relevant information supports audit findings and recommendations and is consistent with the objectives for the audit" (Standard 420).

Answer (A) is incorrect because whether sampling is appropriate and the results are valid are issues related to the determination of sufficiency and competence rather than relevance. Answer (B) is incorrect because objectivity and lack of bias do not assure that information will support audit findings and recommendations and be consistent with the audit objectives. Answer (C) is incorrect because it defines evidence sufficient so that a prudent, informed person would reach the same conclusion as the auditor.

25. What standard of evidence is satisfied by an original signed document?

A. Sufficiency.

B. Competence.

C. Relevance.

D. Usefulness.

The correct answer is (B). *(CIA, adapted)*

REQUIRED: The standard of evidence satisfied by an original signed document.

DISCUSSION: Competent evidence is reliable and the best available through the application of appropriate audit procedures. An original document is the prime example of such evidence.

Answer (A) is incorrect because sufficient evidence is factual, adequate, and convincing. The information contained on the document may be none of those things. Answer (C) is incorrect because relevancy concerns the relationship of the evidence to some objective of the audit. No audit objective is disclosed in the question. Thus, whether the information on the document is relevant to the investigation cannot be determined. Answer (D) is incorrect because usefulness is achieved if the item of evidence helps the organization (the auditor, in this case) to accomplish predetermined goals. No such goals are specified.

Questions 26 through 28 are based on the following information. The director of internal auditing is reviewing some of the basic concepts inherent in the performance of an audit with three auditors who are on a rotation assignment. After 6 months in the department, they will move back to line positions. Each of them has fairly extensive organizational experience and is on a fast track to a high-level management line position. To develop their analytical decision-making abilities, the director pulls some old audit working papers, holding back the review notes and clearing comments. The director asks the team to indicate the evidential criteria that are violated.

26. The organization is required to comply with certain specific standards related to environmental issues. One of these standards requires that certain hazardous chemicals be placed in certified containers for shipment to a federal disposal site. The container must bear an inspection seal signed within the last 90 days by a federal inspector. Based on the following tests, the auditor concluded that the company was in compliance for the audit period:

1. Determine from each chemical loading supervisor that compliance requirements are understood.
2. Inspect sealed containers for evidence of leakage.
3. Ask personnel about procedures performed.

Identify which of the following evidential criteria is(are) violated.

A. Sufficiency.

B. Competency.

C. Relevance.

D. No criteria are violated.

The correct answer is (A). *(CIA, adapted)*

REQUIRED: The evidential standard(s) violated, if any.

DISCUSSION: Sufficient information is defined as factual, adequate, and convincing so that a prudent, informed person would reach the same conclusions as the auditor. These tests are insufficient because the auditor did not determine that each container had an inspection seal signed within the last 90 days.

Answer (B) is incorrect because the information is competent. It is reliable and the best attainable through the use of appropriate audit techniques. Answer (C) is incorrect because the information is relevant. It supports audit findings and recommendations and is consistent with the objectives for the audit. Answer (D) is incorrect because the sufficiency criterion was violated.

27. During the planning stage of an audit, the auditor made an on-site observation of the vehicle maintenance department and included the following statement in a memorandum summary of the results:

> We noted that several maintenance garages were deteriorating badly. Fencing around the property was in need of repair.

Identify which of the following evidential criteria is(are) violated.

A. Sufficiency.

B. Competency.

C. Relevance.

D. No criteria are violated.

The correct answer is (D). *(CIA, adapted)*

REQUIRED: The evidential standard(s) violated, if any.

DISCUSSION: The observations made about the vehicle maintenance department contain sufficient information (factual, adequate, and convincing so that a prudent, informed person would reach the same conclusions) that is competent (reliable and the best attainable through the use of appropriate audit techniques) and relevant (supports audit findings and recommendations and is consistent with the objectives for the audit).

Answer (A) is incorrect because the sufficiency criterion has not been violated. Physical observation by the auditor is sufficient to determine deterioration and need for repairs. Answer (B) is incorrect because the competency criterion has not been violated. On-site observation is an appropriate technique to determine deterioration and needed repairs. Answer (C) is incorrect because the relevance criterion has not been violated. The evidence obtained by the auditor supports findings about the physical condition of the department.

28. In an audit of the effectiveness and validity of a subsidiary's marketing expenditures, the auditor's evidence consists of

1. Analytical comparisons of advertising expenditures and changes in shopping patterns and item sales
2. Direct observation of various advertising media used
3. Review of marketing survey of general public reaction to the marketing plan

Identify which of the following evidential criteria is(are) violated.

A. Sufficiency.

B. Competency.

C. Relevance.

D. No criteria are violated.

The correct answer is (D). *(CIA, adapted)*

REQUIRED: The evidential standard(s) violated, if any.

DISCUSSION: The audit evidence contains sufficient information (factual, adequate, and convincing so that a prudent, informed person would reach the same conclusions) that is competent (reliable and the best attainable through the use of appropriate audit techniques) and relevant (supports audit findings and recommendations and is consistent with the objectives for the audit).

Answer (A) is incorrect because the sufficiency criterion has not been violated. The analytical comparison, direct observation, and review of the market survey provide sufficient evidence of the effectiveness and validity of expenditures. Answer (B) is incorrect because the competency criterion has not been violated. Analysis, observation, and review by the auditor are all methods of obtaining competent, reliable evidence. Answer (C) is incorrect because the relevance criterion has not been violated. The analytical comparisons, direct observations, and review of the marketing survey are all types of evidence relevant to the evaluation of the marketing expenditures.

F. Operational Auditing

29. An example of the subject of an operational audit is

A. The income tax return information of a manufacturer.

B. The performance statistics on the delivery of a city's services.

C. The verification of the dollar amount of royalties due to the developer of a manufacturing process from the user of that process.

D. The 5-year revenue and expenses forecast by an entrepreneur seeking to raise venture capital for his prospective operation.

The correct answer is (B). *(CMA, adapted)*

REQUIRED: The subject of an operational audit.

DISCUSSION: An operational audit is designed to evaluate the efficiency, effectiveness, and economy of managerial organization, performance, and techniques. The only answer choice that would fall into these categories is performance statistics (effectiveness) on the delivery of services.

Answers (A) and (C) are incorrect because each is a typical subject of a financial audit. Answer (D) is incorrect because it concerns prospective financial information. An operational auditor would evaluate such forecasts only after the 5-year period had ended.

30. Operational audits are designed to

A. Produce an opinion on the fairness of the firm's financial statements.

B. Produce an opinion on the accuracy of a firm's financial accounting system.

C. Produce recommendations for improving the accuracy of a firm's financial accounting system.

D. Review performance of an organization or some portion of an organization (e.g., department, function, etc.) using some preestablished standard as the primary evaluation criterion.

The correct answer is (D). *(CMA, adapted)*

REQUIRED: The purpose of operational audits.

DISCUSSION: Operational audits are nonfinancial audits designed to evaluate management efficiency, effectiveness, and economy. Performance within an organization or department is reviewed and recommendations are made for improvements. In any audit, however, standards must exist against which the auditor compares the auditee's performance. These standards may consist of budgets, industry averages, policies, procedures manuals, or common business sense.

Answers (A) and (B) are incorrect because a financial audit, not an operational audit, results in an opinion on financial statements. However, the accounting system may be the subject of an operational audit examination and report, usually by internal auditors. Answer (C) is incorrect because an operational audit is much broader than an evaluation of accounting systems. It embraces administrative as well as accounting controls.

31. Which one of the following items is included in an operational audit but is not required in a financial audit conducted by an external auditor?

A. Planning and control over the work done by an audit team.

B. Supervision of the audit team's activities and output.

C. Fact-finding, analysis, and documentation.

D. Recommendations for improvement.

The correct answer is (D). *(CMA, adapted)*

REQUIRED: The item included in an operational audit but not in a financial audit conducted by an external auditor.

DISCUSSION: An operational audit report includes a statement of findings. If a finding is unfavorable, the report should include recommendations for improvement of the condition. Such is not a requirement of financial audits.

Answers (A), (B), and (C) are incorrect because each is as true of an audit in accordance with GAAS as of an operational audit.

32. In conducting an operational audit, which one of the following activities would not be expected of the internal auditor?

A. Make an objective observation and comprehensive analysis of specific activities.

B. Observe performance of personnel.

C. Assess performance as compared with established policies.

D. Perform the operational activity of the line personnel.

The correct answer is (D). *(CMA, adapted)*

REQUIRED: The activity not expected of an internal auditor in an operational audit.

DISCUSSION: IIA Standard 120 concerns the objectivity of internal auditors. It states that internal auditors should not assume operating responsibilities.

Answer (A) is incorrect because internal auditors are expected not only to be objective but also to collect, analyze, interpret, and document information to support audit results (Standard 420). Answers (B) and (C) are incorrect because internal auditors should ascertain whether results are consistent with established goals and objectives (Standard 350). Observation is a necessary audit procedure for achieving that objective.

G. Internal Audit Reports

33. Which of the following is not a major purpose of an audit report?

A. Inform.

B. Get results.

C. Assign responsibility.

D. Persuade.

The correct answer is (C). *(CIA, adapted)*

REQUIRED: The item that is not a major purpose of an audit report.

DISCUSSION: According to *Sawyer's Internal Auditing* (p. 611), audit reports are intended to inform, persuade, and get results. They explain the auditors' findings, attempt to convince the recipients of the report of the value and validity of those findings, and attempt to foster beneficial change.

Answer (A) is incorrect because informing the audit committee and senior management is a major purpose of an audit report. Answer (B) is incorrect because getting results is a major purpose of an audit report. Answer (D) is incorrect because persuading the audit committee and senior management that certain conditions exist is a major purpose of an audit report.

34. The internal auditing department for a chain of retail stores recently concluded an audit of sales adjustments in all stores in the southeast region. The audit revealed that several stores are costing the company an estimated $85,000 per quarter in duplicate credits to customers' charge accounts. The audit report, published 8 weeks after the audit was concluded, included the internal auditors' recommendations to store management that should prevent duplicate credits to customers' accounts. Which of the following standards for reporting has been disregarded in the above case?

A. The follow-up actions were not adequate.

B. The auditors should have implemented appropriate corrective action as soon as the duplicate credits were discovered.

C. Auditor recommendations should not be included in the report.

D. The report was not timely.

The correct answer is (D). *(CIA, adapted)*

REQUIRED: The reporting standard that was disregarded.

DISCUSSION: According to Standard 430, "Reports should be objective, clear, concise, constructive, and timely." SIAS 2 adds, "Timely reports are those that are issued without delay and enable prompt effective action." The report, which was not published until 8 weeks after the audit was concluded, was not issued in a timely fashion, given the significance of the findings and the need for prompt, effective action.

Answer (A) is incorrect because there is not enough information to evaluate the effectiveness of follow-up. Answer (B) is incorrect because auditors may properly make recommendations for potential improvements but should not implement corrective action. Answer (C) is incorrect because auditor recommendations are an element of an audit finding.

Questions 35 through 37 are based on the following information. An auditor has submitted a first draft of an audit report to an auditee in preparation for an exit interview. The following is an excerpt from that report:

The audit was performed to accomplish several objectives:

- Verify the existence of unused machinery being stored in the warehouse.
- Determine whether machinery had been damaged during storage.
- Review the handling procedures being performed by personnel at the warehouse.
- Determine whether proper accounting procedures are being followed for machinery kept in the warehouse.
- Calculate the current fair market value of warehouse inventories.
- Compare the total value of the machinery to company accounting records.

It was confirmed that, of the 30 machines selected from purchasing records for the sample, 13 were present on the warehouse floor and another five were on the loading dock ready for conveyance to the production facility. Twelve others had already been sent to the production facility at a previous time. An examination of the accounting procedures used at the warehouse revealed the failure by the warehouse accounting clerk to reconcile inventory records monthly, as required by policy. A sample of 25 machines was examined for possible damage, and all but one was in good condition. It was confirmed by the auditors that handling procedures outlined in the warehouse policy manual appear to be adequate, and warehouse personnel apparently were following those procedures, except for the examination of items being received for inventory.

35. When an auditor is communicating with auditees, both situational factors and message characteristics can damage the communication process. An auditor has only limited control over situational factors but has substantial control over message characteristics. Which of the following would seem to be a message characteristic that the auditor who prepared the above report overlooked?

A. Sequence of message.

B. Nature of the audience.

C. Noise.

D. Prior encounters with the auditee.

The correct answer is (A). *(CIA, adapted)*

REQUIRED: The message characteristic the auditor overlooked in preparing the audit report.

DISCUSSION: The auditor neglected to organize the information. Because the information being communicated is complicated, the report's content should be organized in a logical sequence to facilitate understanding and acceptance. For this reason, standard formats are often used in business communications.

Answer (B) is incorrect because the nature of an audience is a situational factor that is outside the control of the auditor. Answer (C) is incorrect because noise is a situational factor that interferes with the effective communication of intended messages. Answer (D) is incorrect because the history of previous encounters is a situational factor that is outside the control of the auditor.

36. The objectives of an audit report are to inform and to influence. Whether these objectives are met depends on the clarity of the writing. Which of the following principles of report clarity was violated in the above audit report?

A. Appropriately organize the report.

B. Keep most sentences short and simple.

C. Use active voice verbs.

D. All of the answers are correct.

The correct answer is (D). *(CIA, adapted)*

REQUIRED: The principle(s) of report clarity violated in the audit report.

DISCUSSION: The report should be well-organized so that the information is given appropriate attention. Also, effective organization enhances understanding by presenting information in an logical order that clarifies the auditor's reasoning. Keeping sentences as short and simple as possible likewise facilitates understanding. Also, active voice verbs are more vivid and concise than passive voice verbs.

Answers (A), (B), and (C) are incorrect because an audit report should be appropriately organized, be concise, and use active voice verbs.

37. The following elements are usually included in final audit reports: purpose, scope, results, conclusions, and recommendations. Which of the following describes all of the elements missing from the above report?

A. Scope, conclusion, recommendation.

B. Purpose, result, recommendation.

C. Result, conclusion, recommendation.

D. Purpose, scope, recommendation.

The correct answer is (A). *(CIA, adapted)*

REQUIRED: The elements missing from the audit report.

DISCUSSION: Although a portion of the scope is discussed, the reader cannot determine the significance of the amount of machines selected without knowing the total amount of machines available and the value of the machinery. Also, the conclusion or auditor's opinion of the operation is not stated, and the report makes no recommendations.

Answers (B), (C), and (D) are incorrect because the purpose of the audit was clearly stated, and the result of the audit was given.

38. In operational audits when fraud is not an issue, the results of the operational audit are ideally exposed initially to

A. The manager in charge of the subject department or function.

B. The supervisor of the manager in charge of the subject department or function.

C. The chief executive officer of the corporation.

D. The divisional controller or corporate controller of the subject department or function.

The correct answer is (A). *(CMA, adapted)*

REQUIRED: The person(s) to whom operational audit results are initially exposed.

DISCUSSION: When fraud is not involved, the initial draft of an operational audit report should be exposed to the manager in charge of the department being audited during what is known as an exit interview. This gives the auditor an opportunity to check his/her findings with the department head before submitting the report to higher management. If the auditor has made a mistake, the department head can rectify the error. If the audit report is accurate, early exposure permits prompt corrective action. Thus, both auditor and department head can benefit from the exit interview.

Answers (B), (C), and (D) are incorrect because higher levels of management should not see the report until it has been reviewed by the manager of the auditee.

39. Which of the following individuals would normally not receive an internal auditing report related to a review of the purchasing cycle?

A. The director of purchasing.

B. The independent external auditor.

C. The general auditor.

D. The chair of the board of directors.

The correct answer is (D). *(CIA, adapted)*

REQUIRED: The person who does not normally receive an internal auditing report related to a review of the purchasing cycle.

DISCUSSION: According to SIAS 2, "Audit reports should be distributed to those members of the organization who are able to ensure that audit results are given due consideration. This means that the report should go to those who are in a position to take corrective action or to ensure that corrective action is taken. The final audit report should be distributed to the head of each audited unit. Higher-level members in the organization may receive only a summary report. Reports may also be distributed to other interested or affected parties such as external auditors and audit committees." As interested or affected parties, the external auditors and the director of purchasing are proper recipients of the report. The board chair would not normally receive a copy. A detailed report, especially one with routine findings, is not usually sent to the board chair.

Answer (A) is incorrect because the director of purchasing should receive a copy. Answer (B) is incorrect because the external auditor should receive a copy. Answer (C) is incorrect because the general auditor should receive a copy.

40. Several levels of management are interested in the results of the marketing department audit. What is the best method of communicating the results of the audit?

A. Write detailed reports for each level of management.

B. Write a report to the marketing management and give summary reports to other management levels.

C. Discuss results with marketing management and issue a summary report to top management.

D. Discuss results with all levels of management.

The correct answer is (B). *(CIA, adapted)*

REQUIRED: The best method of communicating the results of an audit.

DISCUSSION: A written report should be issued after completion of an audit. The report should be addressed to the level of management capable of acting on deficiencies noted in the report. Top management should be aware of internal auditing's activities and any major deficiencies noted. This purpose can be accomplished in an oral or summary report.

Answer (A) is incorrect because each level of management does not need a detailed report. Answer (C) is incorrect because a formal, detailed written report should be addressed to marketing management if that is the level of management able to act on the deficiencies. Answer (D) is incorrect because conclusions and recommendations should be discussed with the appropriate levels of management, but an audit report should still be issued.

H. Auditor Follow-Up

41. The internal auditor should follow up to ascertain that appropriate action is taken on deficiency findings. To accomplish this, the internal auditor should

A. Work closely with the external auditor.

B. Be guided by the wishes of the audit committee.

C. Limit internal audit follow-up to receiving written confirmation from the auditee that appropriate corrective action has been taken.

D. Make any field tests needed to provide assurance that the condition has been corrected.

The correct answer is (D). *(CMA, adapted)*

REQUIRED: The internal audit procedure to ascertain that appropriate action is taken on deficiency findings.

DISCUSSION: The internal auditor is obligated to determine that corrective action is taken and is achieving the desired results or that management has explicitly assumed the risk of not taking corrective action with regard to deficiency findings. Field tests may be needed to obtain adequate assurance.

Answer (A) is incorrect because following up on deficiency findings is an internal auditor's responsibility (only internal auditors issue deficiency findings). Answer (B) is incorrect because, according to the Standards for the Professional Practice of Internal Auditing, the internal auditor must follow up deficiency findings. Answer (C) is incorrect because field testing, not mere auditee confirmation, is required by The IAA Standards to assure that action was taken and the desired results are being achieved.

42. Why should organizations require auditees to promptly reply and outline the corrective action that has been implemented on reported deficiencies?

A. To remove items from the pending list as soon as possible.

B. To effect savings or to institute compliance as early as possible.

C. To indicate concurrence with the audit findings.

D. To ensure that the audit schedule is kept up to date.

The correct answer is (B). *(CIA, adapted)*

REQUIRED: The reason auditees should promptly reply and outline the corrective action that has been implemented on reported deficiencies.

DISCUSSION: Reports should be timely to enable prompt corrective action, and reports should be distributed to those in a position to take corrective action or to ensure that corrective action is taken (SIAS 2). Moreover, Standard 440 requires internal auditors to follow up to ascertain that appropriate action is taken on deficiency findings. The internal auditor should determine that corrective action being taken has the desired results or that management or the board has assumed the risk of not taking corrective action. Consequently, it follows that the objectives of audits and the timely reporting of findings would be defeated if auditees do not promptly implement and report on corrective action.

Answer (A) is incorrect because removing items from the pending list concerns a mechanical and immaterial aspect of the reporting process. Answer (C) is incorrect because the auditee may not concur with the finding. This dispute may or may not be considered in closing the audit. Answer (D) is incorrect because ensuring that the audit schedule is kept up to date is an administrative function of the audit organization.

43. The preliminary survey discloses that a prior audit deficiency was never corrected. Subsequent field work confirms that the deficiency still exists. Which of the following courses of action should the internal auditor pursue?

A. Take no action. To do otherwise would be an exercise of operational control.

B. Discuss the issue with the director of internal auditing. The problem requires an ad hoc solution.

C. Discuss the issue with the person(s) responsible for the problem. (S)he or they should know how to solve the problem.

D. Order the person(s) responsible to correct the problem. (S)he or they have had long enough to do so.

The correct answer is (C). *(CIA, adapted)*

REQUIRED: The auditor's course of action indicated when a prior audit deficiency was never corrected.

DISCUSSION: Standard 440 states, "Internal auditors should follow up to ascertain that appropriate action is taken on reported audit findings. Internal auditors should determine that corrective action was taken and is achieving the desired results, or that management or the board has assumed the risk of not taking corrective action on reported findings." Also, Standard 430 requires discussion of conclusions and recommendations at appropriate levels of management before issuing final reports. Auditee management is at "an appropriate" level. Obtaining auditee cooperation (or at least understanding) is a vital part of the solution of any problem.

Answer (A) is incorrect because a deficiency finding places the firm at risk until the situation changes or the deficiency is corrected. Answer (B) is incorrect because deficiency findings that have not been corrected are not unique and do not require ad hoc solutions. Answer (D) is incorrect because the internal auditor has no line authority over the auditee. To exercise such authority impairs the internal auditor's objectivity.

44. If an internal auditor finds that no corrective action has been taken on a prior audit finding that is still valid, the Standards state that the internal auditor should

A. Restate the prior finding along with the findings of the current audit.

B. Determine whether management or the board has assumed the risk of not taking corrective action.

C. Seek the board's approval to initiate corrective action.

D. Schedule a future audit of the specific area involved.

The correct answer is (B). *(CIA, adapted)*

REQUIRED: The auditor's action when no corrective action has been taken on a prior audit finding that is still valid.

DISCUSSION: Standard 430 states that reports may make recommendations for potential improvements. Also, Standard 440 states, "Internal auditors should follow up to ascertain that appropriate action is taken on reported audit findings. Internal auditors should determine that corrective action was taken and is achieving the desired results, or that management or the board has assumed the risk of not taking corrective action on reported findings."

Answer (A) is incorrect because reporting the matter is unnecessary if management or the board has assumed the risk of inaction. Answer (C) is incorrect because the internal auditor should not assume the operating responsibility of undertaking corrective action. Answer (D) is incorrect because a future audit of the specific area may not be needed if management or the board has assumed the risk of inaction.

45. Upon reviewing the results of the audit report with the audit committee, executive management agreed to accept the risk of not implementing corrective action on certain audit findings. Evaluate the following and select the best alternative for the internal auditing director.

A. Notify regulatory authorities of management's decision.

B. Perform additional audit steps to further identify the policy violations.

C. Conduct a follow-up audit to determine whether corrective action was taken.

D. Determine that internal audit responsibility has been discharged and no further audit action is required.

The correct answer is (D). *(CIA, adapted)*

REQUIRED: The best choice when management agrees to accept the risk of not implementing corrective action.

DISCUSSION: Standard 440 states, "Internal auditors should follow up to ascertain that appropriate action is taken on reported audit findings. Internal auditors should determine that corrective action was taken and is achieving the desired results, or that management or the board has assumed the risk of not taking corrective action on reported findings."

Answer (A) is incorrect because regulatory authorities do not need to be notified. Management has agreed to accept responsibility and no regulatory violations were mentioned. Answers (B) and (C) are incorrect because no further audit action is required.

APPENDIX A: ETHICS AS TESTED ON THE CMA/CFM EXAMS

6 pages of outline
17 multiple-choice questions

A. Ethics
B. Codes of Ethical Conduct
C. IMA Code of Ethics
D. Conflict of Interest
E. Legal Aspects of Social Responsibility

ETHICS

Questions containing ethical issues can appear on any part of the examination, presented within the context of specific subject areas. Candidates should be familiar with:

> *Statement on Management Accounting Number 1C (Revised)*, "Standards of Ethical Conduct for Practitioners of Management Accounting and Financial Management," *Institute of Management Accountants, Montvale, N.J., 1997.*

Current references to business ethics are also found in recent periodicals and newspapers.

A. Ethics

1. Definitions

a. Corporate ethics -- an organization's policies and standards established to assure certain kinds of behavior by its members

b. Individual ethics -- principles of conduct adhered to by an individual

2. Increased Concern for Business Ethics

a. Electrical-equipment conspiracy cases in 1960 caused public concern and creation of the Business Ethics Advisory Council (BEAC) in 1961 under the Secretary of Commerce.

b. BEAC pointed out areas needing self-evaluation by the business community:

1) General business understanding of ethical issues

2) Compliance with laws

3) Conflicts of interest

4) Entertainment and gift expenses

5) Relations with customers and suppliers. Should gifts or kickbacks be given or accepted?

6) Social responsibilities

c. BEAC's recommendations generated business interest, especially from big business, in problems of ethical behavior.

3. **Factors That May Lead to Unethical Behavior**
    a. In any normal population, some people have less than desirable levels of ethics. If these people hold leadership positions, they will adversely influence subordinates.
    b. Organizational factors may lead to unethical behavior.
        1) Pressures for short-run performance in decentralized return on investment (ROI) centers may inhibit ethical behavior.
        2) Emphasis on strict adherence to chain-of-command authority may provide excuses for ignoring ethics when following orders.
        3) Informal work-group loyalties may subvert ethical behavior.
        4) Committee decision processes may make it possible to abstain from or dodge ethical obligations.
    c. External factors may lead to unethical behavior.
        1) Pressure of competition may compromise ethics in the interest of survival.
        2) Unethical behavior of others may force a compromise of ethics.
        3) Definitions of ethical behavior may vary from one culture to another.
            a) Bribes to overseas officials or buyers may be consistent with some countries' customary business practices, but such a practice is not considered ethical among U.S. purchasing agents.
                i) Bribes are now considered illegal under the Foreign Corrupt Practices Act.
            b) The propriety of superimposing our cultural ethical standards (by refusing to bribe) on another culture may be controversial.
4. **General Guides to Ethics**
    a. Golden Rule -- Do unto others as you would have others do unto you.
    b. Maximize good -- Act to provide the greatest good for the greatest number.
    c. Fairness -- Act in ways that are fair or just to all concerned.
    d. Maximize long-run outcomes -- Act to provide the best long-range benefits to society and its resources.
    e. General respect -- Act to respect the planet all humans share and the rights of others because corporate and individual decisions affect them.
5. **Simplified Criteria for Evaluating Ethical Behavior**
    a. Would this behavior be acceptable if people I respect knew I was doing this?
    b. What are the consequences of this behavior for myself, other employees, customers, and society?
6. Ethics are individual and personal, influenced by
    a. Life experiences (rewards for doing right, punishment for doing wrong)
    b. Friendship groups (professional associations, informal groups)
    c. Organizational pressures (responsibilities to superiors and the organization)

B. Codes of Ethical Conduct

1. An organization's code of ethical conduct is the established general value system the organization wishes to apply to its members' activities through
   a. Communicating organizational purposes and beliefs
   b. Establishing uniform ethical guidelines for members
      1) Including guidance on behavior for members in making decisions
2. Laws and written rules cannot cover all situations. However, organizations can benefit from having an established ethical code because it
   a. Effectively communicates acceptable values to all members
      1) Including recruits and subcontractors
   b. Provides a method of policing and disciplining members for violations
      1) Through review panels (formal)
      2) Through group pressure (informal)
   c. Establishes high standards against which individuals can measure their own performance
   d. Communicates to those outside the organization the value system from which the organization's members must not be asked to deviate
3. A typical code for accounting activities (note similarities to the Standards for the Professional Practice of Internal Auditing, GAAP, GAAS, etc.) holds that a financial manager must have
   a. Independence from conflicts of economic interest
   b. Independence from conflicts of professional interest
      1) Responsibility to present information fairly to shareholders/owners and not intentionally protect management
      2) Responsibility to present data to all appropriate managers and not play favorites with information or cover up bad news
      3) Responsibility to exercise an ethical presence in the conduct of professional activities
         a) Ensuring organizational compliance with spirit as well as letter of pertinent laws and regulations
         b) Conducting oneself according to the highest moral and legal standards
         c) Reporting to appropriate internal or external authority any illegal or fraudulent organizational act
   c. Integrity in not compromising professional values for the sake of personal goals
   d. Objectivity in presenting information, preparing reports, and making analyses

C. IMA Code of Ethics

1. The National Association of Accountants (now the Institute of Management Accountants or IMA), through its Management Accounting Practices Committee, issued a revised code of ethics for management accountants in April 1997. This code reflects the official position of the organization. Candidates are urged to study the provisions of the code closely because it is tested. The code is printed below and on the following page in its entirety. (Source: Statement on Management Accounting 1C (Revised), Objectives: Standards of Ethical Conduct for Practitioners of Management Accounting and Financial Management, April 1997, pp. 69-70). The mnemonic CCIO (competence, confidentiality, integrity, and objectivity) is useful. The final section, Resolution of Ethical Conflict, is especially significant.

STANDARDS OF ETHICAL CONDUCT FOR PRACTITIONERS OF MANAGEMENT ACCOUNTING AND FINANCIAL MANAGEMENT

Practitioners of management accounting and financial management have an obligation to the public, their profession, the organizations they serve, and themselves, to maintain the highest standards of ethical conduct. In recognition of this obligation, the Institute of Management Accountants has promulgated the following standards of ethical conduct for practitioners of management accounting and financial management. Adherence to these standards, both domestically and internationally, is integral to achieving the Objectives of Management Accounting. Practitioners of management accounting and financial management shall not commit acts contrary to these standards nor shall they condone the commission of such acts by others within their organizations.

Competence

Practitioners of management accounting and financial management have a responsibility to:

- *Maintain an appropriate level of professional competence by ongoing development of their knowledge and skills.*
- *Perform their professional duties in accordance with relevant laws, regulations, and technical standards.*
- *Prepare complete and clear reports and recommendations after appropriate analyses of relevant and reliable information.*

Confidentiality

Practitioners of management accounting and financial management have a responsibility to:

- *Refrain from disclosing confidential information acquired in the course of their work except when authorized, unless legally obligated to do so.*
- *Inform subordinates as appropriate regarding the confidentiality of information acquired in the course of their work and monitor their activities to assure the maintenance of that confidentiality.*
- *Refrain from using or appearing to use confidential information acquired in the course of their work for unethical or illegal advantage either personally or through third parties.*

Integrity

Practitioners of management accounting and financial management have a responsibility to:

- *Avoid actual or apparent conflicts of interest and advise all appropriate parties of any potential conflict.*
- *Refrain from engaging in any activity that would prejudice their ability to carry out their duties ethically.*
- *Refuse any gift, favor, or hospitality that would influence or would appear to influence their actions.*
- *Refrain from either actively or passively subverting the attainment of the organization's legitimate and ethical objectives.*
- *Recognize and communicate professional limitations or other constraints that would preclude responsible judgment or successful performance of an activity.*
- *Communicate unfavorable as well as favorable information and professional judgments or opinions.*
- *Refrain from engaging in or supporting any activity that would discredit the profession.*

Objectivity

Practitioners of management accounting and financial management have a responsibility to:

- *Communicate information fairly and objectively.*
- *Disclose fully all relevant information that could reasonably be expected to influence an intended user's understanding of the reports, comments, and recommendations presented.*

Resolution of Ethical Conflict

In applying the standards of ethical conduct, practitioners of management accounting and financial management may encounter problems in identifying unethical behavior or in resolving an ethical conflict. When faced with significant ethical issues, practitioners of management accounting and financial management should follow the established policies of the organization bearing on the resolution of such conflict. If these policies do not resolve the ethical conflict, such practitioners should consider the following courses of action:

- *Discuss such problems with the immediate superior except when it appears that the superior is involved, in which case the problem should be presented initially to the next higher managerial level. If satisfactory resolution cannot be achieved when the problem is initially presented, submit the issues to the next higher managerial level. If the immediate superior is the chief executive officer, or equivalent, the acceptable reviewing authority may be a group such as the audit committee, executive committee, board of directors, board of trustees, or owners. Contact with levels above the immediate superior should be initiated only with the superior's knowledge, assuming the superior is not involved. Except where legally prescribed, communication of such problems to authorities or individuals not employed or engaged by the organization is not considered appropriate.*
- *Clarify relevant ethical issues by confidential discussion with an objective advisor (e.g., IMA Ethics Counseling Service) to obtain a better understanding of possible courses of action.*
- *Consult your own attorney as to legal obligations and rights concerning the ethical conflict.*
- *If the ethical conflict still exists after exhausting all levels of internal review, there may be no other recourse on significant matters than to resign from the organization and to submit an informative memorandum to an appropriate representative of the organization. After resignation, depending on the nature of the ethical conflict, it may also be appropriate to notify other parties.*

D. Conflict of Interest

1. Conflict of interest is a conflict between the private and the official responsibilities of a person in a position of trust, sufficient to affect judgment, independence, or objectivity in conducting the affairs of the business.

2. **Examples of Conflict of Interest**
   a. Having a substantial financial interest in a supplier, customer, or distributor
   b. Using privileged information gained from one's official position to enter transactions for personal gain

3. **Methods for Control**
   a. Provide a code of conduct provision applying to conflicts of interest.
   b. Require full financial disclosure by managers.
   c. Require prior notification of any transaction that may raise conflict of interest.
   d. Prohibit financial ties to any supplier, customer, or distributor.
   e. Encourage adherence to strong ethical behavior through corporate actions, policies, and public communications.

E. Legal Aspects of Social Responsibility

1. The **Racketeer Influenced and Corrupt Organization (RICO) Act** was passed in 1970 as an attempt to combat the problem of organized crime and its infiltration of legitimate enterprises.
   a. Its goals were to eliminate organized crime by concentrating on the illegal monies through the use of civil and criminal forfeitures.
   b. Criminal penalties can be levied up to $25,000 and 20 years in jail. Civil penalties include the awarding of treble damages and attorney's fees to the successful plaintiff.
   c. RICO specifically makes the following activities unlawful:
      1) Using income derived from a pattern of racketeering activity to acquire an interest in an enterprise.
      2) Acquiring or maintaining an interest in an enterprise through a pattern of racketeering activity.
      3) Conducting the affairs of an enterprise through a pattern of racketeering activity.
      4) Conspiring to commit any of these offenses.
   d. RICO has been used against white-collar criminals, terrorists, Wall Street insider trading, anti-abortion protesters, local law enforcement agencies, and public accounting firms -- none of which was intended by Congress when the law was passed.
2. The **Foreign Corrupt Practices Act (FCPA) of 1977** regulates payments by U.S. firms operating in other nations.
   a. The act is a reaction to publicity over questionable foreign payments.
   b. The FCPA makes it a criminal offense to make payments to a foreign government or representative thereof to secure or retain business.
   c. It prohibits payments of sales commissions to independent agents, if the commissions are knowingly passed to foreign officials.
   d. Corporations are required to establish internal accounting controls to assure that all overseas payments are proper.
   e. The FCPA applies even if payment is legal in the nation where it is made.
   f. The rationale for the FCPA is that the international reputation of the United States is affected by its international business conduct, which should reflect the best of the United States' ethics.
3. The SEC mandates that the composition of boards of directors include outside directors.
   a. To create diversity and broaden the overview of a company's place in the market and in society
4. Courts are increasingly willing to hold boards of directors and auditors liable for problems.
5. Stop and review! The following 17 multiple-choice questions cover the topic of ethics as it may be tested on the CMA/CFM exams. They are not divided by subunits. We anticipate some "case" questions that describe situations and are followed by one or more questions; e.g., to whom should the financial manager report an ethics violation?

MULTIPLE-CHOICE QUESTIONS

1. If a financial manager/management accountant has a problem in identifying unethical behavior or resolving an ethical conflict, the first action (s)he should normally take is to

A. Consult the board of directors.

B. Discuss the problem with his/her immediate superior.

C. Notify the appropriate law enforcement agency.

D. Resign from the company.

The correct answer is (B). *(Publisher)*

REQUIRED: The proper ethical behavior by a financial manager/management accountant.

DISCUSSION: The Standards of Ethical Conduct for Practitioners of Management Accounting and Financial Management state that the financial manager/management accountant should first discuss an ethical problem with his/her immediate superior. If the superior is involved, the problem should be taken initially to the next higher managerial level.

Answer (A) is incorrect because the board would be consulted initially only if the immediate superior is the chief executive officer and that person is involved in the ethical conflict. Answer (C) is incorrect because unless "legally prescribed, communication of such problems to authorities or individuals not employed or engaged by the organization is not considered appropriate." Answer (D) is incorrect because resignation is a last resort.

2. Sheila is a financial manager who has discovered that her company is violating environmental regulations. If her immediate superior is involved, her appropriate action is to

A. Do nothing since she has a duty of loyalty to the organization.

B. Consult the audit committee.

C. Present the matter to the next higher managerial level.

D. Confront her immediate superior.

The correct answer is (C). *(Publisher)*

REQUIRED: The proper action when a financial manager/management accountant's immediate superior is involved in an ethical problem.

DISCUSSION: To resolve an ethical problem, the financial manager/management accountant's first step is usually to consult his/her immediate superior. If that individual is involved, the matter should be taken to the next higher level of management.

Answer (A) is incorrect because "practitioners of management accounting and financial management have an obligation to the public, their profession, the organization they serve, and themselves, to maintain the highest standards of ethical conduct." Answer (B) is incorrect because the audit committee would be consulted first only if it were the next higher managerial level. Answer (D) is incorrect because if the superior is involved, the next higher managerial level should be consulted first.

3. If a financial manager/management accountant discovers unethical conduct in his/her organization and fails to act, (s)he will be in violation of which ethical standard(s)?

A. "Actively or passively subvert the attainment of the organization's legitimate and ethical objectives."

B. "Communicate unfavorable as well as favorable information."

C. "Condone the commission of such acts by others within their organizations."

D. All of the answers are correct.

The correct answer is (D). *(Publisher)*

REQUIRED: The ethical standard(s) violated by failure to disclose unethical behavior.

DISCUSSION: A financial manager/management accountant displays his/her competence and objectivity and maintains integrity by taking the appropriate action within the organization to resolve an ethical problem. Failure to act would condone wrongful acts, breach the duty to convey unfavorable as well as favorable information, undermine the organization's legitimate aims, discredit the profession, and violate the duty of objectivity owed to users of the subordinate's work product.

Answers (A), (B), and (C) are incorrect because each standard is violated by a financial manager/management accountant who fails to act upon discovering unethical conduct.

4. Corporate social responsibility is

A. Effectively enforced through the controls envisioned by classical economics.

B. Defined as the obligation to shareholders to earn a profit.

C. More than the obligation to shareholders to earn a profit.

D. Defined as the obligation to serve long-term, organizational interests.

The correct answer is (C). *(Publisher)*

REQUIRED: The true statement about corporate social responsibility.

DISCUSSION: The concept of corporate social responsibility involves more than serving the interests of the organization and its shareholders. Rather, it is an extension of responsibility to embrace service to the public interest in such matters as environmental protection, employee safety, civil rights, and community involvement.

Answer (A) is incorrect because a perfectly competitive market was envisioned by classical economics. Answers (B) and (D) are incorrect because the concept embraces the public or societal interest.

5. A common argument against corporate involvement in socially responsible behavior is that

A. It encourages government intrusion in decision making.

B. As a legal person, a corporation is accountable for its conduct.

C. It creates goodwill.

D. In a competitive market, such behavior incurs costs that place the company at a disadvantage.

The correct answer is (D). *(Publisher)*

REQUIRED: The common argument against corporate involvement in socially responsible behavior.

DISCUSSION: Socially responsible behavior clearly has immediate costs to the entity, for example, the expenses incurred in affirmative action programs, pollution control, and improvements in worker safety. When one firm incurs such costs and its competitor does not, the other may be able to sell its products or services more cheaply and increase its market share at the expense of the socially responsible firm. The rebuttal argument is that in the long run the socially responsible company may maximize profits by creating goodwill and avoiding or anticipating governmental regulation.

Answer (A) is incorrect because such behavior may prevent governmental action. Answers (B) and (C) are incorrect because each is an argument for such behavior.

6. The IMA Code of Ethics requires a financial manager/management accountant to follow the established policies of the organization when faced with an ethical conflict. If these policies do not resolve the conflict, the financial manager/management accountant should

A. Consult the board of directors immediately.

B. Discuss the problem with the immediate superior if (s)he is involved in the conflict.

C. Communicate the problem to authorities outside the organization.

D. Contact the next higher managerial level if initial presentation to the immediate superior does not resolve the conflict.

The correct answer is (D). *(Publisher)*

REQUIRED: The proper action when organizational policies do not resolve an ethical conflict.

DISCUSSION: In these circumstances, the problem should be discussed with the immediate superior unless (s)he is involved. In that case, initial presentation should be to the next higher managerial level. If the problem is not satisfactorily resolved after initial presentation, the question should be submitted to the next higher level.

Answer (A) is incorrect because this course of action would be appropriate only for the chief executive officer or for his/her immediate subordinate when the CEO is involved in the conflict. Answer (B) is incorrect because the proper action would be to present the matter to the next higher managerial level. Answer (C) is incorrect because such action is inappropriate unless legally prescribed.

7. Financial managers/management accountants are obligated to maintain the highest standards of ethical conduct. Accordingly, the IMA Code of Ethics explicitly requires that they

A. Obtain sufficient competent evidence when expressing an opinion.

B. Not condone violations by others.

C. Comply with generally accepted auditing standards.

D. Adhere to generally accepted accounting principles.

The correct answer is (B). *(Publisher)*

REQUIRED: The conduct required of financial managers/management accountants.

DISCUSSION: The preamble to the IMA Code of Ethics states, "Practitioners of management accounting and financial management have an obligation to the public, their profession, the organizations they serve, and themselves, to maintain the highest standards of ethical conduct. In recognition of this obligation, the Institute of Management Accountants has promulgated the following standards of ethical conduct for practitioners of management accounting and financial management. Adherence to these standards, both domestically and internationally, is integral to achieving the <u>Objectives of Management Accounting</u>. Practitioners of management accounting and financial management shall not commit acts contrary to these standards nor shall they condone the commission of such acts by others within their organizations."

Answers (A), (C), and (D) are incorrect because each applies to external auditors. The IMA Code of Ethics does not expressly use such language.

8. Integrity is an ethical requirement for all financial managers/management accountants. One aspect of integrity requires

A. Performance of professional duties in accordance with applicable laws.

B. Avoidance of conflict of interest.

C. Refraining from improper use of inside information.

D. Maintenance of an appropriate level of professional competence.

The correct answer is (B). *(Publisher)*

REQUIRED: The aspect of the integrity requirement.

DISCUSSION: According to the IMA Code of Ethics, financial managers/management accountants must "avoid actual or apparent conflicts of interest and advise all appropriate parties of any potential conflict."

Answers (A) and (D) are incorrect because each states an aspect of the competence requirement. Answer (C) is incorrect because it states an aspect of the confidentiality requirement.

9. Under the express terms of the IMA Code of Ethics, a financial manager/management accountant may not

A. Advertise.

B. Encroach on the practice of another financial manager/management accountant.

C. Disclose confidential information unless authorized or legally obligated.

D. Accept other employment while serving as a financial manager/management accountant.

The correct answer is (C). *(Publisher)*

REQUIRED: The action explicitly proscribed by the IMA Code of Ethics.

DISCUSSION: Financial managers/management accountants may not disclose confidential information acquired in the course of their work unless authorized or legally obligated to do so. They must inform subordinates about the confidentiality of information and monitor their activities to maintain that confidentiality. Moreover, financial managers/management accountants should avoid even the appearance of using confidential information to their unethical or illegal advantage.

Answers (A) and (B) are incorrect because the code does not address these matters. Answer (D) is incorrect because other employment may be accepted unless it constitutes a conflict of interest.

10. A financial manager/management accountant discovers a problem that could mislead users of the firm's financial data and has informed his/her immediate superior. (S)he should report the circumstances to the audit committee and/or the board of directors only if

A. The immediate superior, who reports to the chief executive officer, knows about the situation but refuses to correct it.

B. The immediate superior assures the financial manager/management accountant that the problem will be resolved.

C. The immediate superior reports the situation to his/her superior.

D. The immediate superior, the firm's chief executive officer, knows about the situation but refuses to correct it.

The correct answer is (D). *(Publisher)*

REQUIRED: The situation in which the financial manager/management accountant must report to the audit committee and/or board of directors.

DISCUSSION: According to the IMA Code of Ethics, the financial manager/management accountant should "discuss such problems with the immediate superior except when it appears that the superior is involved, in which case the problem should be presented initially to the next higher managerial level. If satisfactory resolution cannot be achieved when the problem is initially presented, submit the issues to the next higher managerial level. If the immediate superior is the chief executive officer, or equivalent, the acceptable reviewing authority may be a group such as the audit committee, executive committee, board of directors, board of trustees, or owners."

Answer (A) is incorrect because, in this situation, the chief executive officer is the next higher managerial level. Answers (B) and (C) are incorrect because the immediate superior has promised or taken action toward satisfactory resolution.

11. In which situation is a financial manager/management accountant permitted to communicate confidential information to individuals or authorities outside the firm?

A. There is an ethical conflict and the board has refused to take action.

B. Such communication is legally prescribed.

C. The financial manager/management accountant knowingly communicates the information indirectly through a subordinate.

D. An officer at the financial manager/management accountant's bank has requested information on a transaction that could influence the firm's stock price.

The correct answer is (B). *(Publisher)*

REQUIRED: The situation in which a financial manager/management accountant may disclose information to those outside the firm.

DISCUSSION: According to the IMA Code of Ethics, financial managers/management accountants are responsible for observing the standard of confidentiality. Thus, the financial manager/management accountant should "refrain from disclosing confidential information acquired in the course of his/her work except when authorized, unless legally obligated to do so."

Answer (A) is incorrect because the IMA Code of Ethics states that "except where legally prescribed, communication of such [ethical conflict] problems to authorities or individuals not employed or engaged by the organization is not considered appropriate." Answer (C) is incorrect because the financial manager/management accountant should "inform subordinates as appropriate regarding the confidentiality of information acquired in the course of their work and monitor their activities to assure the maintenance of that confidentiality." Answer (D) is incorrect because the financial manager/management accountant is required to "refrain from using or appearing to use confidential information acquired in the course of his/her work for unethical or illegal advantage either personally or through third parties."

12. Which ethical standard is most clearly violated if a financial manager/management accountant knows of a problem that could mislead users but does nothing about it?

A. Competence.

B. Legality.

C. Objectivity.

D. Confidentiality.

The correct answer is (C). *(Publisher)*

REQUIRED: The ethical standard most clearly violated when a financial manager/management accountant does nothing about information that is misleading to users.

DISCUSSION: Objectivity is the fourth part of the IMA Code of Ethics. It requires that information be communicated "fairly and objectively," and that all information that could reasonably influence users be fully disclosed.

Answer (A) is incorrect because the competence standard pertains to the financial manager/management accountant's responsibility to maintain his/her professional skills and knowledge. It also pertains to the performance of activities in a professional manner. Answer (B) is incorrect because legality is not addressed in the IMA Code of Ethics. Answer (D) is incorrect because the confidentiality standard concerns the financial manager/management accountant's responsibility not to disclose or use the firm's confidential information.

13. The IMA Code of Ethics includes an integrity standard, which requires the financial manager/management accountant to

A. Identify and make known anything that may hinder his/her judgment or prevent satisfactory completion of any duties.

B. Report any relevant information that could influence users of financial statements.

C. Disclose confidential information when authorized by his/her firm or required under the law.

D. Refuse gifts from anyone.

The correct answer is (A). *(Publisher)*

REQUIRED: The action required of the financial manager/management accountant by the integrity standard.

DISCUSSION: One of the responsibilities of the financial manager/management accountant under the integrity standard is to "recognize and communicate professional limitations or other constraints that would preclude responsible judgment or successful performance of an activity."

Answer (B) is incorrect because the objectivity standard requires the financial manager/management accountant to "disclose fully all relevant information that could reasonably be expected to influence an intended user's understanding of the reports, comments, and recommendations presented." Answer (C) is incorrect because the confidentiality standard requires the financial manager/management accountant to "refrain from disclosing confidential information acquired in the course of his/her work except when authorized, unless legally obligated to do so." Answer (D) is incorrect because the integrity standard requires the financial manager/management accountant to "refuse any gift, favor, or hospitality that would influence or would appear to influence his/her actions."

14. The IMA Code of Ethics includes a competence standard, which requires the financial manager/management accountant to

A. Report information, whether favorable or unfavorable.

B. Develop his/her professional proficiency on a continual basis.

C. Discuss ethical conflicts and possible courses of action with an unbiased counselor.

D. Discuss, with subordinates, their responsibilities regarding the disclosure of information about the firm.

The correct answer is (B). *(Publisher)*

REQUIRED: The action required of the financial manager/management accountant by the competence standard.

DISCUSSION: One of the responsibilities of the financial manager/management accountant under the competence standard is to "maintain an appropriate level of professional competence by ongoing development of his/her knowledge and skills."

Answer (A) is incorrect because the integrity standard requires the financial manager/management accountant to "communicate unfavorable as well as favorable information and professional judgments or opinions." Answer (C) is incorrect because one of the suggestions from the "Resolution of Ethical Conflict" paragraph is to "clarify relevant ethical issues by confidential discussion with an objective advisor (e.g., IMA Ethics Counseling Service) to obtain a better understanding of possible courses of action." Answer (D) is incorrect because the confidentiality standard requires the financial manager/management accountant to "inform subordinates as appropriate regarding the confidentiality of information acquired in the course of their work and monitor their activities to assure the maintenance of that confidentiality."

15. According to Statements on Management Accounting Number 1C (SMA 1C) (revised), *Standards of Ethical Conduct for Practitioners of Management Accounting and Financial Management*, a practitioner has a responsibility to recognize professional limitations. Under which standard of ethical conduct would this responsibility be included?

A. Competency.

B. Confidentiality.

C. Integrity.

D. Objectivity.

The correct answer is (C). *(CMA, adapted)*

REQUIRED: The standard of ethical conduct related to the responsibility to recognize professional limitations.

DISCUSSION: One of the responsibilities of the financial manager/management accountant under the integrity standard is to "recognize and communicate professional limitations or other constraints that would preclude responsible judgment or successful performance of an activity."

Answer (A) is incorrect because the competence standard pertains to the financial manager/management accountant's responsibility to maintain his/her professional skills and knowledge. It also pertains to the performance of activities in a professional manner. Answer (B) is incorrect because the confidentiality standard concerns the financial manager/management accountant's responsibility not to disclose or use the firm's confidential information. Answer (D) is incorrect because objectivity is the fourth part of the IMA Code of Ethics. It requires that information be communicated "fairly and objectively," and that all information that could reasonably influence users be fully disclosed.

16. At Key Enterprises, the controller is responsible for directing the budgeting process. In this role, the controller has significant influence with executive management as individual department budgets are modified and approved. For the current year, the controller was instrumental in the approval of a particular line manager's budget without modification, even though significant reductions were made to the budgets submitted by other line managers. As a token of appreciation, the line manager in question has given the controller a gift certificate for a popular local restaurant. In considering whether or not to accept the certificate, the controller should refer to which section of Statements on Management Accounting Number 1C (SMA 1C) (revised), *Standards of Ethical Conduct for Practitioners of Management Accounting and Financial Management*?

A. Competency.

B. Confidentiality.

C. Integrity.

D. Objectivity.

The correct answer is (C). *(CMA, adapted)*

REQUIRED: The ethical standard relevant to the controller's acceptance of a gift from a line manager.

DISCUSSION: The integrity standard requires the financial manager/management accountant to "refuse any gift, favor, or hospitality that would influence or would appear to influence his/her actions.

Answer (A) is incorrect because the competence standard pertains to the financial manager/management accountant's responsibility to maintain his/her professional skills and knowledge. It also pertains to the performance of activities in a professional manner. Answer (B) is incorrect because the confidentiality standard concerns the financial manager/management accountant's responsibility not to disclose or use the firm's confidential information. Answer (D) is incorrect because objectivity is the fourth part of the IMA Code of Ethics. It requires that information be communicated "fairly and objectively," and that all information that could reasonably influence users be fully disclosed.

17. In accordance with Statements on Management Accounting Number 1C (SMA 1C) (revised), *Standards of Ethical Conduct for Practitioners of Management Accounting and Financial Management*, a management accountant who fails to perform professional duties in accordance with relevant standards is acting contrary to which one of the following standards?

A. Competency.

B. Confidentiality.

C. Integrity.

D. Objectivity.

The correct answer is (A). *(CMA, adapted)*

REQUIRED: The ethical standard violated by a management accountant who fails to perform professional duties in accordance with relevant standards.

DISCUSSION: One of the responsibilities of the financial manager/management accountant under the competence standard is to "maintain an appropriate level of professional competence by ongoing development of his/her knowledge and skills." (S)he must also "perform professional duties in accordance with relevant laws, regulations, and technical standards." The third requirement under this standard is to "prepare complete and clear reports and recommendations after appropriate analyses of relevant and reliable information."

Answer (B) is incorrect because the confidentiality standard concerns the financial manager/management accountant's responsibility not to disclose or use the firm's confidential information. Answer (C) is incorrect because the integrity standard pertains to conflicts of interest, refusal of gifts, professional limitations, professional communications, avoidance of acts discreditable to the profession, and refraining from activities that prejudice the ability to carry out duties ethically. Answer (D) is incorrect because objectivity is the fourth part of the IMA Code of Ethics. It requires that information be communicated "fairly and objectively," and that all information that could reasonably influence users be fully disclosed.

APPENDIX B: TAXES AS TESTED ON THE CMA/CFM EXAMS

6 pages of outline
9 multiple-choice questions

A. Taxation
B. Fundamentals of Corporate Taxation

Taxes are tested in Parts 1, 2, and 4. The intended emphasis is on a user's point of view rather than a preparer's point of view.

Part 1 - The effect of tax provisions on financing and capital structure decisions, such as deductibility of payments for use of capital and after-tax cost of capital (i.e., debt vs. equity issues), is discussed in Study Unit 6, Capital Structure Finance, and Study Unit 7, Risk.

Part 2CMA - Accounting for deferred income taxes is a separate subunit in Study Unit 5, Liabilities.

Part 2CFM - Accounting for deferred income taxes is a subunit in Study Unit 2, Financial Statements: Special Topics.

The income tax code provisions that impact investment decisions are in Study Units 3 through 7 and cover advanced topics in corporate financial management.

- Deductibility of payments for use of capital
- After-tax cost of capital
- Tax shields from property transactions
- Transaction timing for favorable tax consequences

International tax considerations and their effect on international regulations on business are in Study Unit 8, Risk Management.

The tax implications of multinational corporations are in Study Unit 9, External Financial Environment.

Part 4 - The impact of income taxes on operational decisions, including operating income subject to income taxes, gains and losses subject to income taxes, and tax credits apply to Study Unit 1, Decision Analysis.

The impact of income taxes on investment decisions, including tax shields from sunk costs, transaction timing for favorable tax consequences, and tax elections apply to Study Unit 3, Capital Budgeting.

Important background for the above two study units appears in this appendix.

This appendix provides background and overview of taxes as tested on the CMA/CFM exams. Additional detail required in each part is provided as set forth above.

Obviously taxes are a major expense of doing business, and each business attempts to minimize taxes. Thus, financial managers and management accountants must be knowledgeable about taxes.

A. Taxation

1. Two Principles of Taxation

a. Benefits received. Individuals should pay a tax based on the benefits received from the services (e.g., paying for the use of a public park or swimming pool).

b. Ability to pay. Consumers should pay taxes based on their ability to pay them (e.g., taxes on income and wealth).

2. Three Classifications of Taxes Reflecting Ability-to-Pay Principles

a. **Progressive**. With a higher income, individuals pay a higher percentage of their income in taxes (e.g., income tax).

b. **Proportional**. At all levels of income, the percentage paid in taxes is constant (e.g., sales tax).

c. **Regressive**. As income increases, the percentage paid in taxes decreases (e.g., payroll or excise taxes).

1) EXAMPLE: An excise tax is regressive because its burden falls disproportionately on lower-income persons. As personal income increases, the percentage of income paid declines since an excise tax is a flat amount per quantity of the good or service purchased.

3. Taxes also may be classified as either direct or indirect.

a. **Direct taxes** are imposed upon the taxpayer and paid directly to the government, e.g., the personal income tax.

b. **Indirect taxes** are levied against others and thus only indirectly on the individual taxpayer, e.g., sales taxes and Social Security taxes paid by employers.

4. **Incidence of Taxation** -- the parties who actually bear a particular tax. For example, the person who actually bears the burden of an indirect tax may not be the same one who pays the tax to the government.

a. The incidence of taxation becomes important when a government wants to change the tax structure. Because taxation is a form of fiscal policy, the government needs to know who will actually bear the incidence of taxation, not just who will pay the tax.

b. EXAMPLE: Taxes such as the corporate income tax, the corporate property tax, and the excise tax are often shifted to the consumer in the form of higher prices.

c. EXAMPLE: Taxes such as windfall profits taxes are not shifted to the consumer via higher prices. This type of one-time-only tax levied on part of the output produced does not increase the equilibrium price of the taxed good.

5. In recent years, some authorities have supported a **value-added tax**, a tax based on consumption.

a. Many major industrial nations have already adopted a value-added tax.

b. The tax is levied on the value added to goods by each business unit in the production and distribution chain.

1) The amount of value added is measured by the difference between a firm's sales and its purchases.

2) Each firm in the chain collects the tax on its sales, takes a credit for taxes paid on purchases, and remits the difference to the government.

3) The consumer ultimately bears the incidence of the tax through higher prices.

c. A value-added tax encourages consumer saving because taxes are paid on consumption only, not on savings.

1) Because the value-added tax is based on consumption, people in the lower income groups would spend a greater proportion of their income on taxes.

2) The value-added tax is thus regressive.

d. Only those businesses that make a profit have to pay income taxes. Under the value-added tax, however, all businesses have to pay taxes, regardless of income.

6. **Taxes** as a business expense (federal, state, local, and foreign) are an important consideration because they are frequently 25% to 50% of all costs.

a. They include income, use, excise, property, legal document, payroll, and others.

b. Thus, governmental services (national defense, fire, police, etc.) are an important and costly factor of production.

c. Tax planning is very important in investment and financing decisions.

1) Investment tax credits have at times provided direct reduction of taxes when assets were purchased for use in the business.

a) The net effect is to decrease the cost of the asset.

b) The amount of the credit and limitations on the tax credit on used equipment affect investment decisions.

c) Investment tax credit is currently available for solar and geothermal property (business energy credit), for rehabilitation of historic structures, and for certain reforestation property.

2) Accelerated depreciation is permitted on many types of business assets.

a) Accordingly, in the early periods of an asset's life, depreciation is higher, taxable income is lower, and the rate of return on investment is higher.

3) Corporate capital gains are taxed at regular rates, and the capital gains of individuals are taxed at a maximum rate of 28%.

4) Special loss carryforward and carryback rules permit businesses to deduct net operating losses incurred in one period against income earned in other periods.

5) A dividends-received deduction makes tax free 70% to 100% of dividends received by one company from investments in the stock of another company.

a) This deduction prevents or reduces double taxation.

b) It also encourages one company to invest in the stock of another company.

6) Interest is a tax-deductible expense of the debtor company.

a) But dividends on common or preferred stock are not deductible by the issuer.

d. Federal tax policy is fiscal policy that affects the overall economy, which in turn affects production and the finished-goods markets in which the company deals.

e. Government monetary policy determines the availability and cost of capital, which affects financing (and in turn, investing) decisions.

1) Monetary policy also affects overall economic activity.

7. Stop and review! You have completed the outline for this subunit. Study multiple-choice questions 1 through 3 on page 473.

B. Fundamentals of Corporate Taxation

1. A C corporation is subject to tax liability imposed on its income (S corporations are taxed as partnerships).

2. **Gross income** means all income from whatever source derived (unless excluded by statute).

a. Income is considered realized when certain transactions occur.

1) Mere increases in value (appreciation) do not create income.

a) EXAMPLE: Land owned by a corporation increases in value from $5,000 to $20,000 per acre. There is no income until the land is sold or exchanged.

2) Transactions are normally in the form of a sale, an exchange, or the rendering of services.

3) Taxation of realized income may be excluded or deferred, for example, upon exchange of like-kind property.

4) The term "recognized" is used when income is realized and also taxed.

5) Nonresident aliens (including foreign corporations) are, generally, subject to federal income tax on U.S.-source income.

3. **Deductions** are those expenses and other statutorily prescribed items that are subtracted from gross income to determine taxable income.

a. Deductions should be distinguished from **exclusions**. Exclusions are never included in gross income (e.g., interest on tax-exempt bonds).

4. **Taxable income** is gross income minus deductions.

a. Taxable income is analogous to net income.
b. The tax rates are applied to taxable income to determine the tax liability.

5. **Credits** are subtracted from the tax liability.

a. Credits directly reduce taxes, while deductions reduce taxable income upon which the taxes are computed.

6. **Book Income vs. Taxable Income**

a. CMA/CFM candidates (Parts 2) are required to understand the difference between book income and taxable income. Problems may begin with book income, list numerous transactions, and require the adjustments necessary to arrive at taxable income.

b. The following adjustments are necessary to adjust book income to taxable income:

1) Add to book income

a) Federal income tax

b) Excess of capital losses over capital gains

c) Taxable income not reported on books

d) Expenses reported on books but not deducted on tax return (e.g., charitable contributions in excess of limit)

2) Subtract from book income

a) Income reported on books but not included on tax return (e.g., interest income from tax-exempt municipal bonds)

b) Deductions on tax return but not reported on books (e.g., dividends-received deduction)

c. These adjustments are the fundamental differences between net income for financial accounting purposes (book income) and taxable income. The discussion of income and deductions that follows will concentrate on these differences and assume that candidates can already compute book (i.e., accrual accounting) income.

d. According to generally accepted accounting principles (SFAS 109), accounting for income taxes (interperiod tax allocation) is based on the asset and liability method.

7. **Temporary Differences**

a. Depreciation expense arises from using different methods for tax and book purposes.

b. To account for bad debt expense, the direct write-off method is used for tax purposes, and the allowance method (estimating expenses) is used for book purposes.

c. For tax purposes, warranty expense is deductible only when paid, even if reserves based on estimates are used for book purposes.

d. Charitable contributions exceeding 10% of corporate taxable income before charitable contributions and special deductions are deductible during the 5 succeeding tax years.

e. Capital losses in excess of capital gains are not deductible in the current year. Instead, a net capital loss must be carried back 3 years (as a short-term capital loss) and applied towards any capital gains. Any loss remaining can be carried forward for 5 years. Capital losses can be used only to offset capital gains (i.e., ordinary income is unaffected).

f. Prepaid rent income is included in taxable income when received. Prepaid rent expense is deductible in computing taxable income for the period to which it is attributable.

g. The cost of certain intangibles (e.g., goodwill) acquired in connection with the conduct of a trade or business or income-producing activity is amortized over a 15-year period, beginning with the month in which the intangible is acquired.

h. Amortization of organizational costs and business start-up costs can be elected for a period of not less than 60 months for tax purposes.

8. **Permanent Differences**
   a. Life insurance premiums paid on key persons are not deductible if the corporation is the beneficiary. Proceeds from the policy are not taxed.
   b. Seventy percent of dividends received from taxable domestic corporations are deductible by a distributee corporation that owns less than 20% of the stock of the distributing corporation. The percentage deductible is 80 if the distributee owns 20% to 80% of the distributing corporation. A 100% dividend-received deduction is permitted for dividends received among members of an affiliated group of corporations (80% ownership).
   c. Federal income tax expense is not deductible in computing taxable income.
   d. Tax-exempt interest is included in book income. Expense attributable to earning it is not deductible for tax purposes.
   e. Business gifts exceeding $25 are generally not deductible for income tax purposes.
   f. Compensation in excess of $1 million paid to the chief executive officer and certain other employees is not deductible in computing taxable income. Premiums, tax-free benefits, and compensation based on performance goals are excluded from the $1 million limit.
9. **Nonrecognition Transactions**. Certain types of transactions result in deferral or nonrecognition of gains. Examples are like-kind exchanges, involuntary conversions, and tax-free reorganizations.
   a. **Like-kind exchanges**. Sec. 1031 defers recognizing gain or loss to the extent that property productively used in a trade or business or held for the production of income (investment) is exchanged for property of like-kind.
   b. **Involuntary conversions**. A taxpayer may elect to defer recognition of gain, but not losses, if property is involuntarily converted (e.g., by fire, loss, or government condemnation for public use) into money or property that is not similar or related in use. Nonrecognition of gain is contingent on the involuntarily converted property being reinvested in qualified replacement property.
   c. **Tax-free reorganization**. A nontaxable reorganization of one or more corporations is considered a mere change in form of investment rather than a disposition of assets. For this reason, a general rule of nonrecognition of gain or loss applies to qualifying reorganizations. However, gain is recognized to the extent of receipt of cash or other nonqualifying payment.
10. **Taxation on Multinational Corporations**
   a. Multinational corporations frequently derive income from several countries. The government of each country in which a corporation does business may enact statutes imposing one or more types of tax on the corporation.
   b. To avoid double taxation, most countries have adopted treaties to coordinate or synchronize the effects of their taxing statutes.
11. Stop and review! You have completed the outline for this subunit. Study multiple-choice questions 4 through 9 beginning on page 474.

MULTIPLE-CHOICE QUESTIONS

A. Taxation

1. A regressive tax is a tax in which

A. Individuals with higher incomes pay a higher percentage of their income in tax.

B. The burden for payment falls disproportionately on lower-income persons.

C. The individual pays a constant percentage in taxes, regardless of income level.

D. Individuals with lower incomes pay a lower percentage of their income in tax.

The correct answer is (B). *(Publisher)*

REQUIRED: The statement that describes a regressive tax.

DISCUSSION: With a regressive tax, the percentage paid in taxes decreases as income increases. For example, excise taxes and payroll taxes are both regressive taxes. An excise tax is regressive because its burden falls disproportionally on lower-income persons. As personal income increases, the percentage of income paid declines because an excise tax is a flat amount per quality of the good or service purchased.

Answers (A) and (D) are incorrect because a progressive tax is a tax in which individuals with higher (lower) incomes pay a higher (lower) percentage of their income in tax. For example, income taxes are progressive. Answer (C) is incorrect because a proportional tax is a tax in which the individual pays a constant percentage in taxes, regardless of income level. A sales tax is a proportional tax.

2. Two examples of indirect taxes are

A. Taxes on business and rental property and personal income taxes.

B. Sales taxes and Social Security taxes paid by employees.

C. Sales taxes and Social Security taxes paid by employers.

D. Social Security taxes paid by employees and personal income taxes.

The correct answer is (C). *(CMA, adapted)*

REQUIRED: The two forms of indirect taxes.

DISCUSSION: Indirect taxes are those levied against someone other than individual taxpayers and thus only indirectly affect the individual. Sales taxes are levied against businesses and are then passed along to the individual purchaser. Social Security taxes are levied against both the employer and the employee. Those levied against the employee are direct taxes; those levied against the employer are indirect.

Answers (A), (B), and (D) are incorrect because personal income taxes and Social Security taxes levied against the employee are direct taxes.

3. When a fixed plant asset with a 5-year estimated useful life is sold during the second year, how would the use of an accelerated depreciation method instead of the straight-line method affect the gain or loss on the sale of the fixed plant asset?

	Gain	Loss
A.	Increase	Increase
B.	Increase	Decrease
C.	Decrease	Increase
D.	Decrease	Decrease

The correct answer is (B). *(CMA, adapted)*

REQUIRED: The effect of using an accelerated depreciation method instead of straight-line.

DISCUSSION: An accelerated method reduces the book value of the asset more rapidly in the early years of the useful life than does the straight-line method. Hence, the effect of an early sale is to increase the gain or decrease the loss that would have been recognized under the straight-line method.

Answers (A), (C), and (D) are incorrect because the effect is to increase gains and decrease losses.

B. Fundamentals of Corporate Taxation

Questions 4 and 5 are based on the following information. The Hando Communications Corporation (HCC) has just finished its first year of operations. For the year, HCC had $80,000 of income. HCC also earned $11,000 of interest on a tax-exempt bond, a $10,000 depreciation deduction, and an $8,000 tax credit. Assume that HCC has a 30% tax rate.

4. What is HCC's net tax liability?

A. $17,700

B. $15,300

C. $9,700

D. $2,000

The correct answer is (C). *(Publisher)*

REQUIRED: The net tax liability.

DISCUSSION: The first step in calculating HCC's net tax liability is to subtract the exclusion for tax-exempt interest from income, for a gross income amount of $69,000 ($80,000 – $11,000). Next, the deprecation deduction is subtracted from gross income, for a taxable income of $59,000 ($69,000 – $10,000). Then, the taxable income is multiplied by the tax rate, for a gross tax liability of $17,700 ($59,000 x 30%). Finally, net tax liability is computed by subtracting the tax credits from the gross tax liability. Therefore, HCC's net tax liability is $9,700 ($17,700 – $8,000).

Answer (A) is incorrect because $17,700 is the gross tax liability. Answer (B) is incorrect because $15,300 is the net tax liability found when incorrectly treating the tax credit as a direct reduction in taxable income. Answer (D) is incorrect because $2,000 is the net tax liability found when incorrectly treating the exclusion as a credit.

5. Which item reduces HCC's gross tax liability by the largest amount?

A. Gross income.

B. The tax-exempt interest exclusion.

C. The depreciation deduction.

D. The tax credit.

The correct answer is (D). *(Publisher)*

REQUIRED: The item that decreases HCC's gross tax liability by the largest amount.

DISCUSSION: Credits directly reduce taxes, whereas exclusions and deductions reduce income prior to the computation of the gross tax liability. Thus, the credit reduces the gross tax liability on a dollar-for-dollar basis, or $8,000. An exclusion or deduction reduces gross tax liability by the amount of the exclusion or deduction multiplied by the tax rate. Accordingly, the exclusion will reduce HCC's gross tax liability by $3,300 ($11,000 x 30%), and the deduction will reduce HCC's gross tax liability by $3,000 ($10,000 x 30%). Gross income does not reduce the gross tax liability; rather, it increases the gross tax liability by $24,000 ($80,000 x 30%).

Answers (A), (B), and (C) are incorrect because an exclusion or deduction reduces gross tax liability by the amount of the exclusion or deduction multiplied by the tax rate. Therefore, the exclusion will reduce HCC's gross tax liability by $3,300 ($11,000 x .30), and the deduction will reduce HCC's gross tax liability by $3,000 ($10,000 x .30).

6. None of the following items are deductible in calculating taxable income except

A. Estimated liabilities for product warranties expected to be incurred in the future.

B. Dividends on common stock declared but not payable until next year.

C. Bonus accrued but not paid by the end of the year to a cash-basis 90% shareholder.

D. Vacation pay accrued on an employee-by-employee basis.

The correct answer is (D). *(CMA, adapted)*

REQUIRED: The item that is deductible in the calculation of taxable income.

DISCUSSION: Sec. 162(a) states that a deduction is allowed for the ordinary and necessary expenses incurred during the year in any trade or business. A corporation may therefore deduct a reasonable amount for compensation. Accrued vacation pay is a form of compensation that results in an allowable deduction for federal income tax purposes.

Answer (A) is incorrect because warranty expenses are not deductible until paid. Answer (B) is incorrect because dividends on common stock are never deductible by a corporation; they are distributions of after-tax income. Answer (C) is incorrect because amounts accrued by an accrual-basis taxpayer to be paid to a related cash-basis taxpayer in a subsequent period are not deductible until the latter taxpayer includes the items in income. This rule effectively puts related taxpayers on the cash basis.

7. All of the following are adjustments/preference items to corporate taxable income in calculating alternative minimum taxable income except

A. All of the gain on an installment sale of real property in excess of $150,000.

B. Mining exploration and development costs.

C. A charitable contribution of appreciated property.

D. Sales commission earned in the current year but paid in the following year.

The correct answer is (D). *(CMA, adapted)*

REQUIRED: The item that is not an adjustment or preference item in calculating the alternative minimum tax (AMT).

DISCUSSION: Taxable income is adjusted to arrive at alternative minimum taxable income. Some of the common adjustments include gains or losses from long-term contracts, gains on installment sales of real property, mining exploration and development costs, charitable contributions of appreciated property, accelerated depreciation, the accumulated current earnings adjustment, and tax-exempt interest on private activity bonds issued after August 7, 1986. A sales commission accrued in the current year but paid in the following year is not an example of an AMT adjustment.

Answer (A) is incorrect because the gain on an installment sale of real property in excess of $150,000 is an adjustment to taxable income for purposes of computing alternative minimum taxable income. Answer (B) is incorrect because mining exploration and development costs are adjustments to taxable income for purposes of computing alternative minimum taxable income. Answer (C) is incorrect because a charitable contribution of appreciated property is an adjustment to taxable income for purposes of computing alternative minimum taxable income.

8. Which one of the following factors might cause a firm to increase the debt in its financial structure?

A. An increase in the corporate income tax rate.

B. Increased economic uncertainty.

C. An increase in the federal funds rate.

D. An increase in the price-earnings ratio.

The correct answer is (A). *(CMA, adapted)*

REQUIRED: The factor that might encourage a firm to increase the debt in its financial structure.

DISCUSSION: An increase in the corporate income tax rate might encourage a company to borrow because interest on debt is tax deductible, whereas dividends are not. Accordingly, an increase in the tax rate means that the after-tax cost of debt capital will decrease. Given equal interest rates, a firm with a high tax rate will have a lower after-tax cost of debt capital than a firm with a low tax rate.

Answer (B) is incorrect because increased uncertainty encourages equity financing. Dividends do not have to be paid in bad years, but interest on debt is a fixed charge. Answer (C) is incorrect because an increase in interest rates discourages debt financing. Answer (D) is incorrect because an increase in the price-earnings ratio means that the return to shareholders (equity investors) is declining; therefore, equity capital is a more attractive financing alternative.

9. The deferral or nonrecognition of gains is not allowed for tax purposes when the transaction is a(n)

A. Reorganization that is a change in the form of investment.

B. Exchange of property that is used in a business for like-kind property.

C. Reorganization that is considered a disposition of assets.

D. Involuntary conversion of property into qualified replacement property.

The correct answer is (C). *(Publisher)*

REQUIRED: The statement that does not allow for the deferral or nonrecognition of gain.

DISCUSSION: Like-kind exchanges, involuntary conversions, and tax-free reorganizations are examples of transactions that result in the deferral or nonrecognition of gain. A reorganization is nontaxable when it is considered a mere change in investment, not a disposition of assets.

Answer (A) is incorrect because a reorganization that is a mere change in the form of investment is nontaxable. Answer (B) is incorrect because a like-kind exchange allows for the deferral of gain. Answer (D) is incorrect because an involuntary conversion allows for the deferral of gain.

Appendix C: IMA Membership, ICMA Application, and Examination Forms

A. IMA Membership Application
B. Application for Admission to the ICMA
C. Examination Registration Form

You must apply and become an IMA member in order to participate in the IMA Certification programs. The cost is $140 per year for Regular or International membership; $47 for the first year and $93 for the second year for Associate membership (for those within 2 years of completing full-time studies and who reside in the U.S. or Canada); $70 per year for Academic membership (full-time faculty in the U.S. and Canada); and $28 per year for Student membership (must carry at least 6 hours per semester and reside in the U.S. or Canada). The IMA offers three member interest groups at $75 per year: the Controllers Council, the Cost Management Group, and the Small-Business Council. Everyone except students and associates must pay a $15 IMA registration fee. See pages 478 and 479, which can be photocopied and used to apply for IMA membership, or call the IMA at (800) 638-4427 ext. 510 and ask for a CMA/CFM "kit." You may also e-mail the IMA at info@imanet.org to request an information kit.

The application for admission to the ICMA is on pages 480 and 481. This form is required for participation in the IMA Certification Programs. Completion of the registration form on page 482 is required in order to take any and all of the five examination parts.

> Photocopy the following 5 pages and fill them out and mail them today! It will take you less than 10 minutes!

IMA MEMBERSHIP APPLICATION

BATCH NUMBER: ______ ❑ INITIAL APPLICATION ❑ REDINSTATEMENT

PERSONAL INFORMATION

❑ Mr. ❑ Ms. ❑ Miss ❑ Mrs. ❑ Dr. ❑ Male ❑ Female

Social Security Number

Date of Birth / / Month/Day/Year

Send IMA mail to ❑ Home ❑ Business

Last Name/Family Name/Surname | First Name | Middle Initial | Suffix

Home Mailing Address

City | State/Province/Country | Zip Code/Postal Code

COMPANY INFORMATION

Company Name | SIC Code (See Reverse Side)

Business Mailing Address | Job Title Code (See Reverse Side)

City | State/Province/Country | Zip Code/Postal Code | Responsibility Code (See Reverse Side)

Telephone Preference ❑ Business ❑ Home (Include Area/Country/City Codes) | Business Telephone Number with Extension | Home Telephone Number

E-Mail Address | Fax Number: (Include Area/Country/City Codes)

EDUCATIONAL QUALIFICATIONS

College or University	Degree	Major	Date Received/Expected
Undergraduate			
Graduate			

Professional Designations Earned

❑ CPA

❑ CFA

❑ CIA

❑ Other: ______

❑ **If you are applying for the CMA or CFM program, complete side two and pay the $50 credentialing fee.**

❑ **Check here to receive information about the CMA and CFM programs.**

CHAPTER AFFILIATION

Chapter Affiliation (your choice of chapter/student chapter - separate listing) | Chapter Number | ❑ Member-At-Large (check here if no chapter affiliation desired) ❑ International Member-At-Large

If you have ever been convicted of a felony check here. ❑ Explain in an accompanying letter.

PAYMENT INFORMATION - FILL IN AS APPROPRIATE. ALL PAYMENTS MUST BE IN US DOLLARS

❑ REGULAR MEMBERSHIP: (You must reside in the U.S. or Canada) ❑ International Member-At-Large	1 Year–$140.00	$	
❑ ACADEMIC: (You must be a full-time faculty member and reside in the United States or Canada)	$70.00	$	
❑ ASSOCIATE: (You must apply within 2 years of completing full-time studies and reside in the United States or Canada) ❑ 1st Year After Graduation	$47.00	$	
❑ 2nd Year After Graduation	$93.00	$	
❑ STUDENT: (You must have 6 or more hours per semester and reside in the United States or Canada) Expected Graduation Date ______ YEAR	$28.00	$	
MEMBER INTEREST GROUPS: (IMA Membership Required) - 1 Year $75.00 ea. ❑ Controllers Council ❑ Cost Management Group ❑ Small-Business Council No. of MIGs ______	x$75.00	$	
OPTIONAL SERVICES: (IMA Membership Required) ❑ Research Publications Service for 1 year (US members)	$85.00	$	
❑ Research Publications Service for 1 year (International members)	$120.00	$	
CERTIFICATION: (Optional) ❑ U.S. CPA Waiver Fee	$80.00	$	
❑ Credentialing Fee (required of ICMA program applicants only; U.S. college students are exempt)	$50.00	$	
REGISTRATION FEE: Applies to Regular, Academic & International Members *ONLY*. Students and associates are exempt.		$ 15	.00
TOTAL DUE		$	

❑ Check Enclosed (Make check payable in U.S currency to the Institute of Management Accountants, Inc.)

❑ Charge my credit card: ❑ VISA ❑ MasterCard ❑ American Express ❑ Discover Credit Card Number:

Expiration Date: ___/___ MM/YY

I affirm that the statements on this application are correct and agree to abide by the Standards of Ethical Conduct

Signature of Applicant: ______ Date: ______ ID Number from Mailing Label ______

PLEASE READ BOTH SIDES

NOTE: PAYMENT IN FULL MUST ACCOMPANY APPLICATIONS - FEES SUBJECT TO CHANGE

IMA • 10 Paragon Drive • Montvale, New Jersey 07645-1760 • www.imanet.org
(201) 573-9000 • (800) 638-4427 • Fax (201) 573-0559

CERTIFICATION PROGRAMS

If you are applying for admission to the ICMA, please complete the following.

Check the programs you wish to enter: ❑ CMA ❑ CFM

ADDITIONAL EDUCATIONAL INFORMATION

Name on transcript (if different from front of application)

Candidates with U.S. CPA License:	State	Year	License Number

Check the appropriate box(es) and make arrangements for supporting documents to be forwarded to the ICMA.

❑ **College Graduate** - Submit official transcript showing university degree conferred and official university seal or arrange for university to send proof of degree directly to the ICMA. If you have more than one degree, submit only one transcript. Candidates with foreign degrees must have their degree evaluated by an independent evaluation agency.

❑ **GMAT or GRE scores** - Provide copy of scores to ICMA.

❑ **U.S. CPA Exam** - Arrange to have proof of U.S. CPA examination completion sent directly from your State Board of Accountancy to the ICMA.

❑ **U.S. CPA Waiver for Part 2** - Arrange to have proof of U.S. CPA examination completion sent directly from your State Board of Accountancy to the ICMA. Enclose the waiver fee with this application or use space on reverse side to charge this fee to your credit card.

❑ **Applying as a Student** - Upon graduation, be sure to arrange for an official copy of your transcript to be sent to the ICMA.

❑ **Applying as Faculty** - In addition to confirming your educational qualification, please provide a letter on school stationery affirming full-time teaching status.

❑ **Later** - You can sit for the exam without submitting your educational credentials, but you will not be certified until your credentials are verified.

STANDARD INDUSTRY CLASSIFICATIONS (SIC)

AGRICULTURE, FORESTRY, FISHERIES

01 AGRICULTURAL PRODUCTION
07 AGRICULTURAL SVCS/HUNTING/TRAPPING
08 FORESTRY
09 FISHERIES

MINING

10 METAL MINING
11 ANTHRACITE MINING
12 BITUMINOUS COAL/LIGNITE MINING
13 CRUDE OIL/LIGNITE MINING
14 MINING/QUARRY NONMETALLICS

CONTRACT CONSTRUCTION

15 BLDG. CONSTRUCTION - GENERAL CONTRACTORS
16 CONSTRUCTION - OTHER
17 CONSTRUCTION - SPECIAL TRADE CONTRACTORS

MANUFACTURING

19 ORDINANCE/ACCESSORIES
20 FOOD/KINDRED PRODUCTS
21 TOBACCO MANUFACTURERS
22 TEXTILE MILL PRODUCTS
23 APPAREL/FINISHED FABRICS
24 LUMBER/WOOD PRODUCTS
25 FURNITURE/FIXTURES
26 PAPER/ALLIED PRODUCTS
27 PRINTING/PUBLISHING
28 CHEMICALS/ALLIED PRODUCTS
29 OIL REFINING/RELATED INDUSTRIES
30 RUBBER/MISC. PLASTICS PRODUCTS
31 LEATHER PRODUCTS
32 STONE, CLAY, GLASS/CONCRETE PRODUCTS
33 PRIMARY METAL INDUSTRIES
34 FABRICATED METAL PRODUCTS
35 MACHINERY, NONELECTRICAL
36 ELECTRICAL MACHINERY
37 TRANSPORTATION
38 PROFESSIONAL, SCIENTIFIC, CONTROL INSTRUMENTS
39 MISC. MANUFACTURING INDUSTRIES

TRANSPORTATION, COMMUNICATION & UTILITY SERVICES

40 RAILROAD TRANSPORTATION
41 LOCAL AND SUBURBAN TRANSPORTATION
42 MOTOR FREIGHT/WAREHOUSING
44 WATER TRANSPORTATION
46 PIPELINE TRANSPORTATION
47 TRANSPORTATION SERVICES
48 COMMUNICATION
49 ELECTRIC, GAS/SANITARY SERVICES

WHOLESALE & RETAIL TRADE

50 WHOLESALE TRADE
52 BUILDING/HARDWARE/FARM EQUIP DEALERS
53 RETAIL TRADE - GENERAL
54 FOOD STORES
55 AUTO DEALERS/SERVICE STATIONS
56 APPAREL/ACCESSORY STORES
57 FURNITURE/FURNISHINGS/STORES
58 EATING/DRINKING PLACES
59 MISC. RETAIL STORES

FINANCE, INSURANCE & REAL ESTATE

60 BANKING
61 CREDIT AGENCIES NOT BANKS
62 SECURITY/COMMODITY BROKERS, AND SERVICES
63 INSURANCE CARRIERS
64 INSURANCE AGENTS, BROKERS
65 REAL ESTATE
66 COMBINATIONS OF REAL ESTATE, INSURANCE, LOANS, LAW OFFICES
67 HOLDING, INVESTMENT COMPANIES

SERVICES

70 HOTELS/ROOMING HOUSES/CAMPS, ETC.
72 PERSONAL SERVICES
73 MISC. BUSINESS SERVICES
75 AUTO REPAIR, AUTO SERVICES/GARAGES
76 MOTION PICTURES
79 AMUSEMENT/RECREATION SERVICES
80 MEDICAL/HEALTH SERVICES
81 LEGAL SERVICES
82 EDUCATIONAL SERVICES
84 MUSEUM/ART GALLERIES/GARDENS
86 NONPROFIT MEMBERSHIP ORGANIZATIONS
88 PRIVATE HOUSEHOLDS
89 PUBLIC ACCOUNTING

GOVERNMENT

91 FEDERAL GOVERNMENT
92 STATE GOVERNMENT
93 LOCAL GOVERNMENT
94 INTERNATIONAL GOVERNMENT

JOB TITLE

01 OWNER
03 CHAIRMAN OF THE BOARD
05 CHIEF EXECUTIVE OFFICER
06 CHIEF FINANCIAL OFFICER
07 PRESIDENT
09 GROUP PRESIDENT
11 CORPORATE SECRETARY
13 CORPORATE TREASURER
15 EXECUTIVE VICE PRESIDENT
17 SENIOR VICE PRESIDENT
19 VICE PRESIDENT
21 ASSISTANT VICE PRESIDENT
23 GROUP VICE PRESIDENT
25 DIVISIONAL VICE PRESIDENT
27 CORPORATE CONTROLLER
29 ASST. CORPORATE CONTROLLER
31 DIVISIONAL CONTROLLER
33 PLANT CONTROLLER
35 DIRECTOR
37 GENERAL MANAGER
39 MANAGER
41 GENERAL SUPERVISOR
43 SUPERVISOR
45 CHIEF ACCOUNTANT
47 ACCOUNTANT
49 ECONOMIST
51 ANALYST
53 SYSTEMS ANALYST
55 PROGRAMMER
57 ADMINISTRATOR
59 AUDITOR
61 BOOKKEEPER
63 ACCOUNTING CLERK
65 DEAN
67 PROFESSOR
69 ASSOCIATE PROFESSOR
71 ASSISTANT PROFESSOR
73 INSTRUCTOR
75 CONSULTANT
77 PRINCIPAL
79 PARTNER
99 OTHER

STRATEGIC FINANCE MAGAZINE

SUBSCRIPTION RATES PER YEAR:
MEMBERS: $45
(INCLUDED IN DUES, NONDEDUCTIBLE)
STUDENT MEMBERS: $24
(INCLUDED IN DUES NONDEDUCTIBLE)
NON-MEMBER: $140
NONPROFIT LIBRARY: $70

MANAGEMENT ACCOUNTING QUARTERLY

SUBSCRIPTION RATES PER YEAR:
MEMBERS: $10
(INCLUDED IN DUES, NONDEDUCTIBLE)
NON-MEMBER: $60

RESPONSIBILITY AREA

01 GENERAL MANAGEMENT	30 GOVERNMENTAL ACCOUNTING	60 EDUCATION
05 CORPORATE MANAGEMENT	35 FINANCE	65 INFORMATION SYSTEMS
10 PUBLIC ACCOUNTING	40 RISK MANAGEMENT	70 STUDENT
15 GENERAL ACCOUNTING	45 BUDGET AND PLANNING	75 RETIRED
20 PERSONNEL ACCOUNTING	50 TAXATION	80 OTHER
25 COST ACCOUNTING	55 INTERNAL AUDITING	

DEMOGRAPHIC SURVEY (OPTIONAL)

1. Check all that apply:

❑ African American/Black ❑ Asian or Pacific Islander ❑ European/Caucasian ❑ Latino/Latina/Hispanic ❑ American Indian/Native American

❑ Middle Eastern ❑ Mixed Ethnic Identity ❑ Other (Please Specify): ____________ ❑ Physically Challenged

2. Who will pay your IMA Dues? ❑ You ❑ Your Company

3. Why did you become a member of the IMA? ❑ Certification ❑ Networking ❑ Education ❑ Other (Please Specify): ____________

9/00

APPLICATION FOR ADMISSION TO THE ICMA

Institute of Certified Management Accountants
10 Paragon Drive • Montvale, New Jersey 07645-1759
(201) 573-9000 • 1 (800) 638-4427 • FAX: (201) 573-8438
Endorsed by the Institute of Management Accountants

PLEASE CHECK THE PROGRAMS YOU WISH TO ENTER

CMA CFM

☐ CMA PROGRAM ☐ CFM PROGRAM

PERSONAL INFORMATION

TYPE OR PRINT CLEARLY

☐ Mr. ☐ Ms. ☐ Miss ☐ Mrs. ☐ Dr. ☐ Male ☐ Female Social Security Number ☐☐☐-☐☐-☐☐☐☐ Date of Birth ☐☐-☐☐-☐☐

☐ *I would like my mail sent to my home address*

Last Name (Family Name) First Name Middle Name Suffix

Home Mailing Address

City State Zip Code (9 digit)

Home Telephone Number (include area code)
( )

☐ *I would like my mail sent to my business address*

Firm Name

Business Mailing Address

City State Zip Code (9 digit)

Business Telephone Number (include area code) Extension
( )

You must be a member of the Institute of Management Accountants (IMA) to qualify for acceptance into the ICMA (IMA application page 15). Members of the Financial Management Association (FMA) may take the examination but must become IMA members prior to certification

☐ IMA Member No. ______ ☐ FMA Member No. ______ ☐ IMA Membership application submitted.

EDUCATIONAL QUALIFICATIONS

A. College or University	Degree	Date Received/Expected
Undergraduate		
Graduate		
Name on transcript (if different than above)		

	State	Year	License Number
B. Candidates with CPA License:			

Check the appropriate box(es) and make arrangements for supporting documents to be forwarded to the ICMA. Verification must be received for examination authorization.

☐ **College Graduate.** Submit official transcript showing university degree conferred and official university seal or arrange for university to send proof of degree directly to the ICMA. If you have more than one degree, submit only one transcript. Candidates with foreign degrees must have their degree evaluated by an independent agency. Contact the ICMA for details.

☐ **GMAT or GRE scores.** Provide copy of scores to ICMA.

☐ **CPA.** Arrange to have proof of license sent directly from your State Board of Accountancy to the ICMA.

☐ **CPA Waiver for Part 2.** Arrange to have proof of CPA examination completion sent directly from your State Board of Accountancy to the ICMA Enclose the waiver fee with this application or use space on reverse side to charge this fee to your credit card.

☐ **Applying as Student.** Be sure to provide, in the Character Reference Section on the reverse side of this form, the name of a professor who can verify student status. Upon graduation, be sure to arrange for an official copy of your transcript to be sent to the ICMA.

☐ **Applying as Faculty.** In addition to confirming your educational qualification, please provide a letter on school stationery affirming full-time teaching status.

PLEASE COMPLETE BOTH SIDES

CHARACTER REFERENCES

The names, addresses and telephone numbers of two character references must be supplied. One, your current employer, or current professor if you are a student; the other, a person neither employed by your firm nor a member of your family. The ICMA may contact the references as appropriate.

Name - Employer or Professor -(Circle one)			Name - Personal		
Title:			Address:		
Address:			Address:		
City:	State:	Zip Code:	City:	State:	Zip Code:
Phone number:			Phone number:		

Have you ever been convicted of a felony? ☐ Yes ☐ No If yes, please explain in an accompanying letter.

EXPERIENCE REQUIREMENT

Appropriate experience is required before the certification is awarded. The experience listed below will be used by the ICMA to identify candidates whose experience may not meet the requirements. For students and others who do not have accounting experience, but who expect to be employed in the accounting field, please check the following space: ☐ Not currently employed in the required field but expect to be in the future.

DATES (MONTH AND YEAR)	EMPLOYER	JOB TITLE	DUTIES & RESPONSIBILITIES
START DATE: END DATE:			
START DATE: END DATE:			

I agree to comply with the Standards of Ethical Conduct. I declare and affirm that the statements made in the foregoing application, including accompanying statements and transcripts, are true, complete, and correct. I authorize the investigation of all statements contained in this application.

I hereby attest that I will not divulge the content of the examination, nor will I remove any examination materials, notes, or other unauthorized materials from the examination room. I understand that failure to comply with this attestation may result in invalidation of my grades and disqualification from future examinations. For those already certified by the Institute of Certified Management Accountants, failure to comply with the statement will be considered a violation of the Standards of Ethical Conduct for Practitioners of Management Accounting and Financial Management and could result in revocation of the certification.

Signature of Applicant*: X______________________ Date:__________

*Original signature required; do not fax this form. ID Number from mailing label:__________

PAYMENT INFORMATION

☐ I am enclosing a check payable to the Institute of Certified Management Accountants

☐ $50. Credentialing Fee (required)**

☐ Charge my: ☐ VISA ☐ MasterCard ☐ American Express ☐ Discover Card

☐ CPA Waiver Fee $80.

Card Number: ☐☐☐☐☐☐☐☐☐☐☐☐☐☐☐☐

Expiration Date: ☐☐-☐☐

Fees are not refundable

Signature of Applicant: X______________________ Date:__________

FEES SUBJECT TO CHANGE

** Not required for students.

NOTE: A SEPARATE EXAMINATION REGISTRATION FORM IS REQUIRED.

771-08C-40M-998

OFFICE USE ONLY	BATCH NUMBER	PAYMENT BY CHECK # ☐ PERSONAL ☐ COMPANY	AMOUNT

Institute of Certified Management Accountants
10 Paragon Drive • Montvale, New Jersey 07645-1759
(201) 573-9000 • 1 (800) 638-4427 • FAX: (201) 573-8438
Endorsed by the Institute of Management Accountants

CMA® CFM®

EXAMINATION REGISTRATION FORM

PERSONAL INFORMATION

TYPE OR PRINT CLEARLY

☐ Mr. ☐ Ms. ☐ Miss ☐ Mrs. ☐ Dr. ☐ CMA ☐ CFM ☐ Certification Candidate

Social Security Number ☐☐☐-☐☐-☐☐☐☐

Last Name (Family Name) First Name Middle Name Suffix

Preferred Mailing Address

☐ Home
☐ Business

City State Zip Code (9 digit)

Daytime Telephone Number (include area code)

Please Specify ☐ Home ☐ Business

The separate Application for Admission to the ICMA and an application to the IMA must accompany this Examination Registration Form unless they have been previously submitted.

PLACE A CHECK MARK IN THE BLANK SPACES BELOW FOR THE PART(S) YOU WISH TO TAKE AT THIS TIME

☐ PART 1
Economics Finance, and Management

☐ PART 2CMA
Financial Accounting and Reporting

☐ PART 2CFM
Corporate Financial Management

☐ PART 3
Management Reporting, Analysis, and Behavioral Issues

☐ PART 4
Decision Analysis and Information Systems

TOTAL PARTS ______

X $ 80.00 Domestic Fee............... $ ______

X $120.00 International Fee......... $ ______

Less Student/Faculty Discount...... $ ______
(if eligible)

AMOUNT DUE .... $ ______

It should be noted that examination fees are not refundable (see page 9).

CONFIDENTIALITY STATEMENT & PAYMENT INFORMATION

I hereby attest that I will not divulge the content of this examination, nor will I remove any examination materials, notes, or other unauthorized materials from the examination room. I understand that failure to comply with this attestation may result in invalidation of my grades and disqualification from future examinations. For those already certified by the Institute of Certified Management Accountants, failure to comply with the statement will be considered a violation of the Standards of Ethical Conduct for Practitioners of Management Accounting and Financial Management and could result in revocation of the certification.

Signature: X ______ ID Number from mailing label: ______

☐ I am enclosing a check payable to the Institute of Certified Management Accountants

☐ Charge my: ☐ VISA ☐ MasterCard ☐ American Express ☐ Discover Card

Card Number: ☐☐☐☐☐☐☐☐☐☐☐☐☐☐☐☐ Expiration Date: ☐☐-☐☐

Signature: X ______ Date: ______

FEES SUBJECT TO CHANGE

771-08B-40M-898

APPENDIX D: ICMA CONTENT SPECIFICATION OUTLINES

Authors' Note: This is reference material only. The ICMA's CSOs have been carefully analyzed and are incorporated in our Study Units 1 through 10 to provide more systematic and rational coverage of topics within each part of the exam.

We believe we provide comprehensive coverage of the subject matter tested on the CMA/CFM exams. If, after you take the exam, you find topics, concepts, etc., tested that are not covered or are inadequately covered, please call, fax, or e-mail (irvin@gleim.com). We do not want any information about CMA/CFM questions, only information/feedback about our books/software.

The content specification outlines presented on the following pages represent the body of knowledge that will be covered on CMA/CFM Part 4.

Candidates are responsible for being informed on the most recent developments in the areas covered in the outline. This includes understanding of public pronouncements issued by accounting organizations as well as being up-to-date on recent developments reported in current accounting and business periodicals.

The content specification outlines serve several purposes. These purposes include

- Establishing the foundation from which each examination will be developed
- Providing consistent coverage on each examination
- Communicating to interested parties more detail as to the content of each examination part
- Aiding candidates in their preparation for each examination
- Providing information to those who offer courses designed to assist candidates in preparing for the examinations

Important additional information about the content specification outlines and the examinations is listed below.

1. The percentage range given for each major topic within each examination part represents the relative weight range given to that topic in an examination part. A candidate will be required to answer questions that equal the minimum relative weight but do not exceed the maximum relative weight for a major topic.
2. Each examination will sample from the subject areas contained within each major topic area to meet the relative weight specifications. No relative weights have been assigned to the subject areas within each major topic. No inference should be made from the order in which the subject areas are listed or from the number of subject areas as to the relative weight or importance of any of the subjects.
3. Each major topic within each examination part has been assigned a coverage level designating the depth and breadth of topic coverage, ranging from an introductory knowledge of a subject area (Level A) to a thorough understanding of and ability to apply the essentials of a subject area (Level C). Detailed explanations of the coverage levels and the skills expected of candidates are presented beginning on the next page.
4. The topics for each of the examination parts have been selected to minimize the overlapping of subject areas among the examination parts. The topics within an examination part and the subject areas within topics may be combined in individual questions.
5. Questions containing ethical issues can appear on any part of the examination. The ethical issues will be addressed within the context of a specific subject area. For determining the relative weight distribution of an examination, the questions containing ethical issues will be counted in the major topic area in which the ethical issues are raised.
6. Federal income taxation issues will be divided into two categories and be contained in questions that relate to the two categories.
   a. Accounting for income taxes. The financial reporting requirements for income taxes, including the proper treatment of deferred income taxes, will be contained in questions in Part 2CMA, Financial Accounting and Reporting, and Part 2CFM, Corporate Financial Management.
   b. Tax implications for decisions. The tax code provisions that impact decisions will be contained in: (1) questions that require decision analysis regarding debt versus equity issues in Part 1, Economics, Finance, and Management; (2) questions in Part 2CFM, Corporate Financial Management; and (3) the decision analysis questions in Part 4, Decision Analysis and Information Systems.
7. Candidates for the CMA and CFM designations are expected to have a minimum level of business knowledge that transcends all examination parts. This minimum level would include knowledge of basic financial statements, time value of money concepts, and elementary statistics.

In order to more clearly define the topical knowledge required by a candidate, varying levels of coverage for the treatment of major topics of the content specification outlines have been identified and defined. The cognitive skills that a successful candidate should possess and that should be tested on the examination can be defined as follows:

Knowledge: Ability to remember previously learned material such as specific facts, criteria, techniques, principles, and procedures (i.e., identify, define, list).

Comprehension: Ability to grasp and interpret the meaning of material (i.e., classify, explain, distinguish between).

Application: Ability to use learned material in new and concrete situations (i.e., demonstrate, predict, solve, modify, relate).

Analysis: Ability to break down material into its component parts so that its organizational structure can be understood; ability to recognize causal relationships, discriminate between behaviors, and identify elements that are relevant to the validation of a judgment (i.e., differentiate, estimate, order).

Synthesis: Ability to put parts together to form a new whole or proposed set of operations; ability to relate ideas and formulate hypotheses (i.e., combine, formulate, revise).

Evaluation: Ability to judge the value of material for a given purpose on the basis of consistency, logical accuracy, and comparison to standards; ability to appraise judgments involved in the selection of a course of action (i.e., criticize, justify, conclude).

The three levels of coverage can be defined as follows:

Level A: Requiring an appreciation of the broad nature and fundamentals, the ability to recognize the existence of special features and problems in various business transactions, and the ability to apply this knowledge in new and concrete situations, i.e., requiring the skill levels of knowledge and comprehension.

Level B: Requiring a sound understanding of the broad aspects of practices and procedures and awareness of the problems relating to more detailed aspects, the ability to apply such broad knowledge in new and concrete situations, the ability to analyze the components of the material and their relationships, and the ability to reformulate the components into a new whole, i.e., requiring the skill levels of knowledge, comprehension, application, and analysis.

Level C: Requiring a sound understanding of principles, practices, and procedures, the ability to analyze material so that its organizational structure can be understood, the proficiency to apply such knowledge to situations likely to be encountered, the ability to deal with all aspects of the situation without extensive recourse to technical research and assistance, and the ability to judge the value of material for a given purpose, i.e., requiring all six skill levels.

The levels of coverage as they apply to each of the major topics of the Content Specification Outlines are shown on the following pages with each topic listing. The levels represent the manner in which topic areas are to be treated and represent ceilings, i.e., a topic area designated as Level C may contain requirements at the "A," "B," or "C" level but a topic designated as Level B will not contain requirements at the "C" level.

Your authors have organized and presented study outlines and multiple-choice questions in anticipation of the ICMA's questions to appear on the exam. We expect to have 95+% of the questions on your exam explained in the study outlines and/or illustrated in our questions. After you take your exam, please send us your evaluation of this book and our *CMA/CFM Review Test Prep* software. We want to know the topics, concepts, etc., that need to be expanded or added to our CMA/CFM books and software. Remember to check our e-mail updates. See page iv for details.

CONTENT SPECIFICATION
Part 4 - Decision Analysis and Information Systems

A. ***Decision Theory and Operational Decision Analysis (20% - 30%) (Level C)***

1. *Logical steps to reach a decision*
2. *Relevant data concepts in the decision process*
    a. *Future oriented revenues and costs*
    b. *Sunk costs*
    c. *Opportunity costs*
    d. *Importance of cash flows*
3. *Cost/profit/volume analysis for decision making*
    a. *Breakeven analysis*
    b. *Profit performance and alternative operating levels*
    c. *Setting prices*
    d. *Analysis of multiple products*
4. *Marginal analysis*
    a. *Special orders and pricing*
    b. *Make vs. buy*
    c. *Sell or process further*
    d. *Add or drop a segment*
    e. *Select marketing channels*
5. *Cost-based pricing*
    a. *Comparing to market-based prices*
    b. *Submitting bids*
    c. *Setting prices*
    d. *Target costing*
    e. *Government contract pricing and the Cost Accounting Standards Board*

In conjunction with the decision subject areas in this topic, the impact of income taxes on operational decisions

-- Operating income subject to income taxes
-- Gains and losses subject to taxes
-- Tax credits

B. ***Investment Decision Analysis (20% - 30%) (Level C)***

1. *Cash flows*
   a. *Pretax cash flows*
   b. *After-tax cash flows*
2. *Discounted cash flow*
3. *Net present value*
4. *Internal rate of return*
5. *Discounted payback*
6. *Ranking investment projects*
7. *Non-discounting methods*
   a. *Payback period*
   b. *Accounting rate of return*

In conjunction with the decision subject areas in this topic, the impact of income taxes on investment decisions

-- *Tax shields from sunk costs*
-- *Transaction timing for favorable tax consequences*
-- *Property transactions*
-- *Tax elections*

C. ***Quantitative Methods for Decision Analysis (10% - 15%) (Level B)***

1. *Forecasting analysis*
   a. *Regression analysis*
   b. *Learning curve analysis*
   c. *Exponential smoothing*
   d. *Time series analysis*
2. *Linear programming*
   a. *Scarce resource considerations*
   b. *Capacity constraints*
3. *Sensitivity analysis*
4. *Program Evaluation Review Technique (PERT) Analysis*
5. *Probability concepts and expected value*
6. *Statistical concepts*
7. *Decision tree analysis*
8. *Simulation*

D. ***Information Systems (20% - 30%) (Level B)***

1. *Nature of management information systems (MIS)*
   - a. *Purposes and objectives*
   - b. *Components of a MIS*
   - c. *Decision support systems*
   - d. *Hardware concepts*
   - e. *Software concepts*
   - f. *Types of processing*
   - g. *Database management systems*
   - h. *Artificial intelligence and expert systems*
   - i. *Data communication*
2. *Nature of the accounting information system (AIS)*
   - a. *Purposes, activities, benefits of an AIS*
   - b. *Information economics of an AIS (cost/benefit)*
   - c. *Environment of an AIS*
   - d. *Users of an AIS*
   - e. *Components of an AIS*
3. *Systems development and design*
   - a. *Systems development life cycle*
   - b. *Role of the management accountant in systems development*
   - c. *Development tools and techniques*
4. *Systems controls and security measures*
   - a. *General accounting system controls*
   - b. *Application accounting or transaction controls*
   - c. *Network controls*
   - d. *Backup controls*

E. ***Management Controls (10% - 15%) (Level B)***

1. *Internal controls*
   - a. *Control environment*
   - b. *Control procedures*
   - c. *Foreign Corrupt Practices Act*
   - d. *Treadway Report*
   - e. *Committee on Sponsoring Organizations (COSO) report*
2. *Internal Auditing*
   - a. *Responsibility and authority of the internal audit function*
   - b. *Type of audits conducted by internal auditors*
   - c. *Types of assistance provided to management*
   - d. *Types of audit reports*
3. *Assessing the adequacy of the accounting information system*
   - a. *Evaluation of internal controls in an AIS*
   - b. *Evaluation of procedures in an AIS*

The topics and subject areas above may include ethical considerations in decision analysis and information systems.

Statement on Management Accounting No. 1C, "Standards of Ethical Conduct for Practitioners of Management Accounting and Financial Management"

-- Competence
-- Confidentiality
-- Integrity
-- Objectivity

Manipulation of decision factors
Fraud prevention and internal controls

APPENDIX E
ICMA SUGGESTED READING LIST

The ICMA suggested reading list that follows is reproduced to give you an overview of the scope of each part. You will not have the time to study these texts. Our *CMA/CFM Review* books and *CMA/CFM Test Prep* software will be all you need. For all five parts, candidates are expected to stay up-to-date by reading articles from journals, newspapers, and professional publications.

CMA/CFM PART 1 -- ECONOMICS, FINANCE, AND MANAGEMENT

Economics

Lipsey, Richard G., and Courant, Paul, *Economics*, 12th edition, Harper Collins College Publishers, New York, NY, 1998.

McConnell, Campbell R., and Brue, Stanley L., *Economics: Principles, Problems, and Policies*, 14th edition, Irwin/McGraw-Hill, New York, NY, 1999.

International Business Environment

Salvatore, Dominick, *International Economics*, 7th edition, Prentice-Hall, Inc., Upper Saddle River, NJ, 2001.

Madura, Jeff, *International Financial Management*, 6th edition, South Western Publishing Co., Boston, MA, 2000.

Zuckerman, Amy, *International Standards, Desk Reference*, 1st edition, Amacom, New York, NY, 1997.

Root, Franklin R., *International Trade and Investment*, 7th edition, South-Western Publishing Co., Cincinnati, OH, 1993.

Government and Business

Post, James E., Frederick, William C., Lawrence, Anne T., and Weber, James, *Business and Society: Corporate Strategy, Public Policy, Ethics*, 9th edition, McGraw-Hill Inc., New York, NY, 1998.

Weidenbaum, Murray L., *Business and Government in the Global Marketplace*, 6th edition, Prentice-Hall Inc., Upper Saddle River, NJ, 1999.

Greer, Douglas F., *Business, Government, and Society*, 3rd edition, Prentice-Hall, Inc., Upper Saddle River, NJ, 1993.

Business Finance

Ross, Stephen A., Westerfield, Randolph W., and Jaffe, Jeffrey F., *Corporate Finance*, 5th edition, Richard D. Irwin Inc., Chicago, IL, 1999.

Brigham, Eugene F., and Houston, Joel F., *Fundamentals of Financial Management*, 9th edition, Harcourt Brace College Publishers, Orlando, FL, 2000.

Van Horne, James C., and Wachowicz, John M., Jr., *Fundamentals of Financial Management*, 11th edition, Prentice-Hall, Inc., Upper Saddle River, NJ, 2001.

Brealey, Richard A., and Myers, Stewart C., *Principles of Corporate Finance*, 6th edition, McGraw-Hill, Inc., New York, NY, 2000.

Organization and Management Theory and Communication

Mathis, Robert L., and Jackson, John H., *Human Resource Management*, 9th edition, Thomson Learning, St. Paul, MN, 2000.

Tosi, Henry L., Rizzo, John R., and Carroll, Stephen J., *Managing Organizational Behavior*, 4th edition, Blackwell Publishers, Ltd., Cambridge, MA, 2000.

Moorhead, Gregory, and Griffin, Ricky W., *Organizational Behavior*, 6th edition, Houghton Mifflin Company, Boston, MA, 2000.

PART 2CMA -- FINANCIAL ACCOUNTING AND REPORTING

Financial Statements and Reporting Requirements

Nikolai, Loren A., and Bazley, John D., *Intermediate Accounting*, 8th edition, South-Western Publishing Company, Boston, MA, 1999.

Kieso, Donald E., and Weygandt, Jerry J., *Intermediate Accounting*, 10th edition, John Wiley & Sons, Inc., New York, NY, 2000.

Larsen, E. John, *Modern Advanced Accounting*, 8th edition, McGraw-Hill Book Co., New York, NY, 1999.

Afterman, Allan B., *SEC Regulation of Public Companies*, 1st edition, Prentice Hall Inc., Upper Saddle River, NJ, 1995.

Analysis of Accounts and Statements

Gibson, Charles H., *Financial Statement Analysis*, 8th edition, South-Western Publishing Co., Cincinnati, Ohio, 1999.

Bernstein, Leopold A., and Wild, John J., *Financial Statement Analysis, Theory, Application, and Interpretation*, 7th edition, McGraw-Hill Book Co., New York, NY, 2000.

White, Gerald I., Sondhi, Ashwinpaul C., and Fried, Dov, *The Analysis and Use of Financial Statements*, 2nd edition, John Wiley & Sons, New York, NY, 1998.

External Auditing

Arens, Alvin A., and Loebbecke, James K., *Auditing: An Integrated Approach*, 8th edition, Prentice-Hall Inc., Upper Saddle River, NJ, 2000.

Boynton, William C., and Kell, Walter G., *Modern Auditing*, 7th edition, John Wiley & Sons, New York, NY, 2000.

PART 2CFM -- CORPORATE FINANCIAL MANAGEMENT

Financial Management

Ross, Stephen A., Westerfield, Randolph W., and Jaffe, Jeffrey F., *Corporate Finance*, 5th edition, Richard D. Irwin, Inc., Chicago, IL, 1999.

Brigham, Eugene F., and Houston, Joel F., *Fundamentals of Financial Management*, 9th edition, Harcourt Brace College Publishers, Orlando, FL, 2000.

Van Horne, James C. and Wachowicz, John M., Jr., *Fundamentals of Financial Management*, 11th edition, Prentice-Hall, Inc., Upper Saddle River, NJ, 2001.

Brealey, Richard A., and Myers, Stewart C., *Principles of Corporate Finance*, 6th edition, McGraw-Hill, Inc., New York, NY, 2000.

Financial Environment

Weidenbaum, Murray L., *Business and Government in the Global Marketplace*, 6th edition, Prentice-Hall Inc., Upper Saddle River, NJ, 1999.

Post, James E., Frederick, William C., Lawrence, Anne T., and Weber, James, *Business and Society: Corporate Strategy, Public Policy, Ethics*, 9th edition, McGraw-Hill Inc., New York, NY, 1998.

Greer, Douglas F., *Business, Government, and Society*, 3rd edition, Macmillan Publishing, New York, NY, 1993.

Lipsey, Richard G., and Courant, Paul, *Economics*, 12th edition, Harper Collins College Publishers, New York, NY, 1998.

McConnell, Campbell R., and Brue, Stanley L., *Economics: Principles, Problems and Policies*, 14th edition, McGraw-Hill Book Co., New York, NY, 1999.

Madura, Jeff, *International Financial Management*, 6th edition, South Western Publishing Co., Boston, MA, 2000.

Geisst, Charles R., *Investment Banking in the Financial System*, 1st edition, Prentice-Hall Inc., Upper Saddle River, NJ, 1995.

Boynton, William C., and Kell, Walter G., *Modern Auditing*, 7th edition, John Wiley & Sons, New York, NY, 2000.

Rose, Peter S., *Money and Capital Markets, Financial Institutions and Instruments in a Global Market*, 7th edition, Richard D. Irwin Inc., Chicago, IL, 2000.

Williams, C. Arthur, Smith, Michael L., and Young, Peter C., *Risk Management and Insurance*, 8th edition, McGraw-Hill Inc., New York, NY, 1998.

Analysis of Financial Statements

Gibson, Charles H., *Financial Statement Analysis*, 8th edition, South-Western Publishing Co., Cincinnati, OH, 1999.

Bernstein, Leopold A., and Wild, John J., *Financial Statement Analysis, Theory, Application, and Interpretation*, 7th edition, McGraw-Hill Book Co., New York, NY, 2000.

White, Gerald I., Sondhi, Ashwinpaul C., and Fried, Dov, *The Analysis and Use of Financial Statements*, 2nd edition, John Wiley & Sons, New York, NY, 1998.

Financial Statements and Accounting Standards

Kieso, Donald E., and Weygandt, Jerry J., *Intermediate Accounting*, 10th edition, John Wiley & Sons, Inc., New York, NY, 2000.

Nikolai, Loren A., and Bazley, John D., *Intermediate Accounting*, 8th edition, South-Western Publishing Company, Boston, MA, 2000.

Larsen, E. John, *Modern Advanced Accounting*, 8th edition, McGraw-Hill, Inc., New York, NY, 1999.

Afterman, Allan B., *SEC Regulation of Public Companies*, 1st edition, Prentice Hall Inc., Upper Saddle River, NJ, 1995.

CMA/CFM PART 3 -- MANAGEMENT REPORTING, ANALYSIS, AND BEHAVIORAL ISSUES

Management Reporting and Analysis

Welsch, Glenn A., Hilton, Ronald W., and Gordon, Paul N., *Budgeting: Profit Planning and Control*, 5th edition, Prentice-Hall, Inc., Upper Saddle River, NJ, 1998.

Carter, William K., and Usry, Milton F., *Cost Accounting*, 12th edition, Dame Publications Inc., Houston, TX, 1999.

Horngren, Charles, Foster, George, and Datar, Srikant M., *Cost Accounting: A Managerial Emphasis,* 10th edition, Prentice-Hall, Inc., Upper Saddle River, NJ, 2000.

Blocher, Edward J., Chen, Kung H., and Lin, Thomas W., *Cost Management: A Strategic Emphasis*, Irwin/McGraw Hill, New York, NY, 1999.

CMA/CFM PART 4 -- DECISION ANALYSIS AND INFORMATION SYSTEMS

Decision Analysis and Quantitative Methods

Anderson, David R., Sweeney, Dennis J., and Williams, Thomas A., *An Introduction to Management Science-Quantitative Approaches to Decision Making*, 9th edition, West Publishing Company, St. Paul, MN, 2000.

Carter, William K., and Usry, Milton F., *Cost Accounting*, 12th edition, Dame Publications Inc., Houston, TX, 1999.

Horngren, Charles, Foster, George, and Datar, Srikant, *Cost Accounting: A Managerial Emphasis,* 10th edition, Prentice-Hall, Inc., Upper Saddle River, NJ, 2000.

Blocher, Edward J., Chen, Kung H., and Lin, Thomas W., *Cost Management: A Strategic Emphasis*, Irwin/McGraw Hill, New York, NY, 1999.

Information Systems

Romney, Marshall B., and Steinbart, Paul J., *Accounting Information Systems*, 8th edition, Addison-Wesley Publishing Company, Reading, MA, 1999.

Moscove, Stephen A., Simkin, Mark G., and Bagranoff, Nancy A., *Core Concepts of Accounting Information Systems*, 7th edition, John Wiley & Sons, Inc., New York, NY, 2000.

O'Brien, James A., *Introduction to Information Systems*, 10th edition, Irwin/McGraw Hill Publishers, Barr Ridge, IL, 2000.

Internal Auditing

Sawyer, Lawrence B., and Dittenhofer, Mortimer A., *Internal Auditing*, 4th edition, Institute of Internal Auditors, Altamonte Springs, FL, 1996.

Ratliff, Richard L., Wallace, Wanda A., Summers, Glenn E., McFarland, William G., and Loebbecke, James K., *Internal Auditing Principles and Techniques*, 2nd edition, The Institute of Internal Auditors, Altamonte Springs, FL, 1996.

CMA/CFM REVIEW: Part 4, CPE

CONTINUING PROFESSIONAL EDUCATION

FIVE SEPARATE COURSES

Courses Available	Course Level	Average Comp. Time	Approximate CPE Credit
1. Decision Analysis and CVP	Basic	500 min.	5/10* hours
2. Capital Budgeting and Decision Making	Basic	500 min.	5/10* hours
3. Quantitative Methods	Basic	500 min.	5/10* hours
4. Information Systems	Basic	500 min.	5/10* hours
5. Internal Control and Internal Auditing	Basic	500 min.	5/10* hours
	Total CPE Credit		25/50* hours

*Check with your accreditation agency to determine whether it gives credit for half or full completion time.

The Gleim CMA/CFM CPE course is available via the Internet beginning with the Tenth Edition. The cost is only $50 (in addition to the $24.95 for this study manual, which serves as your workbook documentation) for up to 50 hours of CPE credit. Gleim CPE credit is accepted at most agencies and all state boards of accountancy. Visit our web site for details.

To register, simply go to www.gleim.com/Accounting/cpe/cma and sign up to receive complete instructions and access to our online grading program.

QAS Quality Assurance Service

Gleim Publications, Inc. is registered with the National Association of State Boards of Accountancy as a Quality Assurance Service (QAS) sponsor of continuing professional education. Participating state boards of accountancy have final authority on the acceptance of individual courses for CPE credit. Complaints regarding QAS sponsors may be addressed to: NASBA, 150 Fourth Avenue North, Suite 700, Nashville, TN 37219-2417, (615) 880-4200.

Gleim Publications, Inc. is registered with the National Association of State Boards of Accountancy as a sponsor of continuing professional education on the National Registry of CPE Sponsors. State boards of accountancy have final authority on the acceptance of individual courses. Complaints regarding sponsors may be addressed to: NASBA, 150 Fourth Avenue North, Suite 700, Nashville, TN 37219-2417, (615) 880-4200.

NASBA NATIONAL REGISTRY OF CPE SPONSORS

Index

Gleim Publications, Inc.
P.O. Box 12848
Gainesville, FL 32604

TOLL FREE: (800) 87-GLEIM/(800) 874-5346
LOCAL: (352) 375-0772
FAX: (888) 375-6940 (toll free)
INTERNET: http://www.gleim.com
E-MAIL: sales@gleim.com

Customer service is available:
8:00 a.m. - 7:00 p.m., Mon. - Fri.
9:00 a.m. - 2:00 p.m., Saturday
Please have your credit card ready or save time by ordering online!

CPA REVIEW

	Books	Software	Book/Software Package	Audiotapes	Book/Software/ Audio Package	
Auditing	☐ @ $29.95	☐ @ $39.95	☐ @ $62.95	☐ @ $79.95	☐ @ $125.00	$______
Business Law	☐ @ $29.95	☐ @ $39.95	☐ @ $62.95	☐ @ $79.95	☐ @ $125.00	______
TAX-MAN-GOV	☐ @ $29.95	☐ @ $39.95	☐ @ $62.95	☐ @ $79.95	☐ @ $125.00	______
Financial	☐ @ $29.95	☐ @ $39.95	☐ @ $62.95	☐ @ $79.95	☐ @ $125.00	______
A System for Success	☐ FREE with any Gleim *CPA Review* Book					

Save 16% on the Complete Gleim CPA System (add $18 S&H below) . . . . $500.00 ______
[5 books, 4 cassette albums (41 tapes), 4 CPA Test Prep disks, bonus calculator and versatile book bag]

CIA REVIEW

	Books	Software	Book/Software Package	
Part I Internal Audit Process	☐ @ $24.95	☐ @ $39.95	☐ @ $57.95	$______
Part II Internal Audit Skills	☐ @ $24.95	☐ @ $39.95	☐ @ $57.95	______
Part III Mgmt. Ctl. & Info. Tech.	☐ @ $24.95	☐ @ $39.95	☐ @ $57.95	______
Part IV The Audit Environment	☐ @ $24.95	☐ @ $39.95	☐ @ $57.95	______

CPE credit also available (online only). Please visit www.gleim.com/Accounting/cpe/cia

CMA/CFM REVIEW

	Books	Software	Book/Software Package	Audiotapes	Book/Software/ Audio Package	
Part 1 Eco., Fin., & Mgmt.	☐ @ $24.95	☐ @ $44.95	☐ @ $62.95	☐ @ $64.95	☐ @ $114.95	$______
Part 2CMA Fin. Acc. & Rep.	☐ @ $24.95	☐ @ $44.95	☐ @ $62.95	☐ @ $64.95	☐ @ $114.95	______
Part 2CFM Corp. Fin. Mgmt.	☐ @ $24.95	☐ @ $44.95	☐ @ $62.95	☐ @ $64.95	☐ @ $114.95	______
Part 3 Mgmt. Rep./Behav. Iss.	☐ @ $24.95	☐ @ $44.95	☐ @ $62.95	☐ @ $64.95	☐ @ $114.95	______
Part 4 Dec. Anal. & Info. Sys.	☐ @ $24.95	☐ @ $44.95	☐ @ $62.95	☐ @ $64.95	☐ @ $114.95	______

CPE credit also available @ $50 per part. Purchase of corresponding book is required.

EA REVIEW

	Books	Software	Book/Software Package	
Part 1 Individuals	☐ @ $24.95	☐ @ $39.95	☐ @ $57.95	$______
Part 2 Sole Prop. & Prtnrshps.	☐ @ $24.95	☐ @ $39.95	☐ @ $57.95	______
Part 3 Corp./Fid./Est. & Gift Tax	☐ @ $24.95	☐ @ $39.95	☐ @ $57.95	______
Part 4 IRS Adm/Other Topics	☐ @ $24.95	☐ @ $39.95	☐ @ $57.95	______

"THE GLEIM SERIES" EXAM QUESTION AND EXPLANATION BOOKS, SOFTWARE, & CPE

	Books	Software	Book/Software Package	Book & CPE	
Auditing & Systems	☐ @ $19.95	☐ @ $20.00	☐ @ $29.95	☐ $125.00	$______
Business Law/Legal Studies	☐ @ $19.95	☐ @ $20.00	☐ @ $29.95	☐ $125.00	______
Federal Tax	☐ @ $19.95	☐ @ $20.00	☐ @ $29.95	☐ $125.00	______
Financial Accounting	☐ @ $19.95	☐ @ $20.00	☐ @ $29.95	☐ $125.00	______
Cost/Managerial Accounting	☐ @ $19.95	☐ @ $20.00	☐ @ $29.95	☐ $125.00	______

Shipping (nonrefundable): **First item = $5; each additional item = $1** $______

Add applicable sales tax for shipments within Florida. ______

Fax or write for prices/instructions for shipments outside the 48 contiguous states. **TOTAL** $______

1. We process and ship orders daily, within one business day over 98.8% of the time. Call by noon for same-day service!
2. Please PHOTOCOPY this order form for others.
3. No CODs. Orders from individuals must be prepaid. Library and company orders may be purchased on account.
4. Gleim Publications, Inc. guarantees the immediate refund of all resalable texts and unopened software and audiotapes if returned within 30 days. Applies only to items purchased direct from Gleim Publications, Inc. Our shipping charge is nonrefundable.
5. Components of specially priced package deals are nonreturnable.

NAME (please print) ______________________

ADDRESS ______________________ Apt. ______
(street address required for UPS)

____

CITY ______________ STATE ____ ZIP ______

____ MC/VISA/DISC ____ Check/M.O. Daytime Telephone (____)__________

Credit Card No. ______ - ______ - ______ - ______

Exp. ___/___ Mo. / Yr. Signature ______________________

Printed 02/01. Prices subject to change without notice.

020

Please forward suggestions, corrections, and comments concerning typographical errors, etc., to **Irvin N. Gleim • c/o Gleim Publications, Inc. • P.O. Box 12848 • University Station • Gainesville, Florida • 32604**. Please include your name and address so we can properly thank you for your interest.

1. ______________________________

2. ______________________________

3. ______________________________

4. ______________________________

5. ______________________________

6. ______________________________

7. ______________________________

8. ______________________________

9. ______________________________

10. ______________________________

11. ______________________________

12. ______________________________

13. ______________________________

14. ______________________________

15. ______________________________

16. ______________________________

17. ______________________________

18. ______________________________

19. ______________________________

20. ______________________________

Remember for superior service:	Mail, e-mail, or fax questions about our books or software. Telephone questions about orders, prices, shipments, or payments.

Name: ______________________________

Company: ______________________________

Address: ______________________________

City/State/Zip: ______________________________

Phone: ( ) __________ FAX: ( ) __________ E-mail: __________